T0178229

Lecture Notes in Computer Science 13104

More information about this subseries at http://www.springer.com/series/7408

Paolo Ballarini · Hind Castel ·
Ioannis Dimitriou · Mauro Iacono ·
Tuan Phung-Duc · Joris Walraevens (Eds.)

Performance Engineering and Stochastic Modeling

17th European Workshop, EPEW 2021
and 26th International Conference, ASMTA 2021
Virtual Event, December 9–10 and December 13–14, 2021
Proceedings

 Springer

Editors
Paolo Ballarini 🆔
University Paris-Saclay
Gif-sur-Yvette, France

Ioannis Dimitriou 🆔
University of Patras
Patras, Greece

Tuan Phung-Duc 🆔
Faculty of Engineering, Information
and Systems, Division of Policy
and Planning Sciences
University of Tsukuba
Tsukuba, Japan

Hind Castel 🆔
Telecom Sub Paris
Evry, France

Mauro Iacono 🆔
Università degli studi della Campania Luigi
Vanvitelli
Caserta, Italy

Joris Walraevens 🆔
Ghent University
Gent, Belgium

ISSN 0302-9743   ISSN 1611-3349   (electronic)
Lecture Notes in Computer Science
ISBN 978-3-030-91824-8   ISBN 978-3-030-91825-5   (eBook)
https://doi.org/10.1007/978-3-030-91825-5

LNCS Sublibrary: SL2 – Programming and Software Engineering

This Springer imprint is published by the registered company Springer Nature Switzerland AG
The registered company address is: Gewerbestrasse 11, 6330 Cham, Switzerland

Preface

This volume contains the papers presented at the 17th European Performance Engineering Workshop (EPEW 2021) and the 26th International Conference on Analytical & Stochastic Modelling Techniques & Applications (ASMTA 2021) which were held concurrently during December 9–10 and December 13–14, 2021, respectively. Due to the COVID-19 pandemic it was decided to organize them - exceptionally - online.

For the EPEW conference, there were 11 submissions. Each submission was reviewed by three Program Committee members on average. The committee decided to accept nine papers. For the ASMTA conference, there were 26 submissions. Each submission was reviewed by three Program Committee members on average. The committee decided to accept 20 papers.

It was our privilege to invite Dieter Fiems (Ghent University) and Jean-Michel Fourneau (Université de Versailles St Quentin) to give keynote talks at EPEW 2021 and to invite Antonis Economou (National and Kapodistrian University of Athens) and Neil Walton (University of Manchester) to give keynote talks at ASMTA 2021.

We would like to thank all authors of all the papers appearing in this proceedings for their interesting contributions. Special thanks go to the members of the Program Committees for their time and effort in assuring the quality of the selected papers. Finally, we cordially thank the EasyChair and Springer teams for their support in publishing this volume.

See you in 2022!

December 2021

Paolo Ballarini
Hind Castel
Ioannis Dimitriou
Mauro Iacono
Tuan Phung-Duc
Joris Walraevens

Organization EPEW 2021

General Chair

Tuan Phung-Duc   University of Tsukuba, Japan

Program Committee Chairs

Paolo Ballarini   CentraleSupélec, France
Hind Castel   Telecom SudParis, France
Mauro Iacono   Università degli studi della Campania Luigi Vanvitelli, Italy

Logistics and Web Chair

M. Akif Yazici   Istanbul Technical University, Turkey

Program Committee

Paolo Ballarini	CentraleSupélec, France
Enrico Barbierato	Università Cattolica del Sacro Cuore, Italy
Benoit Barbot	Université Paris-Est Créteil, France
Marco Beccuti	Università degli Studi di Torino, Italy
Marco Bernardo	University of Urbino, Italy
Laura Carnevali	University of Florence, Italy
Hind Castel	Telecom SudParis, France
Davide Cerotti	Politecnico di Milano, Italy
Ioannis Dimitriou	University of Patras, Greece
Dieter Fiems	Ghent University, Belgium
Jean-Michel Fourneau	Université de Versailles-Saint-Quentin-en-Yvelines, France
Stephen Gilmore	University of Edinburgh, UK
Marco Gribaudo	Politecnico di Milano, Italy
András Horváth	Università degli Studi di Torino, Italy
Gábor Horváth	Budapest University of Technology and Economics, Hungary
Emmanuel Hyon	Université Paris Nanterre, France
Esa Hyytiä	University of Iceland, Iceland
Mauro Iacono	Università degli studi della Campania Luigi Vanvitelli, Italy
Alain Jean-Marie	Inria, France
William Knottenbelt	Imperial College London, UK
Lasse Leskelä	Aalto University, Finland

Organization ASMTA 2021

General Chair

Tuan Phung-Duc   University of Tsukuba, Japan

Program Committee Chairs

Ioannis Dimitriou   University of Patras, Greece
Joris Walraevens   Ghent University, Belgium

Logistics and Web Chair

M. Akif Yazici   Istanbul Technical University, Turkey

Publicity Chair

Sabine Wittevrongel   Ghent University, Belgium

Program Committee

Nail Akar	Bilkent University, Turkey
Jonatha Anselmi	Inria, France
Konstantin Avrachenkov	Inria, France
Paolo Ballarini	CentraleSupélec, France
Simonetta Balsamo	Università Ca' Foscari di Venezia
Tejas Bodas	IIT Dharwad, India
Hind Castel	Telecom SudParis, France
Dieter Claeys	Ghent University, Belgium
Céline Comte	Eindhoven University of Technology, The Netherlands
Koen De Turck	CentraleSupélec, Belgium
Antonis Economou	University of Athens, Greece
Dieter Fiems	Ghent University, Belgium
Marco Gribaudo	Politecnico di Milano, Italy
Irina Gudkova	People's Friendship University of Russia, Russia
Antonio Gómez-Corral	Universidad Complutense de Madrid, Spain
Yezekael Hayel	LIA/University of Avignon, France
András Horváth	University of Turin, Italy
Gábor Horváth	Budapest University of Technology and Economics, Hungary
Mauro Iacono	Università degli studi della Campania Luigi Vanvitelli, Italy
Yoshiaki Inoue	Osaka University, Japan

Contents

ASMTA 2021

EPEW 2021

Workload Prediction in BTC Blockchain and Application to the Confirmation Time Estimation

Ivan Malakhov$^{(\boxtimes)}$ (iD), Carlo Gaetan (iD), Andrea Marin (iD), and Sabina Rossi (iD)

Università Ca' Foscari Venezia, Via Torino 155, 30173 Venice, Italy
{ivan.malakhov,gaetan,marin,sabina.rossi}@unive.it

Abstract. Blockchains are distributed ledgers storing data and procedures in an immutable way. The validation of the information stored therein as well as the guarantee of its immutability can be achieved without the need of a central authority. Proof-of-work is the maximum expression of the distributed nature of such systems, and requires miners to spend a large amount of energy to secure the blockchain. The cost is mostly paid by the end-users that offer fees to support the validation of their transactions. In general, higher fees correspond to shorter validation delays. However, given the limited throughput of the system and variability of the workload, the fee one needs to offer to satisfy a certain requirement on the validation delay strongly depends on the intensity of the workload that, in turns, is subject to high variability.

In this work, we propose a time series analysis of the workload of Bitcoin blockchain and compare the accuracy of Facebook Prophet model with a ARIMA model. We take into account the periodicity of the workload and show by simulations how these predictions, accompanied with their confidence intervals, can be used to estimate the confirmation delays of the transactions given the offered fees.

Keywords: Blockchain · Confirmation time analysis · Time series analysis

1 Introduction

Blockchains have been attracting more and more attention from the research community from the economical, security and application points of view. More recently, the quantitative analysis of blockchains has also emerged as an important research challenge.

The blockchain distributed ledger has substantially three main roles: (i) verify the information or procedures that end-users wants to store according to some rules, (ii) guarantee the immutability of the stored information and (iii) make the information or procedure publicly available. While there are several ways to achieve these goals, in this paper, we focus on the most popular one, namely the *proof-of-work* (PoW).

© Springer Nature Switzerland AG 2021
P. Ballarini et al. (Eds.): EPEW 2021/ASMTA 2021, LNCS 13104, pp. 3–21, 2021.
https://doi.org/10.1007/978-3-030-91825-5_1

PoW has been introduced by the seminal paper [12] by the pseudonym Satoshi Nakamoto with the aim of creating a distributed ledger for economical transactions based on the cryptocurrency Bitcoin (BTC). Many public blockchains use PoW as consensus algorithm and, recently, it has been proposed also for permissioned blockchains [10]. Therefore, henceforth, we will focus on BTC blockchain since it is, together with Ethereum, the mostly used and known blockchains. However, the methodologies and discussions that we propose can be easily extended to other blockchain systems with similar characteristics.

In BTC, the blockchain stores transactions into blocks. Blocks have a maximum size of 1MB and are generated, on average, every 10 min. This means that the maximum throughput of the system is fixed by design. Transactions are proposed by the end users and are sent to the *miners* for being processed.

Miners maintain a queue called *Mempool* that contains all the pending transactions, i.e., the transactions sent by end users but that have not been added to a block yet. When a transaction is included in a block, we say that it is *confirmed*, i.e., it is permanently stored in the system. The transaction residence time in the Mempool is called *confirmation delay*. This delay is crucial in determining the Quality-of-Service (QoS) of applications based on blockchains.

In order to understand the quantitative dynamics of the transaction conformations, we need to review the procedure implemented by the miners to secure the blockchain system. Each miner selects from the Mempool a set of transactions to fill a block, then he/she checks their integrity (e.g., when there is a transfer of cryptocurrency it verifies that there is not double spending) and finally it works on a computational problem that requires a large amount of energy in order to be solved. This latter step is the PoW. The miner that firstly announces the solution of its computational problem is entitled to add his/her new block to the blockchain after the other peers have verified the correctness of the solution.

In order to cover the energy and hardware costs, miners receive a certain amount of cryptocurrency when they succeed in a block consolidation: some is freshly created by the system and given to the miners and then they receive the fees promised by the owners of the transactions added to new block. These fees are offered by the end users on a voluntary base, i.e., they can even offer 0 BTC. However, the miners tend to select from the Mempool those transactions that offer the highest fee.

From the end-user's point of view, an interesting trade off arises: on the one hand, he/she wishes to offer the lowest possible fee to reduce the running costs of his/her activities, on the other he/she may have some requirements on the QoS, e.g., the need to confirm the transaction within a certain amount of time. For example, the transaction may be associated with a trading speculation and hence must be confirmed in a few minutes, or may be a bid for a certain auction with a deadline.

Blockchain systems can be studied as distributed systems by means of formal methods in the style of [5,6]. Queueing theory allows us to study the relation between the holding time in the Mempool and the arrival intensity of the transactions. Clearly, when the holding times increase, the transaction fees also increase. However, while for high fees consolidated in few blocks it is safe to assume that the arrival intensity is time homogeneous, for transactions offering low fees this is unrealistic.

In this paper, we study the problem of predicting the traffic intensity in BTC blockchain with the aim of parameterising a simulation model that studies the expected confirmation time of transactions. After collecting data about the transaction arrival process at our BTC node, we use these traces to train two possible prediction models: one based on Facebook Prophet model [13], and the other is the well-known Autoregressive Integrate Moving Average (ARIMA) model. Both models provide confidence intervals in the prediction of the arrival process and allow us to consider pessimistic-, average- and optimistic-case scenarios. After comparing the two predictive models, we study by simulation the transaction confirmation time as a function of the offered fees and compare the results obtained with the real trace as input with those obtained by using the predicted trace as input.

The paper is structured as follows. In Sect. 2, we discuss the related work done in similar fields. Section 3 describes the motivation of this paper and gives a brief description of the applied prediction models. In Sect. 4, we examine the ARIMA and Prophet forecasts accuracy after certain hours from the transaction arrival and the accuracy of the predictions on the expected confirmation time using Monte Carlo simulations. Finally, Sect. 5 concludes the paper and provides an insight for future work.

2   Related Work

Statistical analysis on blockchain and in particular BTC system have been widely investigated in the recent years. However, most of the research efforts have been devoted to the prediction of the conversion rate to USD or other currencies (see, e.g., [4,11]).

In our case, we are interested in studying the cost of transaction fees. Most of the previous works assume a time-homogeneous arrival process, as in [1,7,8] which can be reasonable for expensive transactions that are confirmed within one hour from their request. However, when the delay is longer, the fluctuations of the arrival process cause the model with the homogeneity assumption to generate inaccurate predictions.

In addition, [9] provides a similar contribution by demonstrating the stationary analysis of the queueing model and the definition of the customer priority classes. However, the authors focus on a game theoretical framework where they attempt to find correlations between the fee fluctuations and the miners' economical incentive.

Another work [14] analyses the transaction fees in the blockchain networks. However, their research is related to the Ethereum blockchain and particularly the smart contract transactions.

To the best of our knowledge, this is the first study that aims at predicting the intensity of the transaction arrival process by using time series analysis and predicting on the confirmation time based on the offered fee.

3 Background and Motivation

This section describes the goal of the paper and provides background information about the models used for the prediction of the arrival rate of transactions.

3.1 The Problem of Predicting the Minimum Fee for QoS

PoW is a method for both reaching consensus among the miners and guaranteeing the immutability of the blockchain contents. More precisely, it quantifies the expected energy cost required to modify a confirmed transaction. The more computational power (usually called hashpower) the miners invest the more secure the distributed ledger is. For this reason it becomes crucial to incentivize more miners to join the network with some rewards.

In the BTC blockchain, miners are rewarded in two ways: i) for each confirmed block, the miner who created it receives a certain amount of cryptocurrency and ii) for each transaction included in the block, the same miner receives the fee offered by the user who created that transaction.

In BTC the cryptocurrency is the Bitcoin but since its value is high (at the moment, $1\text{BTC} \simeq 35,000$ USD), fees are usually expressed in Satoshi (sat) where $1\text{BTC} = 10^8 \text{sat}$.

While the former reward is going to be dismissed in the next years, the latter plays a crucial role in understanding the QoS of applications that use BTC blockchain. Indeed, miners aim at maximising their profit and thus choose to include in the block the transactions with the highest fee.

Transaction fees are known to be subject to high fluctuations as shown by Fig. 1a. We may notice that the average fee for a transaction can vary from around 4.5 to 9 USD in a month. How to decide which fee to offer to have an expected confirmation delay?

It is important to understand that the answer to this question depends on several state variables of the blockchain. First, we should consider the Mempool occupancy (usually called improperly Mempool size), i.e., the backlog of the transactions that are waiting to be confirmed. Figure 1b shows the trace of the Mempool occupancy in the month of June 2021. The are several bursts that clearly affect the decision on the fee to be offered.

However, the most important factor is the transaction arrival process. Recall that all the transactions arriving after a tagged transaction t offering a fee per byte f will overtake t if they offer more than f. Fee per byte is commonly used to compare the cost of transactions because these may have different sizes. Since block sizes are fixed (i.e., a block can contain approximately $2,300$ transactions at most) and the inter-generation time of blocks is on average 10 min, this implies that the competition among the transactions gets tougher when the traffic is

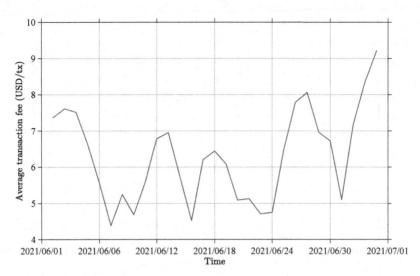

(a) Transaction fee in USD in the Bitcoin network. The data are retrieved from
`http://www.blockchain.com`

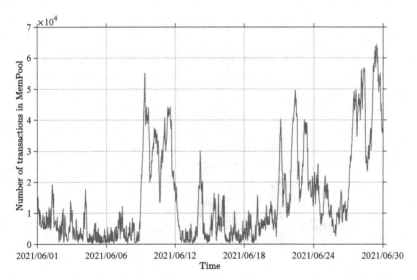

(b) Memory Pool size in transactions in the Bitcoin network. The data are retrieved
from `http://www.blockchain.com`

Fig. 1. Blockchain network indicators.

higher. Figure 2b shows the distribution of the fee per byte offered under heavy-
load conditions as measured by our monitor.

Summarising, the confirmation time of a transaction t arriving at time τ
depends on the following aspects:

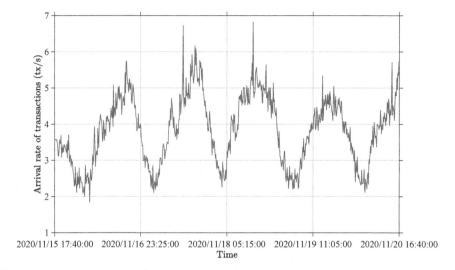

(a) Arrival rate of transactions as a function of time with step of 10 minutes. The data are retrieved from the installed node.

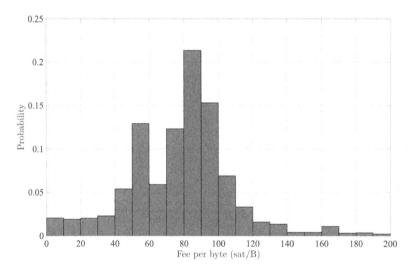

(b) Empirical probability density function of fee per byte in heavy workload conditions. The data are retrieved from the installed node.

Fig. 2. Continue. Blockchain network indicators.

– The arrival rate of the transactions after τ and before the confirmation of t, limited to those whose fees are higher than the fee offered by t;
– The state of the Mempool at τ;
– The distribution of fees offered by the other users.

In general, we measure the confirmation delay in number of blocks rather than in seconds. However, it is well-known that the time between consecutive block consolidations is approximately exponential with mean 600 s [3].

Figure 2a shows the intensity of the arrival process in a period of time. This is subject to high variability and exhibits a clear seasonality. Therefore, in any procedure aimed at predicting the expected confirmation time of transactions, we must implicitly or explicitly deal with the prediction of the arrival intensity at the moment in which the transactions is sent to the ledger.

The goal of this paper is that of evaluating the quality of the predictions of the arrival process given by two popular approaches to time series analysis: the ARIMA and Facebook Prophet models. Moreover, we will use these predictions to assess, by simulation, the quality of the estimations of the expected confirmation time of transactions.

3.2   Background on the ARIMA Model

ARIMA(p, d, q) model [2] one of the most widely used models for statistical forecasting a time series of observations X_t. The ARIMA equation is a linear (i.e., regression-type) equation in which the predictors consist of lags of the dependent variable and/or lags of the forecast errors. The general model can be written as

$$(1 - \phi_1 L - \cdots - \phi_p L^p)(1 - L)^d X_t = c + (1 + \theta_1 L + \cdots + \theta_q L^q)\varepsilon_t$$

where L is the lag-operator, i.e. $L^k a_t = a_{t-k}$ and ε_t is a white noise. The value p refers to the "AutoRegressive" component and represents the number of lagged observations included in the model. The "Integrated part" of the ARIMA model indicates that the data values have been replaced with the difference between their current and previous values, i.e. $(1 - L)x_t = x_t - x_{t-1}$. The value d is a number of times that the raw observations are differenced. In general, differencing refers to the transformation applied to non-stationary time series in order to make them stationary by attempting to remove the deterministic components such as trends or periodicities. The value q, stands for the size of the "Moving Average" window for the forecast errors. Automatic identification of the orders p, d, q and statistical estimation of the parameters $\phi_1, \ldots, \phi_p, \theta_1, \ldots, \theta_q$ can be done easily (see [2]).

The data are collected every 10 min and our time series exhibit seasonality with frequency of $144 = 24 \times 6$ which is exactly 24 h in 10-min terms. In our experiment, using the Akaike Information Criterion, we identify a special instance of the ARIMA model, namely a multiplicative seasonal model [2]:

$$(1 - \phi_1 L - \phi_2 L^p)(1 - L)(1 - L^{144})X_t = (1 + \theta_1 L + \theta_2 L^2)\varepsilon_t.$$

3.3   Background on the Facebook Prophet Model

The Prophet model [13] is a modular regression model with interpretable parameters that can be adjusted in order to optimize the prediction response.

The authors use a decomposable time series model with three key components, namely trend, seasonality, and holidays. The model may be represented as follows:

$$X_t = g_t + s_t + h_t + \varepsilon_t.$$

where g_t refers to the trend function that simulates non-periodic changes in the value of the time series, s_t describes periodic changes of the series, that is any seasonality effects, and h_t stands for the effects of holidays which occur on rather irregular pattern over one or more days. The error term ε_t is still white noise and represents any idiosyncratic changes of the model.

What is more, one of the features of g_t can be changepoint prior scale. The changepoints allow to incorporate trend changes in the growth models and stand for the points in time at which the trend is supposed to change its vector. It can be set manually otherwise it will be done automatically. This feature modulates the flexibility of the automatic changepoint selection. Larger values will allow many changepoints and small ones - few.

The authors frame the forecasting problem as a curve-fitting exercise, which differs from the models that account for the temporal dependence structure in the data. Although they miss some inferential benefits of using a generative model, e.g., the ARIMA model, their approach provides several practical advantages such as the fast fitting, ability to use irregular time data, flexible tuning of the trend, and seasonality behaviour.

4   Evaluation of the Accuracy in Performance Predictions

This section consists of two parts. First, we study the accuracy of the ARIMA and Prophet predictions on the time series of the transaction arrivals in the BTC blockchain. This allows us to obtain a punctual value of the prediction after τ hours from the last considered arrival of transaction and its confidence interval. Thus, for each epoch, we have a predicted expected value, a lower bound that represents the optimistic scenario and an upper bound leading to the pessimistic scenario.

The second contribution of the section is the estimation of the accuracy of the predictions on the expected confirmation time by means of Monte Carlo simulations of the confirmation process. The simulation uses as input three values of the confidence interval (lower, upper and central) to obtain an optimistic, pessimistic and expected estimation of the confirmation time.

It is worthy of notice that, while the expected confirmation delay is monotonic increasing with respect to the arrival rate, the relation between waiting time and intensity of the arrival process is not linear and hence the intervals obtained in the confirmation delay predictions are not symmetric with respect to the prediction obtained using the expected arrival rate.

4.1   Comparison of Time Series Prediction Models

This section describes the accuracy of the estimates as well as their insights obtained by the aforementioned prediction models.

In order to collect the time series, we have installed a BTC mining node and logged the transactions announced at its Mempool. We have collected the data for five days and obtained our dataset that was coherent with the information available on specialised websites but with higher granularity (see Fig. 2a). Additionally, we analysed the distribution of transaction fees of the Bitcoin clients in heavy load conditions (see Fig. 2b).

In order to train the models, we divided our dataset in two parts with the same size: the first one has been used to train the models, while the second part has been used to assess the accuracy of the prediction.

For both the models, we use prediction intervals with a coverage of 95%.

Figure 3 and 4 show predictions of the transaction arrival intensity provided by the Prophet and ARIMA models, respectively. What is more, Fig. 3a and 3b illustrate the prediction deviation due to the choice of different changepoint prior scale values, namely, 0.06 and 0.07 accordingly. Thus, the outcome of the Prophet model at the parameter 0.07 gives the best prediction, according to our experiment. In our assessment, we will use the best results.

For both the plots, we used the first 2.5 days of data to train the model, and then we predicted the future arrivals. We show the test data of our dataset (blue line), the prediction of the model (red line) and the confidence intervals (grey lines). As expected, as the prediction time is moved far in the future, the confidence interval becomes wider. However, for practical applications, predictions are useful when performed within approximately 10 or 12 h, otherwise it is very likely that the transaction is delay tolerant.

Even before formally testing the accuracy of the predictions with an error measure, we may notice that Prophet seems to give a better accuracy in this context.

Now, we consider the predictions of the Prophet and ARIMA model at fixed time intervals. More precisely, given an interval τ, at each time t we use all the data up to t to train the model, and forecast the value of the time series at time $t + \tau$.

Figure 6 shows the comparison of the predictions obtained with the Prophet and ARIMA models for different values of τ. We can see that, although the ARIMA predictions tend to be more noisy, both the models show rather good predictions of the test data.

More precisely, in Table 1 we compute the absolute errors of the Prophet and ARIMA predictions. According to our experiments, Prophet outperforms ARIMA, especially for short term predictions. Henceforth, we will carry out our experiments by using the Prophet model (Figs. 5, 7 and 8).

4.2   Simulations

In this section, we are interested in determining the accuracy of the estimation of the expected confirmation time using the Prophet prediction model to determine the arrival intensity of the transactions.

We resort to Monte Carlo simulations whose structure can be summarised as follows:

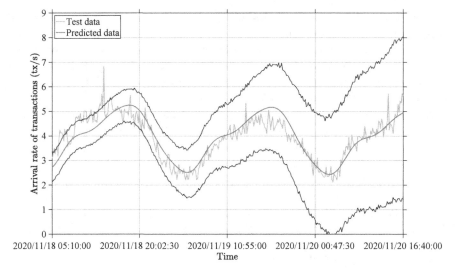

(a) The comparison at changepoint prior scale of 0.06.

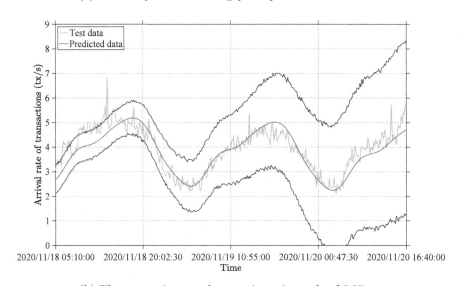

(b) The comparison at changepoint prior scale of 0.07.

Fig. 3. Comparison of the actual arrival rate of transactions and the predicted response based on the Prophet model with different changepoint prior scale.

- We consider a fixed sequence of transaction arrivals. This can be trace-driven by our dataset or obtained by the models (optimistic-, average- or pessimistic-case scenarios).
- The generation of the blocks occurs at random time intervals, exponentially distributed with average 10 min. This follows from the memoryless charac-

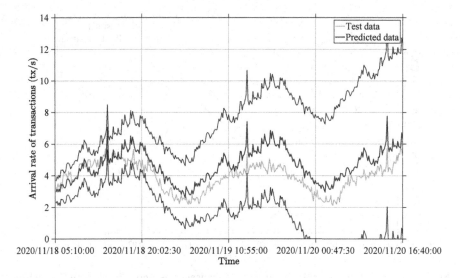

Fig. 4. Comparison of the actual arrival rate of transactions and the predicted response based on the ARIMA model.

Table 1. Mean absolute errors of prophet and ARIMA models with different size of the prediction horizon.

Prediction horizon τ in hours	Prophet error	ARIMA error
1	0.3120	0.3668
2	0.3366	0.3896
4	0.3966	0.4219
12	0.6333	0.6416

teristic of the mining process and from the invariant properties of the BTC blockchain.

- At a block generation instant, the most valuable transactions of the Mempool are confirmed and removed from the queue. We assume that the block contains 2,300 transactions. Transaction fees are chosen probabilistically using the distribution of Fig. 2b.
- Initially, the Mempool is populated with a fixed amount of transactions. These transactions offer a fee per byte according to the distribution of Fig. 2b. Notice that, although this is an approximation since the cheapest transactions tend to remain in the Mempool, the comparison remains fair since the initial Mempool population is the same for all the scenarios.

More precisely, we number the transactions from $-M$ to ∞, where M is the initial Mempool size, transaction 0 is the tagged transaction whose confirmation time is measured, and transaction denoted by $i > 0$ are those arriving after the tagged one.

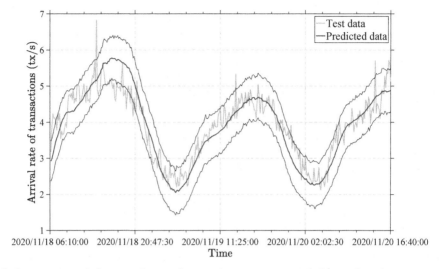

(a) Comparison of the actual arrival rate of transactions and the predicted response for $\tau = 1$ hour ahead based on the Prophet with changepoint prior scale of 0.07.

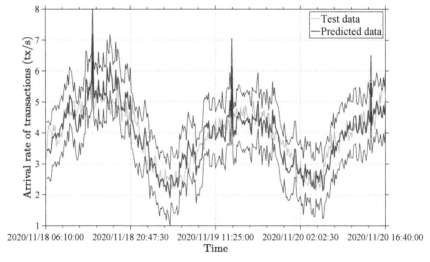

(b) Comparison of the actual arrival rate of transactions and the predicted response for $\tau = 1$ hour ahead based on the ARIMA model.

Fig. 5. Comparison of the Prophet and ARIMA prediction models at prediction horizon $\tau = 1$ h and confidence interval of 0.95.

Transaction t_i is denoted by a pair (τ_i, f_i), where τ_i is the arrival time and f_i the offered fee. For $i \leq 0$, $t_i = 0$. f_i is sampled from the distribution of Fig. 2b independently of τ_i. τ_i, for $i > 0$ are obtained from the real traces or from the predictions of Prophet. Notice that, in practice, the fees may be dependent

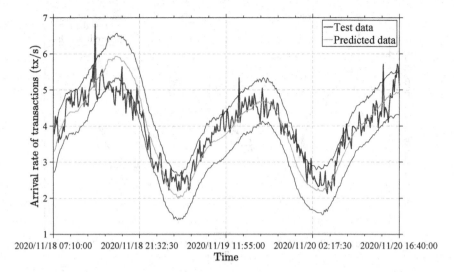

(a) Comparison of the actual arrival rate of transactions and the predicted response for $\tau = 2$ hours ahead based on the Prophet with changepoint prior scale of 0.07.

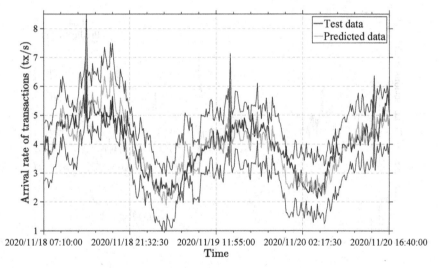

(b) Comparison of the actual arrival rate of transactions and the predicted response for $\tau = 2$ hours ahead based on the ARIMA model.

Fig. 6. Comparison of the Prophet and ARIMA prediction models with prediction horizon $\tau = 2$ h and confidence interval of 0.95.

from the system state (Mempool size, intensity of the arrival process) but in this context we use the simplifying assumption of independence since we mainly focus on the accuracy of the predictive power of the Prophet model.

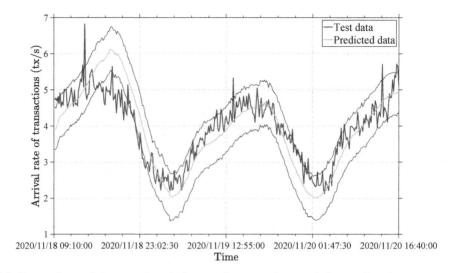

(a) Comparison of the actual arrival rate of transactions and the predicted response for $\tau = 4$ hours ahead based on the Prophet with changepoint prior scale of 0.07.

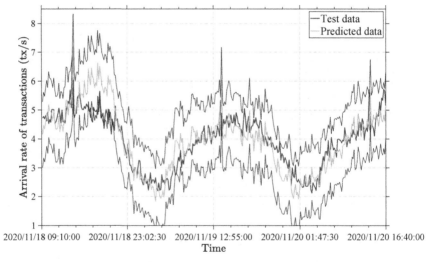

(b) Comparison of the actual arrival rate of transactions and the predicted response for $\tau = 4$ hours ahead based on the ARIMA model.

Fig. 7. Comparison of the Prophet and ARIMA prediction models with prediction horizon $\tau = 4$ h and confidence interval of 0.95.

Let $\mathcal{T}$ be the set of transactions.

Let $X_1, X_2, \ldots$ be the sequence of block consolidation times, and assume $X_0 = 0$. Then, $X_{i+1} - X_i$, $i \geq 0$, are i.i.d. exponential random variables with mean 10 min.

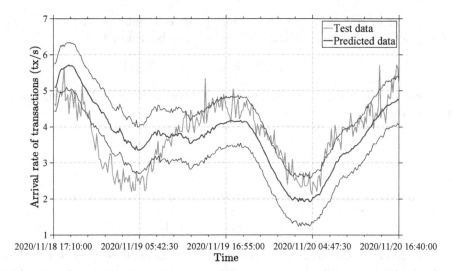

(a) Comparison of the actual arrival rate of transactions and predicted response for $\tau = 12$ hours ahead based on the Prophet prediction approach by Facebook with changepoint prior scale of 0.07.

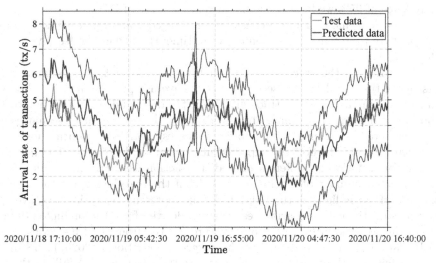

(b) Comparison of the actual arrival rate of transactions and predicted response for $\tau = 12$ hours ahead based on the ARIMA model.

Fig. 8. Comparison of the Prophet and ARIMA prediction models with prediction horizon $\tau = 12$ h and confidence interval of 0.95.

The state of the simulation model is described by a collection of transactions in the Mempool, denoted by $\mathcal{M}_i$, where the subscript i expresses that the state is associated with the instant immediately after the consolidation of block i.

The set of transactions arriving during the consolidation of the $(i + 1)$-th block, but after the consolidation of the i-th, can be denoted by:

$$\mathcal{A}_i = \{t_i \in \mathcal{T} : \tau_i > X_i \land \tau_i \leq X_{i+1}\}$$

Now, let $\mathcal{F}(\mathcal{M})$ the set of at most $2,300$ transactions with the highest fee present in $\mathcal{M}$.

Thus we have the following recursive relation:

- $\mathcal{M}_0 = \{t_i \in \mathcal{T} : \tau_i \leq 0\}$
- $\mathcal{M}_{i+1} = \mathcal{M}_i \cup \mathcal{A}_i \backslash \mathcal{F}(\mathcal{M}_i \cup \mathcal{A}_i)$

Thus, the confirmation time T_c for the tagged transaction is given by:

$$T_c = \min\{i : t_0 \notin M_i\}.$$

The Monte Carlo simulation experiment consists of $10,000$ samples of T_c for a fixed fee f_0. Then, the expected confirmation time is obtained by averaging the sample values. The experiments have been repeated 30 times and the estimates have been used to determine the confidence interval for the expected confirmation delay. To avoid confusion, we omit the confidence interval from the plot. For a confidence of 95% we have a maximum relative error of 7%.

For each scenario that we consider, the tagged transaction offers a certain fee per byte that controls the confirmation time: the higher the fee, the quicker the process.

According to the trace of arrival that we use, we obtain 4 estimates: the first using the real data, the second using the average prediction of Prophet, the third and fourth using the trace given by the lower and upper bounds of the confidence intervals determined by the Prophet. These two latter scenarios can be interpreted as pessimistic and optimistic cases in terms of confirmation delay.

Figure 9a and 9b show the expected number of blocks required for the transaction confirmation for different offered transaction fees. The grey bars refer to the expected number of blocks obtained from the real data while the stack of the second bar in light, normal and dark blue represent the optimistic-, average- and pessimistic- case scenarios, respectively, derived from the predicted data.

In the scenario of Fig. 9a, the data were derived from the time series shown in Fig. 3b, and the arrival time of the tagged transaction is 2020/11/18 05:10. Thus, the first half of the dataset was used to train the model. While in the second scenario (Fig. 9b), the arrival time for the tagged transaction is 2020/11/19 08:30:00, and hence the training data include all the series up to that epoch.

The inspection of Fig. 9a shows that the pessimistic case scenario for 70 sat/B is absent: this happens because the transaction is dropped before its confirmation (usually after 72 h of residence in the Mempool). A second observation is that, especially in heavy-load (70 and 80 sat/B), the distance between the optimistic prediction and the average is smaller than that from the pessimistic and the average. This is due to the non-linearity of expected response time of a queueing system with respect to the arrival intensity. Finally, for this scenario, we notice

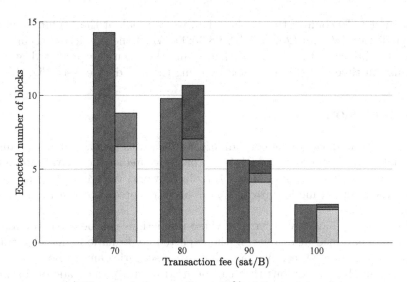

(a) The simulation results at 50% of the training data.

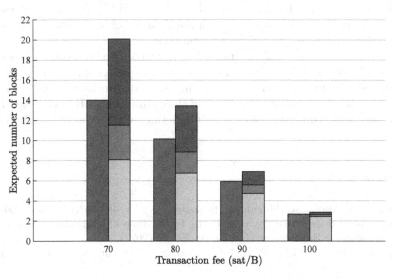

(b) The simulation results at 70% of the training data.

Fig. 9. Simulation results based on the actual data (grey bar) compared to the results of the Prophet predicted response (blue bars) with the optimistic, average and pessimistic cases and the initial Mempool occupancy of 10,000 transactions and different amount of the training data. (Color figure online)

that while the prediction obtained with the dataset is always within the optimistic and pessimistic cases, it seems to be closer to the latter. Indeed, Fig. 3b shows that predicted values for the first period of time are rather underestimated by the model. To confirm this explanation, we can look at the beginning of the next prediction interval (2020/11/19 08:30:00) when the prediction accuracy is higher. In this case, there is a good matching between the predicted average confirmation time and that obtained by using the real dataset (see Fig. 9b).

5   Conclusion

In this paper, we have applied two different time series forecast models, namely the Prophet by Facebook and ARIMA, in order to predict the arrival rate of the transactions at the Mempool of the Bitcoin network. According to our experiments, the Prophet model provides more accurate predictions in terms of the absolute errors.

Moreover, we have investigated if these predictions can be used to parameterise a model aimed at estimating the expected confirmation time of a transaction given its offered fee. We have shown two scenarios and in both cases we obtained valuable predictions that can be used to study the trade off between the blockchain running costs and the quality of service.

Although our study has been carried out for the BTC blockchain, it can be extended to any similar system where transactions are chosen from the Mempool according to an auction (e.g., Ethereum blockchain).

Future works have several directions. First, it would be important to compare the approach proposed here with other forecasting models, e.g., based on machine learning. Second, an analytical model of the queueing processes associated with the transactions should be studied to avoid the computationally expensive Monte Carlo simulations required to obtain the prediction on the expected confirmation time.

References

1. Balsamo, S., Marin, A., Mitrani, I., Rebagliati, N.: Prediction of the consolidation delay in blockchain-based applications. In: Proceedings of International Conference on Performance Engineering (ICPE), pp. 81–92 (2021)
2. Box, G.E., Jenkins, G.M., Reinsel, G.C., Ljung, G.M.: Time Series Analysis: Forecasting and Control. Wiley, Hoboken (2015)
3. Decker, C., Wattenhofer, R.: Information propagation in the bitcoin network. In: IEEE P2P 2013 Proceedings, pp. 1–10. IEEE (2013)
4. Faghih Mohammadi Jalali, M., Heidari, H.: Predicting changes in Bitcoin price using grey system theory. Financ. Innov. 13(6) (2020)
5. Fourneau, J., Marin, A., Balsamo, S.: Modeling energy packets networks in the presence of failures. In: Proceedings of 24th IEEE International Symposium on Modeling, Analysis and Simulation of Computer and Telecommunication Systems, MASCOTS, pp. 144–153. IEEE Computer Society (2016)
6. Gallina, L., Hamadou, S., Marin, A., Rossi, S.: A probabilistic energy-aware model for mobile ad-hoc networks. In: Al-Begain, K., Balsamo, S., Fiems, D., Marin, A. (eds.) ASMTA 2011. LNCS, vol. 6751, pp. 316–330. Springer, Heidelberg (2011). https://doi.org/10.1007/978-3-642-21713-5_23

7. Kasahara, S., Kawahara, J.: Effect of Bitcoin fee on transaction-confirmation process. J. Ind. Manag. Optim. **15**(1), 365–386 (2019)
8. Kawase, Y., Kasahara, S.: Priority queueing analysis of transaction-confirmation time for bitcoin. J. Ind. Manag. Optim. **16**(3), 1077–1098 (2020)
9. Li, J., Yuan, Y., Wang, F.-Y.: Analyzing Bitcoin transaction fees using a queueing game model. Electron. Commer. Res. 1–21 (2020). https://doi.org/10.1007/s10660-020-09414-3
10. Malakhov, I., Marin, A., Rossi, S., Smuseva, D.: Fair work distribution on permissioned blockchains: a mobile window based approach. In: Proceedings of IEEE International Conference on Blockchain, pp. 436–441. IEEE (2020)
11. Mudassir, M., Bennbaia, S., Unal, D., Hammoudeh, M.: Time-series forecasting of bitcoin prices using high-dimensional features: a machine learning approach. Neural Comput. Appl. (2020)
12. Nakamoto, S.: Bitcoin: a peer-to-peer electronic cash system (2009). http://www.bitcoin.org/bitcoin.pdf
13. Taylor, S.J., Letham, B.: Forecasting at scale. Am. Stat. **72**(1), 37–45 (2018)
14. Zarir, A.A., Oliva, G.A., Jiang, Z.M., Hassan, A.E.: Developing cost-effective blockchain-powered applications: a case study of the gas usage of smart contract transactions in the ethereum blockchain platform. ACM Trans. Softw. Eng. Methodol. (TOSEM) **30**(3), 1–38 (2021)

A Petri Net Formalism to Study Systems at Different Scales Exploiting Agent-Based and Stochastic Simulations

M. Beccuti[1], P. Castagno[1], G. Franceschinis[2](✉), M. Pennisi[2], and S. Pernice[1]

[1] Dipartimento di Informatica, Università di Torino, Turin, Italy
{beccuti,castagno,pernice}@di.unito.it
[2] Università del Piemonte Orientale, Alessandria, Italy
{giuliana.franceschinis,marzio.pennisi}@uniupo.it

Abstract. The recent technological advances in computer science have enabled the definition of new modeling paradigms that differ from the classical ones in describing the system in terms of its components or entities. Among them, Agent-Based Models (ABMs) are gaining more and more popularity thanks to their ability to capture emergent phenomena resulting from the interactions of individual entities. However, ABMs lack a formal definition and precisely defined semantics. To overcome this issue we propose a new method exploiting Petri Nets as a graphical meta-formalism for modeling a system from which an ABM model with clear and well-defined semantics can be automatically derived and simulated. We aim to define a framework, based on a PN formalism, in which a system can be efficiently studied through both Agent-Based Simulation and classical Stochastic one depending on the study goal.

Keywords: Agent based modeling and simulation · Stochastic simulation · Extended stochastic symmetric Petri nets

1 Introduction

In the last decades, computational modeling has become increasingly common for studying real-world phenomena, either natural or artificial, by using different modeling perspectives. Indeed, it is possible to focus on the specific mechanisms driving the phenomena of interest (microscopic point of view), or, on the other hand, to model the system's overall behavior, allowing to study the interaction with the external environment (macroscopic point of view). According to this, researchers may exploit different modeling approaches to capture the *macro* or the *micro* behavior of the system.

Roughly speaking, the approaches focusing on the *macro* behavior of systems are further classified into *deterministic* or *stochastic*. Deterministic models are typically formulated as systems of differential equations (in continuous time) or difference equations (in discrete time) and provide an average description of the system evolution at the population scale [10]. Stochastic ones are

© Springer Nature Switzerland AG 2021
P. Ballarini et al. (Eds.): EPEW 2021/ASMTA 2021, LNCS 13104, pp. 22–43, 2021.
https://doi.org/10.1007/978-3-030-91825-5_2

instead formulated as stochastic processes defined on families of random variables (e.g. Discrete-Time Markov Chain - DTMC, Continuous Time Markov Chain - CTMC) and they are particularly suitable in cases where randomness plays an important role [2]. It is important to observe that the system's complexity may prevent the explicit generation and solution of the underlying stochastic process so that Stochastic Simulation Algorithm (SSA) [8] and its approximations [3,9] end up being the only possibility.

Among the approaches for modeling a system in terms of its *micro* behavior, those based on the Agent-Based Modeling (ABM) paradigm are quite powerful: in this case the system is described in terms of interactions among its individuals, namely agents [12]. This approach is more suitable for studying specific spatial aspects thanks to its ability to represent a more realistic environment in which the agents interact, as well as any agent's specific behavior.

In the literature, various works highlight the strengths of ABM over other modeling techniques in terms of their ability to (i) capture emergent phenomena from the interactions of agents; (ii) provide a natural description of the system. However, to the best of our knowledge, ABM suffers from a lack of unique well-defined semantics specifying how the agents and environment behaviors are coupled and scheduled. One aspect concerns the time advance mechanism [11] that can be Fixed Interval Time Advance (FITA) or Next Event Time Advance (NETA). The first advances the current time of a fixed and constant amount, while the latter employs an event-driven approach, increasing the simulation time to that of the first event occurring in time. Another relevant aspect common to many case studies using ABM is how conflicts and concurrency among the agents are handled. Indeed in most cases, the ABM scheduler may introduce implicit (probabilistic or deterministic) rules in the ordering of events that may affect the simulation outcome. These two aspects are deeply discussed in [7], where the authors compare SSA and ABM simulations through four increasingly complex models of the immune system.

To deal with this crucial aspect, we propose to use high-level Stochastic Petri Nets (SPN) as a graphical meta-formalism for modeling a system and derive both an SSA simulator and an ABM one with clear and well-defined semantics based on the same underlying CTMC. This approach has allowed us to define a graphical framework where users can study a system through different modeling perspectives without any intervention on the model description itself. According to the questions of interest, users can exploit SSA simulation to efficiently derive measures on the overall system or ABM one to study the behavior of the single system entities and understand how the global behavior is derived by the interactions of these entities.

2 Background

Before describing how an ABM with clear and well-defined semantics can be automatically derived from the PN model, in this section, we briefly introduce the Extended Stochastic Symmetric Nets (ESSN) and the ABM formalisms.

2.1   The ESSN Formalism

Among the existing SPN formalisms in this paper we focus on Extended Stochastic Symmetric Nets (ESSN) [13], a high-level SPN formalism extending Stochastic Symmetric Net (SSN) [6] with the possibility of easily defining complex rate functions. Such extension has proven its effectiveness in modeling epidemiological and biological systems in a compact, readable and parametric way [4,14,15].

The formal ESSN definition is reported in Appendix 3, while hereafter we describe the main features of this formalism using the ESSN example in Fig. 1.

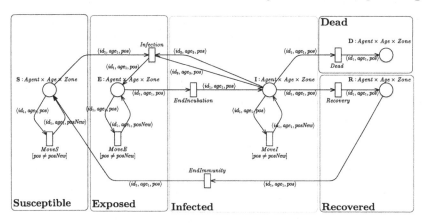

Fig. 1. The SEIRS ESSN model. Initial marking: N agents in place **S** (uniformly distributed among $|Zone|$ positions and 5 agents in place **I** (all in the same position).

An ESSN is a bipartite graph whose nodes are *places*, and *transitions*. Places, graphically represented as circles, denote a system local state while transitions, graphically represented as boxes, model the system events. Places and transitions are connected by directed and annotated *arcs*, which express the relation between states and event occurrences. Let us consider the ESSN model depicted in Fig. 1, which represents an extended version of the Susceptible-Exposed-Infected-Recovered-Susceptible (SEIRS) model considering the population space and age distribution, for more details we refer to Sect. 4. The model is defined by five places **S**, **E**, **I**, **R**, and **D** modeling susceptible, exposed, infected, recovered, and dead individuals, respectively, and eight transitions defining the possible events, for instance transition *Recovery* represents the recovery of infected individuals.

A place p can hold tokens belonging to the place color domain $cd(p)$ where color domains are defined by the Cartesian product of elementary types called *color classes*, $\mathcal{C} = \{C_1, \ldots, C_n\}$, which are finite and disjoint sets, and might be further partitioned into (static) subclasses. For instance, considering the SEIRS example, the color domains of all places is defined by the product of the following three color classes: *Agents*, *Age* and *Zones*. The former represents the color class of agents ids (i.e., individuals) in the model, while the last two classes are the features that are associated with each agent. Specifically, *Age* is subdivided into three subclasses A_0, A_1, and A_2, representing the population age structure, comprising three sub-classes. While *Zones* represents the different positions that each agent might stay in.

Similarly, a color domain is associated with each transition and is defined as a set of typed variables where the variables are those appearing in the functions labeling the transition arcs and their types are the color classes. Thus, *var* assigns to each transition $t \in T$ a set of variables, each taking values in a given element C_i of $\mathcal{C}$ (the variable's type); fixed order on the set of variables, the color domain of t, $cd(t)$, is defined as the Cartesian product of its variables' types. Then, we can define an instance, denoted as $\langle t, c \rangle$, of a given transition t as an assignment (or binding) c of the transition variables to a specific color of a proper type. Arcs are annotated by the functions $I[p, t]$, if the arc connects a place p to a transition t, or $O[p, t]$ for the opposite direction. The evaluation of $I[p, t]$ (resp. $O[p, t]$), given a legal binding of t, provides the multiset of colored tokens that will be withdrawn from (input arc) or added to (output arc) the place connected to that arc by the firing of such transition instance. The set of input/output places of transition t is denoted by $\bullet t/t \bullet$. Considering the *Recovery* transition in Fig. 1, its color domain is defined as $cd(Recovery) = Agent \times Age \times Zone$, and the firing of color instance $\langle Recovery, id_1 = 1, age_1 = young, pos = zone_0 \rangle$ would remove from place $\mathbf{I}$ a token representing the *young* agent with id 1 staying in position $zone_0$, and add it to place $\mathbf{R}$. This is due to the arc expression $\langle id_1, age_1, pos \rangle$ annotating both input and output arcs of transition *Recovery*.

It is possible to associate specific *guards* with transitions: a guard is a logical expression defined on the color domain of the transition, which can be used to define constraints on its legal instances. A transition instance $\langle t, c \rangle$ is enabled and can fire in a marking[1] m, if its guard evaluated on c is true, and for each input place p we have that $I[p, t](c) \leq m[p]$, where $\leq$ is the comparison operator between multisets. The firing of $\langle t, c \rangle$ in m produces a new marking m' such that, for each place p, we have $m'[p] = m[p] + O[p, t](c) - I[p, t](c)$. The set of all instances of t enabled in marking m is denoted by $E(t, m)$.

Finally, each transition is associated with a specific *velocity*, representing the parameter of the exponential distribution modeling its random firing time. Considering the ESSN, the set of transition T is divided into two subsets T_{ma} and T_g depending on the associated velocity. The former subset contains all transitions which fire with a velocity defined by the Mass Action (MA) law [17]. The latter includes all transitions whose random firing times have velocities that are defined as general real functions. Hence, we will refer to the transitions belonging to T_{ma} as standard transitions and as general transitions those in T_g. This allows to easily model events that do not follow the MA law but more complex function. Observe that in the SEIRS example in Fig. 1, we considered only standard transitions for simplicity, but general ones are anyway taken into account in the translation algorithm proposed in Sect. 3. Let us define $\hat{m}(\nu) = m(\nu)_{|\bullet t}$, where the notation $m_{|P_i}, P_i \subset P$—denotes the projection of the marking m on a subset P_i of places; the rate parameter associated with an enabled transition instance $\langle t, c \rangle$ is given by the function

---

[1] A marking m is a P-indexed vector, $m[p]$ is a multiset on $cd(p)$, and $m[p][c]$ is the multiplicity of $c \in cd(p)$ in the multiset $m[p]$; $m(\nu)$ denotes the marking at time ν.

$$F(\hat{m}(\nu), t, c, \nu) := \begin{cases} \varphi(\hat{m}(\nu), t, c), & t \in T_{ma}, \\ f_{\langle t, c \rangle}(\hat{m}(\nu), \nu), & t \in T_g, \end{cases} \tag{1}$$

In particular, $\varphi(m(\nu), t, c)$ is the MA law, i.e.

$$\varphi(m(\nu), t, c) = \omega(t, c) \prod_{\langle p_j, c' \rangle |\ p_j \in {}^\bullet t\ \wedge\ c' \in cd(p_j)} m[p_j][c'](\nu)^{I[p_j, t](c)[c']} \tag{2}$$

with $\omega(t, c)$ the MA constant rate parameter of the enabled transition instance $\langle t, c \rangle$. Observe that $\varphi(\hat{m}(\nu), t, c)$ and $f_{\langle t, c \rangle}(\hat{m}(\nu), \nu)$ can depend only on the time ν and the marking of the input places of transition t at time ν. Stochastic firing delays, sampled from a negative exponential distribution, allow one to automatically derive the underlying CTMC that can be studied to quantitatively evaluate the system behaviour [1]. In details, the CTMC state space, $\mathbb{S}$, corresponds to the reachability set of the corresponding ESSN, i.e., all possible markings that can be reached from the initial marking. The Master equations (MEs) for the CTMC are defined as follows:

$$\frac{d\pi(m_i, \nu)}{d\nu} = \sum_{m_k} \pi(m_k, \nu) q_{m_k, m_i} \qquad m_i, m_k \in \mathbb{S} \tag{3}$$

where $\pi(m_i, \nu)$ represents the probability to be in marking m_i at time ν, and q_{m_k, m_i} the element of the *infinitesimal generator* (i.e., the velocity to reach the marking m_i from m_k), which is defined as follows:

$$q_{m_k, m_i} = \sum_{\substack{t \in T \wedge c' \in cd(t) \wedge \\ \langle t, c' \rangle \in E(t, m_k)_{|m_i}}} F(m_k, t, c', \nu). \tag{4}$$

with $E(t, m_k)_{|m_i}$ is the subset of $E(t, m_k)$ whose firing leads to marking m_i.

In complex systems, the equations (3) are often computationally intractable and several techniques can be exploited to study the system taking into account stochasticity. The Stochastic Simulation Algorithm (SSA) [8] is an exact stochastic method used to simulate systems, whose behaviour can be described by the MEs. Although the SSA was proposed to simulate chemical or biochemical systems of reactions, it can be easily extended to simulate different systems. Since in this work the algorithm is directly applied to ESSN models, here we describe it by using the ESSN notation. Let us consider an ESSN model whose state at time ν is described by marking $m(\nu)$. Thus, given the MEs introduced in Eq. 3 and the exponential distribution properties[2], the time necessary (τ) to the next transition firing is the exponentially distributed random variable whose mean is defined as:

$$\frac{1}{q_{m(\nu)}}, \qquad \text{with} \qquad q_{m(\nu)} \equiv \sum_{m_i \in \mathbb{S},\ m_i \neq m(\nu)} q_{m(\nu), m_i}. \tag{5}$$

---

[2] Let $\{X_i\}_{i=1}^n$ be independent exponentially distributed random variables with parameters $\{\lambda_i\}_{i=1}^n$ respectively. Then 1) $X = \min(X_1, \ldots, X_n)$ is exponentially distributed with parameter $\sum_{i=1}^n \lambda_i$, and 2) $Prob(X_i = \min(X_1, \ldots, X_n)) = \frac{\lambda_i}{\sum_{i=1}^n \lambda_i}$.

To advance the system from $m(\nu)$ to $m(\nu + \tau)$, the *inverse transform sampling* approach is exploited. Specifically, two random numbers r_1 and r_2 are sampled from the uniform distribution in the unit interval, and then the time τ to the next reaction is obtained from the following equation:

$$\tau = \frac{1}{q_{m(\nu)}} ln(\frac{1}{r_1}). \tag{6}$$

While, the next transition instance to fire is the first $\langle t, c \rangle \in E(t, m(\nu))$ (assuming an arbitrary total order of the instances in the set) satisfying

$$\sum_{\langle t',c' \rangle \in E(t,m(\nu)), \langle t',c' \rangle \leq \langle t,c \rangle} F(m(\nu), t', c', \nu) > r_2 \, q_{m(\nu)}. \tag{7}$$

The system state is then updated according to the marking change $O - I$ associated with the transition instance $\langle t, c \rangle$, and this process gets repeated until some final time or condition is reached.

2.2   The ABM Formalism

Agent-based Modeling and Simulation (ABMS) follows a bottom-up approach, in which the global behavior of a system results from the local behaviors of the individual particles.

While there is no clear and universally recognized consensus on the definition of ABM, it is possible to delineate some peculiar characteristics. First, ABMs are based on *agents*: identifiable and discrete objects or individuals that can be heterogeneous in nature. The agents may possess internal features or states used to describe some relevant properties and intrinsic characteristics; the features may be static or dynamic, may differ from an agent to another, and usually influence the behavior of the agents.

Agents act and adapt their internal and external behavior following predefined rules, and according to their internal states (non-interaction-driven dynamics). Furthermore, agents can interact with each other and behave consequently (interaction driven dynamics), according to interaction rules that may define both the requirements (i.e., the required value of the internal features of the agents and/or the type of the agents) enabling the interaction occurrence and the results of the interaction (i.e., state-change, death and birth of the agents).

Interactions can be deterministic or stochastic. In the former case, if all requirements are satisfied the interaction occurs in a deterministic manner. In the latter, a probability $\gamma \in (0, 1)$ is defined, possibly as a function of the agents' internal states and/or types, and the interaction is treated as a Bernoulli event. Rules may be very simple, as if-then rules, or very complex, as rules that make use of complex inner models from, for example, the fields of AI or Bioinformatics.

The way and the protocols used to describe how an agent behaves may largely vary from an implementation to another. This also holds for the mechanisms used to describe non-interaction driven dynamics.

It is worth noting that most ABMs used for the simulation of population dynamics include a description of the spatial domain, that can be either discrete or continuous. In such cases, interactions among agents may depend also on their physical proximity inside the environment. Furthermore, agents may be allowed to move and diffuse throughout the environment, or in parts of it. While the agent position could be considered as an internal feature of the agent itself, and thus managed in the context of the internal-driven dynamics, it is usually represented explicitly and separately from the other features, mainly for the visualization of agents distribution in space.

We can define an ABM as $\langle \mathcal{P}, \mathcal{F}, \mathcal{S}, \mathcal{I}, \Omega(0) \rangle$ where:

- $\mathcal{P} = \{\mathcal{P}_1, \mathcal{P}_2, ... \mathcal{P}_n\}$ is a finite set of agent types;
- $\mathcal{F} = \{\mathcal{F}_1, \mathcal{F}_2, ... \mathcal{F}_m\}$ is a finite set of agent features (i.e., attributes);
- $\mathcal{S}$ ($\mathcal{S}_{\mathcal{P}_i}$) is the set of all the possible configurations of agents (of type $\mathcal{P}_i$);
- $\mathcal{I}$ is the set of interaction/transition rules;
- $\Omega(0)$ defines the initial set of agents and their configuration at time 0.

We define $\mathcal{F} = \{\mathcal{F}_1, \mathcal{F}_2, ... \mathcal{F}_m\}$ as a set of agent features possibly owned by agents. Each feature $\mathcal{F}_i$, can be a limited or unlimited integer or real variable used to represent the specific attributes of an agent type (e.g., position, age, energy or activity state, ...). $\mathcal{S}_{\mathcal{P}_i} \in \mathcal{S}$ is the set of possible attribute configurations of type $\mathcal{P}_i$ agents. More specifically, $\mathcal{S}_{\mathcal{P}_i}$ is a subset of the Cartesian product of the features $\{\mathcal{F}_{k_1}, \mathcal{F}_{k_2}, ..., \mathcal{F}_{k_h}\} \subseteq \mathcal{F}$ associated with the agent type $\mathcal{P}_i$.

Each agent $e_j \in \Omega(t)$ is characterized by a type $\widetilde{\mathcal{P}} \in \mathcal{P}$, and its configuration $s' \in \mathcal{S}_{\widetilde{\mathcal{P}}}$ at time t. Note that even if two agents of the same type $\widetilde{\mathcal{P}}$ have the same identical configuration $s' \in \mathcal{S}_{\widetilde{\mathcal{P}}}$ at time t, they will be identified and treated as distinct entities. We assume that agents cannot change type over time.

The set of interaction/transition rules $\mathcal{I}$ can be partitioned in two sets $\mathcal{I}_{tran}$ and $\mathcal{I}_{int}$, with $\mathcal{I}_{tran} \cap \mathcal{I}_{int} = \emptyset$. $\mathcal{I}_{tran}$ refers to the transition rules describing non-interaction driven dynamics. $\mathcal{I}_{int}$ defines the set of rules describing interaction driven dynamics. In general, a rule defines:

- A list of involved agents (including the newborn) with their type;
- The precondition on the agents' configurations expressed through a boolean function on their state;
- An occurrence probability or rate expressed as a real function that may depend on the internal states of the involved agents;
- The definition of the state-changes caused by the rule execution on the involved agents, possibly creating new agents or eliminating existing ones. It can be expressed as a set of functions, one for each involved agent (including the newborn). Each function is defined on the configuration of all (already existing) involved agents, and its codomain is the set of possible configurations of the corresponding agent type plus a *null* configuration modeling the agent *death*.

It is worth noting here that while discrete-time ABMs generally use γ as a probability for the application of a stochastic transition, continuous-time

ABMs use γ to represent a transition rate characterizing the random variable modeling the waiting time for the application of the rule.

So, when a rule has to be applied, one or more agents $e_i \in \Omega(t)$ are selected, the preconditions on the agents' types and configurations defined by the interaction rule are checked and, if the rule is applied (according to γ), the configuration of the involved agents is changed according to the rule definition. Furthermore, the application of the rule may lead to the death of some of the involved agents, and/or to the introduction of newborn agents.

Let $e_i, [e_j, \ldots] \in \Omega(t)$ be one or more agents, the application of a generic rule to the selected agents can be denoted as follows:

$$\mathcal{I}([e_i(\mathcal{P}_i, s'), e_j(\mathcal{P}_j, s''), \ldots]) \xrightarrow{p}$$

$$\langle [e_i(\mathcal{P}_i, \tilde{s}'), e_j(\mathcal{P}_j, \tilde{s}''), \ldots, e_{new1}(\mathcal{P}_{new1}, s_{new1}), \ldots] \rangle$$

Rules that produce newborn agents without involving any existing agent are also possible: $\mathcal{I} \xrightarrow{p} \langle e_{n+1}(\mathcal{P}_j, s') \rangle$ creates a new agent of type $\mathcal{P}_j$ and configuration s'.

Generally, the evolution of the system is obtained by continuously applying the interaction/transition rules to the agents that satisfy the preconditions defined by rules themselves. To this end, it is fundamental to define a scheduling function **F** that establishes the policies on the order of the agents, on the order of the rules that will be applied on them, and manages possible conflicts and the synchronous or asynchronous updating of agents. However there is no universal consensus on how such a scheduler should behave, and therefore any ABM tool and implementation may use a different approach.

Discrete-time ABMs usually make use of a FITA approach. In this scenario, the evolution proceeds by discrete time-steps of the same length δ. The application of scheduling function **F** to $\Omega(t)$ moves the system from $\Omega(t)$ to $\Omega(t + \delta)$, by executing a series of actions (each action being the application of a rule to a given set of compatible agents), with a given order. However, it is supposed that all the applied actions happen within the same interval $(t, t + \delta)$.

Depending on the scheduling policies of ABM tools (e.g. NetLogo), an agent already involved in a given action may be involved again for the application of another rule within the same time-step (for passively participating in another interaction, or for a rule that requires a different agent configuration).

In Continuous-time ABMs the time is instead continuously updated for each action (i.e., for each applied rule), and the scheduling function considers only one action at a time. Each possible action has an associated time which is a random variable with a given distribution. When this distribution is exponential all actions have an associated rate that is used to calculate a waiting-time, and the time increment $\delta(t)$ depends on the sum of the rates associated with all the possible actions that can be executed at time t.

Anyhow, in both approaches the role of the scheduling function and its implemented policies are fundamental to obtain reliable results. We will present in the next section our approach for the simulation of a continuous-time ABM.

3   Deriving an ABM from the ESSN Representation

The class of ESSN models that can be translated into an ABM following the procedure described in this section is called AB-ESSN, and comprises some additional annotation allowing to identify the agent classes and the space structure where the agents are located. Each place color domain is a Cartesian product of classes including exactly one agent class and possibly other classes used to represent the agent attributes including its location in space. The set of places can thus be partitioned in as many subsets as the number of agent classes in the model, their initial marking must ensure that each agent identity color appears only once in the corresponding subset of places, and the net structure must imply that each such subset identifies a marking invariant of the model. Such invariance property can be checked using automatic structural analysis methods.

The transitions of AB-ESSNs may represent intra-agent events or interactions among two or more agents. The former type of transition causes a state change in only one agent, and it must comprise a variable, in its color domain, corresponding to the agent identity; it typically has one input and one output place from the subset of places for that agent type. The latter type of transitions involves two or more agents, and this is reflected in the corresponding color domain, which must comprise (at least) as many variables as the number of agents participating in the interaction: one of these agents is indicated as the active agent while the others are passive (the explicit choice of the agent with the active role may convey useful information on the model interpretation, however a default selection is also possible). The transition velocities can be defined through general functions but they cannot depend on the specific agent id color.

For each transition that does not involve agents' birth or death, the set of *input agents* is equal to the set of *output agents*, while for the transitions including birth or death of agents this is not true: this balance can be recovered by adding a dummy place which can be seen as a pool of fresh agent ids, from which a new id is withdrawn every time a new agent is generated, and where the ids of dead agents are put for reuse. This ensures the above mentioned marking invariance property.

Hereafter we assume that the modeler explicitly annotates the model to identify the agent classes and those representing the locations; the automatic identification of the agent classes in a not annotated ESSN is not trivial, although it may be pursued by checking which class satisfies the required constraints.

The set of agent types in the ABM coincides with the color classes representing agent ids in the AB-ESSN. For each agent type, its state is represented by a variable corresponding to the id of the place where the "agent token" is located and a data structure representing the position in the modeled space. In case the "agent token" in the AB-ESSN model carries additional information, representing the value of some agent feature, additional variables are included (whose type corresponds to the color class used to represent the attribute in the AB-ESSN: e.g. the age category in the SEIRS model).

The agent behavior must embed the state change rules for the ESSN projection on the corresponding agent type and the mechanism for the continuous time, event-driven simulation algorithm described in Algorithm 1. The following translation algorithm defines how to translate an AB-ESSN into an executable (NETA) agent-based simulation model.

The Translation Algorithm

- For each color class $C_{(i)}, i \in \{1, \ldots, m\}$, identifying an agent type, a Type $\mathcal{P}_i$ will be added to the set of possible agent types $\mathcal{P}$. For each other color class C_j a feature $\mathcal{F}_j$ is added to $\mathcal{F}$.
- Let G_i be the set of places whose color domain includes $C_{(i)}, (C_{(i)} \in cd(p), \forall p \in G_i)$ (an agent id class) then a new feature $\mathcal{F}_{\mathcal{P}_i}$ will be added to the set of possible agents features $\mathcal{F}$, and the set of possible values of $\mathcal{F}_{\mathcal{P}_i}$ is G_i. Moreover, the set of possible configurations for agents of type $\mathcal{P}_i$, $\mathcal{S}_{\mathcal{P}_i}$ will include $\mathcal{F}_{\mathcal{P}_i}$.
  For each place $p \in G_i$, let $cd(p) = C_1^{n_{p,1}}, \ldots, C_m^{n_{p,m}}$ be its color domain (where $n_{p,k}$ denotes the number of repetitions of class C_k in $cd(p)$), there shall be a feature $\mathcal{F}_j$ of type C_j in $\mathcal{F}$. Furthermore, $\mathcal{S}_{\mathcal{P}_i}$ will include as many repetitions of the feature $\mathcal{F}_j$ as the $max_j(n_{p,j})$, i.e., $\mathcal{F}_{j,1} \ldots \mathcal{F}_{j,max_j}$. So, depending on the value of the feature encoding the place identity, the corresponding subset of features will be used in the agent ABM specification. Concluding, $\mathcal{S}_{\mathcal{P}_i}$ will be defined as a subset of the Cartesian product among $\{\mathcal{F}_{\mathcal{P}_i}, \mathcal{F}_{j1,1}, \ldots, \mathcal{F}_{j1,max_{j1}}, \mathcal{F}_{j2,1}, \ldots, \mathcal{F}_{j2,max_{j2}}, \ldots\}$, with $\mathcal{F}_{j1}, \mathcal{F}_{j2}, \ldots \in \mathcal{F}$.
- For each transition t, a new rule $\mathcal{I}_j \in \mathcal{I}$ will be introduced. The color domain of the transition with its guard c, and the input arc expressions define the rule preconditions. The rule $\mathcal{I}_j$ will involve as many agents as the tokens identified by $I[p, t], \forall p \in^{\bullet} \mathbf{t}$. The conditions on the configuration of the involved agents will include a clause checking that the value of the feature corresponding to each agent state be equal to the input place, and other clauses on the features (transition variables) appearing in the transition guard c.
- Let $O[p, t], \forall p \in \mathbf{t}^{\bullet}$ be the expressions corresponding to the output arcs of transition t; the rule $\mathcal{I}_j$ will affect as many agents as the number of tokens produced by $O[p, t], \forall p \in \mathbf{t}^{\bullet}$. The function defining the state change will be derived from the input and output arc expressions of t. We note that if a transition consumes a token with a given color "id" $\in C_{(i)}$, and produces a new token with the same type and "id", but possibly in another place $\in G_i$, and/or with different values on the other features, the corresponding interaction rule $\mathcal{I}^*$ will not destroy the involved agent and will not create a new agent, but it will just change the internal configuration s' of the selected agent e_j of type $\mathcal{P}_i$ to s'', with $s', s'' \in \mathcal{S}_{\mathcal{P}_i}$.
- The rate function of the transition t (described by a Mass Action law function or a general function such as, for example, a Michaelis-Menten equation) will be used to determine the appropriate rate function γ of the corresponding rule.

We introduce now the definition of the scheduling algorithm for the simulation of the translated ABM. The main idea is to use an approach that is in line with the SSA algorithm presented in Sect. 2.1, but that takes also into account the concept of agent. To simplify the implementation in existing ABM frameworks, we suppose that the scheduler function will choose the next event by randomly selecting among all the agents (with a roulette-wheel method) the next agent with a probability proportional to the sum of the rates of the rules that can be applied on it, and thus selecting (again with a roulette-wheel method on the rates of the rules) a rule among all the rules that can be applied on the selected agent. While this process is quite straightforward for transition rules, some assumptions must be made for interaction rules, i.e., rules that involve more than one agent. In this scenario, we suppose that for each interaction rule $\in \mathcal{I}_{int}$ there is an active participant (identified by a given type $\mathcal{P}^*$, and a given configuration $s^* \in \mathcal{S}_{\mathcal{P}^*}$) that will take care of executing the rule, and one or more passive participants that will undergo the rule execution. As a consequence, the interaction rule rates will be only included in the calculation of the cumulative rates of their respective active agents. This also involves that for the calculation of the rule rates any active agent must be aware of the number of possible combinations of passive agents it can interact with.

When a given agent e^* is selected for the next event, and an interaction rule $\mathcal{I}^*$ has been selected among the possible rules associated with e^*, the passive agents will be randomly selected among those that are compatible with $\mathcal{I}^*$ and the rule will be applied.

For transitions with no input places, that can be used to represent environmental and/or global events such as the introduction of novel agents inside the simulation, we introduce, for the sake of the translation, a global "meta-agent" (the environment), that will be in charge of these rules. This meta-agent will have its own rate according to the associated rules, and will always compete with all the other real agents for the selection of the next agent and rule.

To obtain a more efficient implementation of the scheduler function we will take advantage of a common characteristic owned by the majority of ABM tools, i.e., the description of the physical space on which the agents move and behave. We then suppose that all agents own a specific feature $\mathcal{F}^*$ that will represent their position p_i inside a physical space P, with $p_i \in P$. For simplicity, we consider P as composed of a finite set of positions (i.e., discrete space). Multiple agents may be in the same p_i at the same time, but each agent can be in only one position at a given time. Movement is implemented using specific transition rules that change the position of the agents. Moreover, a notion of physical proximity derived from the position may limit the application of the interaction rules to agents that are within a given range (e.g., in the same position).

As a consequence, when an event occurs and an interaction rule is carried on some selected agents, the event will at most influence only the rates of the agents that are in the positions on which the rule had an effect, i.e., in the same or close positions for agents influenced by an interaction rule, or in the starting and ending position for an agent that moved. So, thanks to the memory-less property of the exponential distribution and the locality of rules, we can recalculate the agent rates only for the agents in the positions affected by the last event. The pseudo-code can be found as follows in Algorithm 1.

We note here that, since the state change in the ESSN is *local*, it is possible to introduce further optimizations in Algorithm 1 by limiting the recalculation of the enabled rules and of the corresponding rates exploiting such locality (partially implemented in the proposed algorithm using on agents' position). This can be directly derived from the net structure. Such optimization method is similar to the technique applied in [16].

Algorithm 1: Scheduler function for continuous-time ABM simulation

---

(Initialization);
CurrentTime ← 0;
foreach *position* p_i **do**
   | set the update status of position u_{p_i} ← 1;
end
(Algorithm execution);
while *CurrentTime* ≤ *FinalTime* **do**
   (Calculation of agents' individual rates);
   foreach *agent* e_j *in a position* p_i*, with* u_{p_i} = 1 **do**
      foreach *rule* I_h *compatible with the agent* e_j **do**
        | Calculate, for the agent e_j, the rule rate $\gamma_{(e_j,I_h)}$;
      end
      Set the agent cumulative rate γ_{e_j} ← $\sum \gamma_{(e_j,I_h)}$;
   end
   foreach *position* p_i *with* u_{p_i} = 1 **do**
      | set the update status of position u_{p_i} ← 0 ;
   end
   (Calculation of the global rate);
   γ_{tot} ← $\sum \gamma_{e_j}$;
   (Selection of the next agent and rule according to their rates);
   e^* ← roulette_wheel_selection(e_j,γ_{e_j});
   I^* ← roulette_wheel_selection(I_h,$\gamma_{(e^*,I_h)}$);
   Execute the rule I^* on agent e^* (and on other involved passive agents);
   foreach p_j *influenced by the execution of* I^* **do**
      | u_{p_j} ← 1;
   end
   (Time update);
   CurrentTime ← CurrentTime + $\frac{1}{\gamma_{tot}} \cdot \ln \left(\frac{1}{random\_float(0,1)} \right)$;
end

---

4 Case Study: SEIRS

In this section, we propose an extended version of the Susceptible-Exposed-Infected-Recovered-Susceptible (SEIRS) model considering the population space and age distribution (already introduced in Sect. 2.1). The model is defined by exploiting the AM-ESSN formalism, as depicted in Fig. 1. It is characterized by five modules, each including one place, representing the susceptible (**S**), exposed (**E**), infected (**I**), recovered (**R**), and dead (**D**) individuals.

Let us now briefly describe each module. The **Susceptible module** is characterized by place **S** representing the susceptible individuals, and it models individuals not exposed to the pathogen. Such individuals may get infected if they get in touch with an infected individual, through the *Infection* transition, becoming an exposed individual. The **Exposed module**, defined by place **E**, is characterized by the exposed individuals which do not show symptoms until the end of the incubation period. Through the *EndIncubation* transition the exposed individual becomes an infected individual modeled by place **I** in the **Infected module**. After a certain time period, an infected individual may (1) recover becoming a recovered individual, which is not contagious anymore: in this case, he/she moves from place **I** in the Infected module to place **R** in the **Recovered module** by means of transition *Recovery*; (2) die moving from place **I** in the Infected module to place **D** in the **Dead module**. A recovered individual eventually becomes susceptible again through the *EndImmunity* transition, but while in **R** state he/she can not be infected and remains out of the disease dynamics.

All demographic changes in the population (births, deaths not caused by the infection, and ageing) are explicitly disregarded in our model. Finally, the Susceptible, Exposed and Infected modules include a movement transition (*MoveS*, *MoveE*, and *MoveI* respectively) representing the movement of agents among different positions. It is possible to automatically verify that all places in this model are covered by a P-invariant which implies that if in the initial marking each token has a different agent id, then all reachable markings have this property, moreover each agent can be in only one state (S, E, I, R or D).

Application of the Translation to the SEIRS Example. We implemented the ABM translation of the SEIRS model using the approach described in Sect. 3. The model has been implemented on top of the NetLogo ABM framework [18]. The template examples to code an ABM model with NetLogo using the approach described here are given in the Appendix 1. In NetLogo agents are named *turtles*, space is explicitly represented by means of a particular type of agents, called patches, that do not move: each position is represented by a patch. In this example there is only one type of turtles; they own age (with three possible values) and a value called "pdelta" that is used to store the death rate according to the corresponding age. We also used another variable named "tgamma" to temporary store the cumulative agent rate. Patches own a specific property called "update", that is used for the local optimization described in the scheduling algorithm, and a property called "gamma" for the cumulative rate of the patch. Using the NetLogo "ask" command it is possible to tell the agents (turtles or patches) to do something. While the procedure for the selection of the next event, and thus of the next turtle to act, implements the scheduling Algorithm 1, the result of the rule logic has been instead implemented inside the turtle code using the "ask" command. We note here that, even if the transition rates of our SEIRS ESSN example are defined according to Mass Action law velocities, in principle any general real function can be used and translated. The complete SEIRS NetLogo model with all the code and comments is available at https://github.com/qBioTurin/AM-ESSN.

Simulation Experiments. Some simulation experiments have been performed using the SSA simulator integrated into GreatMod tool [4] on the (unfolded) ESSN model and using the Netlogo model automatically generated from the same ESSN model and implementing Algorithm 1. The first simulator is obtained

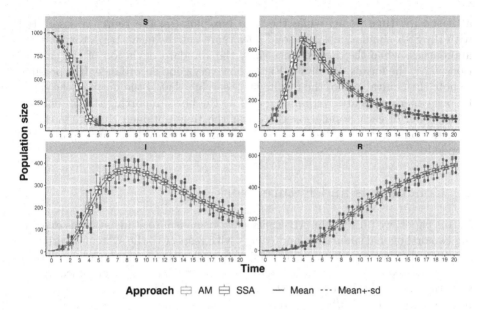

Fig. 2. Comparison among the trajectories (represented exploiting box-plots) generated exploiting the SSA (cyan) and AM (red) approaches considering 5 positions. (Color figure online)

after *decoloring* [5] the model of Fig. 1 w.r.t. the agents' id color class (*Agent*): this can be done without affecting the model behavior (by properly defining the transition MA rates so that the underlying CTMC is an exact lumping of the original one).

Table 1 shows the measured times varying number of positions and susceptible population size: these are obtained performing 1000 simulation runs using the ESSN derived SSA (this includes the code generation time from the unfolded ESSN and the actual simulation) and using the Netlogo simulator (which has been generated by applying the rules described in Sect. 3; the automatic translation is being implemented and some heuristics to speed up the simulation are being experimented). It is easy to see that the first approach is not affected by

Table 1. Execution times (sec.) of the two approaches.

# Pos.	#S	ESSN	ABMS
5	1000	54	328
10	1000	95	315
20	1000	623	332
25	1000	1656	333
5	2000	54	1200
5	4000	54	5259
5	8000	64	25860

the increase in the population size (since the agents id is not maintained) while the time increases with the number of positions; on the other hand, the ABMS time is not affected when increasing the number of positions while it increases when increasing the population size. Furthermore, in Fig. 2 results obtained with both approaches are plotted, considering 5 positions and 1000 simulations for each approach. In details, for each time point two box-plots are plotted: the red one (on the left) shows the variability of the distribution of the AM simulations at a specific time point, while the blue box-plot (on the right) is associated with the SSA simulations. The solid lines represent the mean trajectory of both approaches, while the dashed lines show the standard deviation of the simulations w.r.t. the means. Results in Fig. 2 show very good accordance. It is possible to note the equivalence of the results by observing that *i)* the mean of the simulations experiments are overlapping, and *ii)* the boxes reporting the probability distribution of each population class over the time obtained through AM and SSA are comparable.Further analysis regarding this equivalence are reported the Appendix 2.

5   Conclusions and Future Work

In this paper, we present the AB-ESSN metaformalism to study systems exploiting two orthogonal simulation approaches with respect to the system description detail level (macroscopic or microscopic). Specifically, our methodology derives an agent based or a stochastic simulator starting from a unique graphical representation, provided through the AB-ESSN formalism. The automatic translation from AB-ESSN to ABMS leads to a well defined semantics. The newly proposed scheduling algorithm for agents' actions makes event scheduling independent from any implementation choice about the order of the agents, and the execution of their actions by the underlying ABMS engine. Combining these two aspects provides coherence between the two simulation methodologies.

Closely related to the macro versus micro perspective, the ease of measuring specific aspects of our simulation models is a key aspect in selecting the most suitable computational approach (i.e. SSA or ABMS) for a given study. For instance, as shown by our experiment ABMS is more efficient in handling spatial aspects of the diseases' spread; while SSA scales up better when the population is increased. Moreover, measures taking into account the identities of the components/entities of the system are more easily derived using ABMS. For instance, considering the SEIRS model presented in Sect.4, the individuals' number who fall ill again in the ABM requires only to count, for each agent, the times this event happens. On the other hand, the computation of the same measure using SSA requires updating the ESSN model tracking each reinfection with an additional color, increasing the complexity of the model.

As further work, we will integrate the proposed algorithm in GreatMod (https://qbioturin.github.io/epimod/) a powerful framework developed by our group for the analysis of biological and epidemiological systems. We will work also to implement the proposed translation algorithm on top of other ABM frameworks, as FlameGPU (https://flamegpu.com/). Finally, the ABMS algorithm performance will be improved by taking advantage of the information on the dependency between rules that can be derived from the ESSN structure. Beyond that, we are studying an optimized version of the scheduling algorithm exploiting symmetries automatically derived from the ESSN structure.

Acknowledgements. The research work of G. Franceschinis has been partially supported by Università del Piemonte Orientale (Project MoDRI).

Appendix 1: NetLogo Templates from ESSN to ABM Translation

ESSN to ABM translation step:

> For each color class $C_{(i)}, i \in \{1, \ldots, m\}$, identifying an agent type, a Type $\mathcal{P}_i$ will be added to the set of possible agent types $\mathcal{P}$. For each other color class C_j a feature $\mathcal{F}_j$ is added to $\mathcal{F}$.

Netlogo template code:

```
;this is a comment!
breed [AGENTS_Pi  ANAGENT_Pi]
breed [AGENTS_Pj  ANAGENT_Pj]
breed [AGENTS_Pk  ANAGENT_Pk]
;``breed'' creates a new agent type, AGENTS_Pi refers to the entire
; population of type Pi, while  ANAGENT_Pi will be the keyword
; to indicate an agent of type Pi
```

ESSN to ABM translation step:

> Let G_i be the set of places whose color domain includes $C_{(i)}, (C_{(i)} \in cd(p), \forall p \in G_i)$ (an agent id class) then a new feature $\mathcal{F}_{\mathcal{P}_i}$ will be added to the set of possible agents features $\mathcal{F}$, and the set of possible values of $\mathcal{F}_{\mathcal{P}_i}$ is G_i. Moreover, the set of possible configurations for agents of type $\mathcal{P}_i$, $\mathcal{S}_{\mathcal{P}_i}$ will include $\mathcal{F}_{\mathcal{P}_i}$.

Netlogo template code:

```
; There is no Enum type in NetLogo, so the possible values must be
; initially set into the "setup" procedure, and managed through the
; interaction and transition rules.
AGENTS_Pi-own FPi
AGENTS_Pj-own FPj
AGENTS_Pk-own FPk
;Initial values and populations can be set inside the ``to setup''
;procedure, while the execution code can be set in the ``to go'' procedure
to setup [
create-AGENTS_Pi 10 [  ;; create 10 agents of type Pi
      set color red ;; set initial values...
      set FPi "SOMEINITIALVALUE"
   ]
]
to go [ ;execution code
ask AGENTS_Pi [do-something...]
]
```

ESSN to ABM translation step:

> For each place $p \in G_i$, let $cd(p) = C_1^{n_{p,1}}, \ldots, C_m^{n_{p,m}}$ its color domain (where $n_{p,k}$ denotes the number of repetitions of class C_k in $cd(p)$), there shall be a feature $\mathcal{F}_j$ of type C_j in $\mathcal{F}$. Furthermore, $\mathcal{S}_{\mathcal{P}_i}$ will include as many repetitions of the feature $\mathcal{F}_j$ as the $max_j(n_{p,j})$, i.e. $\mathcal{F}_{j,1} \ldots \mathcal{F}_{j,max_j}$. So, depending on the value of the feature encoding the place identity, the corresponding subset of features will be used in the agent ABM specification. Concluding, $\mathcal{S}_{\mathcal{P}_i}$ will be defined as a subset of the Cartesian product among $\{\mathcal{F}_{\mathcal{P}_i}, \mathcal{F}_{j1,1}, \ldots, \mathcal{F}_{j1,max_{j1}}, \mathcal{F}_{j2,1}, \ldots, \mathcal{F}_{j2,max_{j2}}, \ldots\}$, with $\mathcal{F}_{j1}, \mathcal{F}_{j2}, \ldots \in \mathcal{F}$.

Netlogo template code:

```
;Repeated for each Cj in cd(Gi)
AGENTS_Pi-own Cj
AGENTS_Pi-own Ch
;...
```

```
AGENTS_Pj-own Cl
AGENTS_Pk-own Ck
;...
; As said before,  no Enum type in NetLogo, Variables will be
; initially  set into the "setup" procedure,  and managed
; through the interaction and transition rules
```

ESSN to ABM translation step:

For each transition t, a new rule $\mathcal{I}_j \in \mathcal{I}$ will be introduced. The color domain of the transition with its guard c, and the input arc expressions define the preconditions that will be checked to select the rule $\mathcal{I}_j$. If $I[p,t]$ is the expression function on the input arcs on the transition t (for each $p \in^{\bullet} \mathbf{t}$) the rule $\mathcal{I}_j$ will involve as many agents as the tokens identified by $I[p,t]$, The conditions on the state of the involved agents will be determined according to the guard c of the transition and according to the functions on the input arcs that define conditions on the marking of the input places.Let $O[p,t], \forall p \in \mathbf{t}^{\bullet}$ be the expressions corresponding to the output arcs of transition t, The rule $\mathcal{I}_j$ will affect as many agents $e_i, e_j, \ldots$ as the number of tokens produced by $O[p,t]$. The function defining the state change will be derived from the input and output functions of t.

Netlogo template code:

```
; Let us suppose that NEXT_TO_INTERACT  is one of the active
; agents  of type Pi  that has been selected  according the
; scheduling procedure, and suppose the agent chooses  the next rule
; to be executed with a roulette wheel method.The other passive
; agents participating to the rule can be selected as follows
ask NEXT_TO_INTERACT [
  let FRIEND1 one-of AGENTS_Pj with [Cl = "SOMETHING" ]
  ;...
  let FRIEND2 one-of AGENTS_Pk   with [Ck = "SOMETHINGELSE" ]
; these two last statements are repeated as many times  as
; the number of passive agents required  for  the selected rule.
; With the "with" command we select only agents that satisfy
;the given preconditions on their state established by the rules
; If the rule is used to select another agent  of type Pi
; (i.e., of the same type as the active agent)  it is possible
; to use the statement
let FRIEND3 one-of other AGENTS_Pi with [...]
; To select n agents of a given type it is possible to use
; the statement ``n-of'' instead of ``one-of''.
; To apply the actions required by the rule to change the state
; on passive agents
  ask FRIEND1 [
    ;possible  commands (some examples)
    set Cl "SOMETHINGNEW"  ; changing internal configuration
    hatch_AGENTS_Pj 10 [   ; create 10 agents of type pj with a given configuration
    set Cl "SOMEPREDEFINEDVALUE"]
  ]
  ask FRIEND2 [
    ;possible example commands
    fd 1     ; move the agent ahead of one position
```

```
    set color "green"      ; change agent color
    die      ; kill the agent FRIEND2
]    ; We can also have actions for the active agent
do-something
die
;....
]
```

ESSN to ABM translation step:

> The rate function of the transition t will be used to determine the appropriate rate function p of the corresponding rule.

Netlogo template code:

```
; if the agents of type Pi have, for example, a rate r1 to change  their
; configuration  and rate r2 to interact (as active agents)  with
; agents  of type Pj (passive agents)  in the same position,
; the following code template can be used.
ask patches with [update = 1]  [set totPj count AGENTS_Pj-here  with [...]]
; The rule counts all passive agents Pj in each position (that
; needs recalculation) eventually with  given characteristics.
; totPj can be defined as a patch variable with the command patches-own
ask AGENTS_Pi-on patches with [update = 1] [
; calculation of agent cumulative rate only for agents on positions
; that need to be updated. tgamma is a turtle variable that contains
; its cumulative interaction rate
set tgamma  r1  + totPj  * r2 ]
; here rate calculation for other agents...
; calculation of cumulative gamma per position
; gamma is defined as a patch variable
    ask patches with [update = 1] [
        set gamma sum  [tgamma] of turtles-here
        set update 0 ]
; turtles-here selects all types of agents on the position
set gammatot sum [gamma] of patches ; global rate calculation
; time increment calculation
let increment ((-1 /(gammatot)) * ln(random-float 1))
set time  time + increment
; choice of next agent to interact using  aroulette wheel method.
; In this case we use the NetLogo rnd extension
let NEXT_TO_INTERACT rnd:weighted-one-of turtles [tgamma]
ask NEXT_TO_INTERACT [
; roulette wheel selection among the two possible actions of the agent
let p1 (totPj  * r2 / tgamma)
let pp random-float 1
; interaction rule code here
if-else pp < p1 [do-something... ]
; transition rule code here
[do-something-else...] ]
```

Appendix 2: Simulation Results

The curves in Figs. 2 and 3 show the number of Susceptible, Exposed, Infected and Recovered people in the time period (0, 20) with an initial state comprising

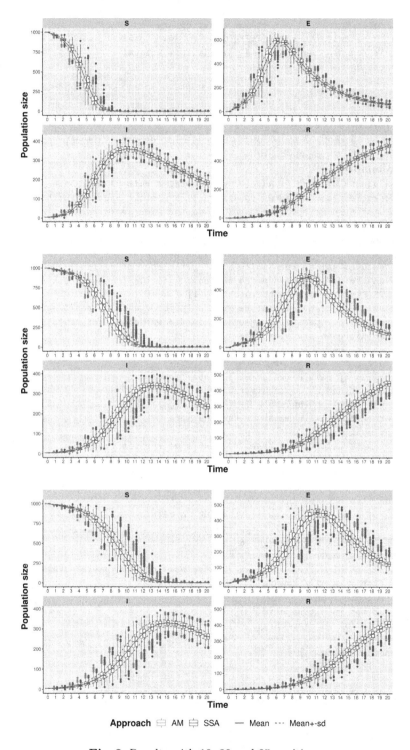

Fig. 3. Results with 10, 20 and 25 positions

1000 susceptible individuals, uniformly distributed among the $K \in \{5, 10, 20, 25\}$ positions, and 5 infected concentrated in one position. People can freely move from one position towards any other position (with equal probability). The results are obtained by performing 1000 runs with the ESSN derived SSA (in cyan) and the NetLogo simulation (in red): the curves show consistent results.

We performed a thorough analysis to check the statistical equivalence between the SSA and the ABM simulation. We wanted to show that curves generated from two independent CTMCs X_1 and X_2 are equally distributed ($X_1 \overset{d}{=} X_2$). Since we are studying data represented by curves (i.e., *functional data*), we apply the functional data analysis (FDA) to test the distribution equivalence. In details, we used the R package *fdatest* (https://cran.r-project.org/web/packages/fdatest) in which the FDA is used to define a statistical hypothesis test with the aim to verify the null hypothesis $X_1 \overset{d}{=} X_2$ against the alternative $X_1 \overset{d}{\neq} X_2$. The approach[3], on which the test is based, is called the **Interval Testing Procedure** (ITP) and it is characterized by projecting the functional data (which belong to an infinite-dimensional space) into a finite-dimensional space with a definition of coefficients of a suitable basis expansion (i.e., B-spline) on a uniform grid. Successively, univariate permutation tests for each basis coefficient are jointly performed, and then combined to obtain the p-values for the test associated with each spline coefficient (the p-values can be further corrected to assure the control of the level of the test for each possible set of true null hypotheses).

The results obtained refer to p = 9 uniformly spaced B-splines of order m = 2, and all the places (**S**, **E**, **I**, **R**) were tested considering all the scenarios with different number of positions. Thus, in each interval characterizing the B-spline no significant differences are found (i.e., all the corrected p-values associated with each B-spline coefficients are greater than .05) between the curves obtained from the SSA and the ABM simulation.

Appendix 3: Formal Definition of the ESSN Formalism

Definition 1 (Extended Stochastic Symmetric Net). *An ESSN is a ten-tuple:*

$$\mathcal{N}_{ESSN} = \langle P, T, \mathcal{C}, I, O, cd, \Theta, \omega, \Omega, m_0 \rangle$$

where

- $P = \{p_i\}$ *is a finite and non empty set of* places, *with* $i = 0, \ldots, n_p$, *where* n_p *is the number of places.*
- T *is the set of transitions and is defined as* $T = T_{ma} \cup T_g$, *with* $T_{ma} \cap T_g = \emptyset$. *Where* $T_{ma} = \{t_i^*\}_{1 \leq i \leq n_{T_{ma}}}$ *is the set of the* $n_{T_{ma}}$ *transitions whose speeds follow the MA law, and* $T_g = \{\bar{t}_i\}_{1 \leq i \leq n_{T_g}}$ *is the set of the* n_{T_g} *transitions whose speeds are defined as continuous functions.*

---

[3] Pini, Alessia, and Simone Vantini. *"The interval testing procedure: inference for functional data controlling the family-wise error rate on intervals"*. MOX-Report 13 (2013): 2013.

- $\mathcal{C} = \{C_1, \ldots, C_n\}$ is the finite set of basic color classes; a color class C_i can be partitioned into subclasses $C_{i,j}$ or can be circularly ordered.
- $I, O[p,t] : cd(t) \rightarrow Bag[cd(p)], \forall p \in P, t \in T$ are the input and output matrices[4], whose elements are functions annotating the directed arc connecting t and p (if the arc does not exist the corresponding function is the empty constant). The arc functions are denoted by expressions that take the form of weighted sums of tuples $\sum_i \lambda_i.\langle f_1, \ldots, f_k\rangle[g]$, where g is a standard predicate (following the same syntax of the transition guards), the tuple elements f_i are class functions [6] built upon a restricted set of basic functions namely projection denoted by a transition variable v_i, successor denoted $!v_i$ (only defined on ordered classes), and a constant function denoted S_i $(S_{i,j})$ returning all the elements of (sub)class C_i $(C_{i,j})$:

$$f_i = \sum_{k=1}^{m} \alpha_k.v_k + \sum_{q=1}^{||\mathcal{C}||}\sum_{j} \beta_{q,j}.S_{C_{q,j}} + \sum_{k=1}^{m} \gamma_k.!v_k; \ \alpha_k, \beta_k, \gamma_k \in \mathbb{Z}$$

- $cd : \bigotimes_{i=1}^{n} \bigotimes_{j}^{e_i} C_i^j$ is a function defining the color domain of each place and transition (where $e_i \in \mathbb{N}$ is the number of occurrences of the class C_i); for places it is expressed as the Cartesian product of basic color classes, for transitions it is expressed as a list of variables with their types. Observe that a place may contain undistinguished tokens or a transition may have no parameters, in this case their domain is neutral.
- Θ is the vector of guards and maps each element of T into a standard predicate. The admissible basic predicates are $v = v'$, $v \in C_{i,q}$, $d(v) = d(v')$ where $v, v' \in var(t)$ have same type C_i and $d(v)$ denotes the static subclass $C_{i,j}$ of the color assigned to v by a given transition instance $c \in cd(t)$ $(\Theta(t)$ may be the constant true, that is also a standard predicate).
- $\omega(t,c)$ is the rate parameter of transition $t \in T_{ma}$ when firing the instance $\langle t, c\rangle$.
- $\Phi = \{f_{\langle t,c\rangle}\}_{t\in T \wedge c \in cd(t)}$ is set of all transition speeds $\forall t \in T$. These must be continuous and derivable functions, and they can depend only on the marking of the input places of transition t at time ν defined in Eqs. 1 and 2.
- $m_0 : P \rightarrow Bag[cd(p)]$ is the initial marking, a P-indexed vector, mapping each place p on a multiset on $cd(p)$.

References

1. Marsan, M.A., et al.: Modelling with Generalized Stochastic Petri Nets. Wiley, Hoboken (1995)
2. Allen, L.: A primer on stochastic epidemic models: formulation, numerical simulation, and analysis. Infect. Dis. Modell. **2**(2), 128–142 (2017)
3. Cao, Y., et al.: Efficient formulation of the stochastic simulation algorithm for chemically reacting systems. J. Chem. Phys. **121**(9), 4059–4067 (2004)

[4] $Bag[A]$ is the set of multisets built on set A, and if $b \in Bag[A] \wedge a \in A$, $b[a]$ denotes the multiplicity of a in the multiset b.

4. Castagno, P., et al.: A computational framework for modeling and studying pertussis epidemiology and vaccination. BMC Bioinf. **21**, 344 (2020)
5. Chiola, G., et al.: A structural colour simplification in well-formed coloured nets. In: Proceedings of the Fourth International Workshop on Petri Nets and Performance Models, PNPM91, pp. 144–153 (1991)
6. Chiola, G., et al.: Stochastic well-formed coloured nets for symmetric modelling applications. IEEE Trans. Comput. **42**(11), 1343–1360 (1993)
7. Figueredo, G., et al.: Comparing stochastic differential equations and agent-based modelling and simulation for early-stage cancer. PLoS One **9**, e95150 (2014)
8. Gillespie, D.T.: Exact stochastic simulation of coupled chemical reactions. J. Phys. Chem. **81**(25), 2340–2361 (1977)
9. Gillespie, D.T.: Approximate accelerated stochastic simulation of chemically reacting systems. J. Chem. Phys. **115**(4), 1716–1733 (2001)
10. Keeling, M., et al.: Modeling Infectious Diseases in Humans and Animals. Princeton University Press, Princeton (2008)
11. Law, A.M.: Simulation Modeling and Analysis, 5th edn. McGraw-Hill, New York (2015)
12. Perez, L., et al.: An agent-based approach for modeling dynamics of contagious disease spread. Int. J. Health Geogr. **8**(1), 1–17 (2009)
13. Pernice, S., et al.: A computational approach based on the colored Petri net formalism for studying multiple sclerosis. BMC Bioinf. **20**(6), 1–17 (2019)
14. Pernice, S., et al.: Computational modeling of the immune response in multiple sclerosis using epimod framework. BMC Bioinf. **21**(17), 1–20 (2020)
15. Pernice, S., et al.: Impacts of reopening strategies for COVID-19 epidemic: a modeling study in piedmont region. BMC Infect. Dis. **20**(1), 1–9 (2020)
16. Reinhardt, O., et al.: An efficient simulation algorithm for continuous-time agent-based linked lives models. In: Proceedings of the 50th Annual Simulation Symposium, Virginia Beach, VA, USA, 23–26 April 2017, pp. 9:1–9:12. SCS Int./ACM (2017)
17. Voit, E., et al.: 150 years of the mass action law. PLoS Comput. Biol. **11**(1), e1004012 (2015)
18. Wilensky, U.: NetLogo (1999). http://ccl.northwestern.edu/netlogo/

State Space Minimization Preserving Embeddings for Continuous-Time Markov Chains

Susmoy Das and Arpit Sharma[✉]

Department of Electrical Engineering and Computer Science,
Indian Institute of Science Education and Research Bhopal, Bhopal, India
{susmoy18,arpit}@iiserb.ac.in

Abstract. This paper defines embeddings which allow one to construct an action labeled continuous-time Markov chain (ACTMC) from a state labeled continuous-time Markov chain (SCTMC) and vice versa. We prove that these embeddings preserve strong forward bisimulation and strong backward bisimulation. We define weak backward bisimulation for ACTMCs and SCTMCs, and also prove that our embeddings preserve both weak forward and weak backward bisimulation. Next, we define the invertibility criteria and the inverse of these embeddings. Finally, we prove that an ACTMC can be minimized by minimizing its embedded model, i.e. SCTMC and taking the inverse of the embedding. Similarly, we prove that an SCTMC can be minimized by minimizing its embedded model, i.e. ACTMC and taking the inverse of the embedding.

Keywords: Markov chain · Behavioral equivalence · Bisimulation equivalence · Stochastic systems · Embeddings

1 Introduction

Continuous-time Markov chains (CTMCs) have a wide applicability ranging from classical performance and dependability evaluation to systems biology. CTMC models are categorized as either state labeled CTMCs (SCTMCs) or action labeled CTMCs (ACTMCs). SCTMCs are primarily used by the model checking community for the formal verification of stochastic systems. Real-time properties can be expressed for SCTMCs using continuous stochastic logic (CSL) [3,5], CSL^{TA} [16], deterministic timed automaton (DTA) [1] and metric temporal logic (MTL) [33] formulas. State-of-the-art model checking algorithms [3,13,16] and tools such as Probabilistic Symbolic Model Checker (PRISM) [27], Markov Reward Model Checker (MRMC) [24] and Storm [15] have been implemented to model check these real-time objectives. To tackle the state space explosion problem [2], several abstraction techniques have been proposed for SCTMCs, e.g. three-valued abstraction [25], symmetry reduction [26] and simulation/bisimulation minimization [23]. In contrast, ACTMCs are more commonly used as semantic model for amongst others stochastic process algebras [8,21,22],

© Springer Nature Switzerland AG 2021
P. Ballarini et al. (Eds.): EPEW 2021/ASMTA 2021, LNCS 13104, pp. 44–61, 2021.
https://doi.org/10.1007/978-3-030-91825-5_3

stochastic Petri nets [28] and stochastic activity networks [29,34]. Various linear-time and branching-time relations on ACTMCs have been defined such as trace and testing equivalences [8,10,39], weak and strong variants of bisimulation equivalence and simulation pre-order [4,9,11,22]. Although both these models are regarded to be on an equal footing, no effort has been made in the past to let the analysis techniques and tools of one community to be utilized by the other. More specifically, these two communities have independently developed their methods and tools without leveraging the advancements made by the other community. For example, performance analysis tools have also been developed for ACTMCs, e.g. mCRL2 [12] and CADP [17]. Similarly, several equivalences have been defined for SCTMCs [3,5,37].

This paper focuses on providing a formal framework which allows one to use the state-of-the-art tools developed in one setting for model minimization in the other setting. For example, backward bisimulation and weak backward bisimulation minimization is not supported by any of the well known stochastic model checking tools, e.g. PRISM, MRMC and Storm. This means that if one of these tools in the state based stochastic setting implements the quotienting algorithm, then using our framework, models in the action labeled stochastic setting can also be minimized and vice versa. To achieve this goal, we define the embeddings slc from ACTMCs to SCTMCs and alc from SCTMCs to ACTMCs. We prove that both these embeddings preserve strong forward and strong backward bisimulation. Next, we define weak backward bisimulation for SCTMCs and ACTMCs, and show that weak forward and weak backward bisimulaion are incomparable. We also prove that weak forward and weak backward bisimulation are preserved by our newly defined embeddings.

In order to reverse the effects of alc and slc, we define the invertibility criterion and the left inverse of these embeddings, i.e. alc^{-1} from ACTMCs to SCTMCs and slc^{-1} from SCTMCs to ACTMCs. We also show that the invertibility is preserved w.r.t. strong forward, strong backward, weak forward and weak backward bisimulation. Finally, we prove that a model can be minimized in one setting by minimizing its embedded model in the other setting and applying the inverse of the embedding. For example, if one applies alc to an SCTMC and minimizes it, then the left inverse embedding, i.e. alc^{-1} will return the minimal SCTMC (for all the strong and weak variants of bisimulation).

Organisation of the Paper. Section 2 presents the related work. Section 3 briefly recalls the main concepts of SCTMCs and ACTMCs. Section 4 defines the embeddings slc and alc. Section 5 proves that these embeddings preserve both forward and backward bisimulation. Section 6 proves that these embeddings also preserve both weak forward and weak backward bisimulation. Section 7 defines inverse embeddings and proves that model minimization is preserved under our framework. Finally, Sect. 8 concludes the paper and provides pointers for future research.

2 Related Work

In [31,32], De Nicola and Vaandrager showed that there are embeddings between Kripke structures (KSs) and labeled transitions systems (LTSs). These results have enabled one to use a process algebra to describe the system behaviour as an LTS and to use Computational Tree Logic (CTL) or CTL* to specify the requirement the system has to comply with. Additionally, authors have shown that stuttering equivalence for KSs coincides with divergence-sensitive branching bisimulation for LTSs. In [35,36], Reniers et al. have extended the above mentioned results by defining two additional translations, i.e. inverse embeddings which enable minimization modulo behavioral equivalences. In this paper, authors have shown that their embeddings can also be used for a range of other equivalences of interest, e.g. strong bisimilarity, simulation equivalence, and trace equivalence. In [30], a tool was developed which takes a process description and an action-based version of CTL (ACTL) formula to be verified, and then translates them into a Kripke Structure and CTL, respectively. In [18,19], author have presented a new proposal to evaluate the relative expressive power of process calculi. Unlike full abstraction, their proposal is more focused on expressiveness issues and is also interesting for separation results. In the software product line setting [6,7], authors have demonstrated that modal transition systems (MTSs) with variability constraints are equally expressive as featured transition systems (FTSs). This has been achieved by defining two transformation functions between these models, and proving the soundness and completeness of both the transformations. In the probabilistic setting, a formal framework for relating the two types of discrete time probabilistic models has been proposed in [14]. This framework preserves strong bisimulation and trace equivalences and allows taking the inverse of the embedding. This framework can not be directly lifted to the stochastic setting as it fails to preserve weak (forward and backward) bisimulation equivalences.

3 Preliminaries

This section presents the necessary definitions and basic concepts related to continuous-time Markov chains (CTMCs) that are needed for the understanding of the rest of this paper.

Definition 1 (SCTMC). *A state labeled continuous-time Markov chain (SCTMC) is a tuple $C = (S, AP, R, s_0, L)$ where:*

- *S is a countable, nonempty set of states,*
- *AP is the set of atomic propositions,*
- *$R : S \times S \to \mathbb{R}_{\geq 0}$ is a rate function,*
- *s_0 is the initial state, and*
- *$L : S \to 2^{AP}$ is a labeling function.*

C is called finite if S and AP are finite. Let $\to = \{(s,r,t) \mid R(s,t) = r > 0\}$ denote the set of all transitions for an SCTMC C. We denote $s \xrightarrow{r} t$ if $(s,r,t) \in \to$.

Let $s \xrightarrow{r}$ denote that $\exists s' \in S$ s.t. $R(s, s') > 0$. Similarly, let $\xrightarrow{r} s$ denote that $\exists s' \in S$ s.t. $R(s', s) > 0$. For $C \subseteq S$, let $R(s, C) = \sum_{t \in C} R(s, t)$. For $C \subseteq S$, $R(C, s) = \sum_{t \in C} R(t, s)$. The exit rate $E(s)$ for the state $s \in S$ be given by $E(s) = \sum_{s' \in S} R(s, s')$. Note that, $E(s) = R(s, S)$. For a state s, $E(s) = 0$ is equivalent to calling s an *absorbing* state.

Definition 2 (ACTMC). *An action labeled continuous-time Markov chain (ACTMC) is a tuple $\mathcal{C} = (S, Act, R, s_0)$ where:*

- *S is a countable, nonempty set of states,*
- *Act is the set of actions which contains the special action τ,*
- *$R : S \times Act \times S \to \mathbb{R}_{\geq 0}$ is a rate function, and*
- *s_0 is the initial state.*

$\mathcal{C}$ is called finite if S and Act are finite. τ is special action used to denote an invisible ccomputation. The analogous concept in SCTMC refers to moving to states with the same label. Let $\to = \{(s, a, r, t) \mid R(s, a, t) = r > 0\}$ denote the set of all transitions for an ACTMC $\mathcal{C}$. We denote $s \xrightarrow{a,r} t$ if $(s, a, r, t) \in \to$. Let $s \xrightarrow{a,r}$ denote that $\exists s' \in S$ and $\exists a \in Act$ s.t. $R(s, a, s') > 0$. If a state s cannot perform a particular action, say a, with a positive rate, we denote it by $s \not\xrightarrow{a}$. Similarly, let $\xrightarrow{a,r} s$ denotes that $\exists s' \in S$ and $\exists a \in Act$ s.t. $R(s', a, s) > 0$. If a state s cannot be reached by performing a particular action, say a, in one step with a positive rate we denote it by $\not\xrightarrow{a} s$. For $C \subseteq S$, let $R(s, a, C) = \sum_{t \in C} R(s, a, t)$. For $C \subseteq S$, $R(C, a, s) = \sum_{t \in C} R(t, a, s)$. The exit rate $E(s)$ for the state $s \in S$ be given by $E(s) = \sum_{s' \in S, a \in Act} R(s, a, s')$. Note that, $E(s) = \sum_{a \in Act} R(s, a, S)$. For a state s, $E(s) = 0$ is equivalent to calling s an *absorbing* state. We assume that our SCTMC and ACTMC models do not have absorbing states.

4   Embeddings

This section defines the embeddings *slc* and *alc*. *slc* allows one to construct an SCTMC from an ACTMC. Similarly, *alc* is useful for constructing an ACTMC from an SCTMC.

Definition 3 (slc). *Let $\mathcal{C} = (S, Act, R, s_0)$ be an ACTMC and $\theta \in \mathbb{R}_{>0}$ be a positive real which is fixed. The embedding slc : $ACTMC \to SCTMC$ is formally defined as $slc(\mathcal{C}) = (S', AP', R', s_0', L')$ s.t.:*

- *$S' = S \cup \{(a, t) \mid R(s, a, t) > 0$ for some $s, t \in S$ and $a \neq \tau\}$,*
- *$AP' = (Act \backslash \{\tau\}) \cup \{\bot\}$ where $\bot \notin Act$,*
- *The rate function R' is defined by:*

$$R'(s, (a, t)) = R(s, a, t) \; \forall s, t \in S \text{ s.t. } R(s, a, t) > 0 \text{ and } a \neq \tau,$$

$$R'(s, t) = R(s, \tau, t) \; \forall s, t \in S \text{ s.t. } R(s, \tau, t) > 0, \text{ and}$$

$$R'((a, t), t) = \theta \; \forall (a, t) \in S' \backslash S,$$

- *$s_0' = s_0$, and*

– $L'(s) = \{\bot\}\ \forall s \in S$ *and* $L'((a, t)) = \{a\}$.

Let $|S|$ denote the cardinality of a given set S.

Example 1. Consider the ACTMC $\mathcal{C}$ shown in Fig. 1 (left). The SCTMC obtained by applying *slc*, i.e. $slc(\mathcal{C})$ is shown in Fig. 1 (right).

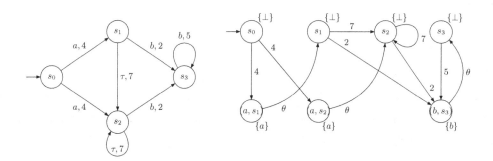

Fig. 1. ACTMC $\mathcal{C}$ and $slc(\mathcal{C})$

Remark 1. Let $\mathcal{C}$ be an ACTMC with $|S| = n$, $|Act| = t$ and let m be the maximum number of different actions (excluding τ) through which any state s can be reached in $\mathcal{C}$ in one step (clearly, $m \leq |Act| = t$). Then for $slc(\mathcal{C})$, $|S'| \leq n + (n \cdot m) = (m+1) \cdot n \leq (t+1) \cdot n$ and $| \rightarrow' | \leq | \rightarrow | + (n \cdot m) \leq | \rightarrow | + (n \cdot t)$. This is due to the fact that each outgoing transition for a particular action ($\neq \tau$) to a state results in only one new state in $slc(\mathcal{C})$, e.g. if $s \xrightarrow{a,r} t$ and $s' \xrightarrow{a,r'} t$, then we only create one new state (a, t) in $slc(\mathcal{C})$. The number of transitions in $slc(\mathcal{C})$ increases by an amount exactly the same as the increase in the number of states, as for each new state, we add exactly one additional transition of the form $(a, t) \xrightarrow{\theta'} t$.

Definition 4 (alc). *Let* $\mathcal{C} = (S, AP, R, s_0, L)$ *be an SCTMC and* $\theta \in \mathbb{R}_{>0}$ *be a positive real which is fixed. The embedding alc : SCTMC $\rightarrow$ ACTMC is formally defined as* $alc(\mathcal{C}) = (S', Act', R', s_0')$ *s.t.:*

– $S' = S \cup \{\overline{s} \mid s \in S\}$,
– $Act' = 2^{AP} \cup \{\tau, \bot\}$,
– R' *is defined by:*

$$R'(\overline{s}, \bot, t) = R(s, t)\ \forall s, t \in S \text{ s.t. } R(s, t) > 0 \text{ and } L(s) \neq L(t),$$

$$R'(\overline{s}, \tau, \overline{t}) = R(s, t)\ \forall s, t \in S \text{ s.t. } R(s, t) > 0 \text{ and } L(s) = L(t),$$

$$R'(s, L(s), \overline{s}) = \theta\ \forall s \in S, \text{ and}$$

– $s_0' = s_0$.

Remark 2. Let $\mathcal{C}$ be an SCTMC with $|S| = n$. Then for $alc(\mathcal{C})$, $|S'| = 2 \cdot n$ and $| \rightarrow' | = | \rightarrow | + n$. The number of transitions increases by an amount exactly the same as the increase in the number of states, as for each state, we add one additional transition of the form $s \xrightarrow{L(s),\theta} \bar{s}$.

Both *alc* and *slc* introduce new transitions with rate θ. Note that, in the embedded CTMC, the steady state probabilities and the timed reachability probabilities are not preserved, as the embedded system will also spend time in the new states. When θ converges to 0, the steady state probabilities converge to that of the original CTMC. The convergence of θ is not a problem because our goal is to minimize the system after applying the embedding and then take the inverse of the embedding to get back the quotient of the original model. Taking the inverse will reverse the effects of applying an embedding and will return a model without θ-labeled transitions which will preserve the steady state and timed reachability probabilities.

Example 2. Consider the SCTMC $\mathcal{C}$ shown in Fig. 2 (left). The ACTMC obtained by applying *alc*, i.e. $alc(\mathcal{C})$ is shown in Fig. 2 (right).

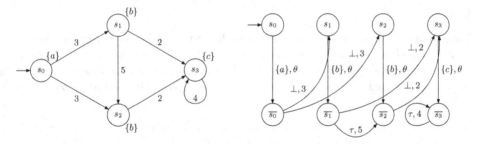

Fig. 2. SCTMC $\mathcal{C}$ and $alc(\mathcal{C})$

Remark 3. From [20], we know that the correctness of the embeddings ('encodings') involves fulfilling the notion of full abstraction. An emebedding is said to satisfy the criterion of 'full abstraction' if and only if it maps every equivalent pair of states to an equivalent pair in the range set and vice versa. As a result, a *'correct'* embedding has to map equivalent source pairs into equivalent target pairs and, conversely, equivalent images of this mapping must have originated from equivalent source pairs.

Next, we establish these claims for our definition of the embeddings.

5   Preservation of Strong Bisimulation

In this section, we show that the embeddings defined in Sect. 4, i.e. *slc* and *alc* preserve both forward and backward notions of strong bisimulation. We

first recall the definitions of forward and backward bisimulation [4,5]. Given
an equivalence relation $\mathcal{R}$, let $S/_{\mathcal{R}}$ denote the set consisting of all $\mathcal{R}$-equivalence
classes. Let $[s]_{\mathcal{R}}$ denote the equivalence class of state s under $\mathcal{R}$, i.e. $[s]_{\mathcal{R}} = \{s' \in S \mid (s, s') \in \mathcal{R}\}$.

Definition 5 (Forward bisimulation for SCTMCs). *Equivalence $\mathcal{R}$ on S
is a forward bisimulation on $\mathcal{C}$ if for any $(s, s') \in \mathcal{R}$ we have: $L(s) = L(s')$, and
$R(s, C) = R(s', C)$ for all C in $S/_{\mathcal{R}}$. s and s' are forward bisimilar, denoted
$s \sim s'$, if there exists a forward bisimulation $\mathcal{R}$ on $\mathcal{C}$ s.t. $(s, s') \in \mathcal{R}$.*

These conditions require that any two forward bisimilar states are equally labeled
and have identical rates to move to any equivalence class $C \in S/_{\mathcal{R}}$.

Definition 6 (Forward bisimulation for ACTMCs). *Equivalence $\mathcal{R}$ on S
is a forward bisimulation on $\mathcal{C}$ if for any $(s, s') \in \mathcal{R}$ we have: $R(s, a, C) = R(s', a, C)$, for all C in $S/_{\mathcal{R}}$ and $a \in Act$. s and s' are forward bisimilar, denoted
$s \sim s'$, if there exists a forward bisimulation $\mathcal{R}$ on $\mathcal{C}$ s.t. $(s, s') \in \mathcal{R}$.*

This condition requires that forward bisimilar states exhibit the same stepwise
behavior. We write $\mathcal{C} \models s \sim s'$ to denote $(s, s') \in \mathcal{R}$ for some forward bisimulation $\mathcal{R}$ on an SCTMC $\mathcal{C}$ (ACTMC $\mathcal{C}$, respectively). Next, we recall the definition
of backward bisimulation for SCTMCs and define it for ACTMCs in a similar
fashion.

Definition 7 (Backward bisimulation for SCTMCs [38]). *Equivalence $\mathcal{R}$
on S is a backward bisimulation on $\mathcal{C}$ if for any $(s, s') \in \mathcal{R}$ we have: $L(s) = L(s')$,
$R(C, s) = R(C, s')$ for all C in $S/_{\mathcal{R}}$, and $E(s) = E(s')$. s and s' are backward
bisimilar, denoted $s \sim_b s'$, if there exists a backward bisimulation $\mathcal{R}$ on $\mathcal{C}$ s.t.
$(s, s') \in \mathcal{R}$.*

These conditions require that any two backward bisimilar states are equally
labeled, have identical incoming rates from any equivalence class $C \in S/_{\mathcal{R}}$ and
have the same exit rates.

Definition 8 (Backward bisimulation for ACTMCs). *Equivalence $\mathcal{R}$ on
S is a backward bisimulation on $\mathcal{C}$ if for any $(s, s') \in \mathcal{R}$ we have: $R(C, a, s) = R(C, a, s')$, for all C in $S/_{\mathcal{R}}$ and $a \in Act$, and $E(s) = E(s')$. s and s' are
backward bisimilar, denoted $s \sim_b s'$, if there exists a backward bisimulation $\mathcal{R}$
on $\mathcal{C}$ s.t. $(s, s') \in \mathcal{R}$.*

We write $\mathcal{C} \models s \sim_b s'$ to denote $(s, s') \in \mathcal{R}$ for some backward bisimulation
$\mathcal{R}$ on an SCTMC $\mathcal{C}$ (ACTMC $\mathcal{C}$, respectively). The definitions of forward and
backward bisimulation can be easily extended to compare the behavior of two
SCTMCs (ACTMCs, respectively). Given an SCTMC $\mathcal{C}$, the function that yields
a minimal quotient of $\mathcal{C}$'s forward or backward bisimulation relation is denoted
by $\sim_{MS}$ ($\sim_{bMS}$, respectively). Similarly, for an ACTMC $\mathcal{C}$, the function that
yields a minimal quotient of $\mathcal{C}$'s forward or backward bisimulation relation is
denoted by $\sim_{MA}$ ($\sim_{bMA}$, respectively). From [38], we know that forward and
backward bisimulation are incomparable for SCTMCs. Next, we show that it is
also true for ACTMCs.

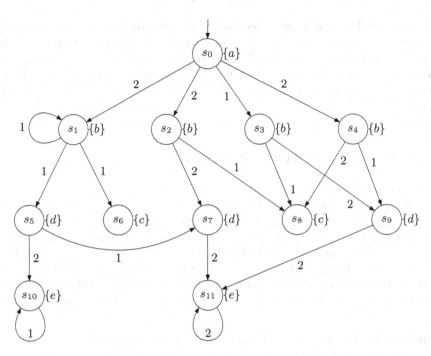

Fig. 3. SCTMC $\mathcal{C}$

Proposition 1. *For ACTMCs,* $\sim \not\Rightarrow \sim_b$ *and* $\sim_b \not\Rightarrow \sim$.

Example 3. Consider the SCTMC in Fig. 3. Here, we find that the states s_2 and s_3 are forward bisimilar as $L(s_2) = L(s_3) = \{b\}$, $R(s_2, C_1) = R(s_3, C_1) = 1$, where $C_1 = \{s_8\}$ and $R(s_2, C_2) = R(s_3, C_2) = 2$, where $C_2 = \{s_7, s_9\}$. But these two states do not satisfy the condition of backward bisimilarity of SCTMC [Definition 7] as $R(C_3', s_2) = 2$ and $R(C_3', s_3) = 1$, where $C_3' = \{s_0\}$, and hence $s_2 \sim s_3$ but $s_2 \not\sim_b s_3$.

On the other hand, in the same figure we have states s_2 and s_4 which are backward bisimilar as $L(s_2) = L(s_4) = \{b\}$ and $R(C_3', s_2) = R(C_3', s_4) = 2$, where $C_3' = \{s_0\}$ and $E(s_2) = E(s_4) = 3$. These two states are not forward bisimilar as $R(s_2, C_1) = 1$ and $R(s_4, C_1) = 2$, where $C_1 = \{s_8\}$, i.e. $s_2 \sim_b s_4$ but $s_2 \not\sim s_4$.

This demonstrates that these relations are incomparable and give different quotients. In a similar way, we can prove that forward and backward bisimulation are incomparable for ACTMCs by showing suitable counterexamples.   □

Next, we discuss the main results of this section.

Theorem 1. *Let* $\mathcal{C} = (S, Act, R, s_0)$ *be an ACTMC. Then, given an equivalence* $\leftrightarrow \in \{\sim, \sim_b\}$, $\forall s, s' \in S$, *we have* $\mathcal{C} \models s \leftrightarrow s' \Leftrightarrow slc(\mathcal{C}) \models s \leftrightarrow s'$.

Theorem 2. *Let* $\mathcal{C} = (S, AP, R, s_0, L)$ *be an SCTMC. Then, given an equivalence* $\leftrightarrow \in \{\sim, \sim_b\}$, $\forall s, s' \in S$, *we have* $\mathcal{C} \models s \leftrightarrow s' \Leftrightarrow alc(\mathcal{C}) \models s \leftrightarrow s'$.

Proof idea for Theorem 1 *and Theorem* 2. We begin by defining a new relation $\mathcal{R}'$ in the embedded system which comprises of all the state pairs from the original system which were related under the forward (backward) bisimulation relation $\mathcal{R}$. Additionally, only those new state pairs are added for which the state components were related under $\mathcal{R}$, e.g. $((a,t),(a,t')) \in \mathcal{R}'$ iff $(t,t') \in \mathcal{R}$ $((\overline{s},\overline{s'}) \in \mathcal{R}'$ iff $(s,s') \in \mathcal{R}$ respectively). Next, we prove that this new relation is indeed a forward (backward) bisimulation equivalence.

6   Preservation of Weak Bisimulation

Unfortunately, bisimulation is too fine, and it is often desirable to obtain a quotient system smaller than bisimulation such that properties of interest are still preserved. To achieve this several variants of weak bisimulation have been defined in the literature. Weak bisimulation relations are important for system synthesis as well as system analysis. Weak bisimulations abstract away from non-observable steps. In this section, we show that the embeddings defined in Sect. 4, i.e. *slc* and *alc* preserve both forward and backward weak bisimulation. We first recall the definition of weak forward bisimulation for SCTMCs and ACTMCs [4,5].

Definition 9 (Weak Forward bisimulation for SCTMCs). *Equivalence $\mathcal{R}$ on S is a weak forward bisimulation on $\mathcal{C}$ if for any $(s,s') \in \mathcal{R}$ we have: $L(s) = L(s')$, and $R(s,C) = R(s',C)$ for all C in $S/\mathcal{R}$ with $C \neq [s]_\mathcal{R}$. s and s' are weak forward bisimilar, denoted $s \approx s'$, if there exists a weak forward bisimulation $\mathcal{R}$ on $\mathcal{C}$ s.t. $(s,s') \in \mathcal{R}$.*

These conditions require that any two weak forward bisimilar states are equally labeled and have identical 'relative rates' to move to any external equivalence class $C \in S/\mathcal{R}$.

Definition 10 (Weak Forward bisimulation for ACTMCs). *Equivalence $\mathcal{R}$ on S is a weak forward bisimulation on $\mathcal{C}$ if for any $(s,s') \in \mathcal{R}$ we have: $R(s,a,C) = R(s',a,C)$, for all C in $S/\mathcal{R}$ and $a \in Act$ with $a \neq \tau$ or $s,s' \notin C$. s and s' are weak forward bisimilar, denoted $s \approx s'$, if there exists a weak forward bisimulation $\mathcal{R}$ on $\mathcal{C}$ s.t. $(s,s') \in \mathcal{R}$.*

We write $\mathcal{C} \models s \approx s'$ to denote $(s,s') \in \mathcal{R}$ for some weak forward bisimulation $\mathcal{R}$ on an SCTMC $\mathcal{C}$ (ACTMC $\mathcal{C}$, respectively). Next, we define weak backward bisimulation and its quotient for SCTMCs and ACTMCs.

Definition 11 (Weak Backward bisimulation for SCTMCs). *Equivalence $\mathcal{R}$ on S is a weak backward bisimulation on $\mathcal{C}$ if for any $(s,s') \in \mathcal{R}$ we have: $L(s) = L(s')$, $R(C,s) = R(C,s')$ for all C in $S/\mathcal{R}$ s.t. $C \neq [s]_\mathcal{R}$, and $E(s) = E(s')$. s and s' are weak backward bisimilar, denoted $s \approx_b s'$, if there exists a weak backward bisimulation $\mathcal{R}$ on $\mathcal{C}$ s.t. $(s,s') \in \mathcal{R}$.*

These conditions require that any two weak backward bisimilar states are equally labeled, have identical incoming rates from any external equivalence class $C \in S/\mathcal{R}$ and have the same exit rates.

Definition 12 (Weak Backward bisimulation for ACTMCs). *Equivalence $\mathcal{R}$ on S is a weak backward bisimulation on $\mathcal{C}$ if for any $(s, s') \in \mathcal{R}$ we have: $R(C, a, s) = R(C, a, s')$, for all C in $S/\mathcal{R}$ and $a \in Act$ with $a \neq \tau$ or $s, s' \notin C$, and $E(s) = E(s')$. s and s' are weak backward bisimilar, denoted $s \approx_b s'$, if there exists a weak backward bisimulation $\mathcal{R}$ on $\mathcal{C}$ s.t. $(s, s') \in \mathcal{R}$.*

We write $\mathcal{C} \models s \approx_b s'$ to denote $(s, s') \in \mathcal{R}$ for some weak backward bisimulation $\mathcal{R}$ on an SCTMC $\mathcal{C}$ (ACTMC $\mathcal{C}$, respectively). The definitions of weak forward and backward bisimulation can be easily extended to compare the behavior of two SCTMCs (ACTMCs, respectively). Given an SCTMC $\mathcal{C}$, the function that yields a minimal quotient of $\mathcal{C}$'s weak forward or weak backward bisimulation relation be denoted by $\approx_{MS}$ ($\approx_{bMS}$, respectively). Similarly, for an ACTMC $\mathcal{C}$, the function that yields a minimal quotient of $\mathcal{C}$'s weak forward or weak backward bisimulation relation be denoted by $\approx_{MA}$ ($\approx_{bMA}$, respectively). We use $\leftrightarrow$ to denote multiple relations for which the result holds true followed by the definition of its scope. The following proposition asserts that weak forward bisimulation and weak backward bisimulation are incomparable.

Proposition 2. *For both SCTMCs and ACTMCs, $\approx \not\Rightarrow \approx_b$ and $\approx_b \not\Rightarrow \approx$.*

Example 4. Consider the SCTMC in Fig. 3. Here, we find that the states s_5 and s_7 are weak forward bisimilar as $L(s_5) = L(s_7) = \{d\}$ and $R(s_5, C_1) = R(s_7, C_1) = 2$, where $C_1 = \{s_{10}, s_{11}\}$. But these two states do not satisfy the condition of weak backward bisimilarity of SCTMC [Definition 11] as $R(C_2', s_5) = 1$ and $R(C_2', s_7) = 2$, where $C_2' = \{s_1, s_2\}$, and hence $s_5 \approx s_7$ but $s_5 \not\approx_b s_7$.

On the other hand, in the same figure we have states s_1 and s_2 which are weak backward bisimilar as $L(s_1) = L(s_2) = \{b\}$ and $R(C_3', s_1) = R(C_3', s_2) = 2$, where $C_3' = \{s_0\}$ and $E(s_1) = E(s_2) = 3$. These two states are not weak forward bisimilar as $R(s_1, C_4) = 1$ and $R(s_2, C_4) = 2$, where $C_4 = \{s_5, s_7\}$, i.e. $s_1 \approx_b s_2$ but $s_1 \not\approx s_2$.

This demonstrates that these relations are incomparable and give different quotients as a result. In a similar way, we can prove that forward and backward bisimulation are incomparable for ACTMCs by showing suitable counterexamples. Table 1 shows the relationship between different bisimulation equivalences discussed in this paper. The table is based on the SCTMC $\mathcal{C}$ shown in Fig. 3. □

From [4,5], we know that $\sim$ is strictly finer than $\approx$ for both SCTMCs and ACTMCs. The following proposition asserts that it is also true for $\sim_b$ and $\approx_b$.

Table 1. Relationship between equivalences

States	$\sim$	$\sim_b$	$\approx$	$\approx_b$
(s_2, s_3)	✓	×	✓	×
(s_1, s_2)	×	×	×	✓
(s_5, s_7)	×	×	✓	×
(s_2, s_4)	×	✓	×	✓

Proposition 3. *For both SCTMCs and ACTMCs, $\sim_b \Longrightarrow \approx_b$ but $\approx_b \not\Longrightarrow \sim_b$.*

Next, we discuss the main results of this section.

Theorem 3. *Let $\mathcal{C} = (S, Act, R, s_0)$ be an ACTMC. Then, given an equivalence $\leftrightarrow \in \{\approx, \approx_b\}$, $\forall s, s' \in S$, we have $\mathcal{C} \models s \leftrightarrow s' \Leftrightarrow slc(\mathcal{C}) \models s \leftrightarrow s'$.*

Theorem 4. *Let $\mathcal{C} = (S, AP, R, s_0, L)$ be an SCTMC. Then, given an equivalence $\leftrightarrow \in \{\approx, \approx_b\}$, $\forall s, s' \in S$, we have $\mathcal{C} \models s \leftrightarrow s' \Leftrightarrow alc(\mathcal{C}) \models s \leftrightarrow s'$.*

These theorems assert that *slc* and *alc* preserve weak forward and backward bisimulation. The proof ideas are similar to that of Theorem 1 and 2 respectively.

7   Inverse of the Embeddings and Minimization

This section defines the invertibility criteria followed by the definitions of the inverse of the embeddings *alc*, i.e. alc^{-1} and *slc*, i.e. slc^{-1}. We also prove that a model in one setting can be minimized by minimizing its embedded model in the other setting and taking the inverse of the embedding. Subsection 7.1 defines alc^{-1} and Subsect. 7.2 defines slc^{-1}.

7.1   Inverse of the Embedding from ACTMCs to SCTMCs

The embedding *alc* defined in Sect. 4 will always create an ACTMC which satisfies the invertibility criterion. The explicit definition of the invertibility criterion is only meant for situations when one wishes to apply the inverse embedding first. Additionally, the unique rate θ will be needed to apply *alc* after applying alc^{-1} to revert back to the original system.

Definition 13 (Invertibility criterion).   *Let $\mathcal{C} = (S, Act, R, s_0)$ be an ACTMC. Then $\mathcal{C}$ is invertible iff:*

- *$Act\backslash\{\tau\} = 2^{AP} \cup \{\bot\}$ for some set AP,*
- *s_0 can only perform $s_0 \xrightarrow{\mathcal{A},\theta}$ for some $\mathcal{A} \in 2^{AP}$,*
- *all the states which can perform an action $\mathcal{A} \in Act\backslash\{\bot,\tau\}$, must have only one outgoing transition with label $\mathcal{A}$. Additionally, all the states which are capable of performing an action from $Act\backslash\{\bot,\tau\}$ must do it with an identical rate, say θ, i.e. $\forall s, s' \in S$ s.t. $s \xrightarrow{\mathcal{A},\theta_1}$ and $s' \xrightarrow{\mathcal{A}',\theta_2}$, then $\theta_1 = \theta_2 = \theta$,*
- *$\forall s, s' \in S$ and $\mathcal{A} \in 2^{AP}$ if $s \xrightarrow{\mathcal{A},\theta} s'$, then all the outgoing transitions from s' have either $\bot$ or τ as their action label,*
- *$\forall s, s' \in S$, if $s \xrightarrow{\bot,r} s'$, then $s' \xrightarrow{\mathcal{A},\theta}$, for some $\mathcal{A} \in 2^{AP}$, and*
- *$\forall s, s' \in S$, if $s \xrightarrow{\tau,r} s'$, then either $s' \xrightarrow{\bot,r}$ or $s' \xrightarrow{\tau,r}$ for some $r \in \mathbb{R}_{>0}$.*

The following lemma asserts that if a given system is invertible, then the minimal quotient of any system equivalent to the given system is also invertible under all the equivalences discussed in this paper.

Lemma 1. *Let $\mathcal{C}$ be an arbitrary invertible ACTMC. For $\leftrightarrow \in \{\sim, \sim_b, \approx, \approx_b\}$, given any ACTMC $\mathcal{C}'$, s.t. $\mathcal{C} \leftrightarrow \mathcal{C}'$, $\leftrightarrow_{MA} (\mathcal{C}')$ is also invertible.*

Now, we are in a position to define the inverse of the embedding alc.

Definition 14 (alc^{-1}). *Let $\mathcal{C} = (S, Act, R, s_0)$ be an invertible ACTMC. Then alc^{-1} is the inverse of the embedding alc, where $alc^{-1} : ACTMC \to SCTMC$ is formally defined as $alc^{-1}(\mathcal{C}) = (S'', AP'', R'', s_0'', L'')$ is an SCTMC s.t.:*

- $S'' = \{s \in S \mid R(s, \mathcal{A}, s') = \theta \text{ where } \mathcal{A} \in 2^{AP''} \text{ and } \mathcal{A} \cap \{\bot, \tau\} = \emptyset\}$,
-

$$AP'' = (\bigcup_{\mathcal{X} \in Act \wedge \bot \notin \mathcal{X}} \mathcal{X}) \backslash \{\tau\},$$

- R'' *is defined as:*

$$R''(s, s'') = r \text{ iff } \exists s', R(s, \mathcal{A}, s') = \theta \text{ and } R(s', \bot, s'') = r, \text{ and}$$

$$R''(s, s''') = r \text{ iff } \exists s', s'', R(s, \mathcal{A}, s') = \theta, \ R(s''', \mathcal{A}', s'') = \theta \text{ and } R(s', \tau, s'') = r$$

  where $\mathcal{A}, \mathcal{A}' \in 2^{AP''} \backslash \{\bot\}$,
- $s_0'' = s_0$, *and*
- $L''(s) = \{x \in \mathcal{A} \mid R(s, \mathcal{A}, s') = \theta \text{ and } \mathcal{A} \in 2^{AP''}\}$.

The following proposition establishes that alc^{-1} is the inverse of embedding of alc.

Proposition 4. *Let $\mathcal{C}$ be an SCTMC. Then, $alc^{-1}(alc(\mathcal{C})) = \mathcal{C}$.*

Proof. Consider an arbitrary SCTMC $\mathcal{C} = (S, AP, R, s_0, L)$. Let $alc(\mathcal{C}) = (S', Act', R', s_0')$ and $alc^{-1}(alc(\mathcal{C})) = (S'', AP'', R'', s_0'', L'')$. We establish the isomorphism by proving that there are isomorphisms between the components of both SCTMCs, i.e. we show isomorphisms between S and S'', R and R'', AP and AP'' and L and L''.

From the definition of alc, when applied to $\mathcal{C}$, we have,

- $S' = S \cup \{\bar{s} \mid s \in S\}$
- $Act' = 2^{AP} \cup \{\tau, \bot\}$
- $R'(\bar{s}, \bot, t) = R(s, t) \ \forall s, t \in S \text{ s.t. } R(s, t) > 0 \text{ and } L(s) \neq L(t)$,
  $R'(\bar{s}, \tau, \bar{t}) = R(s, t) \ \forall s, t \in S \text{ s.t. } R(s, t) > 0 \text{ and } L(s) = L(t) \text{ and}$
  $R'(s, L(s), \bar{s}) = \theta \ \forall s \in S$, and
- $s_0' = s_0$.

Since, $alc(\mathcal{C})$ trivially satisfies all the invertibility criterion, it is invertible. Thus, an application of alc^{-1} on the transformed $alc(\mathcal{C})$ yields:

- $S'' = \{s \in S' \mid R'(s, \mathcal{A}, s') = \theta \wedge \mathcal{A} \cap \{\bot, \tau\} = \phi\}$
  $= \{s \in S' \mid s \in S\}$
  $= S$

- $AP'' = (\bigcup_{\mathcal{X} \in Act' \wedge \perp \notin \mathcal{X}} \mathcal{X}) \backslash \{\tau\}$
  $= (\bigcup_{\mathcal{X} \in 2^{AP} \cup \{\tau, \perp\} \wedge \perp \notin \mathcal{X}} \mathcal{X}) \backslash \{\tau\}$
  $= (\bigcup_{\mathcal{X} \in 2^{AP} \cup \{\tau\}} \mathcal{X}) \backslash \{\tau\}$
  $= (AP \cup \{\tau\}) \backslash \{\tau\}$
  $= AP$
- $R''(s, s') = R'(\overline{s}, \tau, \overline{s'}) = R(s, s') \ \forall s, s' \in S'' = S \ if \ L(s) = L(t)$ and
  $R''(s, s') = R'(\overline{s}, \perp, s') = R(s, s') \ \forall s, s' \in S'' = S \ if \ L(s) \neq L(t)$
  thus establishing $R'' = R$.
- $s_0'' = s_0' = s_0$.
- $L''(s) = \{x \in \mathcal{A} \mid R'(s, \mathcal{A}, s') = \theta \ and \ \mathcal{A} \in 2^{AP''}\}$
  $= \{x \in \mathcal{A} \mid R'(s, \mathcal{A}, s') = \theta \ and \ \mathcal{A} \in 2^{AP}\}$
  $= \{x \in \mathcal{A} \mid R'(s, \mathcal{A}, \overline{s}) = \theta \Leftrightarrow L(s) = \mathcal{A}\}$
  $= \{x \in \mathcal{A} \mid L(s) = \mathcal{A}\}$
  $= L(s)$ □

Note that, the inverse defined is the 'left-inverse' of the embedding alc and not the general inverse, as $alc(alc^{-1}(\mathcal{C}))$ is not defined for any arbitrary ACTMC. Next, we prove that the embeddings alc and its inverse alc^{-1} preserve minimality across ACTMCs and SCTMCs for all the equivalences discussed in this paper. Our goal is to show that if one starts with a minimal invertible ACTMC and applies the inverse embedding alc^{-1} to obtain an SCTMC, then it will also be minimal (w.r.t. $\sim, \sim_b, \approx, \approx_b$). We first prove some auxiliary lemmas which are required for proving the main results of this section.

Lemma 2. *Let alc preserves and reflects through $\leftrightarrow$ where $\leftrightarrow \in \{\sim, \sim_b, \approx, \approx_b\}$. Then, for any SCTMC $\mathcal{C}$,*

$$\leftrightarrow_{MS} (\mathcal{C}) = alc^{-1}(\leftrightarrow_{MA} (alc(\leftrightarrow_{MS} (\mathcal{C})))) \implies \leftrightarrow_{MS} (\mathcal{C}) = alc^{-1}(\leftrightarrow_{MA} (alc(\mathcal{C}))).$$

In the following lemma, we show that the ACTMC obtained after applying alc to a minimal SCTMC is already minimal and we do not need to minimize it further.

Lemma 3. *For any SCTMC $\mathcal{C}$, and for $\leftrightarrow \in \{\sim, \sim_b, \approx, \approx_b\}$, the following holds:*

$$\leftrightarrow_{MA} (alc(\leftrightarrow_{MS} (\mathcal{C}))) = alc(\leftrightarrow_{MS} (\mathcal{C}))$$

We use these lemmas to prove the main theorem which asserts that if we apply alc to an SCTMC and minimize it, then by applying the inverse embedding, i.e. alc^{-1}, we will get the minimal SCTMC (across all equivalences discussed in this paper).

Theorem 5. *For any SCTMC $\mathcal{C}$, and for $\leftrightarrow \in \{\sim, \sim_b, \approx, \approx_b\}$, the following holds:*

$$\leftrightarrow_{MS} (\mathcal{C}) = alc^{-1}(\leftrightarrow_{MA} (alc(\mathcal{C})))$$

Proof. **Case $\leftrightarrow = \sim$:** For any SCTMC $\mathcal{C}$, from Lemma 3 we have:

$$\sim_{MA} (alc(\sim_{MS} (\mathcal{C}))) = alc(\sim_{MS} (\mathcal{C}))$$

Lemma 1 combined with the functionality of alc^{-1} gives us:

$$alc^{-1}(\sim_{MA}(alc(\sim_{MS}(\mathcal{C})))) = alc^{-1}(alc(\sim_{MS}(\mathcal{C})))$$

Finally, using Lemma 2, we have our desired conclusion.

$$\sim_{MS}(\mathcal{C}) = alc^{-1}(\sim_{MA}(alc(\mathcal{C})))$$

The proof is analogous for cases when $\leftrightarrow \in \{\sim_b, \approx, \approx_b\}$.   □

Intuitively, an SCTMC $\mathcal{C}$ can be minimized by simply minimizing its embedded ACTMC and applying the inverse.

7.2   Inverse of the Embedding from SCTMC to ACTMC

Remark 4. The embedding slc defined in Sect. 4 will always create an SCTMC which satisfies the invertibility criterion. The explicit definition of the invertibility criterion is only meant for situations when one wishes to apply the inverse embedding first. Additionally, the unique rate θ will be needed to apply slc after applying slc^{-1} to revert back to the original system.

Definition 15 (Invertibility criterion). *Let* $\mathcal{C} = (S, AP, R, s_0, L)$ *be an SCTMC. Then* $\mathcal{C}$ *is invertible iff:*

- $AP = Act \cup \{\perp\}$ *for some set* Act,
- $|L(s)| = 1 \; \forall s \in S$,
- *all the states, i.e.* $s \in S$ *s.t.* $\perp \notin L(s)$ *must have a single outgoing transition. Additionally, all such states must have an identical rate (say θ) on their single outgoing transition, i.e.* $\forall s, s' \in S$ *s.t.* $\perp \notin L(s)$ *and* $\perp \notin L(s')$, *if we have,* $s \xrightarrow{\theta_1}$ *and* $s' \xrightarrow{\theta_2}$ *then* $\theta_1 = \theta_2 = \theta$, *and*
- *If* $\exists s \in S$ *s.t.* $\perp \notin L(s)$, *then* s *can be reached in one step from a state* s' *s.t.* $L(s') = \{\perp\}$ *and* $s' \xrightarrow{r} s$ *and* s *has a single outgoing transition of the form* $s \xrightarrow{\theta} s''$ *for some* $s'' \in S$ *s.t.* $L(s'') = \{\perp\}$.

The following lemma asserts that if a given system is invertible, then the minimal quotient of any system equivalent to the given system is also invertible under all the equivalences discussed in this paper.

Lemma 4. *Let* $\mathcal{C}$ *be an arbitrary invertible SCTMC. For* $\leftrightarrow \in \{\sim, \sim_b, \approx, \approx_b\}$, *given any SCTMC* $\mathcal{C}'$, *s.t.* $\mathcal{C} \leftrightarrow \mathcal{C}'$, $\leftrightarrow_{MS}(\mathcal{C}')$ *is also invertible.*

Definition 16 (slc^{-1}). *Let* $\mathcal{C} = (S, AP, R, s_0, L)$ *be an invertible SCTMC. Then* slc^{-1} *is the inverse of the embedding* slc, *where* $slc^{-1} : SCTMC \rightarrow ACTMC$ *is formally defined as* $slc^{-1}(\mathcal{C}) = (S'', Act'', R'', s_0'')$ *s.t.:*

- $S'' = \{s \in S \mid L(s) = \{\perp\}\}$,
- $Act'' = (AP \backslash \{\perp\}) \cup \{\tau\}$,

- R'' *is defined as:*

$$R''(s, a, s') = r \text{ iff } R(s, s'') = r, R(s'', s') = \theta \text{ and } \perp \notin L(s''),$$

$$R''(s, \tau, s') = r \text{ iff } R(s, s') = r, \text{ and } L(s) = L(s') = \{\perp\},$$

- $s_0'' = s_0$.

The following proposition establishes that slc^{-1} is the inverse of embedding of slc.

Proposition 5. *Let $\mathcal{C}$ be an ACTMC. Then, $slc^{-1}(slc(\mathcal{C})) = \mathcal{C}$.*

Again, note that, the inverse defined is the 'left-inverse'. Next, we prove that the embeddings slc and its inverse slc^{-1} preserve minimality across ACTMCs and SCTMCs for all the equivalences discussed in this paper.

Lemma 5. *Let slc preserves and reflects through $\leftrightarrow$ where $\leftrightarrow \in \{\sim, \sim_b, \approx, \approx_b\}$. Then, for any ACTMC $\mathcal{C}$,*

$$\leftrightarrow_{MA}(\mathcal{C}) = slc^{-1}(\leftrightarrow_{MS}(slc(\leftrightarrow_{MA}(\mathcal{C})))) \implies \leftrightarrow_{MA}(\mathcal{C}) = slc^{-1}(\leftrightarrow_{MS}(slc(\mathcal{C}))).$$

In the following lemma, we show that the SCTMC obtained after applying slc to a minimal ACTMC is already minimal and we need not minimize it further.

Lemma 6. *For any ACTMC $\mathcal{C}$, and for $\leftrightarrow \in \{\sim, \sim_b, \approx, \approx_b\}$, the following holds:*

$$\leftrightarrow_{MS}(slc(\leftrightarrow_{MA}(\mathcal{C}))) = slc(\leftrightarrow_{MA}(\mathcal{C}))$$

We use these lemmas to prove the main theorem which asserts that if we apply slc to an ACTMC and minimize it, then by applying the inverse embedding, i.e. slc^{-1}, we will get the minimal ACTMC (across all equivalences discussed in this paper).

Theorem 6. *For any ACTMC $\mathcal{C}$, and for $\leftrightarrow \in \{\sim, \sim_b, \approx, \approx_b\}$, the following holds:*

$$\leftrightarrow_{MA}(\mathcal{C}) = slc^{-1}(\leftrightarrow_{MS}(slc(\mathcal{C})))$$

Proof. The proof is similar to that of Theorem 5.   □

Intuitively, an ACTMC $\mathcal{C}$ can be minimized by simply minimizing its embedded SCTMC and applying the inverse.

8   Conclusions

We have proposed a formal framework which allows one to move from ACTMCs to SCTMCs, and, conversely, from SCTMCs to ACTMCs. We have defined two embeddings and proved that strong forward bisimulation, strong backward bisimulation, weak forward bisimulation and weak backward bisimulation are preserved by these embeddings. Next, we have defined the invertibility criteria and the inverse of these embeddings. We have shown that invertibility is preserved with respect to all the four variants of bisimulation. Finally, we have proved that minimization in one setting can be achieved by simply minimizing the embedded model in the other setting and applying the inverse of the embedding. Our framework helps in bridging the gap between two equally important stochastic modeling communities. Additionally, from an application point of view, our results have enabled the practitioners to use the state-of-the-art tools developed in one setting for model minimization and analysis in the other setting. For instance, if one of the stochastic model checking tools, e.g. PRISM or Storm implements the quotienting algorithm for (weak) backward bisimulation, then using our embeddings, models in the action labeled stochastic setting can also be directly minimized. This research work can be extended in several interesting directions which are as follows:

- Implement a tool that allows one to construct an ACTMC from an SCTMC model and vice versa. For example, this tool would allow transforming ACTMCs to SCTMCs and use the state-of-art-the machinery provided by PRISM [27] for minimization and analysis.
- Investigate the possibility to design more advanced embedding techniques to reduce the size of the embedded model.
- Investigate the preservation of (weak) linear-time equivalences by these embeddings [37,39].

References

1. Alur, R., Dill, D.L.: A theory of timed automata. Theor. Comput. Sci. **126**(2), 183–235 (1994)
2. Baier, C., Katoen, J.P.: Principles of Model Checking. MIT Press, Cambridge (2008)
3. Baier, C., Haverkort, B.R., Hermanns, H., Katoen, J.P.: Model-checking algorithms for continuous-time Markov chains. IEEE Trans. Softw. Eng. **29**(6), 524–541 (2003)
4. Baier, C., Hermanns, H., Katoen, J., Wolf, V.: Bisimulation and simulation relations for Markov chains. Electron. Notes Theor. Comput. Sci. **162**, 73–78 (2006)
5. Baier, C., Katoen, J.P., Hermanns, H., Wolf, V.: Comparative branching-time semantics for Markov chains. Inf. Comput. **200**(2), 149–214 (2005)
6. ter Beek, M.H., Damiani, F., Gnesi, S., Mazzanti, F., Paolini, L.: From featured transition systems to modal transition systems with variability constraints. In: Calinescu, R., Rumpe, B. (eds.) SEFM 2015. LNCS, vol. 9276, pp. 344–359. Springer, Cham (2015). https://doi.org/10.1007/978-3-319-22969-0_24
7. ter Beek, M.H., Damiani, F., Gnesi, S., Mazzanti, F., Paolini, L.: On the expressiveness of modal transition systems with variability constraints. Sci. Comput. Program. **169**, 1–17 (2019)
8. Bernardo, M.: Non-bisimulation-based Markovian behavioral equivalences. J. Log. Algebraic Methods Program. **72**(1), 3–49 (2007)
9. Bernardo, M.: Weak Markovian bisimulation congruences and exact CTMC-level aggregations for concurrent processes. In: QAPL, pp. 122–136. EPTCS 85 (2012)

10. Bernardo, M., Cleaveland, R.: A theory of testing for Markovian processes. In: Palamidessi, C. (ed.) CONCUR 2000. LNCS, vol. 1877, pp. 305–319. Springer, Heidelberg (2000). https://doi.org/10.1007/3-540-44618-4_23
11. Buchholz, P.: Exact and ordinary lumpability in finite Markov chains. J. Appl. Prob. **31**, 59–75 (1994)
12. Bunte, O., et al.: The mCRL2 toolset for analysing concurrent systems. In: Vojnar, T., Zhang, L. (eds.) TACAS 2019. LNCS, vol. 11428, pp. 21–39. Springer, Cham (2019). https://doi.org/10.1007/978-3-030-17465-1_2
13. Chen, T., Han, T., Katoen, J., Mereacre, A.: Model checking of continuous-time Markov chains against timed automata specifications. Log. Methods Comput. Sci. **7**(1) (2011)
14. Das, S., Sharma, A.: Embeddings between state and action labeled probabilistic systems. In: Hung, C., Hong, J., Bechini, A., Song, E. (eds.) SAC 2021: The 36th ACM/SIGAPP Symposium on Applied Computing, Virtual Event, Republic of Korea, 22–26 March 2021, pp. 1759–1767. ACM (2021)
15. Dehnert, C., Junges, S., Katoen, J.-P., Volk, M.: A STORM is coming: a modern probabilistic model checker. In: Majumdar, R., Kunčak, V. (eds.) CAV 2017. LNCS, vol. 10427, pp. 592–600. Springer, Cham (2017). https://doi.org/10.1007/978-3-319-63390-9_31
16. Donatelli, S., Haddad, S., Sproston, J.: Model checking timed and stochastic properties with CSL^{TA}. IEEE Trans. Softw. Eng. **35**(2), 224–240 (2009)
17. Garavel, H., Lang, F., Mateescu, R., Serwe, W.: CADP 2010: a toolbox for the construction and analysis of distributed processes. In: Abdulla, P.A., Leino, K.R.M. (eds.) TACAS 2011. LNCS, vol. 6605, pp. 372–387. Springer, Heidelberg (2011). https://doi.org/10.1007/978-3-642-19835-9_33
18. Gorla, D.: Towards a unified approach to encodability and separation results for process calculi. In: van Breugel, F., Chechik, M. (eds.) CONCUR 2008. LNCS, vol. 5201, pp. 492–507. Springer, Heidelberg (2008). https://doi.org/10.1007/978-3-540-85361-9_38
19. Gorla, D.: Towards a unified approach to encodability and separation results for process calculi. Inf. Comput. **208**(9), 1031–1053 (2010)
20. Gorla, D., Nestmann, U.: Full abstraction for expressiveness: history, myths and facts. Math. Struct. Comput. Sci. **26**(4), 639–654 (2016)
21. Hermanns, H., Herzog, U., Katoen, J.: Process algebra for performance evaluation. Theor. Comput. Sci. **274**(1–2), 43–87 (2002)
22. Hillston, J.: A Compositional Approach to Performance Modelling. Cambridge University Press, USA (1996)
23. Katoen, J.-P., Kemna, T., Zapreev, I., Jansen, D.N.: Bisimulation minimisation mostly speeds up probabilistic model checking. In: Grumberg, O., Huth, M. (eds.) TACAS 2007. LNCS, vol. 4424, pp. 87–101. Springer, Heidelberg (2007). https://doi.org/10.1007/978-3-540-71209-1_9
24. Katoen, J., Khattri, M., Zapreev, I.S.: A Markov reward model checker. In: QEST, pp. 243–244. IEEE Computer Society (2005)
25. Katoen, J., Klink, D., Leucker, M., Wolf, V.: Three-valued abstraction for probabilistic systems. J. Log. Algebr. Program. **81**(4), 356–389 (2012)
26. Kwiatkowska, M., Norman, G., Parker, D.: Symmetry reduction for probabilistic model checking. In: Ball, T., Jones, R.B. (eds.) CAV 2006. LNCS, vol. 4144, pp. 234–248. Springer, Heidelberg (2006). https://doi.org/10.1007/11817963_23
27. Kwiatkowska, M., Norman, G., Parker, D.: PRISM 4.0: verification of probabilistic real-time systems. In: Gopalakrishnan, G., Qadeer, S. (eds.) CAV 2011. LNCS,

vol. 6806, pp. 585–591. Springer, Heidelberg (2011). https://doi.org/10.1007/978-3-642-22110-1_47

28. Marsan, M.A., Balbo, G., Conte, G., Donatelli, S., Franceschinis, G.: Modelling with Generalized Stochastic Petri Nets, 1st edn. Wiley, USA (1994)
29. Meyer, J.F., Movaghar, A., Sanders, W.H.: Stochastic activity networks: structure, behavior, and application. In: International Workshop on Timed Petri Nets, Torino, Italy, 1–3 July 1985, pp. 106–115. IEEE Computer Society (1985)
30. Nicola, R.D., Fantechi, A., Gnesi, S., Ristori, G.: An action-based framework for verifying logical and behavioural properties of concurrent systems. Comput. Netw. ISDN Syst. **25**(7), 761–778 (1993)
31. De Nicola, R., Vaandrager, F.: Action versus state based logics for transition systems. In: Guessarian, I. (ed.) LITP 1990. LNCS, vol. 469, pp. 407–419. Springer, Heidelberg (1990). https://doi.org/10.1007/3-540-53479-2_17
32. Nicola, R.D., Vaandrager, F.W.: Three logics for branching bisimulation. J. ACM **42**(2), 458–487 (1995)
33. Ouaknine, J., Worrell, J.: Some recent results in metric temporal logic. In: Cassez, F., Jard, C. (eds.) FORMATS 2008. LNCS, vol. 5215, pp. 1–13. Springer, Heidelberg (2008). https://doi.org/10.1007/978-3-540-85778-5_1
34. Plateau, B., Atif, K.: Stochastic automata network of modeling parallel systems. IEEE Trans. Softw. Eng. **17**(10), 1093–1108 (1991)
35. Reniers, M.A., Schoren, R., Willemse, T.A.C.: Results on embeddings between state-based and event-based systems. Comput. J. **57**(1), 73–92 (2014)
36. Reniers, M.A., Willemse, T.A.C., et al.: Folk theorems on the correspondence between state-based and event-based systems. In: Černá, I. (ed.) SOFSEM 2011. LNCS, vol. 6543, pp. 494–505. Springer, Heidelberg (2011). https://doi.org/10.1007/978-3-642-18381-2_41
37. Sharma, A., Katoen, J.-P.: Weighted lumpability on Markov chains. In: Clarke, E., Virbitskaite, I., Voronkov, A. (eds.) PSI 2011. LNCS, vol. 7162, pp. 322–339. Springer, Heidelberg (2012). https://doi.org/10.1007/978-3-642-29709-0_28
38. Sproston, J., Donatelli, S.: Backward bisimulation in Markov chain model checking. IEEE Trans. Softw. Eng. **32**(8), 531–546 (2006)
39. Wolf, V., Baier, C., Majster-Cederbaum, M.E.: Trace machines for observing continuous-time Markov chains. Electron. Notes Theor. Comput. Sci. **153**(2), 259–277 (2006)

Multi-timescale Fairness for Heterogeneous Broadband Traffic in Access-Aggregation Networks

Szilveszter Nádas[1], Balázs Varga[1], Illés Horváth[2], András Mészáros[3], and Miklós Telek[3(✉)]

[1] Ericsson Research, Budapest, Hungary
{szilveszter.nadas,balazs.a.varga}@ericsson.com
[2] MTA-BME Information Systems Research Group, Budapest, Hungary
[3] Department of Networked Systems and Services, Budapest University of Technology and Economics, Budapest, Hungary
{meszarosa,telek}@hit.bme.hu

Abstract. We propose a throughput value function (TVF) based solution for providing multi time-scale (MTS) fairness for broadband traffic in access-aggregation networks. The primary goal of MTS fairness is a dynamic control of resource sharing that considers the usage history of the broadband connection. We present a flow level description of the multi time-scale throughput value function (MTS-TVF) based resource sharing. We provide dimensioning guidelines in traffic aggregation scenarios and present its simulation-based performance analysis.

In the performance analysis, our focus is on overloaded systems involving both low load users (with temporally active traffic) and high load users (with heavy traffic, e.g. continuous multiple downloads). We find that the Quality of Experience (QoE) of low load users significantly increases when using MTS-TVF and it becomes similar to that of a lightly loaded system, while the change in the QoE is minimal for high load users.

Keywords: Fairness · Multiple timescales · Core stateless · Resource sharing · Throughput value function · QoS · Fluid model

1 Introduction

Resource sharing among traffic flows has remained an area of interest in networking research. Fairness is usually interpreted as equal (or weighted) throughput [1] experienced by flows. By definition, throughput is a measure derived from total packet transmission during a time interval, the length of which is called *timescale*. With the introduction of 5G for mobile and Fiber-To-The-Home (FTTH) for fixed Internet access, the capacity of the last mile has significantly been increased, resulting in much higher load on the access-aggregation

Partially supported by the OTKA K123914 and the TUDFO/51757/2019-ITM grants.

networks than before, thus moving bottlenecks from the edge to routers in the aggregation. In such a network, the congestion controls used by the flows and the (propagation) round trip times (RTTs) are much more heterogeneous than in data centers and other closed enterprise networks. To handle the increased load and to serve these high-speed bottlenecks, a new node functionality is needed, where controlling resource sharing is an important design goal.

Most current resource sharing control methods are based on throughput measured only on a short timescale (e.g. RTT). For bursty traffic, throughput measured on multiple timescales (e.g. RTT, 1 s, 10 s, session duration) usually results in different values. From the end-user perspective, network performance is better described by throughput during the active periods of a source as opposed to the general case when active and inactive periods are both considered. Taking the history of inactivity into account is advantageous for short transmissions like web downloads or initial buffering of adaptive video streaming. A comprehensive recent survey on fairness [1] states that "getting a scheme to instantly serve web flows for improved performance while maintaining fairness between other persistent traffic remains an open and significant design problem to be investigated." For elastic flows, [2] argues that "highly unequal flow rates have led to flow completion times considerably better than with equal flow rates, indeed nearly as good as they were before the contending long-running flow was introduced".

The literature on these two main concepts, resource sharing control methods based on multi-timescale (MTS) throughput measurement and throughput value function (TVF) is limited. A solution for providing MTS fairness, referred to as Multi-Timescale Bandwidth Profile (MTS-BWP), was introduced in [3]. It defines and implements multi-timescale fairness for a network scenario with few sources with well-defined traffic behaviour. MTS-BWP applies several token buckets per Drop Precedence representing increasing timescales of throughput measurements. MTS-BWP implementation complexity increases with the number of drop precedences, which may out-weigh its advantages when fine grained control is needed. The concept of using TVFs for fine-grained resource sharing based on short timescale throughput measure was introduced in [4], and the MTS extension of the TVF is discussed in [5]. The packet level behaviour of multi time-scale throughput value function (MTS-TVF) is considered and evaluated in [5], using a packet level simulation tool. Due to the inherent complexity of the packet level behaviour, the applicability of the packet level analysis is restricted to rather simple scenarios, much smaller than the aggregation scenario considered in this paper. Practically, only the evaluation of the initial transient of a small network scenario is feasible with the packet level simulator, which is a single jump in the fluid simulator. A contribution of the paper is the introduction of the fluid model of MTS-TVF resource sharing and an associated fluid simulation tool, which makes it possible to evaluate the performance of MTS-TVF based resource sharing in real networking aggregation scenarios.

To really utilize the advantages of MTS-TVF, flexible but explicit dimensioning guidelines are needed. The main contribution of this paper is a design approach for MTS-TVF resource sharing, that provides such dimensioning guidelines to achieve MTS fairness goals for heterogeneous broadband traffic in access-aggregation network. The benefit of using the proposed MTS-TVF resource sharing is evaluated by comparing its behaviour with the single timescale TVF (STS-TVF) based one, and the TCP fairness based one.

The rest of the paper is organized as follows. Section 2 gives an overview of TVF based resource sharing. Section 3 introduces multi-timescale fairness. Section 4 describes a fluid model of the proposed resource sharing method that will be used for dimensioning. Section 5 provides dimensioning guidelines for specific goals. Section 6 provides an approximate analysis of the system based on analytic calculations. Section 7 provides numerical results, and Sect. 8 concludes the work.

2   Overview of STS-TVF Resource Sharing

In a very high level view, TVF determines how the resources are shared between users with different bandwidths. The STS-TVF resource sharing [4] extends the idea of core stateless resource sharing solutions like [6,7] by marking each packet with a continuous value called Packet Value (PV). The main goal in a network element is to maximize the total aggregate PV of delivered packets. The resource sharing procedure is composed of two phases: 1) Packet marking at network edge; 2) Packet scheduling and dropping based on the PV in the middle of the network.

1) The goal of packet marking is to assign a PV to each packet based on the operator policy and the traffic rate R of the traffic source node (represents e.g. a subscriber and referred to as *node* in the sequel). The PV represents the potential of the packet to get through the network, but the transmission probability also depends on the congestion level of the network. If the network is highly congested packet with high PVs might be dropped, while in case of moderate congestion even packets with low PV get through.

To achieve this goal, packets are marked at the edge of the network by using the resource sharing policy of the operator described by a TVF (denoted by $TVF(.)$). The marker assigns *random* PVs to packets from a proper TVF and bandwidth dependent distribution, such that the rate of packets of the given node with PV larger than x is $TVF(x)$.

The packet marker is implemented as follows. Generated traffic is measured on a single time scale: when the measured rate is R, the assigned PV is $TVF(x)$, where x is a uniformly distributed sample in $[0, R]$. The same packet marking algorithm is applied in all nodes.

We note that operators might have different TVFs for different user classes (e.g. Gold, Silver, Background, Voice), but in this paper we restrict our attention to a single user class.

2) Resource nodes in the middle of the network treat packets solely relying on the carried PVs. Each such node aims at maximizing the total amount of PV transmitted over the shared bottleneck. To this end, scheduling algorithms of different complexity can be used, including algorithms that drop the packet with the smallest PV (even from the middle of the buffer) when the buffer length is too long [4] or using proportional integral controllers (PI-controllers) to determine a PV threshold for packet dropping [8].

Accordingly, at high congestion only packets with high PVs are transmitted, more precisely packets with PV above a given Congestion Threshold Value (CTV) that reflects the actual congestion level. Note that the amount of high and low PV packets in different flows determines the resource share between them. As a result, flows with larger share of high PV packets receive more throughput.

3 Multi-timescale Fairness

For bandwidth profiling, bitrate is typically measured on a short timescale in the order of RTT. It expresses the instantaneous resource usage and it can even capture short bursts. STS-TVF resource sharing uses only this short timescale bitrate to share bandwidth. In other words, the history of nodes is not considered in STS-TVF resource sharing. When our goal is to ensure long-term fairness (or network usage service level agreement) among flows with largely different profiles, bitrates on longer timescales are far more expressive. We assume n timescales $(TS_1, \ldots, TS_n)$ with different lengths: $TS_1 \approx \text{RTT} < TS_2 < \ldots < TS_n$ (e.g. RTT, 1 s, 10 s, session duration). For a flow with an equally spaced, stable traffic, after the time associated with the largest timescale has elapsed, we expect all those rate measurements to be the same, i.e., $R_i \approx R$, $\forall TS_i$, where R_i is the measured rate on timescale TS_i. However, in transient situations, e.g. when transmission starts for a previously silent flow, we expect small rate measurements of long timescales, while R_1 (of $TS_1 \approx \text{RTT}$) may be high (i.e. $R_1 > R_2 > \ldots > R_n$). Rate measurements at shorter timescales react faster to the changes of network conditions, while at longer timescales temporal changes may remain invisible. Similar behavior with the opposite ordering can be seen for a case when a flow stops transmission (or its rate decreases) after a long active period. Implementation of rate measurement algorithms is detailed in [5].

3.1 Multi-timescale Throughput Value Functions (MTS-TVF)

The goal of MTS-TVF [5] is to control the resource sharing between users with different rates on different timescales. E.g. such that after an inactive period of a subscriber it gets an advantage for its new session compared to subscribers with long time transmission.

Figure 1 depicts an example with four TVFs: $TVF_4() \ldots TVF_1()$ (disregard the other notations for the time being). The actual throughput value of the packet is derived from the four TVFs based on the actual $R_4, \ldots, R_1$ throughput measurements as follows.

3.2 MTS Rate Measurement-Based Marker

The packet marking based on MTS-TVF is a two steps procedure.

Algorithm 1. CTVF$(n, R_i, TVF_i(), i \in \{1, \ldots, n\})$

$R'_n = R_n,$
for $i = n - 1, i > 0, i = i - 1$ **do**
$\quad R'_i = \max(R'_{i+1}, R_i),$ $\quad\quad$ //largest of $R_j, j \geqslant i$
$\quad PV_i = TVF_{i+1}\left(R'_{i+1} + \sum_{j=i+1}^{n-1} \Delta_j\right),$
$\quad \Delta_i = TVF_i^{-1}(PV_i) - \left(R'_{i+1} + \sum_{j=i+1}^{n-1} \Delta_j\right),$
end for
$CTVF =$
$$\begin{cases} TVF_n(x) & \text{if } x < R'_n, \\ TVF_{n-1}(x + \Delta_{n-1}) & \text{if } R'_n \leq x < R'_{n-1}, \\ \vdots & \vdots \\ TVF_1\left(x + \sum_{j=i}^{n-1} \Delta_j\right) & \text{if } R'_2 \leq x, \end{cases}$$
Return($CTVF$),

- First, a composite TVF (CTVF) is computed based on the actual R_i measurements and the $TVF_i(), i \in \{1, \ldots, n\}$ functions.
- In the second step, the computed CTVF function is used as the (single) TVF in STS-TVF resource sharing and the PV is randomly assigned as follows. When the measured rates are $R_1, \ldots, R_n$, the assigned PV is $CTVF(x)$, where x is a uniformly distributed random sample in $[0, R_1]$.

Algorithm 1 (from [5]) implements the marking procedure, where the **for** loop goes downward and the $\sum_{j=i+1}^{n-1}$ summation is idle for $i = n - 1$. In a high level description of the procedure, the first step is to compile a single TVF referred to as CTVF, which is sensitive to the R_i rates and the second step is to apply the "single time scale" packet marking approach from [4]. Figure 1 and 2 demonstrate the composition of the CTVF using a 4 TS example. The algorithm constructs the CTVF by properly shifting sections from each of the $TVF_i()$ functions to form a single monotone decreasing function. Intuitively, the idea is that a given $TVF_i()$ determines the resource share when the instantaneous rate is between R_{i+1} and R_i (when $R_{i+1} < R_i$). For a detailed explanation of Algorithm 1, we refer to [5], while here we provide some further remarks.

The algorithm does not utilize R_1 (it is only used for computing a PV). When $R_{i+1} < R_i$ holds for $i \in \{1, \ldots, n - 1\}$, as it is in Fig. 1, $R'_i = R_i$ and the **for** cycle computes Δ_i values and PV_i values for $i \in \{1, \ldots, n - 1\}$.

The composition of CTVF is demonstrated in Fig. 2 (where this shifting changes the appearance of the function in log-log scale). Essentially, the R_i values determine which part of $TVF_i()$ plays role in the CTVF. The higher $R_i - R_{i+1}$ is, the more significant the role of $TVF_i()$ in the CTVF is. When the $R_{i+1} < R_i$ relation is violated for some i, the procedure compiles the CTVF without using the $TVF_i()$ function.

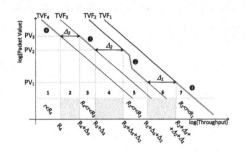

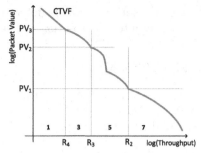

Fig. 1. Example of TVFs for 4-timescales

Fig. 2. CTVF composed from the Example in Fig. 1

4 Fluid Simulation of MTS-TVF Resource Sharing

With MTS-TVF we can implement fine grained resource sharing policies. The appropriate evaluation and validation are crucial elements of developing such policies. However, these are not trivial tasks for MTS-TVS. In [5], the behaviour of the MTS-TVF resource sharing is investigated with a packet level simulator, unfortunately this approach, while precise, restricts the analysis to simple scenarios over a rather short time period. In order to gain dimensioning level information, in this section, we introduce a fluid model [9] of the MTS-TVF resource sharing method. Our fluid model assumes idealized resource sharing characteristics, namely instantaneous bandwidth adaptation and no bottleneck buffer: RTT is equal to 0, there is no packet loss, and packets are infinitesimally small. These assumptions correspond to a fluid model where the throughput of each flow adapts instantly to varying conditions. Although at packet level, congestion results in packets lost and re-sent, for dimensioning purposes there is no need for such level of detail and fluid models work properly.

4.1 Fluid Model of Packet Marking and Forwarding

In the fluid model the rate measurements are maintained on all timescales $(R_1 \ldots R_n)$ and based on that the CTVF is computed for all nodes using Algorithm 1.

In the fluid model the PV computation of the nodes and the associated packet dropping at bottleneck link is replaced by the calculation of the ideal resource sharing using the concept of Congestion Threshold Value (CTV), which is computed from

$$\sum_{u \in \{\text{all nodes}\}} CTVF_u^{-1}(CTV) = C, \tag{1}$$

where C is the capacity of the bottleneck link. Intuitively, (1) states that the instantaneous bandwidth allocated to node u is $CTVF_u^{-1}(CTV)$, and the allocated bandwidth sums up to C. (1) is an implicit equation for the unknown

CTV, whose solution is unique due to the strict monotonicity of $CTVF_u(x)$, the simulator computes the solution of (1) by binary search.

4.2 Fluid Simulator

In our model, a node can generate multiple flows with different characteristics (e.g. web download, video).

Our fluid simulator keeps track of the state of the system:

- the arrival time and finishing time of each flow;
- the list of all active flows at all nodes along with the remaining flow size;
- the bitrate history of each node (from which $R_1, \ldots, R_n$ is known).

Based on the above information, the simulator calculates the CTV according to (1) and the bandwidth rate allocation for each node and for each flow in the system. The simulator recalculates all information at regular small time intervals Δt, and whenever a flow arrives or leaves the system.

5 Dimensioning Guidelines

MTS-TVF is a powerful tool to control resource sharing. However, to fully utilize its capabilities, properly founded dimensioning rules are needed. In this section we propose resource sharing guidelines for providing MTS fairness for heterogeneous broadband traffic in an access-aggregation network. In the dimensioning we only consider the resource sharing for congested system states, because throughput goals are more critical in these cases. Specifically, we consider the following scenario: There are two kinds of nodes, high load nodes (HLNs) and low load nodes (LLNs), competing for the bandwidth (C) of a common bottleneck link. The numbers of HLNs, LLNs and all nodes are N_H, N_L, and $N = N_H + N_L$, respectively. The N_H HLNs are constantly active, resulting in a fully utilized bottleneck link. Consequently, the traffic history of HLNs is the same with relatively high measured throughput on the largest timescale and the load of a single LLN is low enough that a newly active LLN has negligible measured throughput on the largest timescale.

In this scenario we aim to achieve the following dimensioning goals (DGs):

DG1: We want each HLN to achieve at least BW_1 throughput in long-term average.
DG2: If a LLN with inactive history becomes active we aim to allocate it approximately ρ times as much bandwidth as HLNs get, and this allocated bandwidth has to be high enough so that the LLN is able to download *ibs* Mbit in t_1 seconds.
DG3: To avoid extreme fluctuations in the bandwidth allocated, ρ set to be the lowest value satisfying DG2.

DG2 can correspond to, e.g. downloading a web page in t_1 time. In the simulations in Sect. 7 we consider video downloads. The video starts only when a buffer is filled in the video player (hence the name *ibs*, initial buffer size).

5.1 The Proposed MTS-TVF

For DG1 to hold, it is necessary that

$$N_H BW_1 + \ell_L \cdot C < C, \tag{2}$$

where ℓ_L is the total load of LLNs relative to C (i.e. $\ell_L \cdot C$ is the total load of LLNs). To satisfy the DGs we propose to use two timescales with the following TVFs (shown in Fig. 3):

$$TVF_1(x) = \begin{cases} 1/x, & \text{if } x < BW_1, \\ \frac{(BW_2-x)}{(BW_2-BW_1)BW_1} + \frac{\rho(x-BW_1)}{(BW_2-BW_1)\rho BW_2}, & \text{if } BW_1 \le x < BW_2, \\ \rho/x, & \text{otherwise,} \end{cases} \tag{3}$$

$$TVF_2(x) = 1/x, \tag{4}$$

where BW_2 is set to $BW_2 = \rho(BW_1 + \epsilon)$, thus $TVF_1(x)$ is strictly monotone decreasing (i.e., invertible, which is needed for the Algorithm 1 to work) in the (BW_1, BW_2) interval, where ϵ is a small positive value ($0 < \epsilon << BW_1$) and $TVF_1(x)$ is linear between BW_1 and BW_2. Cf. Algorithm 1, as the number of timescales is two

$$\Delta_1 = TVF_1^{-1}(TVF_2(R_2)) - R_2. \tag{5}$$

The first timescale is the RTT. The second timescale and ρ are set to satisfy DG2 and DG3:

$$TS_2 = \frac{ibs}{BW_1} \quad \text{and} \quad \rho = \frac{ibs}{BW_1 t_1}. \tag{6}$$

5.2 Intuitive Behaviour of the Proposed MTS-TVF

For $x < BW_1$, $TVF_1(x) = TVF_2(x)$, and consequently $\Delta_1 = 0$ (see (5) and also Fig. 3). For any node with $R_2 \le BW_1$ (where R_2 is its bitrate on TS_2) the CTVF (purple/dashed curve in Fig. 4) is

$$\widetilde{CTVF}(x) = TVF_1(x). \tag{7}$$

For any node with $R_2 > BW_1$, we assume that ϵ is a rather small value to make TVF_1 invertible (with the given numerical precision) thus we avoid the discussion of the case when $BW_1 < R_2 < BW_1 + \epsilon$. In the case when $R_2 > BW_1 + \epsilon$, $\Delta_1 = (\rho - 1)R_2$ and the CTVF is (brown/solid curve in Fig. 4)

$$\widetilde{CTVF}(x) = \begin{cases} 1/x & \text{if } x \le R_2 \\ 1/(x + \Delta_1)\rho & \text{otherwise.} \end{cases} \tag{8}$$

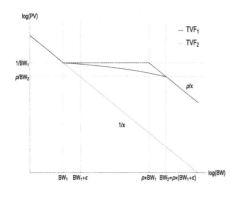

Fig. 3. The proposed TVFs

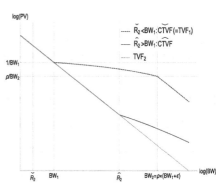

Fig. 4. CTVFs for $R_2 < BW_1$ and $R_2 > BW_1$

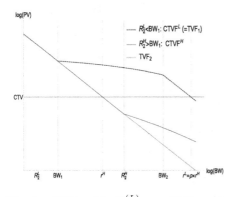

Fig. 5. CTV with $R_2^{(L)} < BW_1$ and $R_2^{(H)} > BW_1$

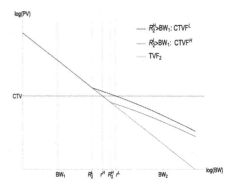

Fig. 6. CTV with $R_2^{(L)} > BW_1$ and $R_2^{(H)} > BW_1$

Let $R_2^{(H)}$ denote the bitrate of HLNs measured on TS_2, $R_2^{(L)}$ denote the bitrate of a chosen active LLN measured on TS_2 and $r^{(H)}$ and $r^{(L)}$ denote the instantaneous bitrates of HLNs and active LLNs, respectively.

Assuming that the system is always close its stationary behaviour and the number of active LLNs is relatively stable, $r^{(H)}$ is close to constant, consequently $R_2^{(H)} \approx r^{(H)}$, and due to (2), $R_2^{(H)} \approx r^{(H)} > BW_1$ the associated CTVF at $r^{(H)}$ is $\widehat{CTVF}(r^{(H)}) \approx 1/r^{(H)} \approx 1/R_2^{(H)}$.

Furthermore, assuming that HLNs with identical history compete for the bandwidth remaining for HLNs, $(1 - \ell_L)C$, the CTV is obtained from (1) as follows

$$\sum_{u\in\{\text{HLNs}\}} CTVF_u^{-1}(CTV) = (1 - \ell_L)C,$$
$$\Downarrow$$
$$\widehat{CTVF}^{-1}(CTV) = (1 - \ell_L)C/N_H = R_2^{(H)} > BW_1,$$
$$\Downarrow$$
$$CTV = \widehat{CTVF}(R_2^{(H)}) = 1/R_2^{(H)} < 1/BW_1.$$

where $R_2^{(H)} > BW_1$ comes from (2).

Let us assume that a formerly inactive LLN becomes active at a given point in time. Then $R_2^{(L)} = 0$, and the CTVF of the node is $\widehat{CTVF}() = TVF_1()$. As this LLN is added to the competition for the bandwidth, the CTV increases a bit and the bandwidth of the HLNs decreases a bit, but the CTV remains below $1/BW_1$ and the bandwidth allocated to LLNs is $r^{(L)} = \widehat{CTVF}^{-1}(CTV) = \rho/CTV$ (from the "otherwise" option of $TVF_1()$), while the bandwidth allocated to HLNs is $r^{(H)} = 1/CTV$ (as it is exemplified in Fig. 5). From this point $R_2^{(L)}$ starts increasing monotonically such that and $r^{(L)} \approx \rho r^{(H)}$ for as long as $R_2^{(L)} < BW_1$. If ibs megabits are downloaded in time $t_{ibs} \le TS_2$ then at t_{ibs} after the LLN becomes active

$$R_2^{(L)} \geqslant \frac{ibs}{TS_2} = BW_1, \tag{9}$$

where we used TS_2 from (6) in the second step. According to (9) and (6), ibs megabits are downloaded with rate

$$r^{(L)} \approx \rho r^{(H)} > \rho BW_1 = \frac{ibs}{t_1}, \tag{10}$$

therefore DG2 will be fulfilled and ibs megabits will be downloaded in less than t_1 seconds. At the limit of the inequality (2), $r^{(L)} = \rho BW_1 = ibs/t_1$, which is just enough to download ibs megabits in t_1 seconds, therefore we set ρ according to (6), which is the lowest ρ to satisfy DG3.

When, due to the high throughput ($r^{(L)} \approx \rho r^{(H)}$) at the beginning of the active period of the LLN, $R_2^{(L)}$ increases above BW_1 the associated CTVF becomes $\widehat{CTVF}(x)$ and the bandwidth allocation modifies according to Fig. 6.

6   Approximate Analysis of the Stationary Behaviour

In general, we expect the system to converge to some stationary behaviour, but calculating the stationary distribution explicitly is infeasible. Instead, we focus on the mean number of active LLNs in the stationary distribution. In this setting, all HLNs are active, and one of the dimensioning parameter is the number of active LLNs.

In this section, we present an approximate calculation based on intuitive assumptions which allow to compute the stationary behaviour and later we evaluate the accuracy of the approximation. The approximate analysis is based on the assumption that all active LLNs have a single flow which started from a perfect node history. Note that even apart from this assumption, the calculations only provide an approximation due to the non-linearity and long memory of the system.

The following calculations are specific to the TVF designed in Sect. 5, which sharply distinguish the nodes with high and low measured R_i rates, referred to as good and bad history. The active LLNs are divided into the following categories:

- $N_L^{(1)}$ is the mean number of active LLNs with a web flow with good history;
- $N_L^{(2)}$ is the mean number of active LLNs with a web flow with bad history;
- $N_L^{(3)}$ is the mean number of active LLNs with a video flow with good history;
- $N_L^{(4)}$ is the mean number of active LLNs with a video flow with bad history.

Using the notations also from Table 1, we approximate the system behaviour with the following equations:

$$C = (N_L^{(1)} + N_L^{(3)})\rho R_{st} + (N_L^{(2)} + N_L^{(4)} + N_H)R_{st}, \tag{11}$$

$$\ell_L \cdot C = (N_L^{(1)} + N_L^{(3)})\rho R_{st} + (N_L^{(2)} + N_L^{(4)})R_{st}, \tag{12}$$

$$\frac{N_L^{(1)}}{N_L^{(2)}} = \frac{ibs_2}{(fs_1 - ibs_2)\rho}, \tag{13}$$

$$\frac{N_L^{(3)}}{N_L^{(4)}} = \frac{ibs_2}{(fs_2 - ibs_2)\rho}, \tag{14}$$

$$\frac{20\%}{80\%} = \frac{\rho N_L^{(1)} + N_L^{(2)}}{\rho N_L^{(3)} + N_L^{(4)}}, \tag{15}$$

where R_{st} is the mean bandwidth allocated to a node with bad history (identical to all nodes with bad history).

(11) corresponds to the fact that the system is always used at full capacity (due to $\ell_L + \ell_H > 1$). According to (12), the entire load of the LLNs is serviced (no discarding at LLNs). (13) and (14) set the ratio of *time* spent in good/bad node history for LLNs, taking into account that the node history changes from good to bad after downloading initial buffer size ibs_2. Finally, (15) sets the web/video ratio of the incoming data.

(11)–(15) leads to a system of linear equations describing the mean stationary behaviour of the system, which can then be compared with actual simulations, done in the next section.

Table 1. Model parameters

C	1000 Mbps	Total capacity
N_L	900	Number of LLNs
N_H	100	Number of HLNs
	80%/20%	LLNs' video/web data ratio
fs_1	5 MB	Web download file size
fs_2	18.75 MB	Video download file size
ibs_2	3.125 MB	Initial video buffer size
t_1	2 s	Initial buffer download time
t_2	30 s	Video download time
	30	Number of HLN flows per node
	10	Maximal number of flows at a LLN
BW_1	5 Mbps	Guaranteed throughput of HLNs
ℓ_L	$0.1, 0.2, 0.3, 0.4$	Load of LLNs proportional to C

Table 2. Number of active LLNs according to simulation and approximate calculation

	$\ell_L = 0.2$		$\ell_L = 0.4$	
	Approx.	Sim.	Approx.	Sim.
Slow LLN $(N_L^{(2)} + N_L^{(4)})$	19.5	17.3	49.4	51.4
Fast LLN $(N_L^{(1)} + N_L^{(3)})$	2.6	3.9	6.9	6.6

7 Simulation Results

7.1 Simulation Setup

The parameters of the considered heterogeneous Broadband traffic scenario of the Access-Aggregation Network is summarized in Table 1. The HLNs have 30 continuously active flows with data to transmit and the LLNs initiate web and video flows according to Poisson arrival processes, whose arrival rate can be obtained from ℓ_L, fs_1, fs_2 and the video/web data ratio of LLNs. The number of flows at a LLN is at most 10. Flows arriving when this limit is reached are dropped.

Based on the simulation runs we check the following requirements:

- if the long-term average throughput of HLNs is larger than BW_1 (DG1);
- if the video flows can fill up the initial buffer of size ibs_2 in time t_1.
- if the full video of size fs_2 is downloaded in time t_2. (The throughput required for this is fs_2/t_2, which is equal to BW_1, which is provided for the nodes even with bad history (see Table 1). So as long as a LLN has exactly one active flow which is a video download, then it is guaranteed to finish downloading in t_2 time).

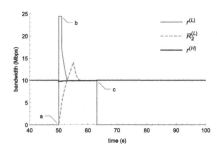

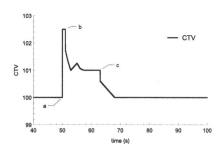

Fig. 7. Node bandwidth time series with 100 bad history nodes

Fig. 8. CTV time series with 100 bad history nodes

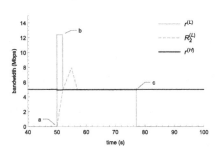

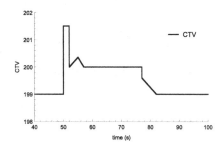

Fig. 9. Node bandwidth time series with 199 bad history nodes

Fig. 10. CTV time series with 199 bad history nodes

We note that web downloads have been included for a more realistic traffic model, but no criteria or dimensioning is included for web downloads in the present paper.

7.2 Numerical Analysis of the Mean Stationary Behaviour

To validate the approximate calculations of Sect. 6, the number of active LLNs was also evaluated by simulation. Table 2 displays the results, grouped according to fast nodes (good history, $N_L^{(1)} + N_L^{(3)}$ with the notation of Sect. 6) and slow nodes (bad history, $N_L^{(2)} + N_L^{(4)}$). The relative load of LLNs is $\ell_L = 0.2$ or 0.4.

Table 2 shows that the approximate calculation holds up nicely, even for $\ell_L = 0.4$. The maximal value of ℓ_L for which (2) holds is 0.5, as this limit is approached the assumption on the single active flow per LLN is violated with higher and higher probability (hence the larger error for $\ell_L = 0.4$).

7.3   Time Series Examples

We show two sample realizations. The first system assumes the parameters from Table 1. The second system differs in the number of HLNs and LLNs. In the example shown in Figs. 7 and 8, the system has 100 active nodes, all with bad history. A video flow arrives at 50 s (point (a)) at a LLN with perfect history.

In the setup of Sect. 7.1, the 100 active nodes with bad history correspond to the 100 HLNs, and the single active LLN corresponds to $\ell_L = 0.01$ as calculated from (11)–(15).

Figure 7 displays the instant bitrate of the LLN $(r^{(L)})$, its bitrate on the 5 s timescale $(R_2^{(L)})$, and the instant bitrate of a node with bad history $(r^{(H)})$, while Fig. 8 displays the associated CTV of the system. The TVF related reasons for this evolution of the bandwidth sharing and the CTV are discussed in relation with Fig. 5 and 6 in Sect. 4.

We note that DG1 (or, equivalently, (2)) allows 200 active nodes at most, so the $100 + 1$ active nodes is well below this limit, and as a result, all nodes are allocated a relatively high bandwidth. Nodes with bad history get 10 Mbps (instead of the required $BW_1 = 5$ Mbps), and due to the dimensioning of the TVF, the single LLN gets $2.5 \cdot 10 = 25$ Mbps until the initial buffer is filled up (point (b)), which takes less than the required $t_1 = 2$ s, and the total download also takes much less time than the required 30 s.

On the other hand, in the example shown in Figs. 9 and 10, the system has 199 active nodes with bad history when a video flow arrives (marked with (a) in the figures) at a LLN with perfect history. Figure 9 displays the instant bitrate of the LLN $(r^{(L)})$, its bitrate on the 5 s timescale $(R_2^{(L)})$, and the instant bitrate of a node with bad history $(r^{(H)})$, while Fig. 10 displays the CTV.

In the setup of Sect. 7.1, the 200 active nodes correspond to 100 HLNs, 99 LLNs with bad history and 1 LLN with good history corresponding to $\ell_L = 0.5$ as calculated from (11)–(15).

This system is critical in the sense that DG1 and (2) hold with equality now. From (6),

$$\rho = \frac{ibs_2}{BW_1 t_1} = \frac{25\,\text{Mbit}}{5\,\text{Mbps} \cdot 2\,\text{s}} = 2.5,$$

and the single LLN has bandwidth $\rho BW_1 = 12.5$ Mbps allocated until the initial buffer size is reached at point (b) (exactly $t_1 = 2$ s after (a)); after that, its history reverts back to bad and its bandwidth allocation drops to $BW_1 = 5$ Mbps until the video is finished at point (c) (27 s after (a)).

7.4   Statistical Results

In this section, we make a statistical comparison of three congestion control principles: the MTS-TVF is compared with TCP-fair and node-fair. For TCP-fair, each active node is allocated bandwidth proportional to its number of active flows, while for node-fair, each active node is allocated equal bandwidth. Actually, node-fair can be realized by using STS-TVF, see also [8].

Based on the simulator output, we compute the following statistics:

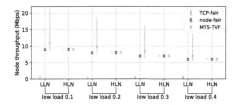

Fig. 11. Node throughput statistics

Fig. 12. Video throughput statistics

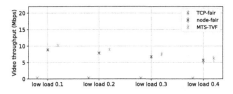

Fig. 13. Web download throughput statistics

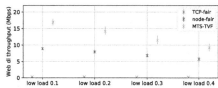

Fig. 14. Video initial buffer criterion at LLNs

- the node throughput for active periods (periods when there is no traffic at the respective node are excluded) for LLNs and HLNs;
- the flow throughput for video and web download flows at LLNs;
- ratio of flows where the time-to-play (TTP) criterion and total download time criterion is satisfied.

Figure 11 compares the node throughput of both LLNs and HLNs for the three congestion controls and various load setups according to Sect. 7.1, with the total low load varying. The figure depicts the node throughput average with a × symbol and the 10% best – 10% worst interval with bars. The main advantage of MTS-TVF is that it offers better performance for LLNs without hurting the long term performance of HLNs. TCP-fair provides flow count proportional throughput, resulting in very poor performance for LLNs. Node-fair and MTS-PPV provide proper prioritization for LLNs at no cost in the performance of HLNs.

Figure 12 and 13 display video throughput and web download throughput at LLNs. The significantly better throughput provided to web downloads by MTS-TVF is due to prioritizing nodes with good history, which applies to LLNs as the load of an individual LLN is so small that rare arrivals occur mostly at good history. The initial buffer size of video downloads is 3.125 MB = 25 Mbit, and the applied TVF is dimensioned so that the first 25 Mbit of any flow at a node with good history is allocated a high bandwidth (2.5 times larger than for HLNs). The effect on the overall flow throughput is more pronounced for web downloads (5 MB = 40 Mbit) than video downloads (18.75 MB = 150 Mbit) as a relatively larger portion of the flow is downloaded at a high bandwidth.

Figure 14 displays the ratio of video download flows which meet the initial buffer criterion of downloading 25 Mbit in 2 s. Note that this criteria is included

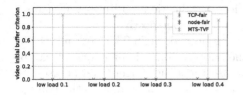 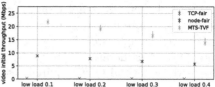

Fig. 15. Video initial buffer throughput at LLNs

Fig. 16. Throughput criterion for the whole video at LLNs

in the dimensioning guidelines of the MTS-TVF, and accordingly, for MTS-TVF, over 90% of all video download flows meet this criterion even for low load $\ell_L = 0.4$, while for TCP-fair and node-fair, the ratio of flows meeting this criterion is practically zero. This is one of the major advantages of using a properly dimensioned MTS-TVF. The flows that do not meet the criterion for MTS-TVF are due to a flow arriving shortly after another flow at a LLN, with the node history still bad when the second flow arrives. As the number of LLNs increases, the probability of this goes to 0; for 900 LLNs, it still occurs with a small probability (also depending on the total load of LLNs).

Figure 15 displays the throughput statistics for the initial buffer of video downloads at LLNs, that is, the throughput the flow until the initial buffer is full. Again, MTS-TVF vastly outperforms the other two resource sharing methods. Figure 16 displays the ratio of video download flows which meet the total download criterion. MTS-TVF was dimensioned so that this criterion is met, and accordingly, for MTS-TVF, it is met for the vast majority of flows. For TCP-fair, this criterion is failed entirely, while for node-fair, it is met for a smaller but still relatively high portion of the flows.

8 Conclusion

The MTS-TVF based resource sharing introduced in [5], extends the advantages of Multi-Timescale Bandwidth Profile to a wide range of traffic scenarios from only a well defined scenario. It formalizes Multi-Timescale fairness and describes ideal time-series behaviour of resource sharing. However, to utilize the potential benefits of MTS-TVF resource sharing, we need flexible, but explicit dimensioning rules.

In this work we provided a dimensioning method for an access-aggregation network scenario and illustrated the advantages of MTS-TVF using heterogeneous broadband traffic model in access-aggregation network. Using an idealized fluid system model, we showed the time-series behaviour for the working point (CTV) and we showed how the system behaves for several dynamic workloads.

In the studied system, the QoE (assumed based on experienced bandwidth) of low load users significantly increased when using MTS-TVF, effectively making the QoE similar to that of a lightly loaded system, while the effect on the QoE of high load users was minimal. MTS-TVF uses the same policy for both heavy and light loaded users, does not require service identification, and uses well defined policies, therefore it is ideal from a net neutrality perspective.

As an example we showed how the proposed MTS-TVF optimizes the video QoE of moderate loaded users. The current dimensioning concept can be extended for several QoE requirements. Also the same concept can be used for other traffic aggregates, e.g. services, network slices. Additionally, the concept can be combined with the multi-layer virtualization concept, when MTS-TVF is applied for different traffic aggregates simultaneously, e.g. for services, subscribers and network slices at the same time.

References

1. Abbas, G., Halim, Z., Abbas, Z.H.: Fairness-driven queue management: a survey and taxonomy. IEEE Commun. Surv. Tutor. **18**(1), 324–367 (2016)
2. Briscoe, B.: Per-flow scheduling and the end-to-end argument. Technical report (2019). http://bobbriscoe.net/projects/latency/per-flow_tr.pdf
3. Nádas, S., Varga, B., Horváth, I., Mészáros, A., Telek, M.: Multi timescale bandwidth profile and its application for burst-aware fairness. In: IEEE Wireless Communications and Networking Conference (WCNC 2020) (2020)
4. Nadas, S., Turanyi, Z.R., Racz, S.: Per packet value: a practical concept for network resource sharing. In: 2016 IEEE Global Communications Conference (GLOBECOM), pp. 1–7 (2016)
5. Nádas, S., Gombos, G., Fejes, F., Laki, S.: Towards core-stateless fairness on multiple timescales. In: Proceedings of the Applied Networking Research Workshop, pp. 30–36 (2019)
6. Cao, Z., Zegura, E., Wang, Z.: Rainbow fair queueing: theory and applications. Comput. Netw. **47**(3), 367–392 (2005)
7. Stoica, I., Shenker, S., Zhang, H.: Core-stateless fair queueing: a scalable architecture to approximate fair bandwidth allocations in high-speed networks. IEEE Trans. Netw. **11**(1), 33–46 (2003)
8. Laki, S., Gombos, G., Nádas, S., Turányi, Z.: Take your own share of the pie. In: Proceedings of the Applied Networking Research Workshop, ANRW 2017, pp. 27–32. ACM (2017)
9. Figueiredo, D., Liu, B., Guo, Y., Kurose, J., Towsley, D.: On the efficiency of fluid simulation of networks. Comput. Netw. **50**, 1974–1994 (2006)

DiPS: A Tool for Data-Informed Parameter Synthesis for Markov Chains from Multiple-Property Specifications

Matej Hajnal[1,3]($\boxtimes$), David Šafránek[3], and Tatjana Petrov[1,2]

[1] Department of Computer and Information Sciences, University of Konstanz,
Konstanz, Germany
374185@mail.muni.cz
[2] Centre for the Advanced Study of Collective Behaviour, University of Konstanz,
78464 Konstanz, Germany
[3] Systems Biology Laboratory, Faculty of Informatics, Masaryk University,
Botanická 68a, 602 00 Brno, Czech Republic

Abstract. We present a tool for inferring the parameters of a Discrete-time Markov chain (DTMC) with respect to properties written in probabilistic temporal logic (PCTL) informed by data observations. The tool combines, in a modular and user-friendly way, the existing methods and tools for parameter synthesis of DTMCs. On top of this, the tool implements several hybrid methods for the exploration of the parameter space based on utilising the intermediate results of parametric model checking – the symbolic representation of properties' satisfaction in the form of rational functions. These methods are combined to support three different parameter exploration methods: (i) optimisation, (ii) parameter synthesis, (iii) Bayesian parameter inference. Each of the available methods makes a different trade-off between scalability and inference quality, which can be chosen by the user depending on the application context. In this paper, we present the implementation, the main features of the tool, and we evaluate its performance on several benchmarks.

1 Introduction

Modelling stochastic dynamical systems such as a biological cell, epidemic spread in a population, or a randomised communication protocol is challenging, especially when parameters are not available, subject to uncertainty, and when exper-

TP's research is supported by the Ministry of Science, Research and the Arts of the state of Baden-Württemberg, the DFG Centre of Excellence 2117 'Centre for the Advanced Study of Collective Behaviour' (ID: 422037984), and AFF (Committee on Research, University of Konstanz), MH's research is supported by Young Scholar Fund (YSF), project no. P83943018FP430_/18, and Max Planck Institute of Animal Behaviour. DŠ's research is supported by the Czech Grant Agency grant no. GA18-00178S. The authors acknowledge Denis Repin, Nhat-Huy Phung, and Stefano Tognazzi for functional testing, feedback, and solving code issues.

© Springer Nature Switzerland AG 2021
P. Ballarini et al. (Eds.): EPEW 2021/ASMTA 2021, LNCS 13104, pp. 79–95, 2021.
https://doi.org/10.1007/978-3-030-91825-5_5

imental data measurements are scarce. *Parameter synthesis* is particularly useful in this context, as it determines the regions of parameter space, for which a high-level property holds. Such high-level property is typically a functional specification (e.g. 'error states are reached with small probability'). Parameter synthesis of discrete-time Markov chains (DTMCs) is supported by several existing tools for probabilistic verification [8,9,12,22]. Most of these implementations specialise in the case when a single qualitative or quantitative property is of interest. In practice, there is an emerging need to reason about multiple properties at the same time. One such situation is when multiple functional properties should be satisfied simultaneously. Another scenario occurs when a functional property (specification) is additionally constrained by properties derived from experimental data, typical for modelling biological systems or in the context of grey-box system verification and testing [1]. For instance, a qualitative summary of experimental observations at a steady-state such as 'a certain group of states is eventually reached as a terminal state with probability greater than a threshold', can be used to additionally constrain the parameter synthesis procedure. Data observations alone can be encoded in the form of multiple temporal properties, e.g. steady-state observations in a chain with more than one bottom strongly connected component (BSCC) [14].

In this paper, we present DiPS[1] – a tool for data-informed parameter synthesis for *parametric discrete time Markov chains* (pMC) from multiple-property specifications. For a single property expressed in Probabilistic Computation Tree Logic (PCTL) [17], the standard parameter synthesis procedures provide a symbolic representation of satisfaction probability in the form of *rational functions*, which will evaluate exactly to the satisfaction probability for that single property in the given chain. We leverage existing tools PRISM [22] and Storm [9] to obtain the rational functions characterising satisfaction probability of each among the multiple properties in the specification and before incorporating threshold constraints available from the data measurements. Resulting rational functions are the cornerstone of the tool as all further analyses are based on them. In the next step, the (experimental) data are used as thresholds for constraining the rational functions, for given confidence level and based on frequentist statistics interpretation. The resulting algebraic constraints are finally employed to explore the parameter space for which the chain behaviour agrees with the observations.

To explore parameter values respecting given specification supported with data, DiPS employs several different methods working with the synthesised algebraic constraints – rational functions and confidence intervals obtained from experimentally observed satisfaction of the specification. The computational workflow utilises the following methods: optimisation, parameter space refinement, parameter space sampling, and Metropolis-Hastings. In Fig. 1, it is shown how these methods are combined in the tool to tackle the complex data-informed specification-driven procedures including optimisation, parameter synthesis, and Bayesian inference. In particular, the tool implements the following tasks:

---

[1] https://github.com/xhajnal/DiPS.

- marking single points in the parameter space sat (green) or unsat (red) wrt. the algebraic constraints satisfaction [8] (*space sampling*), or
- marking entire regions (hyper-rectangles) in the parameter space safe (green), unsafe (red) wrt. the algebraic constraints satisfaction with SMT solver [8, 11,19,20] or interval arithmetics (*space refinement*), or
- identifying a single point in the parameter space with the least distance wrt. data (*optimisation*), or
- identifying a distribution over possible parametrisations based on their relative likelihood wrt. data using Bayesian inference (*Metropolis-Hastings*).
- providing a novel hybrid method that combines Bayesian inference with space refinement (*HMH*),
- facilitating a user-friendly interface allowing to visualise the results and adapt the workflow by combining the tasks above,
- exploring the potential of the methods for efficient parallel processing on a multi-core hardware.

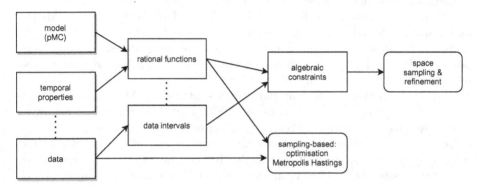

Fig. 1. The main workflow of DiPS. Parametric model checking produces a rational function, encoded as a symbolic expression representing the satisfaction probability of temporal properties, which can be observed at execution time (in the data). The data can be used to compute confidence intervals and set thresholds to rational functions, resulting in a set of algebraic constraints. The space of parameters satisfying the algebraic constraints is computed by sampling and refinement technique that partitions the space into rectangular regions. The data can be applied directly (without computing confidence intervals) with rational functions to find parameter points minimising the distance between these two inputs (optimisation) or to approximate the posterior distribution of the parameters using Bayesian inference (Metropolis-Hastings).

We start with briefly introducing the key theoretical concepts of the implemented methods (Sect. 2) and follow by stating the original features of DiPS in detail (Sect. 3). The implementation of the tool is described in Sect. 4 including the evaluation of the key tool features conducted on several models from different domains. The tool is available as open source (see footnote 1) including a ready to run virtual machine with instructions on how to run the experimental evaluation. A tutorial containing detailed information on the tool's functionality accompanied with a running example is bundled with the tool.

1.1 Related Work

Parametric model checking has been continuously developed, starting with state elimination similar to finite-state automata reduction to regular expressions [6, 15], later enhanced with set-based state elimination simplifying the intermediate results [13], all the way to the SCC decomposition technique with a special structure to store individual factors [18] that improved the speed and memory efficiency.

PRISM [22] is a well-established tool for modelling and model checking DTMCs, Continuous-Time Markov Chains (CTMCs), Markov decision processes (MDPs), and Probabilistic Timed Automata (PTA). As PRISM is easy to be installed, we use it as the first option to obtain rational functions - parametric model checking. PRISM also provides space partitioning using sampling. We leverage this functionality and add a visualisation of the result.

Storm [9] is a command-line tool for analysis of DTMCs, CTMCs, MDPs, and Markov automata (MA). It improves memory efficiency, speed, and output usability of parametric model checking by implementing efficient methods proposed in [18]. In DiPS, one can use Storm instead of PRISM to improve the performance of parametric model checking.

Storm also provides efficient parameter synthesis of Markov chains with multi-affine parametrisations – *parameter lifting* [25]. DiPS can call Storm to refine the parameter space. Storm output consists of separate results for each property while considering the lower and upper bound of the interval of the respective algebraic constraint separately. To that end, DiPS can merge these partial results to obtain (and visualise) the overall result for the conjunction of properties.

Parameter lifting technique was updated with monotonicity checking in [27].

PARAM [12] is another tool for parametric model checking of DTMCs employing state elimination and state-lumping techniques, however, it is not that efficient as Storm – see benchmarks in [8].

PROPhESY [8] supports discrete-time models with safety and liveness properties. It provides space sampling and refinement employing SMT-solvers; however, the usability is limited to properties with exactly two parameters and by the dependencies/VM environment.

In [24], Bayesian inference ideas were used to constrain the parameter values directly from data (without using rational functions). It improves the results of Statistical Model Checking (SMC), especially in the case of sparse data.

PRISM-PSY [5] implements parametric uniformisation to explore parameter space for parametrised CTMCs with Continuous Stochastic Logic (CSL) specification and employs GPU hardware. It was reused for robust design synthesis in RODES [4].

U-check [2] employs Bayesian statistical algorithm and smoothed model checking for CTMCs with Metric Interval Temporal Logic (MiTL) specification.

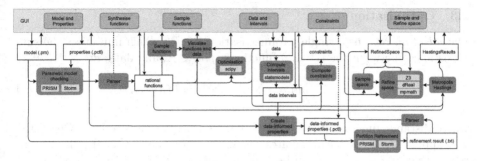

Fig. 2. The architecture of DiPS. Main GUI components, six tabs (in green), main functionality components (in blue), and leveraged tools and libraries (in red). (Color figure online)

2   Methods

In this section, we briefly recall incorporated methods introduced in former works and explain the methods and concepts in detail. Moreover, we describe a novel method based on a combination of Monte Carlo and refinement-based approaches. Additionally, we add information on the parallelisation potential of the individual methods. More information about methods' outputs and their settings in DiPS can be seen in the tutorial which is a part of the tool package.

2.1   Model Checking

Model checking verifies whether a given model satisfies a given specification, while the specification is often formalised in the form of temporal property. Probabilistic operators within the property answer questions such as whether a probability of reaching target state is higher than a given threshold, e.g. 0,4: $P_{>0.4}[F\ Target]$. In the second form of temporal properties, quantitative properties, we can ask for the value of probability itself: $P_{=?}[F\ Target]$.

When the values of probabilities within the models are unknown, a parameter can be used to address this uncertainty. Model checking parametrised models using quantitative property, in the form $P_{=?}$, results in a symbolic expression over model parameters. For pMCs, the expressions are in the form of rational functions, f_i. To obtain rational functions, we leverage already existing tools PRISM and Storm. All the methods implemented in DiPS build upon calculated rational functions.

2.2   Data

$Data$, $[d_1, \ldots, d_m]$, represents empirical estimates of satisfaction probabilities or reward for each of the respective properties. Data points are used to compute intervals constraining the rational function or directly within optimisation and Metropolis-Hastings.

2.3 Optimisation

Optimisation returns a single parametrisation $\hat{\theta} \in \mathbb{R}^n$ which minimises the sum of distances between the rational functions, $f(\hat{\theta}) \in \mathbb{R}$, and data, d:

$$\hat{\theta} := \arg\min_{\theta \in \Theta} \sum_{i \in \{1,\ldots,m\}} w_i \cdot dist(f_i(\theta), d_i) \qquad (1)$$

where distance function $dist : [0,1] \times [0,1] \to \mathbb{R}_{\geq 0}$ can be redefined. Currently, we support the least mean squares distance as provided by scipy library. To reflect that an observation can be more influential or desired to achieve, w_i is a weight term to scale the distance of the respective data point.

2.4 Data Intervals

Data intervals, $[I_1, \ldots, I_m]$, are confidence intervals with given number of measurements, N, and confidence level, C. We currently support six methods for confidence intervals for proportions: standard (CLT/Wald), Agresti-Coull (default) [3], Wilson [3], Jeffreys [3], and Clopper-Pearson [3] implemented in library stats [26], and Rule of three [16]. The standard confidence intervals, defined as

$$I_k = d_k \pm \left(z_{\alpha/2} \sqrt{\frac{d_k(1-d_k)}{N}} \right) \qquad (2)$$

where d_k is the k-th data point, and $\alpha = 1 - C$ is the chosen alpha level, are not generally reaching expected coverage of selected confidence level [3,7]. Therefore, we provide more suitable options with Agresti-Coull method as the default one. When the number of observation is low (below 40), Wilson or Jeffreys method may be more suitable [3]. All the methods are directly applicable when the data estimates probability, the observed value of a quantitative probabilistic property - $P_{=?}$.

Data intervals are then used to constrain the rational functions, $\forall k \in \{1, \ldots, m\}, f_k \in I_k$, and the algebraic constraints are used in the space sampling and refinement methods.

2.5 Sampling

The decision problem whether the instantiation of pMC model satisfies a given PCTL property can be answered by model checking the instantiated DTMC. With the knowledge of rational functions, this problem boils down to evaluating algebraic constraints.

In the sampling, a uniform grid of points is created and in each point we evaluate the constraints to mark the point *sat*(green)/*unsat*(red). Where for the sat point, all the constraints are satisfied and for unsat point at least one of the constraints is violated.

For one or two parameters, the result is visualised as green and red dots in phase space - see Fig. 4a, c. In the multidimensional case, each parametrisation satisfying the properties is visualised as a scatter-line plot where each parameter is plotted against its index in the parameter space, i.e. parametrisation $\hat{\theta} \in \Theta$ is plotted as a function: $i \mapsto \hat{\theta}_i$ for each parameter index $i \in \{1, 2, \ldots, n\}$.

Trivial parallelisation of sampling is based on the independence of algebraic constraint evaluation in the points to be sampled.

2.6   Quantitative Sampling

This method is very similar to ordinary sampling. The only difference is that instead of checking satisfaction, L1 distance to violate each of the algebraic constraints is summed to give a numeric value. For each pair of algebraic constraints derived from lower and upper bound of intervals, a lower distance is used for the pair. Positive values in the sum represent that the respective algebraic constraint is satisfied (in the given point). Note that this assumption does not hold for the whole sum. In each of the sampled points, the sum of distances is visualised by a colour spectrum.

This method hence provides quantitative estimation to better describe the satisfaction landscape of the parameter space. The parallelisation of this method benefits from the same fact as sampling: the algebraic constraints are evaluated in parallel for each point to be sampled.

2.7   Space Refinement

Here we address the problem of inferring parameter values for quantitative properties globally, not only in separate points. This problem is usually solved by space partitioning [20]. For multi-affine parametrisations, Storm implements a efficient method, *parameter lifting* [25]. Prophesy uses SMT solvers and PRISM uses sampling of the partitions to solve the problem approximatively. We provide similar methods for partitioning of space with aim to solve multiple properties in a CEGAR like style, while providing an option to run PRISM or Storm for multiple properties as well.

DiPS supports two SMT solvers, z3 and dreal, and interval arithmetics as proposed in [10] (implemented by library mpmath) to solve the satisfaction of individual regions[2]. This is done in two steps, check safe, verifying whether all the points within the region are satisfying, and check unsafe, verifying whether all the points within the region are not satisfying. If neither of these holds, we split the region (in the longest dimension into two rectangles with equal volume). As verifying of the region can be expensive, we provide an option to sample the region before calling the solver - *sampling-guided refinement*. In all cases of sampling result, one of the solver calls can be skipped and if both sat and unsat samples are found, both solver calls are skipped[3] and the region is

---

[2] In comparison with SMT solvers, interval arithmetics provide faster iterations for price of higher probability to mark a region unknown.

[3] If the sampling contains an unsat point, it is a counterexample of safeness and vice versa if the sampling contains a sat point, it is a counterexample of unsafeness of the region under consideration.

`split` based on the position of sat vs unsat points. In more detail, we calculate rectangular hulls of the sat and unsat points. If the two hulls have no overlap there is a single line/plane dividing these two hulls and we split the region along the line/plane. If one of the two hulls is inside the other hull, we cut the space along the borders of the smaller hull. And finally, if none of two previous holds, we cut the space in all dimensions. As we use two sample points in each dimension the cutting lines/planes are always in the middle of dimension(s).

To choose a new region to check, we select all unknown regions with the biggest volume. Refinement parallelisation relies on the independence of refining these selected regions.

For one and two parameters a phase space of safe (green), unsafe (red), and unknown (white) rectangles is shown - see Fig. 4b, c. For more dimensions, over-approximation of safe or unsafe space as a projection to each of dimensions is visualised.

2.8   Metropolis-Hastings

Metropolis-Hastings [23] is a Markov chain Monte Carlo (MCMC) algorithm for approximating the posterior distribution over model parametrisations wrt. available data. For a given number of iterations, it walks through the parameter space and compares the posterior probability of the current and the next parameter point. It results in a sequence of accepted points predicting the true parameter value.

Importantly, the rational functions $f_i(\theta)$ allow us to evaluate the data likelihood $P(D \mid \theta)$ for each parametrisation and data outcome exactly. Without the rational functions, we would have to hypothesise a class of distributions proportional to the likelihood or simulate the chain to approximate the likelihood which is computationally expensive and/or imprecise.

For one or two parameters, posterior distribution is visualised as rectangularised space where the number of accepted points within each of the rectangle is visualised by a colour gradient - see Fig. 4d. For more dimensions DiPS shows scatter-line plot connecting values of parameters for each of the accepted points.

This visualisation is accompanied by two metadata visualisations. In the first one, the sequence of the accepted point with a histogram is shown for each parameter. In the latter, the sequence of all points, accepted and rejected, is shown as a projection for each of the parameters. For one or two dimensions also the sequence is shown in phase space for both accompanying visualisations.

2.9   Metropolis-Hastings-guided Refinement – HMH

The newly proposed method for parameter synthesis, Hajnal-Metropolis-Hastings (HMH), combines two interconnected methods: Metropolis-Hastings and space refinement. Posterior distribution as the result of Metropolis-Hastings serves to split the space into rectangles with marking of the number of accepted points within each of the rectangles. The discrepancy of this value and the expected number of accepted points in each of the rectangles serves to quantify expectation of probability of the rectangle to be safe. Safe rectangles are expected to contain more accepted points than unsafe rectangles. This aids the refinement procedure to select rectangles: 1. with a higher probability to be either safe or unsafe and 2. to find a safe region faster, as on many occasions, one is interested in finding a safe area rather than validate the whole parameter space.

3 Key Tool Features and Contributions

The main contribution of this paper is a tool offering a palette of parameter inference methods for pMCs. In comparison with the state-of-art we extend existing tools and workflows in the following aspects:

1. DiPS provides a fully automated computation of the confidence intervals serving as thresholds for probability satisfaction of the specified property. The user can pick one of the six methods, with the default option, Agresti Coull method [3], performing much better than the standard (Wald) method.
2. DiPS allows two satisfaction probability thresholds (a lower and an upper bound of the satisfaction probability, e.g., bounds of the confidence interval) per each single temporal property. These bounds constrain the rational functions provided that both of the respective inequalities must be satisfied. DiPS supports the conjunction of multiple constrained rational functions – algebraic expressions of satisfaction probability of *multiple properties*. Multiple properties further allow the experts to maximise the predictive power of sparse data in order to find satisfactory parameter values. We have demonstrated the necessity of such a setting in our case study of the population model of honey bee mass stinging [14] containing several different BSCCs. In that case, it was necessary to constrain reachability probabilities for each of the BSCCs.
   Native support for multiple properties allows DiPS to reach desired coverage of space refinement, while PRISM and Storm tends to reach lower than desired coverage without the possibility of continuation of the refinement to enhance the coverage.
3. We extend the palette of methods with optimisation, Metropolis-Hastings, and HMH allowing to work with large model instances. For instance, Bayesian inference will always give some information within the available time frame, even though it cannot provide a global partitioning of parameter space, as is the case with the space refinement method. In addition, Metropolis-Hastings gives the quantitative result providing more information than the qualitative sat/unsat answer.
4. Precision and efficiency of optimisation and Metropolis-Hastings is enhanced by the knowledge of rational functions. To obtain the probability of satisfaction of a PCTL formula without having rational functions, one needs to run the pMC. As this has to be done in each parameter point to be analysed, the estimation becomes imprecise and/or expensive.

5. Modularity of DiPS provides an option to begin the procedure in any phase of the workflow, starting with:
   - model, properties, and data/intervals,
   - (rational) functions and data/intervals,
   - algebraic constraints, or
   - refined space.

   This allows using manually computed confidence intervals instead of data and generalisation of input functions, adding more rational functions and/or algebraic constraints or even to load previously refined parameter space, and continue refinement with a different setting and/or algebraic constraints.
6. Modularity of the design and its multiple methods allow interconnecting the results; Metropolis-Hastings can be initialised from the optimised point, space refinement can start with initial partitioning based on sampling results (pre-sampled refinement) or Metropolis-Hastings result (HMH), etc. – see Fig. 2.
7. Finally, DiPS is able to analyse and visualise output even for models with more than two parameters - more details with example in tutorial.

4   Implementation and Experiments

4.1   Implementation

DiPS is an open source Python project, which is capable to communicate with and leverage PRISM and Storm. The command-line interface (CLI) serves for optimal performance and fast development. It is supplied with a GUI, divided in 6 functionally different tabs. The GUI provides user-friendly access to the workflow implemented by the CLI.

In the main workflow - see Fig. 1, *models* (PRISM models in .p format) and *properties* (in .pctl format) are fed to PRISM or Storm to run parametric model checking resulting in *rational functions*. *Data* are used directly with rational functions to search for parameter point minimising the distance between the inputs (*optimisation*) or in *Metropolis-Hastings* to compute posterior distribution. For other methods (*space sampling* and *space refinement*), *data intervals* (e.g. *confidence intervals*) are computed from the data to either create *data-informed properties* or combine with rational function to create *algebraic constraints*. Moreover, *data-informed properties* (combination of properties and data intervals) and the model are used for the partitioning using PRISM or Storm, while DiPS uses algebraic constraints directly. DiPS's functional units and their connections are depicted in more details in Fig. 2.

Modularity of DiPS allows starting the workflow at any given point, allowing to adjust or to create a new input. Visualisations provide information on results of implemented methods as well as the output of partitioning results of PRISM and Storm. To parallelise the methods, we use **multiprocessing** Python library using Pools with pool.map.

4.2 Experiments

We have evaluated the performance of our tool on a variant of the well-known Knuth's die [21] and a model of stinging bees presented in [14]. The evaluation consists of three parts: (1) runtimes of parallelisation results of sampling, (2) sampling guided refinement vs regular refinement and its parallelisation, and (3) a comparison of refinement methods implemented in DiPS with PRISM and Storm implementations. Shown results were obtained using a tower PC, Skadi, with 64 bit Ubuntu 20.04.2, i9-9900K CPU, 32 GB RAM, SSD disk.

The first case study is Knuth's die [21] which emulates a 6-sided die with a coin. To obtain the result, at least three flips of the coin are used, where we used three biased coins, one for each of the flips, to generate the data – the probability of rolling a side of the die. The parameters are probabilities of tossing heads with each coin – p_1, p_2, p_3. We scale this model using a version with single parameter ($p_1 = p_2 = p_3$), two parameters ($p_1, p_2, p_3 = 0.5$), and a version with all three parameters.

In the second case study, we look at the population model of honeybees [14]. Honeybees protect their hive against vertebrates by mass-stinging. This action costs a bee life. Collective decision which bee stings and which does not is crucial for the vitality of the colony. When the hive is attacked, a bee decides to sting with an unknown probability p. A stinging bee releases an alarm pheromone, which promotes the stinging of other bees. A bee that initially decided not to sting can sense this pheromone and opt for stinging with probability q. In the refined version of the model (multiparam), this parameter, q_i, is modulated by the number of already stinging bees, i. Here we investigate the semisynchronous version of the model, which means that in the first transition (before sensing alarm pheromone) all bees make a decision to sting or not to sting - synchronous update. Afterwards, in each transition only a single bee makes a decision - asynchoronous update. To generate synthetic data, we simulate the chain and obtain probabilities of reaching every possible number of stinging bees from zero to m: $d_0, d_1, \ldots d_m$, where d_i is the fraction of simulations that ended up with i stinging bees. In the model, this is equal to the probability of reaching the respective BSCC and it can be encoded in terms of a PCTL property, $\varphi_i = P_{=?}F(BSCC_i)$. The distribution of the number of stinging bees is reflected as a conjunction of respective probabilities, $\varphi_0 = d_0 \wedge \varphi_1 = d_1 \wedge \cdots \wedge \varphi_m = d_m$.

For easier comparison with results in [14] we use the same settings for confidence intervals – the Wald method with a correction term. All intervals were computed using N - number of samples: 100, C - confidence level: 0.95.

A script reproducing experiments, examined models, properties, and data is included in the tool package.

Sampling Parallelisation. Results visualised in Fig. 3a show that for smaller models the overhead of parallelisation is higher than the benefits, but in absolute values, the speedup for bigger models is much more dominant. Moreover, the overhead of more cores diminishes as the model and hence rational functions increase in size.

Refinement Parallelisation. In Fig. 3b, c we can see that the most advantage gaining solver is z3. Refinement with z3 requires fewer but more exhaustive calls; hence the overhead of creating processes is minimised. The overhead of more cores diminishes as the model and hence rational functions increase in size.

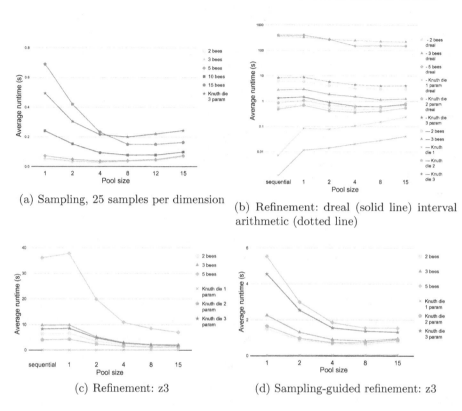

(a) Sampling, 25 samples per dimension

(b) Refinement: dreal (solid line) interval arithmetic (dotted line)

(c) Refinement: z3

(d) Sampling-guided refinement: z3

Fig. 3. Runtimes of sampling and refinement. Time in seconds (vertical axis), the sequential version and the number of processes - Pool size (horizontal axis). The curves display the average of 300 runs (sampling) and 20 runs (refinement).

Sampling-Guided Refinement Parallelisation. We show the benefits of sampling-guided refinement using z3 solver in Fig. 3 (a vs b). In all instances the sampling-guided refinement performs better. The same effect with slightly lower amplitude can be observed using dreal. Finally, calls of interval arithmetic are so fast that the overhead of sampling overweights the benefits. We recommend using the standard (not sampling-guided) parallel refinement in this case.

Comparison of Refinement Using DiPS, Storm, and PRISM. Refinement implemented in Storm does not reach selected coverage because the merging of the refinements for the respective property may create more unknown

Table 1. Runtimes of refinement (fastest setting). Space refinement using SMT solver (z3 and dreal) and interval arithmetic. The fastest method shown in bold. Times in seconds. Timeout (TO) 1 h. In all experiments, a property specifying the reachability of all BSCCs in the model is employed (formulated as a conjunction of reachability of individual BSCCs). Number of samples, $N = 100$, confidence level, $C = 0.95$. Not shown models timed out for all three methods.

Model	Refinement SMT solver: z3, dreal	Refinement interval arithmetic
Knuth die, true point $p_1 = 0.4, p_2 = 0.7, p_3 = 0.5$ data, $[0.208, 0.081, 0.1, 0.254, 0.261, 0.096]$		
# states: 13, # transitions: 20, # BSCCs: 6		
Knuth unfair 1-param	0.01177, **0.001163**	0.007269
Knuth unfair 2-param	0.741, **0.3656**	0.5874
Knuth unfair 3-param	1.326, **0.6033**	3.963
Honeybee model of m agents, 2 parameters, dataset 1 # states: 9,13,24, # transitions: 12,19,39, # BSCCs: m+1		
semisyn 2	0.6514, **0.3593**	3.46
semisyn 3	0.8337, **0.6943**	201.3
semisyn 5	**1.552**, 63.55	TO
Honeybee model of m agents, n parameters, dataset 1 # states: 13, # transitions: 19, # BSCCs: m+1		
semisyn 3	0.3419, 0.1206	**0.1155**

regions, e.g., merging a safe and an unknown rectangle. Surprisingly, we have been unable to obtain the desired coverage for multiple properties input with PRISM as well. Hence manual tweaking of the coverage value and rerunning the analysis is necessary to obtain the desired coverage. Moreover, parameter lifting is limited to multi-affine transition functions. In conclusion, in Table 1 we show the fastest average runtimes (number of processes, sampling-guided vs regular) of refinements that reached the desired coverage. DiPS tackles the scalability problem of refinement with many parameters or large rational functions by using other methods – optimisation, sampling, and Metropolis-Hastings.

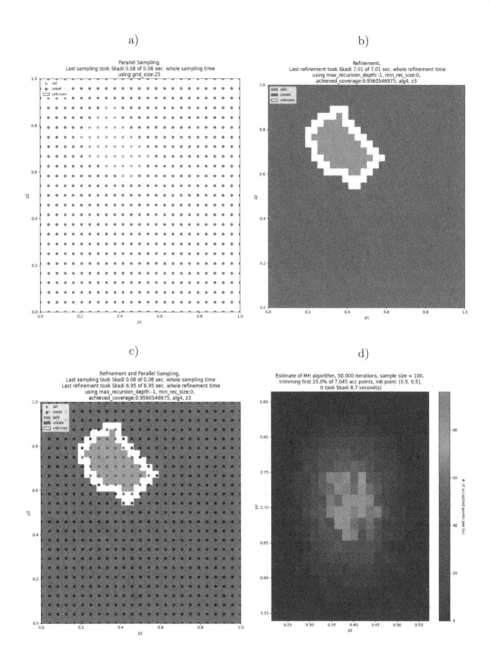

Fig. 4. Visualisation (screenshot from GUI of DiPS) of sampling (a), refinement (b), sampling and refinement (c), and Metropolis-Hastings (d) as a result of 2-param Knuth model. Examples of multidimensional visualisations are shown and explained in the tutorial. (Color figure online)

5   Conclusions and Future Work

We presented a new open source tool, DiPS, dedicated to parameter exploration for pMCs. It focuses on multiple temporal logic properties informed by data. To this aim, we automatically compute rational functions – symbolic representations of satisfaction of each property, by leveraging the existing parameter synthesis tools, as well as their respective probability thresholds, through the confidence intervals derived from data following frequentist statistics interpretation. These two elements are coupled into algebraic constraints over unknown parameters. DiPS solves the algebraic constraints by partitioning the parameter space and can leverage PRISM or Storm for parameter synthesis as well. We add the visualisation of the synthesis results including merging of the Storm bound-wise results. The tool implements two additional methods for parameter exploration, optimisation and Metropolis-Hastings, to tackle the scalability problem of Space Refinement. Moreover, we proposed a new method, HMH, for parameter synthesis combining Metropolis-Hastings and refinement. Finally, parallelisation, modularity, and interconnections of the methods provide further advantages.

In comparison with the mentioned tools, DiPS improves analysis for multiple observations using the conjunction of properties, visualisation of functions in selected points to compare with data, and it complements the analyses with optimisation and Bayesian inference. The possibility to apply different approaches to explore the parameters is especially useful because these analyses have a different trade-off between computational efficiency and the type of information they provide, and the modeller may want to explore different approaches. For example, Bayesian inference will always give some information within the available time frame, but it cannot provide a global partitioning of parameter space, as it is the case with the space refinement method.

We illustrate the applicability of the tool on a variation of Knuth's die[21] and a case study of honeybee mass-stinging behaviour [14].

References

1. Ashok, P., Daca, P., Křetínský, J., Weininger, M.: Statistical model checking: black or white? In: Margaria, T., Steffen, B. (eds.) ISoLA 2020. LNCS, vol. 12476, pp. 331–349. Springer, Cham (2020). https://doi.org/10.1007/978-3-030-61362-4_19

2. Bortolussi, L., Milios, D., Sanguinetti, G.: U-check: model checking and parameter synthesis under uncertainty. In: Campos, J., Haverkort, B.R. (eds.) QEST 2015. LNCS, vol. 9259, pp. 89–104. Springer, Cham (2015). https://doi.org/10.1007/978-3-319-22264-6_6

3. Brown, L.D., Cai, T.T., DasGupta, A.: Interval estimation for a binomial proportion. Stat. Sci. **16**, 101–117 (2001)

4. Calinescu, R., Češka, M., Gerasimou, S., Kwiatkowska, M., Paoletti, N.: RODES: a robust-design synthesis tool for probabilistic systems. In: Bertrand, N., Bortolussi, L. (eds.) QEST 2017. LNCS, vol. 10503, pp. 304–308. Springer, Cham (2017). https://doi.org/10.1007/978-3-319-66335-7_20

5. Češka, M., Pilař, P., Paoletti, N., Brim, L., Kwiatkowska, M.: PRISM-PSY: precise GPU-accelerated parameter synthesis for stochastic systems. In: Chechik, M., Raskin, J.-F. (eds.) TACAS 2016. LNCS, vol. 9636, pp. 367–384. Springer, Heidelberg (2016). https://doi.org/10.1007/978-3-662-49674-9_21

6. Daws, C.: Symbolic and parametric model checking of discrete-time Markov chains. In: Liu, Z., Araki, K. (eds.) ICTAC 2004. LNCS, vol. 3407, pp. 280–294. Springer, Heidelberg (2005). https://doi.org/10.1007/978-3-540-31862-0_21

7. Dean, N., Pagano, M.: Evaluating confidence interval methods for binomial proportions in clustered surveys. J. Surv. Stat. Methodol. 3(4), 484–503 (2015)

8. Dehnert, C., et al.: PROPhESY: a PRObabilistic ParamEter SYnthesis tool. In: Kroening, D., Păsăreanu, C.S. (eds.) CAV 2015. LNCS, vol. 9206, pp. 214–231. Springer, Cham (2015). https://doi.org/10.1007/978-3-319-21690-4_13

9. Dehnert, C., Junges, S., Katoen, J.-P., Volk, M.: A STORM is coming: a modern probabilistic model checker. In: Majumdar, R., Kunčak, V. (eds.) CAV 2017. LNCS, vol. 10427, pp. 592–600. Springer, Cham (2017). https://doi.org/10.1007/978-3-319-63390-9_31

10. Gainer, P., Hahn, E.M., Schewe, S.: Accelerated model checking of parametric Markov chains. In: Lahiri, S.K., Wang, C. (eds.) ATVA 2018. LNCS, vol. 11138, pp. 300–316. Springer, Cham (2018). https://doi.org/10.1007/978-3-030-01090-4_18

11. Hahn, E.M., Han, T., Zhang, L.: Synthesis for PCTL in parametric Markov decision processes. In: Bobaru, M., Havelund, K., Holzmann, G.J., Joshi, R. (eds.) NFM 2011. LNCS, vol. 6617, pp. 146–161. Springer, Heidelberg (2011). https://doi.org/10.1007/978-3-642-20398-5_12

12. Hahn, E.M., Hermanns, H., Wachter, B., Zhang, L.: PARAM: a model checker for parametric Markov models. In: Touili, T., Cook, B., Jackson, P. (eds.) CAV 2010. LNCS, vol. 6174, pp. 660–664. Springer, Heidelberg (2010). https://doi.org/10.1007/978-3-642-14295-6_56

13. Hahn, E.M., Hermanns, H., Zhang, L.: Probabilistic reachability for parametric Markov models. Int. J. Softw. Tools Technol. Transf. 13(1), 3–19 (2011)

14. Hajnal, M., Nouvian, M., Petrov, T., Šafránek, D.: Data-informed parameter synthesis for population Markov chains. In: Bortolussi, L., Sanguinetti, G. (eds.) CMSB 2019. LNCS, vol. 11773, pp. 383–386. Springer, Cham (2019). https://doi.org/10.1007/978-3-030-31304-3_32

15. Han, Y.S.: State elimination heuristics for short regular expressions. Fund. Inform. 128(4), 445–462 (2013)

16. Hanley, J., Lippman-Hand, A.: If nothing goes wrong, is everything all right? Interpreting zero numerators. JAMA 249(13), 1743–1745 (1983)

17. Hansson, H., Jonsson, B.: A logic for reasoning about time and reliability. Formal Aspects Comput. 6(5), 512–535 (1994). https://doi.org/10.1007/BF01211866

18. Jansen, N., et al.: Accelerating parametric probabilistic verification. In: Norman, G., Sanders, W. (eds.) QEST 2014. LNCS, vol. 8657, pp. 404–420. Springer, Cham (2014). https://doi.org/10.1007/978-3-319-10696-0_31

19. Junges, S., et al.: Parameter synthesis for Markov models. CoRR abs/1903.07993 (2019). http://arxiv.org/abs/1903.07993

20. Katoen, J.P.: The probabilistic model checking landscape. In: Proceedings of the 31st Annual ACM/IEEE Symposium on Logic in Computer Science, pp. 31–45. ACM (2016)

21. Knuth, D., Yao, A.: The complexity of nonuniform random number generation. In: Algorithms and Complexity: New Directions and Recent Results. Academic Press (1976)

22. Kwiatkowska, M., Norman, G., Parker, D.: PRISM 4.0: verification of probabilistic real-time systems. In: Gopalakrishnan, G., Qadeer, S. (eds.) CAV 2011. LNCS, vol. 6806, pp. 585–591. Springer, Heidelberg (2011). https://doi.org/10.1007/978-3-642-22110-1_47
23. Metropolis, N., Rosenbluth, A.W., Rosenbluth, M.N., Teller, A.H., Teller, E.: Equation of state calculations by fast computing machines. J. Chem. Phys. **21**(6), 1087–1092 (1953)
24. Polgreen, E., Wijesuriya, V.B., Haesaert, S., Abate, A.: Data-efficient Bayesian verification of parametric Markov chains. In: Agha, G., Van Houdt, B. (eds.) QEST 2016. LNCS, vol. 9826, pp. 35–51. Springer, Cham (2016). https://doi.org/10.1007/978-3-319-43425-4_3
25. Quatmann, T., Dehnert, C., Jansen, N., Junges, S., Katoen, J.-P.: Parameter synthesis for Markov models: faster than ever. In: Artho, C., Legay, A., Peled, D. (eds.) ATVA 2016. LNCS, vol. 9938, pp. 50–67. Springer, Cham (2016). https://doi.org/10.1007/978-3-319-46520-3_4
26. Seabold, S., Perktold, J.: Statsmodels: econometric and statistical modeling with Python. In: 9th Python in Science Conference (2010)
27. Spel, J., Junges, S., Katoen, J.P.: Finding provably optimal Markov chains. Tools Algorithms Const. Anal. Syst. **12651**, 173 (2021)

Modelling a Fair-Exchange Protocol in the Presence of Misbehaviour Using PEPA

Ohud Almutairi[✉] and Nigel Thomas

Newcastle University, Newcastle upon Tyne, UK
{o.m.m.almutairi2,nigel.thomas}@newcastle.ac.uk

Abstract. This paper explores the performance costs introduced by a security protocol known as an anonymous and failure resilient fair-exchange e-commerce protocol. The protocol guarantees customer anonymity and fair exchange between two parties in an e-commerce environment. In this paper, the protocol is studied and modelled when misbehaviour between participants occurs. Models are formulated using the PEPA formalism to investigate the performance overheads introduced by the security properties and behaviour of the protocol when a dispute between the parties exists. This study uses a PEPA Eclipse plug-in to support the creation and evaluation of the proposed PEPA models.

Keywords: PEPA · Security protocol · Misbehaviour

1 Introduction

Computing systems are becoming increasingly complex and consist of multiple interactive components. Performance has been seen as an important aspect for evaluating computing systems. Many computing systems are connected to the network, either privately or publicly, and this can pose vulnerability issues, exposing systems to threats and attacks. Security protocols can add an extra overhead to a system, directly influencing its performance. This is a problem for many different domains. For example, the performance of a web server is reduced in response to the implementation of the secure sockets layer protocol [2,3].

Performance and security are essential aspects for almost all systems. Thus, it is important to develop a system that affords an optimal balance between security and performance concerns. Therefore, the extra cost that security aspects contribute to the system's performance has attracted widespread attention. As a result, developers have conducted explorations and taken measurements with the aim of developing a secure system that gives satisfactory performance [3,8,9].

This paper explores a type of non-repudiation e-commerce security protocol when misbehaviour and disputes between the customer and merchant occur. This protocol is called an anonymous and failure resilient fair-exchange e-commerce protocol. It was proposed by Ray et al. [6]. It guarantees a fair exchange between

© Springer Nature Switzerland AG 2021
P. Ballarini et al. (Eds.): EPEW 2021/ASMTA 2021, LNCS 13104, pp. 96–114, 2021.
https://doi.org/10.1007/978-3-030-91825-5_6

two parties and satisfies the following features: first, fairness – no party can have any advantages over the other party during the course of the exchange; second, the anonymity of the parties – a customer can interact without disclosing any personal information; third, no manual dispute resolution; fourth, not relying on the service of a single trusted third party (TTP) – instead, multiple TTPs are available to provide services; fifth, offline TTP – the involvement of such a party must be at a minimum level, only when any problem occurs; and finally, any type of digital merchandise can be exchanged. Moreover, the protocol is based on an approach called 'cross-validation', which allows the customer to validate the encrypted electronic product without decrypting it.

Based on the description provided by Ray et al. [6], the protocol has two versions: with and without an anonymity feature. In this paper, we consider the protocol version that ensures customer privacy is protected from any other parties. The customer does not need to share any personal information with a merchant to buy. Thus, the customer's true identity is hidden from the merchant. Ray et al. modified the basic failure resilient fair-exchange protocol to prevent the customer's personal information from being known by the merchant by following the electronic cash system [5]. In the basic version, the payment token that the customer sends to the merchant contains some personal information, such as the customer's identity and bank account information. Therefore, the merchant will have detailed personal information about the customer once it receives the payment token. However, in this modified version, the customer uses digital base money to buy from merchants. By using this method, merchants can not obtain any personal information from the customer or create a customer profile without permission.

The protocol relies on TTPs but does not need them to be active at any time except if a problem occurs. Therefore, the protocol has two main descriptions depending on the type of TTP involvement: offline TTP (basic) and online TTP (extension) [6]. With offline TTP involvement, there is no TTP active involvement as no parties misbehave or prematurely terminate the protocol. However, with online TTP, when parties misbehave or prematurely terminate the protocol, the TTP must be involved in resolving the problem and ensuring fair exchange. In [1], we studied the protocol's performance without dispute between parties. In this paper, we study the protocol's performance when a dispute between the parties occurs. In addition, the discussion focuses on the behaviour aspects of the protocols in order to analyse their performance.

The approach is used to model the protocol is Performance Evaluation Process Algebra (PEPA). PEPA is a well-known implementation of the Stochastic Process Algebra (SPA). A system is modelled in the PEPA formalism as a set of components which interact and engage individually or with other components in activities in order to evaluate its performance [4]. Thus, the components represent the active parts in the system and the behaviour of each part is represented by its activities. The creation and performance evaluation of a PEPA model is supported by the PEPA Eclipse plug-in [7]. This tool has been developed to support the Markovian steady-state analysis, Stochastic Simulation Algorithms (SSA) analysis, and Ordinary Differential Equations (ODE) analysis of PEPA models in the Eclipse Platform [7].

The paper is organized as follows. Section 2 provides the protocol specification. In Sect. 3, the proposed PEPA model of the protocol is presented. Section 4 presents the protocol's evaluation and results. Finally, Sect. 5 concludes the paper by providing an overview of the study findings and future work.

2   Protocol Specification

2.1   An Anonymous and Failure Resilient Fair-Exchange E-Commerce Protocol

This subsection provides an informal description of the protocol. The formal description of the protocol and security-related details are provided in [6]. Before the protocol is initiated, the environment needs to be set up with the same steps detailed in [1,6]. The customer (C) uses a pseudo identifier C' when starting a new transaction with the merchant (M) to preserve the anonymity of C. Thus, no parties in the protocol except the customers themselves have sufficient information to link the C' used in the transaction with C, which is the real customer identity. B is the bank or the financial institution, and TTP is the Trust Third Party. The following are the main nine interaction steps (Fig. 1). A more detailed description of the steps is provided in [1,6]. All texts in bold indicate the names of the actions used in the PEPA model presented in this paper:

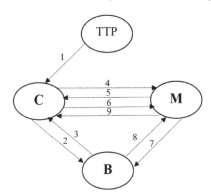

Fig. 1. An anonymous and failure resilient fair-exchange e-commerce protocol.

1. ***download*** (TTP ⇒ C): C visits the TTP website and downloads the encrypted electronic product from the TTP server. This encrypted electronic product can be used to validate the product received from M.
2. ***requestBDigitalCoins*** (C ⇒ B): C sends a request to B for digital coins.
3. ***sendCDigitalCoins*** (B ⇒ C): B sends the signed blinded coin to C.
4. ***sendMPO*** (C' ⇒ M): C' sends a message containing the purchase order (PO) to M.
5. ***sendCEP*** or ***sendCAbort*** (M ⇒ C'): M sends the encrypted product to C' or sends an abort statement to end the transaction if M is not satisfied.

6. **sendMCoinDk** or **sendMAbort** (C' ⇒ M): C' sends the decryption key of the digital coin to M or sends an abort message to end the transaction. If M sends the encrypted electronic product, then C' validates it with the encrypted electronic product received from TTP (step 1). If the product is valid, C' sends the decryption key for the digital coin to M and then waits for the product decryption key by setting a timer. If C' does not receive the key within the time set, they will require TTP involvement. If the product is not valid, C' sends an abort statement to M.

7. **sendBCoinByM** or **sendCAbort** (M ⇒ B or M ⇒ C'): M sends B the digital coin for validation, or M sends C' an abort message to terminate the transaction. If M is unsatisfied for any reason, M sends C' an abort message to terminate the transaction.

8. **sendMyes** or **sendMno** (B ⇒ M): B sends M either 'yes' or 'no'. If the coin has been spent, B sends M 'no'. If the coin has not been spent, B credits M's account with the same amount of money as the digital coin and then sends M 'yes'.

9. **sendCPDk** or **sendCAbort** (M ⇒ C'): M sends the electronic product decryption key to C' after receiving 'yes' from B, or ends the transaction by sending an abort message to C' after receiving 'no' from B.

2.2 The Extended Protocol for Handling Misbehaviours and Communication Problems

This subsection presents scenarios to solve any dispute between the merchant and the customer. When misbehaviours and/or communication problems occur, the extended protocol is initiated, and TTP status is changed to online during the protocol execution. The execution of the extended protocol is started when the customer's timer expires (after Step 6 in the protocol), and the protocol does not reach completion status or when the customer receives an abort message or an invalid product decryption key in Step 9.

A dispute resolution is initiated when a customer sends TTP an initiation message that contains evidence of misbehaving. The misbehaviour scenarios solved by the extended protocol are illustrated as follows (all texts in bold indicate the names of the actions used in the PEPA models presented in this paper):

Merchant Behaves Improperly

Scenario 1: M sends an invalid product decryption key (Step 9 in the protocol steps). The interaction actions for resolving this dispute are as follows:

1. **seekingHelpFromTTP** and then **sendTTPinfo**: C initiates the execution of the extended protocol by seeking help from TTP. Then C sends an initiation message to TTP to seek a resolution of the problem.
2. **validateCoinToB**: TTP receives the customer's dispute resolution request. Then it starts to contact B to validate the coin.

3. **sendTTPyes**: B confirms that the coin is valid, which means that the customer play fairly.
4. **askMForValidK**: TTP orders M to send a valid product decryption key.
5. **sendTTPvalidK** or *timeoutTTP*: M responds within the timeout period by sending the valid product decryption key to TTP, or M does not respond and the timeout expires.
6. **forwardKtoC** or **sendCkByTTP** and **takeActionAgainstM**: If TTP receives the valid product decryption key from M, it forwards it to C. However, if TTP does not receive the valid product decryption key within a specified timeout period, TTP sends C the preserved product decryption key and then takes action against M.

Scenario 2: M sends an invalid product decryption key (Step 9 in the protocol steps). The coin in this scenario is invalid. The interaction actions for resolving this dispute are as follows:

1. **seekingHelpFromTTP** and then **sendTTPinfo**: same as step 1 in the first scenario.
2. **validateCoinToB**: same as step 2 in the first scenario.
3. **sendTTPno**: B confirms that the coin is invalid, which means that TTP needs to investigate who spent the coin.
4. **investigationInvalidCoinToB**: TTP contacts B to investigate who spent the coin.
5. **mspentTheCoinToTTP**: B confirms that the coin is spent by M.
6. **askMForValidK**: same as step 4 in the first scenario.
7. **sendTTPvalidK** or **timeoutTTP**: same as step 5 in the first scenario.
8. **forwardKtoC** or **sendCkByTTP** and **takeActionAgainstM**: same as step 6 in the first scenario.

Scenario 3: M disappears without sending a valid product decryption key (Step 9 in the main protocol steps). The interaction actions for resolving this dispute are as follows:

1. **cTimeoutExpired** and then **sendTTPinfo**: C initiates the extended protocol by sending an initiation message after the timeout period for receiving the decryption key has expired.
2. **validateCoinToB**: same as step 2 in the first scenario.
3. **sendTTPyes**: same as step 3 in the first scenario.
4. **askMForValidK**: same as step 4 in the first scenario.
5. **sendTTPvalidK** or **timeoutTTP**: same as step 5 in the first scenario.
6. **forwardKtoC** or **sendCkByTTP** and **takeActionAgainstM**: same as step 6 in the first scenario.

Scenario 4: M claims that a valid product decryption key has not been sent because an invalid coin's decryption key has been received from C. The interaction actions for resolving this dispute are as follows:

1. **sendTTPreason**: M responds to TTP by identifying the reason for not sending the valid product decryption key to C after TTP contacts it to send a valid decryption key.
2. **sendTTPvalidK**: M must still send TTP the valid product decryption key.
3. **sendMpKbyTTP** and **forwardKtoC**: When TTP receives the valid product decryption key from M, it sends M the valid coin decryption key and forwards the valid product decryption key to C.

Customer Behaves Improperly. In this case, after C sends TTP an initiation message for the extended protocol, TTP starts contacting B to validate the coin. If the coin is invalid (**sendTTPno**), TTP then starts contacting B to investigate who spent the coin (**investigationInvalidCoinToB**). If B confirms that the customer is who spent the coin (**cspentTheCoinToTTP**), TTP will not forward the valid product decryption key to C (**discoverMisbehavingC**).

3   PEPA Models of the Extended Protocol

This section proposes a PEPA model for the protocol extension for handling misbehaviour. The model is an extended protocol to solve the misbehaving event between M and C parties with a probability of M misbehaving. The PEPA model comprises four main components. The four components are Merchant (M), Customer (C), Trust Third Party (TTP) and Bank (B). M, C and TTP are sequential components in the PEPA model, whereas B is a static component. The extended PEPA model is formulated as follows:

Merchant Component

$$M_0 \stackrel{def}{=} (sendMPO, r_{sendMPO}).M_1$$

$$M_1 \stackrel{def}{=} (sendCEP, r_{sendCEP}).M_2 + (sendCAbort, r_{sendCAbort}).M_8$$

$$M_2 \stackrel{def}{=} (sendMCoinDk, r_{sendMCoinDk}).M_3 + (sendMAbort, r_{sendMAbort}).M_6$$

$$M_3 \stackrel{def}{=} (startContactB, r_{startContactB}).M_{3a} + (sendCAbort, r_{sendCAbort}).M_6$$

$$M_{3a} \stackrel{def}{=} (sendBCoinByM, r_{sendBCoinByM}).M_4$$

$$M_4 \stackrel{def}{=} (sendMyes, r_{sendMyes}).M_5 + (sendMno, r_{sendMno}).M_7$$

$$M_5 \stackrel{def}{=} (sendCPDk, r_{sendCPDk}).M_6 + (cTimeoutExpired, r_{cTimeoutExpired}).M_6$$

$$M_6 \stackrel{def}{=} (complete, r_{complete}).M_0 + (askMforValidK, r_{askMforValidK}).M_9$$

$$M_7 \stackrel{def}{=} (sendCAbort, r_{sendCAbort}).M_6$$

$$M_8 \stackrel{def}{=} (sendMAbort, r_{sendMAbort}).M_6$$

$$M_9 \stackrel{def}{=} (sendTTPvalidK, r_{sendTTPvalidK}).M_6$$
$$+ (timeoutTTP, r_{timeoutTTP}).M_{10}$$
$$+ (sendTTPreason, r_{sendTTPreason}).M_{11}$$

$$M_{10} \stackrel{def}{=} (takeActionAgainstM, r_{takeActionAgainstM}).M_6$$

$$M_{11} \stackrel{def}{=} (sendTTPvalidK, r_{sendTTPvalidK}).M_{12}$$

$$M_{12} \stackrel{def}{=} (sendMpkbyTTP, r_{sendMpkbyTTP}).M_6$$

The above model component specifies M's different behaviours, moving from M_0 to M_{12}. It has fourteen behaviours to reflect the protocol's steps for M. M moves sequentially between the different behaviours based on the activities specified in the PEPA component. The actions presented reflect the protocol's interaction steps related to M.

Customer Component

$C_0 \stackrel{def}{=} (download, r_d).C_1$

$C_1 \stackrel{def}{=} (requestBDigitalCoins, r_{requestBDC}).C_2$

$C_2 \stackrel{def}{=} (sendCDigitalCoins, r_{sendCDC}).C_3$

$C_3 \stackrel{def}{=} (sendMPO, r_{sendMPO}).C_4$

$C_4 \stackrel{def}{=} (sendCEP, r_{sendCEP}).C_5 + (sendCAbort, r_{sendCAbort}).C_8$

$C_5 \stackrel{def}{=} (sendMCoinDk, r_{sendMCDk}).C_6 + (sendMAbort, r_{sendMAbort}).C_7$

$C_6 \stackrel{def}{=} (sendCPDk, p * r_{sendCPDk}).C_7 + (sendCPDk, (1-p) * r_{sendCPDk}).C_9$
$\quad + (sendCAbort, p * r_{sendCAbort}).C_7$
$\quad + (sendCAbort, (1-p) * r_{sendCAbort}).C_9$
$\quad + (cTimeoutExpired, p * r_{cTimeoutExpired}).C_7$
$\quad + (cTimeoutExpired, (1-p) * r_{cTimeoutExpired}).C_9$

$C_7 \stackrel{def}{=} (complete, r_{complete}).C_0$

$C_8 \stackrel{def}{=} (sendMAbort, r_{sendMAbort}).C_7$

$C_9 \stackrel{def}{=} (sendTTPinfo, r_{sendTTPinfo}).C_{10}$

$C_{10} \stackrel{def}{=} (forwardKtoC, r_{forwardKtoC}).C_7 + (sendCkByTTP, r_{sendCkByTTP}).C_7$
$\quad + (discoverMisbehavingC, r_{discoverMisbehavingC}).C_7$

The above model component specifies C's different behaviours, moving from C_0 to C_{10}. It has eleven behaviours to reflect the protocol's steps for C. C moves sequentially between the different behaviours based on the activities specified in the PEPA component. The actions presented reflect the protocol's interaction steps related to C. In C_6, we introduced a probability of M misbehaving in the actions rates as the customer is the one who initiates contact with TTP seeking dispute resolution. p indicates the probability that C will receive an honest or satisfactory response from M (assuming M is honest), and $(1-p)$ indicates the probability that C will receive an invalid or no response from M (assuming M is misbehaving). When C receives an invalid or no response from M, C will initiate contact with TTP to resolve the dispute. Therefore, in C_6, there are 6 actions could happen either $sendCPDk$ at rate $r_{sendCPDk} * p$ moving to C_7, $sendCPDk$ at rate $r_{sendCPDk} * (1-p)$ moving to C_9 to seek a dispute resolution form TTP, $sendCAbort$ at rate $r_{sendCAbort} * p$ moving to C_7, $sendCAbort$ at rate $r_{sendCAbort} * (1-p)$ moving to C_9, $cTimeoutExpired$ at rate $r_{cTimeoutExpired} * p$ moving to C_7 or $cTimeoutExpired$ at rate $r_{cTimeoutExpired} * (1-p)$ moving to C_9.

TTP Component

$$TTP_0 \overset{def}{=} (download, r_d).TTP_1 + (sendTTPinfo, r_{sendTTPinfo}).TTP_1$$
$$TTP_1 \overset{def}{=} (validateCoinToB, r_{vr}).TTP_2$$
$$TTP_2 \overset{def}{=} (sendTTPyes, r_{yes}).TTP_3 + (sendTTPno, r_{no}).TTP_7$$
$$TTP_3 \overset{def}{=} (askMforValidK, r_{askMforValidK}).TTP_4$$
$$TTP_4 \overset{def}{=} (sendTTPvalidK, r_{sendTTPvalidK}).TTP_{5a}$$
$$+ (timeoutTTP, r_{timeoutTTP}).TTP_6$$
$$+ (sendTTPreason, r_{sendTTPreason}).TTP_{4a}$$
$$TTP_{4a} \overset{def}{=} (sendTTPvalidK, r_{sendTTPvalidK}).TTP_5$$
$$TTP_5 \overset{def}{=} (sendMpkbyTTP, r_{sendMpkbyTTP}).TTP_{5a}$$
$$TTP_{5a} \overset{def}{=} (forwardKtoC, r_{forwardKtoC}).TTP_0$$
$$TTP_6 \overset{def}{=} (sendCkByTTP, r_{sendCkByTTP}).TTP_8$$
$$TTP_7 \overset{def}{=} (investigationInvalidCoinToB, r_{investigationInvalidCoinToB}).TTP_9$$
$$TTP_8 \overset{def}{=} (takeActionAgainstM, r_{takeActionAgainstM}).TTP_0$$
$$TTP_9 \overset{def}{=} (cspentTheCoinToTTP, r_{cspentTheCoin}).TTP_{10}$$
$$+ (mspentTheCoinToTTP, r_{mspentTheCoin}).TTP_3$$
$$TTP_{10} \overset{def}{=} (discoverMisbehavingC, r_{discoverMisbehavingC}).TTP_0$$

The above model component specifies TTC's different behaviours. TTP has thirteen states. TTP moves from states TTP_0 to TTP_{10} to solve the dispute between C and M. The actions are preformed based on the specified rates in order for TTP to involve in the interaction and provide a fair resolution for the disputed parties. TTP's main actions are *download, validateCoinToB, askMforValidK, timeoutTTP, sendMpkbyTTP, forwardKtoC, sendCkByTTP, investigationInvalidCoinToB, takeActionAgainstM* and *discoverMisbehavingC*. TTP controls the rates of those actions.

Bank Component

$$B \overset{def}{=} (requestBDigitalCoins, r_{requestBDC}).B$$
$$+ (sendCDigitalCoins, r_{sendCDC}).B + (sendBCoinByM, r_{sendBCByM}).B$$
$$+ (sendMyes, r_{sendMyes}).B + (sendMno, r_{sendMno}).B$$
$$+ (cspentTheCoinToTTP, r_{cspentTheCoin}).B$$
$$+ (mspentTheCoinToTTP, r_{mspentTheCoin}).B + (validateCoinToB, r_{vc}).B$$
$$+ (sendTTPyes, r_{yes}).B + (sendTTPno, r_{no}).B$$
$$+ (investigationInvalidCoinToB, r_{investigationInvalidCoinToB}).B$$

The last part of the model is for B component. B has one state. The B's main actions to support the purchase processes between the components C and M are *sendCDigitalCoins, sendMyes* and *sendMno* and to support the dispute resolution are *sendTTPyes, sendTTPno, cspentTheCoinToTTP* and *mspentTheCoinToTTP* as described in the scenarios of the extended protocol specification (Subsect. 2.2). The rates of these actions are controlled by B.

The System Equation. The system equation and complete specification are given by

$$System \stackrel{def}{=} TTP[K] \underset{\mathcal{R}}{\bowtie} (B[S] \underset{\mathcal{M}}{\bowtie} (C_0[N] \underset{\mathcal{L}}{\bowtie} M_0[N]))$$

where the cooperation sets $R = \{download,\ sendTTPinfo,\ validateCoin-ToB,\ sendTTPyes,\ sendTTPno,\ askMforValidK,\ sendTTPvalidK,\ timeoutTTP,\ sendTTPreason,\ sendMpkbyTTP,\ forwardKtoC,\ sendCkByTTP,\ investigation-InvalidCoinToB,\ takeActionAgainstM,\ cspentTheCoinToTTP,\ mspentTheCoin-ToTTP,\ discoverMisbehavingC\}$, $M = \{requestBDigitalCoins,\ sendCDigital-Coins,\ sendMno,\ sendBCoinByM,\ sendMyes\}$ and $L = \{sendMPO,\ sendCEP,\ sendCAbort,\ sendMCoinDk,\ sendMAbort,\ sendCPDk,\ complete,\ cTimeoutEx-pired\}$, any action in the lists R, L and M is shared action between the components specified in the system equation. N is the number of clients and merchant copies on the system, K is the number of TTPs, S is the number of Bs. The four components are initially in the states TTP_0, C_0, M_0 and B.

Furthermore, M has a number of copies depending on the number of customers in the system; each copy is associated with one C in order to serve it. This indicates that the rates of all the main actions carried out by M depend on the number of Cs interacting with M. The rates of M's main activities are divided by the number of Cs that interact with it. The M's main actions are $sendCEP$, $sendCAbort$, $startContactB$, $sendBCoinByM$, $sendCPDk$, $sendTTPvalidK$ and $sendTTPreason$. For example, the rate of $sendCEP$ action is calculated as follows:

$$r_{sendCEP} = \frac{r_{sendCEP1}}{N}$$

Additionally, the service rates of all the main actions of B are calculated based on the number of C and M's copies and the number of Bs involved in the interaction. The B's main actions are $sendCDigitalCoins$, $sendMno$, $sendMyes$, $mspentTheCoinToTTP$, $cspentTheCoinToTTP$, $sendTTPyes$ and $sendTTP-no$. One, two or more Bs can be involved in the protocol to serve C and M. For example, the rate of $sendCDigitalCoins$ action is calculated as follows:

$$r_{sendCDC} = \left(\frac{r_{sendCDigitalCoins1}}{N} \right) * S$$

Further, the service rates of all TTP's main actions depend on the number of both Cs and TTPs interacting with each other. TTP's main actions are $forwardKtoC$, $download$, $sendCkByTTP$, $askMForValidK$, $takeActionAgai-nstM$, $discoverIncorrectPTK$, $sendMpKbyTTP$, $validateCoinToB$, $sendMp-kbyTTP$ and $investigationInvalidCoinToB$. One, two or more TTPs can be involved in the protocol [6]. For example, the rate of $download$ action is calculated as follows:

$$r_d = \left(\frac{r_{download}}{N} \right) * K$$

4   Performance Evaluation of the Extended Protocol

We seek to calculate the average response times of TTPs when they serve C to solve the dispute. We assigned 1 as a value for all rates. The main actions of M are calculated based on the number of customers in the system, and the main actions of TTP are calculated based on the number of customers and TTP in the system, as mentioned in Sect. 3.

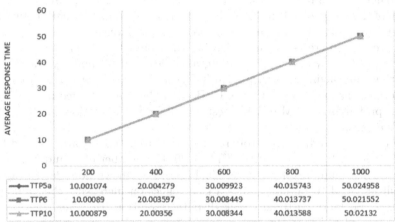

	200	400	600	800	1000
TTP5a	10.001074	20.004279	30.009923	40.015743	50.024958
TTP6	10.00089	20.003597	30.008449	40.013737	50.021552
TTP10	10.000879	20.00356	30.008344	40.013588	50.02132

NUMBER OF CUSTOMERS IN THE SYSTEM

(a) Number of TTPs (K) is 20

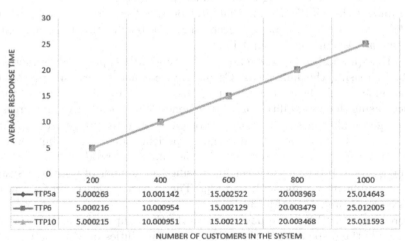

	200	400	600	800	1000
TTP5a	5.000263	10.001142	15.002522	20.003963	25.014643
TTP6	5.000216	10.000954	15.002129	20.003479	25.012005
TTP10	5.000215	10.000951	15.002121	20.003468	25.011593

NUMBER OF CUSTOMERS IN THE SYSTEM

(b) Number of TTPs (K) is 40

Fig. 2. The average response time of TTP_{5a}, TTP_6 and TTP_{10} using ODE when $S = 20$ and the probability of M to be honest is 0.1.

In Fig. 2(a), the number of TTP involved in the system is 20 and we change the number of customers from 200 to 1000 to show how increasing the number of the customer seeking help from TTP would impact the performance of the protocol. The average response time of TTP for all main states (TTP_{5a}, TTP_6 and TTP_{10}) to solve the dispute is the same. Having a large number of customers significantly increases the average response time of TTPs, which creates more performance overhead. However, in Fig. 2(b), the number of TTP is increased to 40 which causes a decrease in the response time in relation to the number of customers in the system compared to Fig. 2(c). Therefore, having a larger number of TTPs involved in the protocol when the dispute occurs between C and M mitigates the security protocol's performance overhead.

We are also interested in investigating the population level analysis of C_4, C_6, C_9 and C_{10} (C_4 and C_6 for having a service from M and C_9 and C_{10} for interacting and having a service from TTP) and throughput analysis of some main actions that provide service to C. Both analyses are studied in relation to different probabilities of M to be honest and different population numbers of Cs and M's copies.

In Figs. 3(a) and 4(a), the probabilities for M to be honest are changed from $p = 0.1$ to $p = 0.9$, the number of TTP is 20, the number of C and M's copies is 200 and the number of banks is 20. Figure 3(a) shows how decreasing the probabilities of M to be honest has a clear effect on increasing the number of C_{10} copies. So more customers seek dispute resolution. However, the number of C in C_4 and C_6 does not experience any change but there are more Cs in C_6 waiting to get a decryption key for the encrypted product than in C_4 to get an encrypted product. Figure 4(a) illustrates the significant increase in the throughput of the TTP's actions that provide the dispute resolution when the probabilities of M to be honest is decreasing. We believe this is because more customers are seeking help from TTP.

In Figs. 3(b) and 4(b), the number of C and M's copies (N) is increased to 600. Increasing the number of Cs in a system has a clear impact on the population level and the actions' throughputs. Just like Fig. 3(a), Fig. 3(b) shows how decreasing the probabilities of M to be honest has a clear effect on increasing the average number of C_{10} as more customers seek a dispute resolution. Also, the number of Cs in C_4 and C_6 does not experience any change, but more Cs in C6 waiting to get a decryption key for the encrypted product than in C4 to get an encrypted product. However, increasing the number of Cs in the system does not significantly impact C_9 and C_{10}. The impact is clearly on C_4 and C_6, compared to C_4 and C_6 in Fig. 3(a).

Moreover, Fig. 4(b) shows the increase in the throughput of the TTP's actions that provide the dispute resolution when the probabilities of M to be honest is decreasing. We believe this is because more customers are seeking help from TTP. Moreover, all actions have a clear reduction on their throughput when we increased the number of C. The throughput values of the TTP's actions are less than the throughput of the TTP's actions in Fig. 4(a) when the number of Cs in a system is 200. Therefore, a larger number of customers in the system will have an effect on the TTP's responses.

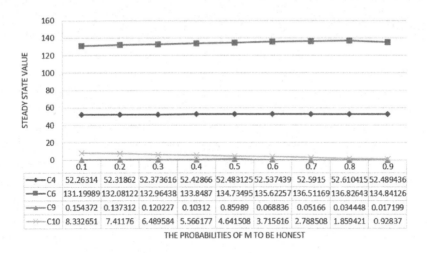

(a) $N = 200$

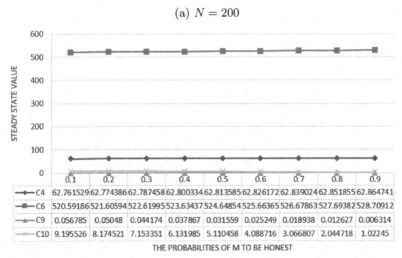

(b) $N = 600$

Fig. 3. The population level analysis using ODE with $K = 20$ and $S = 20$ in relation to different probabilities of M to be honest.

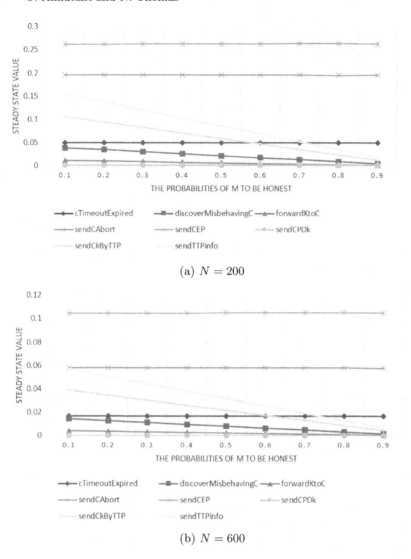

(a) $N = 200$

(b) $N = 600$

Fig. 4. The throughput analysis of actions using ODE with $K = 20$ and $S = 20$ in relation to different probabilities of M to be honest.

Figure 5 shows how faster the system settled in relation to the population number of Cs and M's copies and the probabilities of M being honest. The larger the population number is and the less probability of M to be honest is, the longer time will be taken for the system to settle.

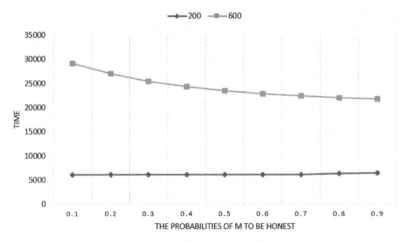

Fig. 5. The steady-state detection time in relation to the population number ($N =$ 200 and $N = 600$).

Now we are interested in changing the rates of the shared actions between M and B. The shared actions between M and B are $rsendBCoinByM$, $rsendMyes$ and $rsendMno$. We increased and decreased the rates to show how these would have an impact on the system performance. First, the shared actions rates between M and B are decreased to $r = 0.2$ and $r = 0.5$. Then they are increased to $r = 2$ and $r = 4$. The rates are calculated depending on the number of N and S involved in the system, as follows:

$$rsendBCoinByM = r/N$$
$$rsendMyes = (r/N) * S$$
$$rsendMno = (r/N) * S$$

Where N is the number of Cs and M's copies and S is the number of banks.

The following figure shows the population level analysis of C_4 and C_6 for having a service from M and C_9 and C_{10} for interacting and having a service from TTP. The probability for M to be honest is changed from $p = 0.1$ to $p = 0.9$. Figures 6(a) and 6(b) illustrate that when the shared actions rates between M and B to check the C's digital coin's validity are slow, C experiences a big delay in receiving a product decryption key. There are large waiting Cs in C_6. In Figs. 6(a) and 6(b), you can notice that the population levels of C_9 and C_{10} are slightly decreased. We believe this is because more Cs waiting in C_4 and C_6 to be served before moving to C_9 and C_{10}. So the faster the rate, the less delay would be, as shown in Figs. 6(c) and 6(d). Moreover, Figs. 6(c) and 6(d) show a significant increase in the population levels of C_{10}.

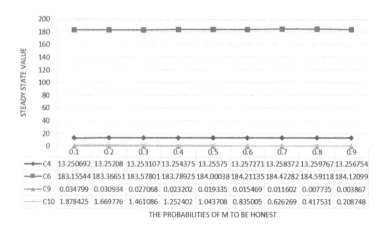

	0.1	0.2	0.3	0.4	0.5	0.6	0.7	0.8	0.9
C4	13.250692	13.25208	13.253107	13.254375	13.25575	13.257271	13.258372	13.259767	13.256754
C6	183.15544	183.36651	183.57801	183.78925	184.00038	184.21135	184.42282	184.59118	184.12099
C9	0.034799	0.030934	0.027068	0.023202	0.019335	0.015469	0.011602	0.007735	0.003867
C10	1.878425	1.669776	1.461086	1.252402	1.043708	0.835005	0.626269	0.417531	0.208748

THE PROBABILITIES OF M TO BE HONEST

(a) $r = 0.2$

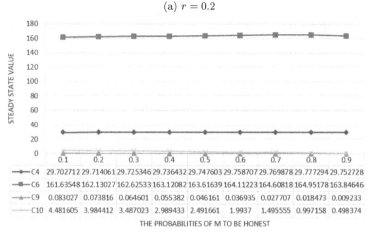

	0.1	0.2	0.3	0.4	0.5	0.6	0.7	0.8	0.9
C4	29.702712	29.714061	29.725346	29.736432	29.747603	29.758707	29.769878	29.77294	29.752728
C6	161.63548	162.13027	162.62533	163.12082	163.61639	164.11223	164.60818	164.95178	163.84646
C9	0.083027	0.073816	0.064601	0.055382	0.046161	0.036935	0.027707	0.018473	0.009233
C10	4.481605	3.984412	3.487023	2.989433	2.491661	1.9937	1.495555	0.997158	0.498374

THE PROBABILITIES OF M TO BE HONEST

(b) $r = 0.5$

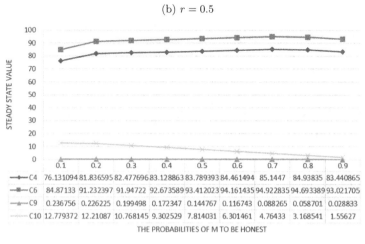

	0.1	0.2	0.3	0.4	0.5	0.6	0.7	0.8	0.9
C4	76.131094	81.836595	82.477696	83.128863	83.789393	84.461494	85.1447	84.93835	83.440865
C6	84.87133	91.232397	91.94722	92.673589	93.412023	94.161435	94.922835	94.693389	93.021705
C9	0.236756	0.226225	0.199498	0.172347	0.144767	0.116743	0.088265	0.058701	0.028833
C10	12.779372	12.21087	10.768145	9.302529	7.814031	6.301461	4.76433	3.168541	1.55627

THE PROBABILITIES OF M TO BE HONEST

(c) $r = 2$

Fig. 6. The population level analysis using ODE with $K = 20$, $N = 200$ and $S = 20$ and with different rates for the shared actions between M and B.

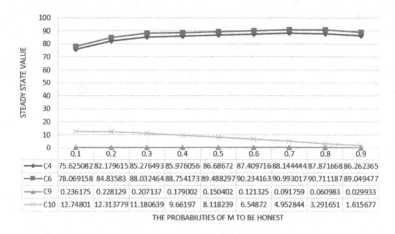

	0.1	0.2	0.3	0.4	0.5	0.6	0.7	0.8	0.9
C4	75.625082	82.179615	85.276493	85.976056	86.68672	87.409716	88.144444	87.871668	86.262365
C6	78.069158	84.83583	88.032464	88.754173	89.488297	90.234163	90.993017	90.711187	89.049477
C9	0.236175	0.228129	0.207137	0.179002	0.150402	0.121325	0.091759	0.060983	0.029933
C10	12.74801	12.313779	11.180639	9.66197	8.118239	6.54872	4.952844	3.291651	1.615677

THE PROBABILITIES OF M TO BE HONEST

(d) $r = 4$

Fig. 6. (*continued*)

Figure 7 shows the throughput of some main actions to serve a customer. These actions are $discoverMisbehavingC$, $forwardKtoC$, $sendCkByTTP$ and $sendTTPinfo$ to interact and seek a help form TTP and $sendCAbort$, $sendCEP$ and $sendCPDk$ to get a service form M and $cTimeoutExpired$. In Fig. 7, the probabilities for M to be honest are changed from $p = 0.1$ to $p = 0.9$, the number of TTP is 20, the number of C and M's copies is 200 and the number of banks is 20. The shared actions rates between M and B are decreased to $r = 0.2$ and $r = 0.5$ and then increased to $r = 2$ and $r = 4$. The rates are calculated based on the number of N and S involved in the system, as mentioned in this Section.

Figure 7 illustrates a significant increase in the throughput of the TTP's actions that provide the dispute resolution when M's probabilities of being honest are decreasing in all Figures. The TTP's actions are $discoverMisbehavingC$, $forwardKtoC$, $sendCkByTTP$ and $sendTTPinfo$. We believe this is because more customers are seeking help from TTP when M's probabilities of being honest are lower. Moreover, there is a considerable improvement in all actions' throughputs when the shared actions rates between M and B to check the C's digital coin's validity are higher.

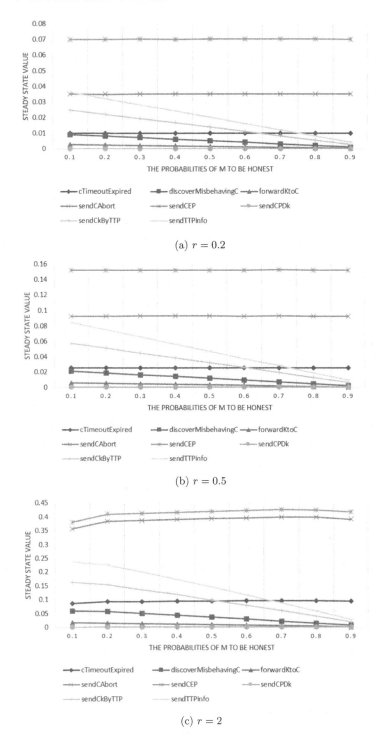

(a) $r = 0.2$

(b) $r = 0.5$

(c) $r = 2$

Fig. 7. The throughput analysis using ODE with $K = 20$, $N = 200$ and $S = 20$ and with different rates for the shared actions between M and B.

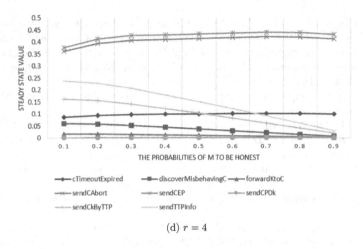

(d) $r = 4$

Fig. 7. (*continued*)

5 Conclusion

This investigation explores an anonymous and failure resilient fair-exchange e-commerce protocol when the misbehaviour and connections problem occurs. The involvement of TTP becomes essential to resolving disputes between partici-pants. We propose the PEPA model for the protocol. The evaluation results indicated that when the protocol preserved customer anonymity, this introduced extra performance costs. Furthermore, when there is a dispute between partic-ipants, the TTP involvement is active, and this would introduce extra load on TTP and/or Bs, which would influence the system performance. Moreover, we showed how scaling up TTP and B resources to handle escalating misbehaving parties mitigates the negative impact on the system performance. In this paper, we concentrate on the misbehaviour of the merchant. In our future work, we will consider models and scenarios when misbehaviour occurs from customers. Also, we will consider PEPA models of malicious misbehaviour where an adversary's behaviour changes over time and the system needs to respond in kind in order to remain secure and to provide a sustainable level of performance to the legiti-mate users by, for example, scaling of the resources to handle escalating threats without a negative impact on the rest of the system.

References

1. Almutairi, O., Thomas, N.: Performance modelling of an anonymous and fail-ure resilient fair-exchange e-commerce protocol. In: Proceedings of the 2019 ACM/SPEC International Conference on Performance Engineering, pp. 5–12 (2019)
2. Apostolopoulos, G., Peris, V., Pradhan, P., Saha, D.: Securing electronic commerce: reducing the SSL overhead. IEEE Netw. **14**(4), 8–16 (2000)

3. Apostolopoulos, G., Peris, V., Saha, D.: Transport layer security: how much does it really cost? In: IEEE INFOCOM 1999. Conference on Computer Communications. Proceedings. Eighteenth Annual Joint Conference of the IEEE Computer and Communications Societies. The Future is Now (Cat. No. 99CH36320), vol. 2, pp. 717–725. IEEE (1999)
4. Hillston, J.: A Compositional Approach to Performance Modelling, vol. 12. Cambridge University Press, Cambridge (2005)
5. Okamoto, T., Ohta, K.: Universal electronic cash. In: Feigenbaum, J. (ed.) CRYPTO 1991. LNCS, vol. 576, pp. 324–337. Springer, Heidelberg (1992). https://doi.org/10.1007/3-540-46766-1_27
6. Ray, I., Ray, I., Natarajan, N.: An anonymous and failure resilient fair-exchange e-commerce protocol. Decis. Support Syst. **39**(3), 267–292 (2005)
7. Tribastone, M., Duguid, A., Gilmore, S.: The PEPA eclipse plug-in. Perform. Eval. Rev. **36**(4), 28 (2009)
8. Zeng, W., Chow, M.Y.: A trade-off model for performance and security in secured networked control systems. In: 2011 IEEE International Symposium on Industrial Electronics, pp. 1997–2002. IEEE (2011)
9. Zhao, Y., Thomas, N.: Performance modelling of optimistic fair exchange. In: Wittevrongel, S., Phung-Duc, T. (eds.) ASMTA 2016. LNCS, vol. 9845, pp. 298–313. Springer, Cham (2016). https://doi.org/10.1007/978-3-319-43904-4_21

Performance Evaluation of a Data Lake Architecture via Modeling Techniques

Enrico Barbierato[1]([✉]) [iD], Marco Gribaudo[2] [iD], Giuseppe Serazzi[2] [iD],
and Letizia Tanca[2] [iD]

[1] Dip. di Matematica e Fisica, Università Cattolica del Sacro Cuore,
Via dei Musei 41, 25121 Brescia, Italy
`enrico.barbierato@unicatt.it`
[2] Dip. di Elettronica, Informazione e Bioingegneria, Politecnico di Milano,
via Ponzio 34\5, 20133 Milano, Italy
`{marco.gribaudo,giuseppe.serazzi,letizia.tanca}@polimi.it`

Abstract. Data Lake is a term denoting a repository storing heterogeneous data, both structured and unstructured, resulting in a flexible organization that allows Data Lake users to reorganize and integrate dynamically the information they need according to the required query or analysis. The success of its implementation depends on many factors, notably the distributed storage, the kind of media deployed, the data access protocols and the network used. However, flaws in the design might become evident only in a later phase of the system development, causing significant delays in complex projects. This article presents an application of queuing networks modeling technique to detect significant issues, such as bottlenecks and performance degradation, for different workload scenarios.

Keywords: Data lake · Queuing networks · JMT

1 Introduction

The concept of *Data Lake* emerged quite recently and relatively to Big Data, aiming at complementing, or even replacing, more conventional approaches such as *Data Warehouses* and Business Intelligence applications[1]. A Data Lake provides a most efficient architecture supporting data science applications projecting the effective value of business performance; as a result, it is of paramount importance to be able to predict the performance of a Data Lake architecture. Unfortunately, environments of this sort do not adhere to a standardised design yet, making difficult to establish valuable metrics capable of determining the effective benefit of using a Data Lake. In principle, the performance of massively distributed environments relies upon a few indicators, most notably *event throughput*, which

---

[1] https://infocus.delltechnologies.com/william_schmarzo/why-do-i-need-a-data-lake-for-big-data/.

© Springer Nature Switzerland AG 2021
P. Ballarini et al. (Eds.): EPEW 2021/ASMTA 2021, LNCS 13104, pp. 115–130, 2021.
https://doi.org/10.1007/978-3-030-91825-5_7

in turn depends on storage performance, the topology of the deployed resources, the co-location of data compared to the computing resources, and the possibility to access data locally against a remote distribution.

Modeling represents a powerful technique to assess system performance. In this respect, the literature offers a wide range of approaches, including stochastic paradigms such as Queueing Networks, Fault Trees and Petri Nets to cite the most notable. The advantage of this approach emerges in an early stage of a project identifying bottlenecks or scenarios causing an impact on performance, resulting in deploying a more efficient and cost-saving architecture.

As a contribution, this work presents a novel performance model of an application querying a Data Lake storage, considering utilization as a performance measure.

The remaining part of the article is organized as follows: Sect. 2 introduces the concept of Data Lake. Section 3 presents some notable examples of implementations, while Sect. 4 discusses the scenario of a Data Lake application modeled by using a queuing system. Section 5 presents an overview regarding the related work, and finally Sect. 6 draws some conclusions proposing the directions of future work.

2   Data Lakes

Typically, data fall into different categories, depending on whether they are structured (e.g., in the form of a relational schema), semi-structured (e.g., JSON or XML format) or unstructured (such as *pdf* files, *YouTube* videos or *Facebook* posts). Due to the growing heterogeneity of information, new paradigms and technologies have been developed; most notably, the Big Data phenomenon has brought innovative ways to analyze and extract significant value in terms of use. In this sense, several enterprises dedicate most of their efforts to mining the vast amount of data at their disposal, to increase their volume of business. Many innovations have been introduced by Big Data, starting from unlimited storage volume, due to the way used to represent data, which are segmented on different clusters. This innovative approach requested the development of a new type of a software framework, called HDFS (Hadoop Distributed File System), providing the capability of storing, retrieving and processing huge data sets. The term Data Lake, introduced in 2010 by J. Dixon[2], is not well defined, though it can be understood[3] as a storage repository, which is massive and scalable, holding raw data, with an associated engine capable of processing the information without altering its structure.

However, if on one side the lack of a rigid data schema is a strength, it can be tough to maintain order when new data are received, making data lake governance very critical. Put in other words, the mere data ingestion does not take into account the semantics and context of the data. It is therefore important to

---

[2] https://jamesdixon.wordpress.com/.

[3] https://searchcio.techtarget.com/feature/Data-lake-governance-A-big-data-do-or-die.

consider, besides traditional dictionaries and indexes, the introduction of a rich set of metadata, as they play a very important role by giving data a meaning. For example, it should be possible to add semantic information to a Data Lake, in order to guarantee that data have a context and ultimately, a governance. In [9] the authors cite a few properties and requirements characterizing a Data Lake, such as i) Architecture scalability; ii) Governance, Cataloging and indexing; iii) Tracking data transformation; iv) Storing the data sharing the content to applications; v) Providing the ability to access the data by using different formats; vi) Granting data access from every sort of device; vii) Allowing different data analytics methods and finally, viii) Data compression and absence of duplication.

Supporting fast data analytics[4] is another crucial aspect of Data Lakes ecosystems. Both relational and NoSQL databases assume that data are physically retrieved from a disk (with slow access time), which requires a significant amount of optimization. However, these techniques turn out to be redundant when flash devices are deployed, and anyway they might not be suited for cloud environments. As a result, new approaches have emerged, such as in-memory databases and data grid on flash units (or even a combination of the two). Since identifying which data need to be mined immediately is of paramount importance, fast data - corresponding to smaller data sets in real-time to solve a specific problem - fit the need to apply fast data analytics. The latter requires two specific approaches: on the one side, since data are received as a stream, the system must be able to deliver the events with the same speed as they were received. On the other hand, the received data must be processed by a data store as they come. This process is articulated in three distinct steps: an *ingest* phase (where massive amounts of data are received in a very small amount of time), a *decision* regarding an event taken on the basis of the data and *real-time analysis*, where an automatic decision-making policy can be implemented.

2.1 Data Lake, Data Warehouse and Data Hub

According to a recent Gartner assessment[5], terms such as *Data Lake, Data Warehouse* and *Data Hubs* are used interchangeably, resulting in a growing confusion about the roles of separate architectures aiming at different purposes.

At first sight, the concept of Data Lake may overlap the definition of Data Warehouse. However, even if these concepts share a common goal, they are distinct. A Data Warehouse can be regarded as a highly structured repository, conceived mainly for (typically already envisaged) business intelligence analysis on structured data. Usually, warehouse data are extracted from other repositories - typically transactional Databases -, cleaned and reorganized according to the business intelligence tasks that are considered as the most frequent, indexed and loaded into tables, and finally related to the remaining data in the warehouse.

---

[4] https://venturebeat.com/2014/06/25/the-next-big-disruption-in-big-data/.
[5] https://www.gartner.com/en/documents/3980938-data-hubs-data-lakes-and-data-warehouses-how-they-are-di.

However, this kind of organization doesn't fit well the adaptability requested by heterogeneous applications, being constrained by data management procedures. Furthermore, a Data Warehouse is not optimized to handle a high volume of data represented in a variety of formats. By contrast, a Data Lake can be hosted by Hadoop, with a significant reduction of ROI (Return of Investment) and ownership, Hadoop being a open source product where data can be structured or not, matches the intrinsic agility of Data Lake, while in a Data Warehouse any structural change must be carefully evaluated.

Summarizing, as well explained in [7], the Data Warehouse performs an *Extract, Transform and Load* (ETL) cycle during the *ingestion* phase, typically works on Relational Data Bases and uses SQL to access the data, while the Data Lake ingests data as is, works on different structures than relational data bases and considers a broad range of methods (such as programming languages or different query techniques than SQL).

To conclude, it can be noted that a Data Hub differs deeply with respect to both Data Lakes and Data Warehouses, as it is not oriented to analytical data investigation. In the case of Data Hubs, the point is to connect data producers and consumers by means of governance controls, rather than running Business Intelligence reports (that are quite rare).

3   Data Lake Architectures

As per [10], the life cycle of a dataset that has entered a Data Lake can be described as composed of the following stages: i) *Ingestion*; ii) *Data extraction*; iii) *Data cleaning*; iv) *Data discovery*; v) *Metadata management* and vi) *Dataset versioning*.

Within the *Ingestion* phase, data are injected into the Data Lake, checked against duplicates, indexed and versioned; the produced output consists of raw datasets in the shape of either text or binary data.

During *Data extraction*, the raw datasets are transformed into an abstract data representation model (e.g., relational).

With the *Data cleaning* step, the organized data obtained earlier are verified against a set of integrity constraints.

The *Data Discovery* process aims at finding similar datasets, for example by using exploration-by-graph techniques.

The creation of a catalog, in the *Metadata management* phase - indispensable to prevent the Data Lake from becoming a *Data Swamp* -, is followed by the *Dataset versioning* stage, where data redundancy, rather typical in Data Warehouses, is discouraged because of the cost (especially in case of large datasets), preferring instead *git-like* approaches.

From an architectural perspective, in [7] the authors recall two classic approaches, i.e., *pond* and *zone*. As described in [8], the former envisages to move the raw data into a specific data pond named *raw*. After a transformation process has been executed, data are moved again to a different pond called *analog*. From this moment, data are ready to be analytically processed. Architectures based on the notion of *zone* are described, for example, in [4]: the Data Lake space is subdivided into zones suited for different stages of the dataset life-cycle (loading, storing, validating and so forth).

More in general, the latter approach suits specific domains such as healthcare. In fact, in these applications[6], very often the Data Lake is organized in different zones, which can be physical or logical ways to separate the data. A *raw* data zone contains data in original format, analyzed by data scientists; a *trusted* data zone is regarded by an organization as the *universal truth*; similarly to the case of Data Warehouses, a *refined* data zone regroups data into *Subject Area Marts* to produce a data view equivalent to a trusted zone but dedicated to a specific domain; finally, the *exploration* zone can be used as a sandbox just by moving data into it from any other zone for private use.

The procedure of data-refining is suitable to present information viewed from different angles. For example, the data in raw format received by a streaming process can be stored in the so-called *bronze* zone. Data of this type are not immediately available, as they need be processed to allow a query to be deployed. As a result, normalized data are stored in a *silver* area. A further, refining step is taken by aggregating the most used metrics. The output of this process constitutes a *gold* zone. From this point onward, Machine Learning and reporting techniques can be applied.

3.1 Smart Data Lake

In [2], the authors present the architecture of a *Smart Data Lake* or SDL, as opposed to classic approaches such as HDFS and S3. A SDL is oriented to self-tuning, and comparable to a middleware operating between two type of nodes: *storage* and *computing*. It provides direct access to raw data and intermediate data representations on-the-fly being, at the same time, aware of the hardware used (local disks, SSDs, NVMs, etc.). A SDL architecture consists of an interface offering a set of layers granting data access, a shared messaging queueing system to swap messages between *RAW* (a commercial package to manage data in heterogeneous formats) and *Proteus* (a tool performing the execution of the analytical processes), a local catalog, and finally a policy for the tiers. Data are subdivided in segments, which are located in storage tiers by a *Storage allocator*, specific for each tier. The authors state that the tier policy algorithm is still ongoing work. However, it is possible to predict access patterns by using tracking techniques, and consequently to manage cache mechanisms more efficiently. A Storage Manager (SM) is located between the storage and the processing layers in order to handle calls for read or write operations. An instance of SM is running on all the computing nodes. When a query is executed, SM is contacted to find the data location and, more importantly, whether the data are stored in a cache or remotely.

---

[6] See, for example, https://www.healthcatalyst.com/four-essential-zones-healthcare-data-lake.

Read and write operations occur by exploiting an inter-process technique. Notably, SDL uses a component to handle the unpredictability of the arrival rate of I/O requests. The authors review two techniques called respectively *polled I/O* and *interrupt-driven I/O*, and propose an adaptive system using both. When SM works in interrupt-driven fashion, it waits for new requests in a sleeping status, but when a request is added to the queue, the event is detected and a signal is sent to SM. From this moment on, SM switches to polled mode, checking if new requests have been sent; after a certain amount of time, SM returns to the initial state. In other words, SM can work in two ways according to the frequency of the times it is invoked.

3.2 Azure Data Lake Store

In [12], the authors introduce Azure Data Lake Store (ADLS), defined as a file system that is scalable and secure at the same time. Its strong point is the support to Hadoop Distributed File System (HDFS) and Cosmos, a file system project developed by Microsoft in 2006. When data are processed, a query invokes ADLS to determine where the data are stored, generating a plan necessary to guarantee that the tasks be executed nearby. As data can be stored remotely, ADLS copies them on the node where the task is requested to run. An ADLS file is composed of a sequence of *extents* (blocks whose size can be up to 4Mb). Any part of a file can be located in one or more storage *tiers*, which in turn can be local (on the same node where processing has been scheduled) or remote (outside an ADLS cluster). Being a file possibly spanning over different tiers, ADLS supports the concept of *partial* file that is regarded as a *sub-sequence*. An ADLS cluster is composed of the following nodes: *back-end*, dedicated to storing data and executing tasks on local tiers; *front-end*, acting as a gateway and checking the access, and *micro-service*, hosting special services. The latter are orchestrated by a *Secure Storage Service*, essentially an entry point to ADLS. A set of metadata services are based on a *RSL-HK*, a component providing an efficient and persistent state.

3.3 Google File Storage

The architecture of *Google File Storage* (GFS) is discussed in [6]. The authors present its design by saying that it is based on the following assumptions: component monitoring, efficient management of multi-GB files, subdivision into large and small streams of *read* workload, a sound semantics to allow users to append data on a file in a concurrent way, and finally the capability to sustain a high bandwidth. A GFS is organized into clusters, each one consisting of one *master* and many replicated *chunkservers*. A file is composed of *chunks*, identified by a *handle* and stored on local disk by *chunkservers*, while the master preserves the metadata. The latter polls periodically the *chunkservers* to gather information about their state. Cache memory is not strictly necessary, as usually the applications tend to stream a high volume of data.

Google claims[7] that such a cloud storage architecture can serve a Data Lake, performing efficiently data ingestion, analytics and data mining.

3.4 Amazon S3

The Amazon S3 storage architecture[8] consists of two-level *namespaces*. The top level contains a set of *buckets* (typically 100) denoted by a unique global name. They can contain users data and consequently allow users to be identified when they are loaded. The following level is the *data object*: identified by a name and a set of metadata, it is the basic information block contained in a bucket. An object can be altered by a user, according to a set of security restrictions. Within a bucket, the object's name is used to perform queries. With regard to data access, S3 supports the SOAP, REST and BitTorrent protocols. Security is assigned to a user when the initial contract is formalized by a public/private key scheme. Amazon S3 storage provides several features, such as object locking, storage classes, versioning, batch operations and replication.

3.5 A Generic Data Lake Architecture

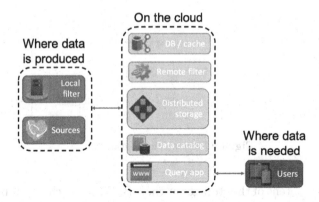

Fig. 1. High level architecture

According to the objective of the paper, we derive a general Data Lake architecture from the descriptions of the previous sections, which will be used to implement the models. The proposed system is shown in Fig. 1 and consists of the following components: i) several Analytics applications querying the data storage; ii) a Data Base (DB) or a cache memory containing the output of an application; iii) a Data Catalog, containing a set of metadata, which is updated

---

[7] https://cloud.google.com/architecture/build-a-data-lake-on-gcp.

[8] https://docs.aws.amazon.com/AmazonS3/latest/userguide/managing-storage.html.

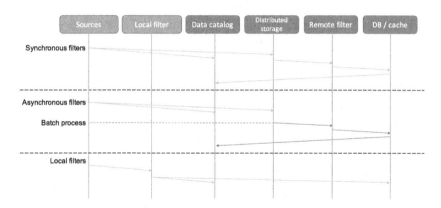

Fig. 2. Data ingestion sequence diagram

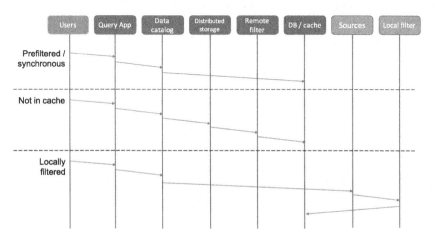

Fig. 3. Query sequence diagram

every time the content of the storage (or the DB/Cache) is changed; iv) a *remote filter*[9], whose output is stored in the DB/cache and finally v) a storage. Note that data can be pre-processed locally by a different type of filter.

In the following sections, three scenarios have been identified and described as per the *sequence diagram* in Fig. 2.

Case 1: Synchronous Filters. A filter is called *Synchronous* when the sources generate data and inject them into the Data Lake, updating the Data Catalog and then the storage. Furthermore, when an instance of remote filtering is executed, the resulting output is stored in the DB/Cache component and the Data Catalog is updated one more time.

---

[9] The term *filter* denotes a type of data processing, producing a result.

Case 2: Asynchronous Filters and Batch Process. A filter is called *Asynchronous* when, after the data have been injected, the Data Catalog and the Storage are updated and the process exits. Another batch process, running in the background, is able to detect that some data have been uploaded and consequently, processes them. Again, the results are stored in the DB/Cache and Data Catalog is updated.

Case 3: Local Filters. In this case, after the process has started and the Data Catalog has been checked out, one or more filters process the data at local level. In the next stage, both Data Catalog and DB/Cache are updated.

Queries. As per the sequence diagram shown in Fig. 3, a query can be invoked on *pre-filtered* or *locally filtered* data. In the former, a user executes an application, causing the corresponding query to look for previously produced results in the Data Catalog. If there is a *hit*, the output is retrieved from the DB/Cache area, otherwise data are retrieved from the Storage and processed by the remote filter. Finally, the db/Cache is updated. In case the data are locally filtered, the application performs first a check against the Data Catalog: as the data have to be locally filtered, they are processed and the resulting output is stored in the DB/Cache component.

4   Case Study

4.1   Introduction

This chapter is organized as follows: Subsect. 4.2 introduces the background of a case study, while Subsect. 4.3 explains how the considered scenario has been modelled by using JMT. Finally, Subsect. 4.4 discusses the experiment settings and the results.

4.2   Considered Scenario

The case study discussed in this work consists of a medical research centre investigating the capability of predicting the presence of cancer in patients at an early stage. Oncologists and data scientists need a shared deposit to store data originated from different sources, such as radiographs, computerized tomographs and magnetic resonance images. While the different data are fed into a Data Lake as a daily routine, the researchers are focusing on the development of efficient Machine Learning classifiers, whose algorithms need to query the stored data providing meaningful results in a short time. The output of the classifiers is used to determine the best treatment for the patients.

It is supposed that the Data Lake architecture reflects the general principles introduced in Sect. 3.5. For the sake of simplicity, the study takes into account a synchronous scenario and the vertical scaling of Application Server and DB/Cache in a high load scenario.

Note that specific Data Lakes areas of interest, such as raw data zone, trusted data zone, refined data zone, Subject Area Marts, exploration zone, bronze, silver and gold zones will be studied in future work.

4.3   The Model

Architecture. The model, created by the JMT suite (see [1]), and shown in Fig. 4 is a queueing network that simulate the resources utilized during the execution of the queries with synchronous filtering. The five queue stations are devoted respectively to the simulation of the following operations: i) *DataCat*, update of the Data Catalog; ii) *Storage*, update of the data in the storage; iii) *Rem. Filter*, perform the operations required by the Remote Filter processing; iv) *DB/Cache*, store and retrieve the data in the DB and cache; and v) *Query App*, simulate the component that manage the application queries.

The workload of the model consists of four classes of requests referred to as i) *Data Raw*, ii) *Meta Data*, iii) *QIC* (Query In Cache) and iv) *QNC* (Query Not in Cache). The requests of each class can follow different paths within the resources. To increase the possible scenarios that can be simulated (e.g., cache hit or miss), the requests can change class during their execution within `Class-Switch` stations. A request entering a `Class-Switch` station in `class-i` can leave the same station in `class-j` according to the probability set in a matrix associated with the station. Three `Class-Switch` stations are in the model: *CSSto*, *ReDCat*, and *CSUsers*.

The requests of the open class *Data Raw* are generated by the `Source1` station with rate λ_{src} and exponentially distributed interarrival times, and are routed to `Data Catalog` server and then to `CSSto` class-switch station. Other types of distributions, e.g., burst, hyperexponential, or phase-type, can be easily generated by the tool if needed. The delay station `Users` simulate the physicians that submit queries to the system. Their number N_{users} is constant and the requests are initially of QIC class, that is a closed class, and sent to the `Query App` server for their management. From the `Query App` server are routed to the `Data Cat` server and then to the `CSSto` class-switch station which split them into two streams: those that have a cache hit (of class QIC) and those that have a cache miss (of class QNC). While QIC requests are sent directly to the `DB/Cache` server, the QNC are routed to `Storage` and `Remote Filter` server before reaching the `DB/Cache` server.

Once a QIC query has been executed by `DB/Cache` it is routed to the `Users` station and leaves the system. Then, since the QIC class is closed, a new query will be generated after a mean time Z_{users}, exponentially distributed, and sent to the `Query App` server. Otherwise, a QNC query that exits the `DB/Cache` is routed to the `Data Cat` server to update the data in the *catalog* and leaves the system after changing its class to QIC in the `CS Users` station. This last operation is necessary to allow the simulator to compute the global performance indices such as *System Response Time* and *Throughput*.

The data generated by the oncological centre, e.g., medical images, etc., are of the open class *Data Raw* and are sent from the `Source1` to the `DataCat` server.

After their analysis in the `Data Cat`, they are routed through CSSto to Storage, Rem.Filter, and `DB/Cache` servers. Their processing in these servers generate some metadata and thus a new access to update the Data Catalog is required. When they exit the `DB/Cache` are sent to `RetDCat` class-switch station where their class is changed from `Data Raw` to `Meta Data` and then are routed to `Data Cat` server. When their processing is completed, they leave the system trough `Sink 1` station.

The service demands of the various servers are exponentially distributed.

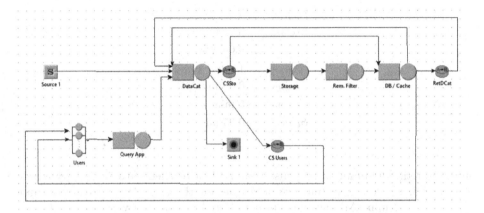

Fig. 4. The JMT model

4.4 Simulation Results

In this section, results have been computed with a 99% confidence interval, with the simulation stopping when a 3% relative error is reached. To simplify the following discussion, only the average of the confidence interval is presented.

It is assumed that the oncological centre produces λ_{src} medical images per minute, where N_{users} doctors query the system on the average every Z_{users}. *Data catalog* requires $S_{DC\text{-read}}$ for reading, and $S_{DC\text{-write}}$ for writing entries. The *storage* requires on the average $S_{storage}$, while *filtering* requires $S_{new\text{-filter}}$ to extract features from a new image being inserted, or $S_{re\text{-filter}}$ to re-analyze a set of images to perform a query requiring results not in cache. Due to the high load of these components, they are respectively replicated on $K_{storage}$ and K_{filter} instances. Requests of data which have already been extracted and are still available in the cache require instead S_{cache}. Table 1 shows the basic parameters used in our test scenario.

As it can be seen from Fig. 5, with the initial settings the bottleneck is the storage. Adding a second storage server $K_{storage} = 2$, the bottleneck moves to the filtering. To remove this secondary bottleneck, at least $K_{filter} = 6$ nodes

Table 1. System parameters

Parameter	Value
λ_{src}	0.5 job/min
N_{users}	20 users
Z_{users}	1 min
S_{app}	0.1 min
K_{storage}	1 servers
K_{filter}	3 servers
p_{inCache}	0.8
$S_{\text{DC-read}}$	0.01 min
$S_{\text{DC-write}}$	0.05 min
S_{storage}	1 min
$S_{\text{new-filter}}$	0.3 min
$S_{\text{re-filter}}$	5 min
S_{cache}	0.05 min

are required. The minimal response time is then reached with $K_{\text{storage}} = 3$ and $K_{\text{filter}} = 8$, where the system response time for the N_{users} stabilizes at around $R_{\text{users}} = 2.72$ min. All experiments considered a probability $p_{\text{inCache}} = 0.8$ of finding the results of the filtering stage already in cache.

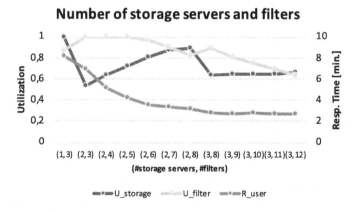

Fig. 5. Determining the optimal number of servers

Figure 6 shows the effect of cache for the filtering data, as function of its hit/miss probability on the optimal scenario previously outlined with $K_{\text{storage}} = 3$ and $K_{\text{filter}} = 8$. As expected, caching has a big impact on the load of both the storage and the filtering. In particular, while storage is also affected by data sources, filtering is mainly conditioned by caching, as shown by the swap of the two curves for $p_{\text{inCache}} = 0.95$. In this scenario, the caching service is still capable of supporting the application, even if as expected its workload increases with the probability of finding data in cache. What is less obvious, is that the increase in performance of the system due to caching, shifts the bottleneck to the query application server, a component that probably would have been undersized without adequate modeling analysis.

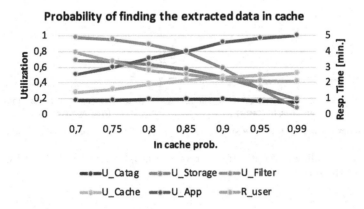

Fig. 6. Effect of cache hit probability

Finally we focus on a case with a larger load, where sources upload images at the rate $\lambda_{\text{src}} = 5$ job/min, and with a number of users in the range $20 \leq N_{\text{users}} \leq 50$, and a probability of finding a result in cache $p_{\text{inCache}} = 0.8$. Both storage and filtering have been properly sized to handle the considered load, respectively with $K_{\text{storage}} = 10$ and $K_{\text{filter}} = 20$; moreover, the data catalog, since it requires only very short queries at the beginning of each transaction, does pose particular threats to the system and its initial configuration can be maintained for all the considered workload range. The application server and the cache are the two components that need to be addressed in this case (see Fig. 7): this time we consider *vertical scaling*. Instead of requesting a larger number of virtual machines to support these components, we imagine to migrate them to server with higher capacity, ranging from 1.33× to 2×. As the application server power is increased, the load to the cache becomes higher, and the bottleneck switches when it is migrated to a system 1.67× faster. To increase the performance of the system then, the cache increased at least to a system 1.33×, moving the bottleneck back to the application server. A subsequent improvement to the latter (2×), has then the effect of moving the bottleneck to the filtering component. This complex sequence of bottleneck switches among the resources shows the importance of addressing Data Lake architectures with a modelling approach, to follow workload evolution which would be otherwise very hard to handle. Figure 7 also shows the effect of the vertical scaling both on the response time and on the throughput of the users.

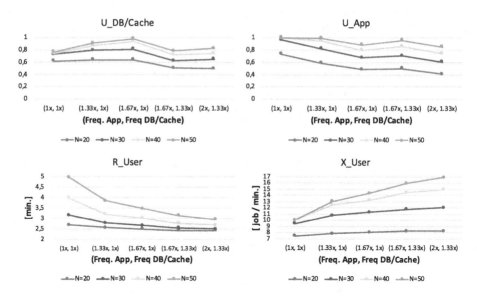

Fig. 7. Vertical scaling of Application Server and DB/Cache with high load workload. The speed of the servers are in the (application server, DB/Cache) order.

5 Related Work

To the best of our knowledge, scientific literature discuss a few approaches regarding Data Lakes performance, mostly from a qualitative perspective. In [13], the authors provide an interesting abstraction by claiming that a data chunk corresponds to a mass and density, two quantities that make difficult to move data from one location to another, therefore introducing an interesting metaphor with the concept of gravity. The notion of *Personal Data Lake* is described as data mass with gravitational pull; the performance of such a system depends on a fast data base system and ultimately on the collections of the corresponding metadata. A qualitative approach measurement of a Data Lake performance, is discussed in [5], where data quality, flexibility, acquisition, access time to raw data, preservation and agility are considered the main parameters to monitor.

In [3], the architecture of a Data Lake is presented, including the performance study of a prototype. The analysis is performed by measuring the data access time on different workflows (CPU and I/O bound respectively, the latter performing worse).

With regard to modeling, the work in [11] presents a metadata *vault* model, i.e. a data model facilitating schema evolutions, to manage Data Lake metadata. The conceptual metadata vault is translated by the authors into logical and physical models. Interestingly, the authors seek to measure the robustness of the metadata model in case the source data scale up, showing that their approach is rather promising.

6 Conclusions and Future Work

Modeling techniques are of paramount importance to recognize flaws and bottlenecks within a proposed system architecture.

With regard to the Data Lake model discussed in Sect. 4.4, it has been shown that caching determines a relevant impact on the load of both storage and filtering. Furthermore, an increase in performance of the system, explainable by caching, shifts the bottleneck to the query application server. Furthermore, another scenario has shown an interesting sequence of bottleneck switches more difficult to determine without the deployment of modeling techniques.

Future work will address the performance analysis of more complex aspects of Data Lake architectures with respect to different domains, such as banking and air traffic control.

References

1. Bertoli, M., Casale, G., Serazzi, G.: JMT: performance engineering tools for system modeling. SIGMETRICS Perform. Eval. Rev. **36**(4), 10–15 (2009). https://doi.org/10.1145/1530873.1530877
2. Bian, H., Chandra, B., Mytilinis, I., Ailamaki, A.: Storage management in smart data lake. In: EDBT/ICDT Workshops (2021)
3. Bird, I., Campana, S., Girone, M., Espinal, X., McCance, G., Schovancová, J.: Architecture and prototype of a WLCG data lake for HL-LHC. In: EPJ Web of Conferences, vol. 214, p. 04024. EDP Sciences (2019)
4. Chessell, M., Scheepers, F., Nguyen, N., van Kessel, R., van der Starre, R.: Governing and managing big data for analytics and decision makers (2014)
5. Derakhshannia, M., Gervet, C., Hajj-Hassan, H., Laurent, A., Martin, A.: Data lake governance: Towards a systemic and natural ecosystem analogy. Future Internet **12**(8), 126 (2020)
6. Ghemawat, S., Gobioff, H., Leung, S.T.: The Google file system. In: Proceedings of the Nineteenth ACM Symposium on Operating Systems Principles, pp. 29–43 (2003)
7. Hai, R., Quix, C., Jarke, M.: Data lake concept and systems: a survey. CoRR abs/2106.09592 (2021). https://arxiv.org/abs/2106.09592
8. Inmon, B.: Data Lake Architecture: Designing the Data Lake and Avoiding the Garbage Dump. Technics Publications (2016)
9. Miloslavskaya, N., Tolstoy, A.: Big data, fast data and data lake concepts. Procedia Comput. Sci. **88**, 300–305 (2016). https://www.sciencedirect.com/science/article/pii/S1877050916316957, 7th Annual International Conference on Biologically Inspired Cognitive Architectures, BICA 2016, held July 16 to July 19, 2016 in New York City, NY, USA
10. Nargesian, F., Zhu, E., Miller, R.J., Pu, K.Q., Arocena, P.C.: Data lake management: challenges and opportunities. Proc. VLDB Endow. **12**(12), 1986–1989 (2019). https://doi.org/10.14778/3352063.3352116
11. Nogueira, I.D., Romdhane, M., Darmont, J.: Modeling data lake metadata with a data vault. In: Proceedings of the 22nd International Database Engineering & Applications Symposium, pp. 253–261 (2018)

12. Ramakrishnan, R., et al.: Azure data lake store: a hyperscale distributed file service for big data analytics. In: Proceedings of the 2017 ACM International Conference on Management of Data, pp. 51–63 (2017)
13. Walker, C., Alrehamy, H.: Personal data lake with data gravity pull. In: 2015 IEEE Fifth International Conference on Big Data and Cloud Computing, pp. 160–167. IEEE (2015)

Computing Bounds for Delay
in a Stochastic Network

Jean-Michel Fourneau[1(✉)], Edouard Longueville[1], Yann Ben Maissa[2],
Loubna Echabbi[2], and Houda Lotfi[2]

[1] DAVID, UVSQ, Univ. Paris-Saclay, Versailles, France
Jean-Michel.Fourneau@uvsq.fr
[2] Telecommunications Systems, Networks and Services Lab, National Institute
of Posts and Telecommunications, INPT, Rabat, Morocco
{benmaissa,echabbi}@inpt.ac.ma

Abstract. We consider a stochastic network where the arcs are associated to discrete random variables which represent the delay. We need to compute the shortest delay (or equivalently the distance) from the source to the sink in the network. Due to the randomness, this problem is known to be hard while it has many polynomial algorithms when the arcs have deterministic lengths (or durations). We provide three approaches and we prove algorithms to obtain stochastic bounds for the distribution of the distance. We present several examples to compare the precision and the time. The approach based on the association of random variables gives very accurate results on the examples and has the smallest complexity.

1 Introduction

In the transportation systems in smart cities, due to the large number of sensors available, we collect a huge volume of data. The data could not be seen as deterministic anymore and we have to deal with the apparent randomness of our measures due to noise, contention, incidents. Here we propose a method to deal with this randomness for a classical problem: the computation of the distance between two nodes.

We consider a directed graph $G = (V, E)$ (digraph) which does not contain any directed cycle, such that each arc (i, j) is associated with a random delay (or distance) $W(i, j)$ to join j from i. These random variables (r.v. in the following) will be denoted as W_m where m is the arc label. We assume that these random variables are discrete and their supports S_m (m is the index of the r.v.) are finite subsets of $\mathbb{R}^+$. We also assume that these random variables are independent. As the digraph does not contain any directed cycle, it is associated to a topological ordering of the nodes. The graph contains N nodes and M directed edges (or arcs). Among these nodes, we distinguish the first node (labelled s and numbered 1). We want to compute the distance (or delay) between s and any node t in the graph. Let $X = (X_1, .., X_t, .., X_N)$ be the random variables associated with the distance or the delay from s to all the nodes t. By construction, X_s has a distribution with a single atom in 0 associated with a probability equal to 1.

© Springer Nature Switzerland AG 2021
P. Ballarini et al. (Eds.): EPEW 2021/ASMTA 2021, LNCS 13104, pp. 131–147, 2021.
https://doi.org/10.1007/978-3-030-91825-5_8

Let $\mathcal{P}_G(s,t)$ be the set of paths $P(t)$ from s to t in the graph G. We assume without loss of generality that $\mathcal{P}_G(s,t)$ is not empty. As G is a Directed Acyclic Graph (or DAG), there exists a finite number of paths from s to t and these paths have a finite number of edges. Let $L(P(t))$ be the delay to reach t departing from s, we can define:

$$L(P(t)) = \sum_{(a,b) \in P(t)} W(a,b), \tag{1}$$

and

$$d(s,t) = Min_{P \in \mathcal{P}_G(s,t)} L(P). \tag{2}$$

Computing $d(s,t)$ is a difficult problem due to the randomness of $W(i,j)$ while many polynomial algorithms exist in the deterministic case. The two reasons are the size of the resulting distributions and the dependence of the path lengths when they share an arc. Indeed, even if the arcs lengths are supposed to be independent, the paths lengths are not independent. Therefore a simple computation of addition and minimum requires conditioning. Furthermore, a convolution of two distributions with size S (associated with the addition of the independent r.v.) may lead to a distribution with size S^2. Thus each new arc added in a path may geometrically increase the number of atoms in the resulting distribution. Such problem was ignored in a recent approach [6] where distributions were modeled by polynomials: during the computation the number of monomials increases geometrically and the approach quickly becomes intractable.

A simple algorithm (with non polynomial complexity) can however be designed, using conditioning on the random variables to solve the problem for small instances when the discrete variables take values in very small sets. It is sufficient to use the Total Probability Theorem after conditioning on the states of all the random variables. Clearly

$$Pr(d(s,t) = k) = \sum_{k_1, k_2, ..., k_M} 1_{D(s,t,k_1,k_2,...,k_M)=k} \prod_{m=1}^{M} Pr(W_m = k_m)$$

where $D(s,t,k_1,k_2,...,k_M)$ is the distance from s to t when the length of arc m is k_m. $D(s,t,k_1,k_2,...,k_M)$ can be obtained by any deterministic algorithm (with complexity $O(M)$) to compute the distances in a directed graph. Clearly the complexity of this approach is $M \prod_{m=1}^{M} |S_m|$ (see [2] for a survey on the complexity for various delays and flow problems for networks or graphs with random discrete costs or durations).

Here we develop several algorithms to derive stochastic bounds on the distribution of the distance. These bounds are easier to compute and they provide guarantee on the expectation and for some of them on the tail of the distribution. The technical part of the paper is as follows. In Sect. 2, we introduce strong stochastic bounds and increasing concave bounds. Based on monotonicity of $d(s,t)$, we propose two algorithms relying on a reduction of the size of the supports of $W(i,j)$. This first approach based on the fact that the distance is an increasing and concave function was already used to bound max flow problems for stochastic networks [3]. Section 3 is devoted to associated random variables to present a new and more efficient approach. We prove that the distances between nodes are associated as some paths share arcs. Then we propose an algorithm to obtain lower bounds on the distance based on this property. We also develop an upper bound algorithm which relies on arc disjoint paths. To the best of our knowledge associated random variables have been considered for PERT networks (Program Evaluation and Review Technic) but their application to the shortest distance problem is original. These algorithms are then numerically compared on some examples in Sect. 4. The algorithms will be integrated in the next version of the XBorne tool [5].

2   Stochastic Bounds Based on Stochastic Monotonicity

The complexity of the exact calculation of the distribution comes from the number of atoms. Therefore it is appealing to derive bounds when we decrease the number of atoms. In [1] we have proposed some methods which keep some quantitative and qualitative information on the results after a reduction of the number of atoms. This is obtained through the use of stochastic orderings. We begin with the definition of the orders we will use in this paper (see [7] for more information).

Definition 1 (strong stochastic ordering). *Let X and Y be two random variables, $X <_{st} Y$ if for all increasing functions Φ, $E[\phi(X)] \leq E[\phi(Y)]$ if the expectations exist.*

Remark that for all r.v. X we have $X <_{st} X$. Let us denote by $X =_{st} Y$ the equality in distribution of random variables X and Y. The stochastic comparison of random variables also implies an inequality between their expectations as seen below.

Property 1. *Let X and Y be two random variables, such that $X <_{st} Y$. Then $E[X] \leq E[Y]$. Furthermore, if $E[X] = E[Y]$ then $X =_{st} Y$.*

We also use some orders based on the variability of the random variables to obtain tighter bounds. Let us first consider convex order $\preceq_{cx}$ which are defined as follows.

Definition 2 (stochastic convex ordering). *Let X and Y be two random variables, $X \preceq_{cx} Y$ if $E[X] = E[Y]$ and for all convex functions ϕ, $E[\phi(X)] \leq E[\phi(Y)]$ if the expectations exist.*

Here we will use the concave ordering $\preceq_{cv}$ which is easily derived from the convex ordering.

Definition 3 (stochastic concave ordering). *Let X and Y be two random variables, $X \preceq_{cv} Y$ if $Y \preceq_{cx} X$*

And finally,

Definition 4 (increasing concave ordering). *Let X and Y be two random variables, $X \preceq_{icv} Y$ if for all increasing concave functions ϕ, $E[\phi(X)] \le E[\phi(Y)]$ if the expectations exist.*

The distance from s to t in the network is define as the minimum of the path lengths from s to t. And the length of a path is the sum of the length of the arcs in the path. Therefore the distance is defined using operators "Min" and "+". And both operators are increasing and concave. More formally, we can define:

$$d(s,t) = f(W_1, W_2, ..., W_M),$$

and we know that f is increasing and concave. We now define the monotonicity for some ordering and we mention the key property for this approach.

Definition 5 (Ψ–Monotony). *A function f is Ψ–monotone if for all random variables X and Y such that $X \preceq_\psi Y$, then $f(X) \preceq_\psi f(Y)$.*

Due to the definitions of the orderings we considered by sets of functions, the following property holds:

Property 2. *If function f is increasing, then it is st – monotone. Similarly, if function f is increasing and concave then it is monotone for the increasing concave ordering.*

Algorithms for st-bounds: Computing st-bounds of $d(s,t)$ is very simple (see [3] for more details). It is sufficient to replace the distributions for the length of an arc by an "st" bound of this distribution as stated in Algorithm 2. If we consider a smaller support, the bound will be easier to compute. The first step consists in building "st" bounds (upper and lower for the input distributions of W_m. This can be done with a very simple algorithm we now describe:

Algorithm 1. Simple "st" Bounds for input discrete distributions

---

1: Choose the size of the bounding distribution (say K) for W_m. Of course $K < S_m$.
2: Divide the set S_m into K proper subsets $S_m^{(i)}$ (i between 1 and K) such that all the atoms in $S_m^{(i)}$ are smaller than atoms in $S_m^{(i+1)}$. Let $l_m^{(i)}$ (resp. $u_m^{(i)}$) be the smallest (resp. largest) atom in $S_m^{(i)}$.
3: The distribution with K atoms $l_m^{(i)}$ with probability $\sum_{a \in S_m^{(i)}} Pr(W_m = a)$ is a lower st-bound of W_m.
4: Similarly the distribution with K atoms $u_m^{(i)}$ and the same probability vector is an upper st-bound of W_m.

In Algorithm 2, we give a short presentation of the algorithm for st-lower bounds. The upper bounds are obtained by similar arguments. We also present in Table 1 and Table 2, the results for the first example we give in Sect. 4. Remark that upper and lower bounds share the same probability vector but they do not have the same support.

Algorithm 2. St Bounds for the distance distributions.

1: Derive L_m stochastic lower bound for W_m, for all $m \in [1..M]$ with the former algorithm.

2: Compute $f(L1, ..., L_M)$ by conditioning and the total probability theorem. The stochastic monotonicity implies that $f(L1, ..., L_M) <_{st} d(s, t)$.

Algorithms for Increasing Concave Bounds: The approach is similar. It is sufficient to build "icv"-bounds of the input distributions. However the derivation of these bounds differs significantly from the st-bound algorithm (see [3] for the details).

Algorithm 3. "icv" Bounds for the distance distributions.

1: **while** The number of atoms in the bound of distribution W_m is larger than the objective **do**

2:    To obtain an upper bound, consider two atoms in the actual distribution, replace these atoms by a new atom which is the barycenter of the atoms.

3:    To obtain a lower bound, consider a subset of at least three atoms in the actual distribution and replace them with the two extreme atoms keeping the same expectation.

4: **end while**

Again, for the sake of conciseness we just state the algorithm for "icv"-lower bounds.

Algorithm 4. "icv" lower bounds for the distance distributions.

1: Derive V_m increasing concave lower bound for W_m, for all $m \in [1..M]$.

2: Compute $f(V1, ..., V_M)$ by conditioning and the total probability theorem. The increasing concave monotonicity implies that $f(V1, ..., V_M) \preceq_{icv} d(s, t)$.

Corollary 1. *Replacing all random variables $W(i, j)$ by their expectation provides an increasing concave upper bound for the distance (like in Jensen inequality). Thus, this strategy to eliminate the randomness has a systematic bias. In the following, this bound will be denoted as Fulkerson bound.*

Property 3 (Complexity). *Let M be the number of arcs in the graph. Assuming that all input distributions have size $|S|$, that we compress the distributions to obtain input bounds with size K, then the algorithm needs $O(M |S|)$ operations to derive the input bounds and $O(M 2^K)$ to get the bounds on the distance.*

Note that both approaches simplify the computation to obtain bounds and that we obtain distinct bounds when we change the partition (finding the most

Table 1. Input Distributions, model 1.

Arcs	Atoms	Probability vector
e0, e2, e4	{2 5 8 10 12 15 18 20}	{0.1 0.1 0.1 0.2 0.1 0.1 0.1 0.2}
e1, e11	{5 10 15 20 25 30 35 40}	{0.1 0.2 0.1 0.2 0.1 0.1 0.1 0.1}
e3	{1 2 3 4 7 8 9 12}	{0.2 0.1 0.2 0.1 0.1 0.1 0.1 0.1}
e5, e10	{2 3 7 8 10 11 15 16}	{0.1 0.1 0.2 0.1 0.1 0.2 0.1 0.1}
e6	{1 2 3 4 8 9 10 13}	{0.1 0.2 0.1 0.1 0.2 0.1 0.1 0.1}
e7	{1 2 3 4 6 9 11 16}	{0.1 0.2 0.1 0.1 0.1 0.1 0.1 0.2}
e8, e9	{3 10 15 20 22 25 30 35}	{0.1 0.3 0.1 0.1 0.1 0.1 0.1 0.1}

Table 2. Stochastic Bounds with two atoms.

Arc	Atoms "st"	lower	Probability		Atoms "st"	upper	Atoms "icv"	lower	Probability		Atoms "icv"	upper	Probability	
e0, e2, e4	2	12	0.5	0.5	10	20	2	20	0.444	0.556	10	15	0.6	0.4
e1, e11	5	25	0.6	0.4	20	40	5	40	0.543	0.457	15	30	0.6	0.4
e3	1	7	0.6	0.4	4	12	1	12	0.636	0.364	2	7	0.4	0.6
e5, e10	2	10	0.5	0.5	8	16	2	16	0.5	0.5	8	13	0.8	0.2
e6	1	8	0.5	0.5	4	13	1	13	0.583	0.417	6	11	0.8	0.2
e7	1	6	0.5	0.5	4	16	1	16	0.6	0.4	3	13	0.6	0.4
e8, e9	3	22	0.6	0.4	20	35	3	35	0.531	0.469	15	30	0.8	0.2

accurate partition and bound is still an open problem). However the complexity remains exponential if we consider distributions which have more than one atom. The theory of associated random vectors will be used to derive less complex bounds. In many cases they are also much more accurate (at least for the lower bound).

3 Bounds Based on Associated Random Vectors

For a more detailed presentation of associated random variables, see [4].

Definition 1. *The random variables $X_1, .., X_n$ are associated if, given two coordinatewise nondecreasing functions f and g: $\mathbb{R}^n \to \mathbb{R}$,*

$$Cov(f(X_1, .., X_n), g(X_1, .., X_n)) \geq 0.$$

Remark 1. One can also consider non increasing functions, as $Cov(f(X), g(X)) = Cov(-f(X), -g(X))$ and f is non increasing implies that $-f$ is non decreasing.

Remark 2. As $Cov(f(X), g(X)) = \mathbb{E}[f(X)g(X)] - \mathbb{E}[f(X)]\mathbb{E}[g(Y)]$, one can also define an associated random vector as $\mathbb{E}[f(X)g(X)] \geq \mathbb{E}[f(X)]\mathbb{E}[g(X)]$.

As a large part of the theory came from reliability, one has more results when the r.v. are Boolean.

Property 4 (Barlow et Proschan). *Let $X_1, ..., X_n$ be n Boolean r.v., then*

$$Pr[(\prod_i X_i) = 1] \geq \prod_i Pr[X_i = 1]$$

and

$$Pr[(\max_i X_i) = 1] \geq \max_i Pr[X_i = 1]$$

Taking into account that the r.v. are boolean, we have: $(\max_i X_i) = 1 - \prod_i(1 - X_i)$. This elementary property is used to derive useful inequalities.

Property 5. *Let X a random vector with n associated random variables. We have:*

$$Pr(X_1 > x_1, X_2 > x_2, .., X_n > x_n)) \geq \prod_k Pr(X_k > x_k),$$

and

$$Pr(X_1 \leq x_1, X_2 \leq x_2, .., X_n \leq x_n)) \geq \prod_k Pr(X_k \leq x_k).$$

Proof: we define $T_i(t) = 1_{X_i > t}$ for all i. T_i is increasing in X_i and is a boolean. The random vector (X_i) is associated. Thus random vector (T_i) is also associated. We apply both inequalities of Property 4 on T_i to prove inequalities on X.

Corollary 2. *Thus, $Pr(min_i(X_i) > x) \geq \prod_k Pr(X_i > x)$.*

Proof: it is sufficient to remark that $Pr(min_i(X_i) > x) = Pr(X_1 > x, X_2 > x, .., X_n > x))$.

To the best of our knowledge, proving an algorithm to check that discrete random variables are associated is still an open problem. Therefore association is proved using the following properties (see the next section on distance in a stochastic network). Starting with a given set of associated random variables (independence is useful here), it is rather simple to obtain new families of associated random variables with increasing transformation.

Property 6. *Every random variable X is associated with itself.*

Property 7. *If $X = (X_1, X_2, ..., X_N)$ is a random vector such that all the X_i are independent random variables, then X is associated.*

Property 8. *Every subset of an associated random vector is associated.*

Property 9. *Let $X_1, X_2, ... X_n$ be independent of the variables $Y_1, Y_2, ..., Y_m$. Assume that $X_1, X_2, ... X_n$ are associated random variables. Assume also that $Y_1, Y_2, ..., Y_m$ are associated random variables. Then $X_1, X_2, ... X_n, Y_1, Y_2, ..., Y_m$ are associated random variables.*

Property 10. *Let $X_1, X_2, ... X_n$ be n associated random variables, and consider coordinatewise nondecreasing functions $f_1, .., f_K \colon \mathbb{R}^n \to \mathbb{R}$. Then random variables $Y_1 = f_1(X_1, X_2, ...X_n), .., f_K(X_1, X_2, ...X_n)$, are associated.*

3.1   Links with $<_{st}$ Ordering and Independence

Notation 1. *Let $X = (X_1, .., X_n)$ an associated random vector. For all i between 1 and n, let us denote by $\overline{X_i}$ a random variable such that $\overline{X_i} =_{st} X_i$ while $\overline{X_i}$ and $\overline{X_j}$ are independent for all $j \neq i$. $\overline{X_i}$ will be denoted as the independent version of X_i.*

We begin with a well-known property which was used in [8,9] to obtain bounds for PERT networks. As we do not use "Max" operator in this paper we do not give the proof of this property.

Property 11. *Let $X = (X_1, .., X_n)$ an associated random vector. Then,*

$$Pr(max_k(X_k)) \geq Pr(max_k(\overline{X_k})).$$

Equivalently $max_k(X_k) <_{st} max_k(\overline{X_k})$. The independent versions of the random variables provide a strong stochastic upper bound of the max (a guarantee).

Here, the operator we use for the path length is the "Min" operator. Thus we derive a similar result for the distance.

Property 12. *Let $X = (X_1, .., X_n)$ an associated random vector. Then, $min_k(\overline{X_k}) <_{st} min_k(X_k)$. The independent versions of the random variables provide a strong stochastic lower bound of the min.*

Proof: We consider the first relation of Property 5, taking for all i $x_i = x$.

$$Pr(X_1 > x, X_2 > x, .., X_n > x)) \geq \prod_k Pr(X_k > x).$$

The left part of the inequality is $Pr(min_k(X_k) > x))$ while the right part is:

$$\prod_k Pr(X_k > x) = \prod_k Pr(\overline{X_k} > x) = Pr(\overline{X_1} > x, \overline{X_2} > x, .., \overline{X_n} > x) = Pr(min_k(\overline{X_k}) > x).$$

Thus,

$$Pr(min_k(X_k) > x)) \geq Pr(min_k(\overline{X_k}) > x).$$

or equivalently

$$min_k(\overline{X_k}) <_{st} min_k(X_k).$$

3.2   Distance and Association

We now prove that in a stochastic network with independent random variables for the length, the distances between nodes are associated random variables. We begin with a technical lemma which also provides an intuition about associated random variables.

Lemma 1. *Let Y, $Z1$ and $Z2$ three independent r.v., then $Y + Z1$ and $Y + Z2$ are associated.*

Proof: Y, $Z1$ and $Z2$ are independent. Thus $(Y, Z1, Z2)$ is an associated random vector. We consider functions $f1$, $f2$ and $f3$: $\mathbb{R}^3 \to \mathbb{R}$ defined by:

$$f1(Y, Z1, Z2) = Y, \quad f2(Y, Z1, Z2) = Y + Z1, \quad f3(Y, Z1, Z2) = Y + Z2.$$

Clearly these three functions are increasing. Thus according to Property 10, random vector $(f1(Y, Z1, Z2), f2(Y, Z1, Z2), f3(Y, Z1, Z2)) = (Y, Y + Z1, Y + Z2)$ is associated. Due to Property 8, $(Y + Z1, Y + Z2)$ is also associated.

Property 13. *Let $Y_P = L(P(t))$ a r.v. equal to the length of path $P(t)$ from s to t in digraph G, $Y = (Y_P)_{P \in \mathcal{P}_G(s,t)}$ is an associated random vector.*

Proof: Let $P1$ et $P2$ two paths from s to t. We have two cases to consider:

1. $P1$ and $P2$ are arc-disjoint.
2. The intersection of $P1$ and $P2$ contains some arcs.

In the first case, $L(P1)$ and $L(P2)$ are independent as they are summations of distinct independent random variables (the length of the arcs which belong to the path). As they are independent, they are also associated according to Property 7.

Now assume that the intersection of $P1$ and $P2$ contains some arcs and let $Q = P1 \cap P2$. As the "+" operator in the definition of the path length is commutative, one can separate each path into two subsets of arcs such that

$$L(P1) = L(Q) + L(P1 \setminus Q) \quad and \quad L(P2) = L(Q) + L(P2 \setminus Q)$$

As $Q = P1 \cap P2$, we have $(P1 \setminus Q) \cap (P2 \setminus Q) = \emptyset$, thus $L(P1 \setminus Q)$ and $L(P2 \setminus Q)$ are independent random variables. Similarly $L(Q)$ is independent of these two r.v. and we can apply Lemma 1 to prove that $L(P1)$ and $L(P2)$ are associated.

3.3   Algorithm for a Lower Bound

Let $d^i(j)$ represent the distance from node 1 to node j where all the nodes between 1 and i have been taken into account to find paths. In general we cannot compute the distribution of $d^i(j)$ because of Property 13 without conditioning and using an exponential complexity algorithm. Instead we compute a stochastic bound $l^i(j)$. The algorithm proceeds by iteration adding one node at each iteration. Each node allows to add new paths and to decrease the shortest distance computed so far. Remember that the directed graph is a DAG and we use this property to consider the nodes in the topological order associated to the DAG. Note that $d(s,t) = d^N(t)$ (i.e. we have considered all the nodes to build paths from 1 to t).

By definition $\overline{l^{k-1}(u)}$ has the same distribution and is independent. Furthermore we know how to numerically compute the minimum of two independent distributions.

Theorem 1. *$l^k(u)$ is an "st" lower bound for the distance: for all node indices k and u, $l^k(u) <_{st} d^k(u)$. And $l^N(t) <_{st} d(s,t)$.*

Algorithm 5. St lower bounds based on association of paths.

1: Init $l^1(u) = W(1, u)$ for all nodes u which are neighbor of node $s = 1$.
2: **for all** k (node number) from 2 to N **do**
3:     **for all** node u neighbors of node $k - 1$ **do**
4:         let $l^k(u) = min\left(\overline{l^{k-1}(u), l^{k-1}(k-1) + W(k-1, u)}\right)$
5:     **end for**
6: **end for**

Proof: by induction on k.

- $k = 1$. We consider the neighbors (say u) of node 1. The distance to reach node u is exactly the length of the arcs $W(1, u)$. Therefore $d^1(u) = W(1, u)$ and thus $l^1(u) <_{st} d^1(u)$ for all these nodes u.
- $k \Rightarrow k + 1$. Adding new arc from k to u allows to decrease the distance (see Fig. 1). By construction we have:

$$d^k(u) = min\left(d^{k-1}(u), d^{k-1}(k-1) + W(k-1, u)\right).$$

By Property 13, we know that the path lengths are associated. Thus for all u,

$$min\left(\overline{d^{k-1}(u)}, \overline{d^{k-1}(k-1) + W(k-1, u)}\right) <_{st} d^k(u)$$

As "min" and "+" are increasing functions, if $x <_{st} y$ and z independent of y and x, then $min(z, x) <_{st} min(z, y)$ and $x + z <_{st} y + z$. By induction, we have: $l^{k-1}(k-1) <_{st} d^{k-1}(k-1)$. Therefore:

$$min\left(\overline{l^{k-1}(u), l^{k-1}(k-1) + W(k-1, u)}\right) <_{st}$$
$$min\left(\overline{d^{k-1}(u), d^{k-1}(k-1) + W(k-1, u)}\right).$$

By transitivity: $min\left(\overline{l^{k-1}(u), l^{k-1}(k-1) + W(k-1, u)}\right) <_{st} d^k(u)$, and finally, $l^k(u) <_{st} d^k(u)$.

The two basic operations in Algorithm 5 is the addition and the minimum of two independent random variables. These operations obviously depend on the size of the discrete distributions. Note that each convolution operation leads to an increase of the size of the distributions and this size has to be controlled to avoid that the last steps of the algorithm deals with very large distributions.

Corollary 3. *In Algorithm 5, after each computation of $l^k(u)$, we replace $l^k(u)$ by a "st" lower bound of $l^k(u)$ with at most K atoms. The algorithm still compute a "st" lower bound of $d(s, t)$ by transitivity.*

Assume that the distributions are represented by sorted lists with size $d1$ and $d2$ and let $d = max(d1, d2)$. The minimum can be computed in $O(d)$ operations and the output distribution has a size smaller than d. The addition of independent random variables is associated with the convolution of their distributions. Efficient algorithms exist with a complexity in $O(d \, log(d))$ and the output has size which is at most d^2.

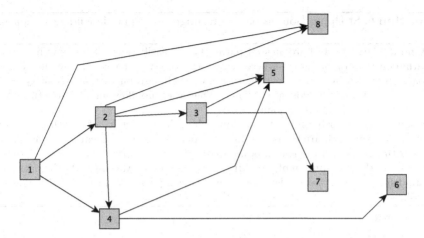

Fig. 1. Considering node 4 allows to modify the distance distribution to node 5 and to initialize the distance distribution to node 6.

Property 14 (Complexity of Algorithm 5**).** *Assume that at each step of the algorithm, the distributions have size K. Each step requires $O(K \, logK)$ operations and after calculation, $l^k(u)$ has size at most K^2 (due to the convolution). The extra step of lower bounds introduced in Corollary 3 has a linear complexity in K^2. And the number of iterations in the nested loops of the algorithm is M. Therefore the complexity is $O(M \, K^2)$.*

3.4   Algorithms for an Upper Bound

We combine two arguments:

- When the two paths are arc disjoint, the lengths of the paths are independent (see the first part of the proof of Property 13) and we know how to compute them.
- If we only consider a subset of paths, we compute an "st" upper bound. More formally, if $\mathcal{A}(s,t) \subset \mathcal{P}(s,t)$, then

$$d(s,t) = Min_{P \in \mathcal{P}(s,t)} L(P) <_{st} m(s,t) = Min_{P \in \mathcal{A}(s,t)} L(P).$$

Thus, the algorithm consists in finding arc-disjoint paths (the largest set) and computing the distance associated with this set of paths.

Property 15 (Complexity). *First step requires $O(F \, M)$ operations to obtain F arc disjoint paths and $F < M$. During step 2, the distribution of the distance is obtained by the convolution of independent random variables of size smaller than K. Each convolution is followed by a compression of the resulting distribution (whose size is at most K^2). Let D the diameter of the graph, the complexity of this step is $O(F \, D \, K^2)$ using the same arguments as in Property 14. The last step consists in computing the minimum of F distributions with size K and this*

Algorithm 6. St Upper Bounds based on subgraphs and independence of paths.

1: Find the largest set of arc-disjoint paths. This is easily done with a max flow algorithm on the original graph where all the arcs receive capacity 1. The augmenting paths provided by a max flow algorithm, on such a graph with arc capacity equal to 1, are by construction arc disjoint. Thus the max-flow algorithm returns the maximal number of arc disjoint paths.
2: Compute the distribution of length for each path. The length of a path is the sum of arc length which are by assumption independent. Thus one can use convolution algorithm to obtain the length of any path.
3: Compute the distribution of the "Min" of these random variables. As the paths are arc-disjoint, the r.v. are independent and the computation of the "Min" is an easy task.

---

can be done with $O(F\ K\ ln(K))$ operations. Therefore, the total complexity is $O(F\ D\ K^2)$.

4 Numerical Results

We begin with a toy example where it is feasible to obtain the exact solution to check the accuracy of the methods. Even if it is possible, the exact algorithm is really time consuming. The graph has 8 nodes and 12 edges (see Fig. 2). All the distributions have 8 atoms. The input distributions are given in Table 1. The "st" bounds and "icv" bounds with two atoms for these distributions are given in Table 2. These distributions with 4 atoms will be given in an appendix of the full paper.

Table 3. Expectation of the distributions (Model1).

Method		Association	St Monotonicity		"icv" Monotonicity		Fulkerson	Exact
			2 atoms	4 atoms	2 atoms	4 atoms		
Exp.	Lower bound	19.35	10.88	16.19	15.02	18.80	.	19.385
	Upper bound	26	28.29	21.99	21.23	20.83	23	.

We present in Fig. 3 the cdf of the distributions. We only present the exact result, the "st" bounds based on association and the "st" upper and lower bounds with 2 and 4 atoms in the input bounds. We do not draw the "icv" bounds to avoid confusion. Note that as the "st" bounds are not all based on the same strategy, the cdf may cross (for instance the purple and the red curves). The most important point to remark here is the tightness of the lower bound based on association.

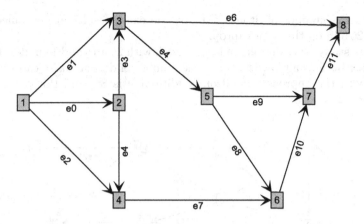

Fig. 2. Graph for toy example.

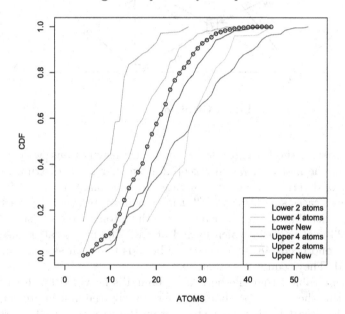

Fig. 3. Stochastic bounds and exact results for toy model. The exact results are depicted as black dots. Note that the new lower bounds (the blue curve) are close to the exact results

To compare all these strategies, we compute the expectation of the distance (in Table 3). The algorithms are very fast for this small graph: computing the distribution with the algorithms based on association need 0.01 s on an ordinary laptop, while the bounds based on bounding input distributions atoms needs 0.02 s (resp. 14s) for bounds with 2 (resp. 4) atoms. Note that computing "st" bounds or "icv" bounds requires the same time as we deal with distributions with the same number of atoms. Finally, the exact results are obtained after 17 h and 26 min on the same laptop.

We now study two examples: a larger graph with 26 arcs and 14 nodes (Fig. 4). As the distributions all have 8 atoms, the number of deterministic cases we have to solve with the approach based on conditioning is 8^{26} and this precludes to give the exact solution.

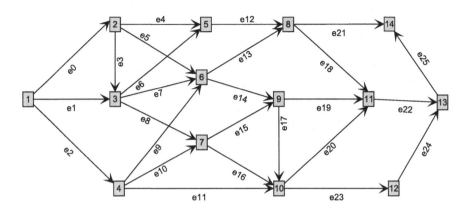

Fig. 4. Graph for Model 2.

For the first method which bounds the input distributions, we consider two strategies. In the first one, we only keep 2 atoms per input distributions. Thus the number of deterministic cases generated by the conditioning is 2^{26}. To keep the number of cases smaller than 2^{30}, the second strategy consists in bounding all the input distributions by distributions with 2 atoms, except distributions for arcs $e22$, $e23$, $e24$ and $e25$ which have 4 atoms. The first strategy needs 3 min while the second needs 48 min. Stochastic bounds based on association are very easily solved. They require less than 1s.

Remark in Table 4 that the best lower bound is provided by the association algorithm. But the best upper bound is obtained by the first approach with 2^{30} deterministic cases. Clearly we have to improve the strategy based on arc-disjoint paths. This point will be developed in the full version of the paper (Fig. 5).

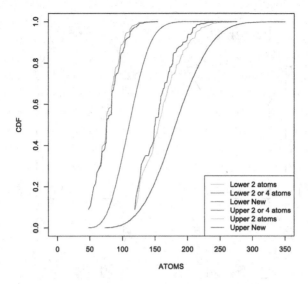

Fig. 5. Probability distribution for the bounds of the distance (Model 2).

Table 4. Bounds of the expectation for Model 2.

Method		Association	Input Bounds: "st"		Input Bounds: "icv"	
			2^{26}	2^{30}	2^{26}	2^{30}
Expectation	Lower bound	111.660	76.868	78.596	92.652	94.198
	Upper bound	183.596	160.149	153.877	132.016	131.109

Now we consider a Series-Parallel (SP for short) digraph (see Fig. 6) with 48 nodes and 54 arcs. Such a graph has a recursive construction and this provides a recursive algorithm to compute the distribution of the distance. Note however that the distributions which appear during the execution of the recursive algorithm still suffer of the size explosion problem due to the convolution operation and we have to bound them if we want to keep their size smaller than K. The arcs all have the same distribution of delay with 8 atoms $(1, 2, 5, 6, 12, 19, 28, 34)$ and probability vector $(0.1, 0.1, 0.05, 0.3, 0.05, 0.1, 0.2, 0.1)$. As the number of arcs is too large, we only report the results based on association. The exact algorithm based on the SP structure only requires 66 s, while the bounding algorithms are faster (1.7 s for the upper bound and 3.8s for the lower bound). The exact expectation is 191.66 while the lower (resp. upper) bounding distribution has an expectation equal to 174.15 (resp. 234.38). This last example (and many others we cannot report here due to the lack of space) shows that the lower bound is still quite accurate. The quality of the upper bound may depend of other graph properties and may be very bad (Fig. 7).

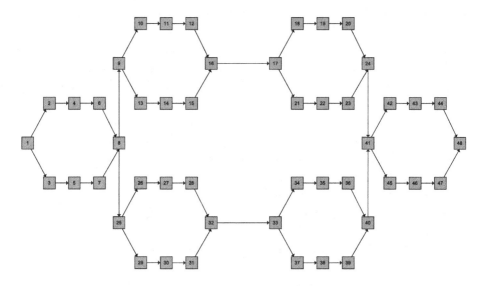

Fig. 6. SP-Graph for Model 3.

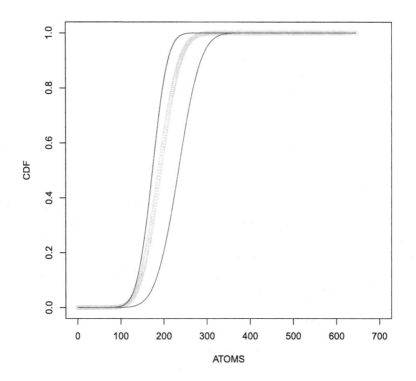

Fig. 7. Probability distribution for the bounds of the distance (Model 3).

5   Concluding Remarks

These two approaches provide some tradeoff between complexity and accuracy of the numerical results. They also suggest new approaches combining association of random variables as well as graph properties that we will present in the extended version of this paper. We have now derived new algorithms based on recursive construction of the graph or a subgraph to combine the "st" bounds approach with the associated r.v. results. We have also generalized some of our results to networks where the delays to cross the arcs are not independent. This allows to study new models of traffic.

Acknowledgments. H. Lotfi was financially supported by CAMPUS FRANCE (PHC TOUBKAL 2019 (French-Morocco bilateral program) Grant Number: 41562UA. This work was partially supported by the Paris Ile-de-France Region through a DIM RFSI grant (EPINE Project).

References

1. Ait Salaht, F., Cohen, J., Castel Taleb, H., Fourneau, J.M., Pekergin, N.: Accuracy vs. complexity: the stochastic bound approach. In: 11th International Workshop on Discrete Event Systems (WODES2012), no. 8 (2012)
2. Ball, M., Colbourn, C., Provan, J.: Network reliability. In: Handbooks in Operation Research and Management Science, vol. 7, pp. 673–762 (1995)
3. Echabbi, L., Fourneau, J., Gacem, O., Lotfi, H., Pekergin, N.: Stochastic bounds for the max flow in a network with discrete random capacities. Electron. Notes Theoret. Comput. Sci. **353**, 77–105 (2020). International Workshop on the Practical Application of Stochastic Modelling (PASM)
4. Esary, J.D., Proschan, F., Walkup, D.W.: Association of random variables, with applications. Ann. Math. Stat. **38**(5), 1466–1474 (1967)
5. Fourneau, J.M., Mahjoub, Y.A.E., Quessette, F., Vekris, D.: XBorne 2016: a brief introduction. In: Czachórski, T., Gelenbe, E., Grochla, K., Lent, R. (eds.) ISCIS 2016. CCIS, vol. 659, pp. 134–141. Springer, Cham (2016). https://doi.org/10.1007/978-3-319-47217-1_15
6. Hastings, K.: Algebraic Approaches to Stochastic Optimization. Ph.D. Clemson University (2012). https://tigerprints.clemson.edu/all_dissertations/935
7. Shaked, M., Shantikumar, J.G.: Stochastic Orders and Their Applications. Academic Press, San Diego (1994)
8. Shogan, A.W.: Bounding distributions for a stochastic pert network. Networks **7**, 359–381 (1977)
9. Yazici-Pekergin, N., Vincent, J.M.: Stochastic bounds on execution times of parallel programs. IEEE Trans. Software Eng. **17**(10), 1005–1012 (1991)

Mixture Density Networks as a General Framework for Estimation and Prediction of Waiting Time Distributions in Queueing Systems

Hung Quoc Nguyen[1] and Tuan Phung-Duc[2(✉)]

[1] Graduate School of Science and Technology, University of Tsukuba,
1-1-1 Tennodai, Tsukuba, Ibaraki 305-8573, Japan
s2030113@s.tsukuba.ac.jp
[2] Faculty of Engineering, Information and Systems, University of Tsukuba,
1-1-1 Tennodai, Tsukuba, Ibaraki 305-8573, Japan
tuan@sk.tsukuba.ac.jp

Abstract. We employ Mixture Density Networks (MDNs) as a general approach for estimation of customer waiting times in queueing systems based on system states that customers observe upon their arrival. We generate a large amount of data by using discrete-event simulation. Part of the generated dataset is used to train the model, and we utilize the whole dataset for the evaluation of the model. Finally, we illustrate this application in a real-world dataset.

Keywords: Queueing systems · Waiting times · Mixture density networks · Discrete-event simulation

1 Introduction

Queueing is a part of everyday life; however, queueing customers are impatient. Psychological studies revealed that waiting in uncertainty over the waiting time seems to become longer [14]. The expected waiting time is also an important measure that affects the decision-making process in most frameworks modeling queues with strategic customers [5,17]. As a matter of fact, it is desirable to have an estimation of waiting times for the sake of not only customers but also platform managers. Queueing theory suggests that when queueing systems are under stationary settings, the waiting time of each customer is a random variable following a probability distribution whose parameters depend on the system state the customer observes upon his arrival. The more information is added to the system states as features, the narrower waiting time distributions become. In other words, it is not really feasible to predict a specific value for

---

Supported by JSPS KAKENHI Grant Number 18K18006 and 21K11765.

P. Ballarini et al. (Eds.): EPEW 2021/ASMTA 2021, LNCS 13104, pp. 148–161, 2021.
https://doi.org/10.1007/978-3-030-91825-5_9

queueing times, especially when less information about system states and fewer data are available.

Under simple settings, waiting time distributions can be handily computed by mathematical procedures if processes have stationary distributions. Whitt [20] proposed the Queue-theory based predictor using Laplace transform and illustrated the method in several examples of systems with exponentially distributed service times. This technique was then followed in multiple later studies and applied in more complex systems, such as service systems with time-varying demand and capacity [12], call centers [11] or call centers with and without reneging customers [13].

In more complicated systems, for example, a $G/G/c$ queue or a queueing system with multiple types of agents, this traditional approach may not be applicable due to the presence of numerous parameters. On the other hand, it is also challenging to fit real-world systems into the framework of queueing theory for two main reasons. First, the queueing theory depends on various assumptions, some of which are usually unrealistic. Second, the information about the system (such as the number of servers or queueing discipline) is not always fully reported. This is when the learning-based approach comes in handy. Several attempts have been made to give a prediction for the queueing time in different systems, such as airport [4], banks [7] or call centers [19]. However, that being mentioned, since the waiting time is stochastic, we would not like to obtain an exact value for the prediction of the queueing time. On the one hand, an overestimation of the actual waiting time may reduce the incentive of customers to join the queue from the beginning. On the other hand, if customers are informed of an underestimation of their actual queueing time, customers may become more impatient during the extra waiting time. Therefore, in many cases, it is more reasonable to derive a distribution of waiting times. Practically, a probability distribution of the waiting time can provide information about a reasonable interval of the actual waiting time.

Motivated by the idea of queueing theory and the power of the learning approach, we employ a machine learning approach called Mixture Density Networks (MDNs) [2] as a general approach for estimation and prediction of customer waiting time in queueing systems based on system states. An MDN is a combination of a neural network and a mixture model output, which outputs conditional probability distributions of the data. Several noticeable applications of MDNs can be named, such as speech generation in voice assistants on smartphones [1], artificial handwriting [9], games [10] and trajectory predictions [15]. Prediction of waiting time distributions in queueing systems using MDNs was once conducted on simulated data with chosen parameters in [18]; however, this study does not leverage the information on system states to give prediction but uses historical waiting times instead. This approach may not be applicable if the time series are subject to discrete breaks in the structure.

In this study, we generate 5 (out of 6) datasets to see how MDNs work on simulated data and illustrate the application of MDNs in different scenarios.

We focus on finding the best-fitted model for a real-world dataset by tuning parameters and designing a proper evaluation method.

The rest of this paper is organized as follows. Section 2 describes datasets and feature selections. In Sect. 3, we summarize the structure of Mixture Density Networks. In Sect. 4, we design evaluation metrics for the model. In Sect. 5, results are presented. Lastly, Sect. 6 contains some concluding remarks.

2   Descriptions of Datasets and Feature Selections

In this section, we describe three datasets that are used to illustrate the performance of MDNs. Three of the four datasets are simulated using Discrete-event simulations, and one is a real-world dataset. All parameters of simulated queueing systems are set in advance for the purpose of simulation only and assumed *unknown* when training models.

Dataset 1: An M/M/1 Queueing System. We simulate an M/M/1 queueing system where both interarrival times and service times follow exponential distributions with rate parameters $\lambda = 4$ and $\mu = 5$, respectively. The simulated dataset includes three features as follows.

- Arriving time: The time when customers arrive at the system.
- Entering time: The time when customers enter the server.
- Leaving time: The time when customers finish and leave the system.

From the above three features, we further process the data. The processed dataset includes two columns: *System state* and *Waiting time*. We try to predict the distributions of waiting times of customers based on the system state they observe upon arrival. In this case, the system state is one-dimensionally represented by the queue length. The simulated dataset contains about $40,000$ observations. The simple setting of this system allows us to derive the ground truth of waiting time distributions, which are Erlang distributions with density functions given as follows

$$f(x) = \frac{\mu^k x^{k-1} e^{-\mu x}}{(k-1)!}, \tag{1}$$

where k denotes the queue length observed by customers upon their arrival.

Dataset 2: M/M/c Queueing Systems. We simulate seven M/M/c queueing systems with the number of servers c ranging from 1 to 7. In this simulation, we fix the service rate at $\mu = 5$. The arrival rates λ are set higher in systems with more servers. Suppose that we are considering employing more servers due to the increasing demand (reflected by arrival rates λ). The simulated dataset includes four features as follows.

- Arriving time: The time when customers arrive at the system.
- Entering time: The time when customers enter the server.

- Leaving time: The time when customers finish and leave the system.
- The number of servers in the system.

From the above three features, we further process the data. The processed dataset includes three columns: *System state*, *Number of servers* and *Waiting time*. We attempt to predict waiting time distributions in the new system from the queue lengths observed by customers upon their arrival (i.e., system states) and the number of servers of the system they enter.

Theoretically, waiting times of customers who arrive at an M/M/c queueing system and observe a queue length of k customers follow an Erlang distribution given as below

$$g(x) = \frac{(c\mu)^k x^{k-1} e^{-c\mu x}}{(k-1)!}. \tag{2}$$

We will use this to evaluate the performance of the model.

Dataset 3: An Erlang-Type G/G/c Queueing System. We simulate an queueing system where interarrival times follow an Erlang distribution with shape parameter $k_1 = 6$ and rate parameter $\lambda_1 = 15$, and service times follow an Erlang distribution with shape parameter $k_2 = 4$ and rate parameter $\mu = 5$. There are 3 identical servers. The simulated dataset includes three features as follows.

- Arriving time: The time when customers arrive at the system.
- Entering time: The time when customers enter the server.
- Leaving time: The time when customers finish and leave the system.

From the above three features, we further process the data. The processed dataset includes two columns: *System state* and *Waiting time*. We try to predict the distributions of waiting times of customers based on the system state they observe upon arrival. In this case, the system state is one-dimensionally represented by the queue length. The simulated dataset contains about $40,000$ observations.

Dataset 4: A Weibull-type G/G/c Queueing System. Similarly to *Dataset 3*, we simulate a queueing system with 3 identical servers but replace the Erlang distributions of interarrival times and service times by Weibull distributions. Interarrival times follow a Weibull distribution with shape parameter $k_1 = 3$ and rate parameter $\lambda_1 = 12$, and service times follow a Weibull distribution with shape parameter $k_2 = 4$ and rate parameter $\mu = 5$. There are 3 identical servers.

Dataset 5: A Preemptive Queueing System with two Classes of Customers. This dataset simulates a cognitive wireless network system considered in [16]. There are two classes of users arriving at a system with $c = 3$ servers: *primary users* and *secondary users*. Interarrival times and service times of both classes of users follow exponential distributions with rate parameters $\lambda_1 = 3, \mu_1 = 6$ (primary users' mean arrival rate and service rate), and $\lambda_2 = 4, \mu_2 = 7$ (secondary users' mean arrival rate and service rate). Behaviors of users are subject to the following discipline.

- If a secondary user arrives and there is at least one available server, the user will occupy the server.
- If a secondary user arrives and there is no available server, the user will wait in a queue.
- If a primary user arrives and there is at least one available server, the user will occupy the server.
- If a primary user arrives and there is at least one available server, the user will occupy the server.
- If a primary user arrives and all servers are occupied by at least one secondary user, a secondary user's session will be interrupted for the primary user to enter and occupy the server. The interrupted secondary user goes back to the queue and waits.
- If a primary user arrives and all servers are occupied by other primary users, that user will leave the system immediately

The simulated dataset includes the following information.

- Arriving time: The time when users arrive at the system.
- Entering time: The time when users enter the server.
- Leaving time: The time when users finish and leave the system.
- Whether a user is a primary or secondary one.

From the above three features, we further process the data. The processed dataset includes three columns: *The number of SUs in the system upon arrival*, *The number of PUs in the system upon arrival* and *Waiting time*.

In this system, the setting is highly complicated and it is not trivial to derive waiting time distributions. The ground truth of distributions, based on the law of large numbers, is obtained by simulating the system within a large amount of time. We extract $90,000$ observations as a part of the dataset to train the model and use the whole large dataset to compare and evaluate the training results.

Dataset 6: Waiting Times and Service Times at Three Banks in Nigeria. This dataset, which is publicly available in [3], includes time recording of customers' activities at three selected banks in Ogun State, Nigeria. Data was collected during four weeks for each bank, containing a total of $52,499$ observations. The dataset includes three features as follows.

- Arriving time: The time when customers arrive at the system.
- Entering time: The time when customers enter the server.
- Leaving time: The time when customers finish and leave the system.

In this dataset, other information such as the number of servers is not reported. From the given information in the dataset, we process and obtain two features describing the system state upon arrivals of each customer: *The number of customers being served* and *The current number of customers waiting in the queue*. Since the data is noisy, we rescale waiting times into the range between 0 and 1, using the following formula

$$z_i = \frac{t_i - \min(\boldsymbol{t})}{\max(\boldsymbol{t}) - \min(\boldsymbol{t})},$$

where $t = (t_1, ..., t_N)$ containing all N observations of waiting times in the dataset, and z_i is now the ith normalized data. This normalization does not ruin the structure of distributions.

We split the available dataset into a training set and a test set containing 50% of data in each.

3 Mixture Density Networks

We are interested in deriving distributions of waiting times of customers. In this study, system parameters are assumed latent and used for the purpose of data generation only. What we really observe are the system state s at each time when a customer arrives and the actual waiting time of the corresponding customer t. Our purpose is to learn the probability densities of t given that s takes a specific value, denoted by $p(t|s)$, which is assumed to be a mixture of k distributions. Note that the type of probability distributions can be arbitrarily chosen. In this paper, we select a Gaussian mixture model because a mixture of Gaussian distributions with enough components is able to universally approximate any probability distribution [8, pp. 65]. The estimation takes the form given as below.

$$p(t|s) = \sum_{k=1}^{K} \pi_k(s)\phi(t, \mu_k(s), \sigma_k(s)),$$

where each $\pi_k(s)$ acts as the weight of each component distribution such that $\sum_k \pi_k = 1$; μ_k, σ_k are the mean and standard deviation parameters for the kth Gaussian; and ϕ denotes the Gaussian density function, given by

$$\phi(t, \mu_k, \sigma_k) = \frac{1}{\sigma_k\sqrt{2\pi}}exp\left(-\frac{(t-\mu_k)^2}{2}\right).$$

The loss function is given by

$$\mathcal{L} = -log\left(\prod_{i=1}^{N}\left(\sum_{k=1}^{K}\pi_k(s_i)\phi(t, \mu_k(s_i), \sigma_k(s_i))\right)\right),$$

where N denotes the number of observations. In any G/G/c queue, the state s is 1-dimensionally represented by the number of customers in the system at the time of each arrival. In the multiserver preemptive queue, the state is 2-dimensional, represented by the number of primary users and the number of secondary users in the system. In the problem of multiple M/M/c queues, s is 2-dimensional, represented by the queue length and the number of servers of the system. Lastly, in the problem of queueing at the three banks, s is 2-dimensional, represented by the queue length and the number of customers being served.

4   Evaluation Metrics

As for the first four datasets, because the ground truth probability distributions are known, we can intuitively compare the prediction for any unobserved or barely observed data in the training set. By plotting the estimated distribution and the known ground truth on the same graph, we can visually compare how well the estimation fits the ground truth.

As for *Dataset 6*, since the ground truth probability distributions are unknown, we measure the goodness of fit of the estimated probability density function on the test dataset by a proper scoring rule [6]. Furthermore, as the data is continuous, Continuous Ranked Probability Score (CRPS) is chosen to measure the error of the estimation. Let t be an observation of the random variable of waiting times and F be its corresponding estimated cumulative distribution function (CDF), then the CRPS between t and F is defined as

$$crps(F,t) = -\int_{-\infty}^{+\infty} (F(t) - \mathbb{H}(y,t))^2 \, dy,$$

where $\mathbb{H}$ denotes the Heaviside step function, which is defined as follows.

$$\mathbb{H}(y,t) = \begin{cases} 1 & \text{if } y \geq t, \\ 0 & \text{otherwise.} \end{cases}$$

We take the mean of CRPSs over all observations to evaluate the performance of MDNs.

$$CRPS = \frac{1}{N} \sum_{i=1}^{N} crps(F_i, t_i).$$

5   Results

In this section, we report the results of experiments that fit mixture models in each dataset described in the earlier section. Note that the number of hidden layers, the number of nodes in each hidden layer and the number of Gaussian components in each mixture are hyperparameters and set manually. Ideally, we want a large enough number of Gaussian components to precisely estimate waiting time distributions; however, with a limited amount of data, too many densities in the mixture may lead to overfitting and trigger bad results. Such hyperparameters are experimentally tuned to yield a reasonable result.

5.1   The M/M/1 Queueing System (*Dataset 1*)

We train the model with about $30,000$ observations with an MDN with 1 hidden layer containing 20 nodes, to output a mixture of 6 normal distributions as the estimation of desired distributions. The following figures illustrate how the estimated distributions fit the data in the test set in some specific examples (Fig. 1).

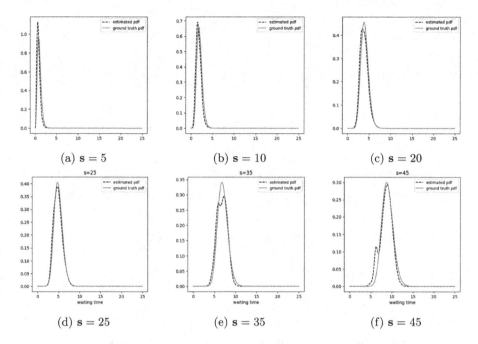

Fig. 1. Estimation of density functions of waiting times in the M/M/1 queueing system.

As **s** becomes larger, the number of corresponding observations in the training set becomes smaller. From the figures, we can see that the estimations fit well until **s** reaches about 25 (which is observed only 3 times in the training set). The accuracy of the prediction is likely to decrease with rarer observations such as **s** = 35 or **s** = 45.

5.2 The M/M/c Queueing Systems (*Dataset 2*)

We train the model with about 50, 000 observations with an MDN with 3 hidden layers containing 12 nodes in each to output a mixture of 10 normal distributions (Fig. 2).

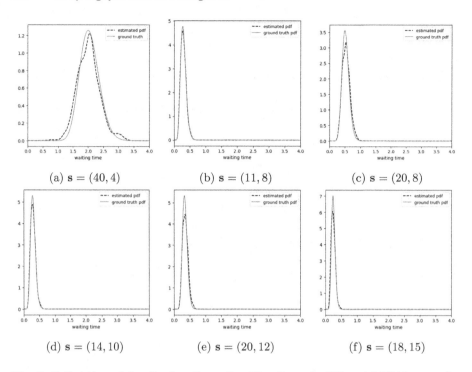

(a) **s** = (40, 4)   (b) **s** = (11, 8)   (c) **s** = (20, 8)

(d) **s** = (14, 10)   (e) **s** = (20, 12)   (f) **s** = (18, 15)

Fig. 2. Estimation of density functions of waiting times in different M/M/c queueing systems.

The above figures visually show how the predicted distributions fit the ground truth in several examples. Note that all of the observations chosen in those examples are not present in the training set. We can observe that MDNs give good predictions for observations which are not too "far" from the existing data. The fitted predictions deviate more from the ground truth with farther new observations such as **s** = (18, 15) (intuitively, it seems more difficult to predict waiting time distributions in a system with 15 servers than in a system with 8 servers, given the data on the systems with the number of servers ranging from 1 to 7).

5.3   The G/G/c Queueing Systems (*Dataset 3* and *Dataset 4*)

For both datasets, we train the modelith an MDN with 1 hidden layer containing 20 nodes, to output a mixture of 6 normal distributions as the estimation of desired distributions. *Dataset 3* contains about 350,000 observations, while *Dataset 4* contains about 110,000 observations. The two datasets are highly imbalanced with no observations of queue lengths which are larger than 7, and fewer than 20 observations of queue lengths equal to 5 and 6 (Figs. 3 and 4).

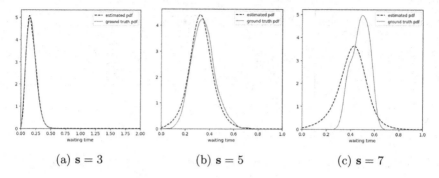

(a) s = 3   (b) s = 5   (c) s = 7

Fig. 3. Estimation of density functions of waiting times in the Erlang-type $G/G/c$ queueing system.

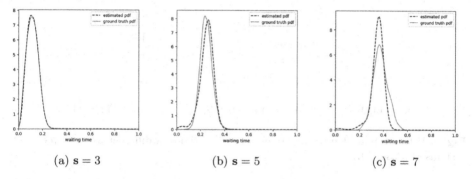

(a) s = 3   (b) s = 5   (c) s = 7

Fig. 4. Estimation of density functions of waiting times in the Weibull-type $G/G/c$ queueing system.

Similarly to the results on *Dataset 1*, we can see that the precision decreases with observations with fewer occurrences in the dataset.

5.4   The Preemptive Queueing System (*Dataset 5*)

We train the model with about $90,000$ observations with an MDN with 2 hidden layers containing 12 nodes and 10 nodes in each, to output a mixture of 10 normal distributions as the estimation of desired distributions. In this case, as the actual density functions are not derived, the ground truth is obtained by simulating the system within a large amount of time. The system parameters used for simulation are $\lambda_1 = 3, \mu_1 = 6, \lambda_2 = 4, \mu_2 = 7$ and $c = 3$. We illustrate the estimation and prediction results in some specific examples of system states (Fig. 5).

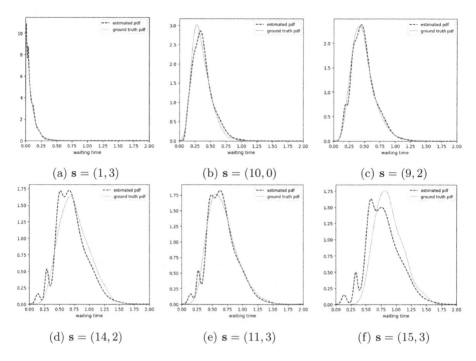

(a) $\mathbf{s} = (1, 3)$ (b) $\mathbf{s} = (10, 0)$ (c) $\mathbf{s} = (9, 2)$

(d) $\mathbf{s} = (14, 2)$ (e) $\mathbf{s} = (11, 3)$ (f) $\mathbf{s} = (15, 3)$

Fig. 5. Estimation of density functions of waiting times in different M/M/c queueing systems (*Dataset 5*).

Except for the first example of $\mathbf{s} = (1, 3)$ which has many observations in the training set, all of the remaining examples show the fitted distributions on the observations of system states which barely show up in the training set (less than 5 observations). We can see that as the states become rarer (for example, in case $\mathbf{s} = (14, 2)$ or $\mathbf{s} = (15, 3)$), the predicted distributions become spiky, and the predicted means are more likely to deviate from the true means.

5.5   Three Banks in Nigeria

We train the model with 26, 249 observations in the training set, experimenting with 3 different settings of MDNs. The results are summarized in Table 1.

Table 1. *CRPS* of estimation by MDNs with different architectures.

#Gaussians	#Hidden layers	#Nodes	CRPS
4	2	5 nodes in each hidden layer	0.06844
6	2	5 nodes in each hidden layer	0.06606
8	3	5 nodes in each hidden layer	0.06621

The experimental results show that the MDN containing 2 hidden layers and outputting a mixture of 6 Gaussian distributions yields the best fit for the data with the smallest $CRPS$ (Fig. 6).

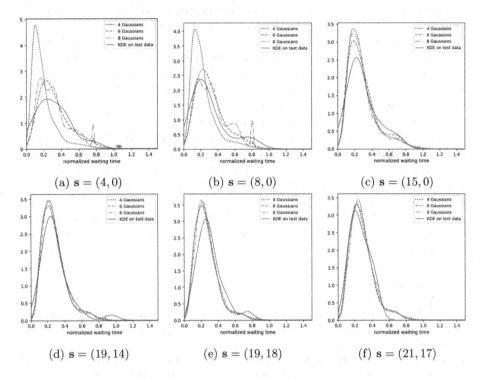

(a) $\mathbf{s} = (4, 0)$ (b) $\mathbf{s} = (8, 0)$ (c) $\mathbf{s} = (15, 0)$

(d) $\mathbf{s} = (19, 14)$ (e) $\mathbf{s} = (19, 18)$ (f) $\mathbf{s} = (21, 17)$

Fig. 6. Estimation of density functions of waiting times in queues in three banks in Nigeria.

The above figures illustrate how the estimated distributions fit the data in the test set in some specific examples. We choose the observations with high frequency in the test set (more than 30) and compare the estimation results with probability density functions estimated on testing data by the Kernel Density Estimation (KDE) method. It can be seen that the estimated distributions become spikier as the number of output mixture distributions is larger. Also, the mixture of 6 Gaussian distributions seems to be closest to the estimated distribution using KDE performed on the test set.

6   Concluding Remarks

This paper illustrated a potential application of Mixture Density Networks (MDNs) in the estimation and prediction of waiting time distributions in queueing systems. This method allows us to derive the relationship between waiting times and system states, and also gives a prediction of waiting time distribution when a customer enters a rarely occurring state, in the case that we barely know information on the system parameters and can only observe historical data.

While the traditional approach using queueing theory relies heavily on assumptions of stationary distributions over processes, MDNs have shown the power in the real-world system. Also, compared to most of the other statistical methods (such as Maximum Likelihood Estimation or Kernel Density Estimation) which can only give estimations for existing observations, MDNs have the ability to make predictions. In this sense, MDNs are particularly meaningful when it comes to estimations of the waiting times corresponding to the queue length with few observations in the dataset.

A shortcoming of this method, as seen in the implementations on *Dataset 3* and *Dataset 4*, is that it does not perform well on highly imbalanced datasets where the feature contains too few different values. Also, predictions become more imprecise with new observations which are distant from existing observations in the dataset. One possible solution could be to instantly feed new data and restrain the model.

Future works may consider learning and outputting a mixture of exponential distributions (as a general phase-type distribution) instead of a conventional Gaussian mixture.

References

1. Deep learning for siri's voice: on-device deep mixture density networks for hybrid unit selection synthesis. https://machinelearning.apple.com/research/siri-voices. Accessed 25 Aug 2021
2. Bishop, C.M.: Mixture density networks. Department of Computer Science and Applied Mathematics, Aston University, UK, Technical report (1994)
3. Bishop, S.A., Okagbue, H.I., Oguntunde, P.E., Opanuga, A.A., Odetunmibi, O.A.: Survey dataset on analysis of queues in some selected banks in Ogun state, Nigeria. Data Brief **19**, 835–841 (2018)
4. Demir, E., Demir, V.B.: Predicting flight delays with artificial neural networks: case study of an airport. In: 2017 25th Signal Processing and Communications Applications Conference (SIU), pp. 1–4 (2017)
5. Edelson, N.M., Hildebrand, K.: Congestion tolls for Poisson queueing processes. Econometrica **43**(1), 81–92 (1975)
6. Gneiting, T., Raftery, A.E.: Strictly proper scoring rules, prediction, and estimation. J. Am. Stat. Assoc. **102**(477), 359–378 (2007)
7. Gomes, D., Nabil, R.H., Nur, K.: Banking queue waiting time prediction based on predicted service time using support vector regression. In: 2020 International Conference on Computation, Automation and Knowledge Management (ICCAKM), pp. 145–149 (2020)
8. Goodfellow, I., Bengio, Y., Courville, A.: Deep Learning. MIT Press, Cambridge (2016)
9. Graves, A.: Generating sequences with recurrent neural networks. arXiv preprint arXiv:1308.0850 (2013)
10. Ha, D., Schmidhuber, J.: Recurrent world models facilitate policy evolution. arXiv preprint arXiv:1809.01999 (2018)

11. Ibrahim, R., Whitt, W.: Real-time delay estimation based on delay history. Manuf. Serv. Oper. Manag. **1**(3), 397–415 (2009)
12. Ibrahim, R., Whitt, W.: Wait-time predictors for customer service systems with time-varying demand and capacity **59**(5), 1106–1118 (2011)
13. Jouini, O., Dallery, Y., Akşin, Z.: Queueing models for full-flexible multi-class call centers with real-time anticipated delays. Int. J. Prod. Econ. **120**(2), 389–399 (2009). Special Issue on Introduction to Design and Analysis of Production Systems
14. Maister, D.: The psychology of waiting lines. In: Czepiel, J., Solomon, M.R., Surprenant, C.F. (eds.) The Service encounter: managing employee/customer interaction in service businesses. D. C. Heath and Company, Lexington Books, Lexington (1985)
15. Makansi, O., Ilg, E., Cicek, O., Brox, T.: Overcoming limitations of mixture density networks: a sampling and fitting framework for multimodal future prediction. In: Proceedings of the IEEE/CVF Conference on Computer Vision and Pattern Recognition, pp. 7144–7153 (2019)
16. Morozov, E.V., Rogozin, S.S., Nguyen, H.Q., Phung-Duc, T.: Modified erlang loss system for cognitive wireless networks (2021). https://arxiv.org/abs/2103.03222, to appear in Journal of Mathematical Sciences
17. Naor, P.: The regulation of queue size by levying tolls. Econometrica **37**(1), 15–24 (1969)
18. Raeis, M., Tizghadam, A., Leon-Garcia, A.: Predicting distributions of waiting times in customer service systems using mixture density networks. In: 2019 15th International Conference on Network and Service Management (CNSM), pp. 1–6 (2019)
19. Senderovich, A., Weidlich, M., Gal, A., Mandelbaum, A., et al.: Queue mining – predicting delays in service processes. In: Jarke, M. (ed.) CAiSE 2014. LNCS, vol. 8484, pp. 42–57. Springer, Cham (2014). https://doi.org/10.1007/978-3-319-07881-6_4
20. Whitt, W.: Predicting queueing delays. Manag. Sci. **45**(6), 870–888 (1999)

ASMTA 2021

Performance Evaluation and Energy Consumption for DVFS Processor

Youssef Ait El Mahjoub[1,2]($\boxtimes$), Jean-Michel Fourneau[1], and Hind Castel-Taleb[2]

[1] DAVID-UVSQ, Université Paris-Saclay, Paris, France
{youssef.ait-el-mahjoub2,jean-michel.fourneau}@uvsq.fr
[2] SAMOVAR, Télécom SudParis, Institut Polytechnique de Paris, Paris, France
hind.castel@telecom-sudparis.eu

Abstract. Dynamic Voltage and Frequency Scaling (DVFS) involves adjusting the speed to match power consumption and performance requirements of the system. This technology has been shown to provide major power and energy savings in many system components (processor cores, memory system ...). In this work, we study the DVFS mechanism for a processor that implements various levels of speed and power (we call these levels "Pstates"). We model the system as a birth-death process and we compare different configurations, using stochastic comparison, in order to evaluate the impact on the response time and power (and energy) consumption. In the case of two Pstates, we proposed a closed form for the steady state distribution and we derived a cost function that uses both performance and energy consumption. Finally, we derive an algorithm that suggests a threshold which minimizes the cost function.

Keywords: DVFS · Performance evaluation · Energy consumption · Stochastic comparison

1 Introduction

The IT sector has a very high contribution on the increase of the overall energy consumption. Many methods to reduce consumption in other industries or services result in more IT and telecommunications (the "Green by IT" approach [1]). Approaches which are used to minimize the power/energy consumption of ICT equipment are known as power management techniques. At processing level, in order to minimize the energy, either the device needs less power to operate, or it is simply switched off when it is not in use. However, a switched off device is unavailable to perform any task and might take considerable time to become available. Therefore, hardware designers implement other capabilities to devices such as Dynamic Voltage and Frequency Scaling (DVFS). This technique allows the processor to adapt its frequency according to energy or performance constraints. Higher frequencies leads to higher performance but more power consumption. DVFS is a promising approach to prevent power wastage

© Springer Nature Switzerland AG 2021
P. Ballarini et al. (Eds.): EPEW 2021/ASMTA 2021, LNCS 13104, pp. 165–180, 2021.
https://doi.org/10.1007/978-3-030-91825-5_10

in distributed large data centers (such as cloud environment [2]), also in the high performance computing (in Graphics Processing Units context [3]). Queuing models have been widely used in the literature to model DVFS schemes. In [4], an M/M/1 queue is considered for the reduction of energy consumption by switching the frequency and voltage, also in [5] authors employs a GI/G/N queue to model server clusters of cloud data centers with the respect of Service Level Agreements (SLA) between users and service providers. In [6] authors model a k-server farm with a fixed power budget which can be split among k servers with Processor Sharing (PS) discipline with the aim of minimizing the mean response time. In this paper, we model a DVFS processor by a queue system to have analytical formulas for performance and energy consumption measures. We established a stochastic comparison for the performance between different systems with one and two Pstates. We also provide sufficient condition for the comparison of power (and energy) consumption. In the case of two Pstates, analytical formulas are used to derive an optimization algorithm that investigates the best threshold that minimizes the cost function.

The objective of this work is to evaluate the response time, power and energy per job consumed in two distinct processor configurations. In Sect. 2, we present the features of a Pstate processor. In Sect. 3, we focus on a configuration where the processor uses a single Pstate, we compare the performance and power consumption of the sixth configurations. In Sect. 4, the processor considers two Pstates. We have proposed a closed form for steady-state distribution and we derived a cost function that uses both performance and energy consumption. We also show, through an algorithm we suggest, the best combination of Pstates to use and the corresponding minimizing threshold.

2   DVFS Pstate Processor

Many manufacturers of CPUs and GPUs (Intel, AMD, ...) implement the DVFS concept. In this document, only numerical values comes from the AMD (Advanced Micro Devices) Opteron processor [7]. The analysis we propose is more general. The Pstate processor operates at several frequencies and voltages denoted as "Pstates" (see Table 1). The highest Pstate runs the processor at full rate and full voltage. But during off-peak periods, the clock can go down to 1 GHz, saving up to 75 percent of the power at full speed.

Table 1. Pstates support in AMD opteron processor [7].

States	P1	P2	P3	P4	P5	P6
Frequency (GHz)	1	1.8	2	2.2	2.4	2.6
Power (W)	≈32	≈55	≈65	≈76	≈90	≈95

Table 1 represents the frequency (in GHz) that corresponds to the number of processor's instruction per time unit "t". We notice that the power consumption increases with the frequency in a non-linear approach. The Pstates power is a super-linearly function of the frequency. We model the Pstate processor as a multi-server queue. Cores of the processor represents servers in the queue. We also assume that

- The processor contains C cores.
- External arrivals of tasks follow independent Poisson process with rate λ.
- Service rates are distributed according to exponential distributions.
- Task scheduling discipline is FCFS (First Come, First Served).
- Tasks arriving into a processor may require "b" instructions (with $b > 0$). So the service rate (i.e. the number of tasks executed per time unit by a processor's core), is $\mu_i = \frac{f(Pstate_i)}{b}$ where "f" is the frequency.

Note that, service rate will depend on the model considered. In one Pstate model, service rate is fixed to the frequency of the Pstate considered, while in two Pstates system, service rate changes and depends on the current workload in the processor.

3   A Multi-core Processor with One Pstate

In this section, we consider that the Pstate processor has only one Pstate "i". Thus, a classical M/M/C queue can represent this model. The servers speed rate $\mu_i > 0$ is fixed and only depends on the chosen Pstate "i". We denote such a system S_i. We will study the performance and power consumption in each Pstate separately. Steady-state distribution and rewards as mean response time, mean number of tasks in the system are well known in the literature. Here, we show how to calculate, analytically, the mean power consumption and energy per job/task in the system. We also compare the performance and the power and energy consumption for the different systems.

We first recall some definitions about the "st" comparison between CTMCs processes and birth-death processes.

Lemma 1. *(Comparison of steady-state distributions* [8]*)*
Let $\{X_1^t, \ t \geq 0\}$ and $\{X_2^t, \ t \geq 0\}$ be two CTMCs. Hence, if the steady-state distribution Π_1 (resp. Π_2) exists, then

$$\{X_1^t, \ t \geq 0\} \leq_{st} \{X_2^t, \ t \geq 0\} \ \Rightarrow \ \Pi_1 \leq_{st} \Pi_2. \tag{1}$$

Theorem 1. *(Comparison of birth-death processes* [8]*)*
Stoyan theorem states that: considering two homogeneous birth-death processes $\{X_1^t, \ t \geq 0\}$ and $\{X_2^t, \ t \geq 0\}$ with Poisson arrivals rate $\lambda_x^{(S_1)}$ and $\lambda_x^{(S_2)}$, and Exponential service rates $\mu_x^{(S_1)}$ and $\mu_x^{(S_2)}$. Note that arrivals rate and service rate depends on the state x. If

$$\forall \ state \ x \geq 0 \quad \mu_x^{(S_1)} \geq \mu_x^{(S_2)} \quad and \quad \lambda_x^{(S_1)} \leq \lambda_x^{(S_2)},$$

then

$$\{X_1^t, \ t \geq 0\} \leq_{st} \{X_2^t, \ t \geq 0\}.$$

3.1   Mean Number of Jobs and Response Time

Under the classical assumptions we mention in Sect. 2, $\{X_i^t, t \geq 0\}$ is the CTMC that describes the system S_i. If the system is stable, then the steady-state distribution exists. Let Π_i be this distribution, and $\mathbb{E}[X_i]$ (resp. $\overline{T}_i$) be the mean number of jobs (resp. the mean response time). So

$$\mathbb{E}[X_i] = \sum_{x=0}^{\infty} x\Pi_i(x) \quad and \quad \overline{T}_i = \frac{\mathbb{E}[X_i]}{\lambda} \quad (Little's\ law). \tag{2}$$

Corollary 1. *We consider six stable systems (corresponding to the sixth Pstates in Table 1). Let $\mu_1, \mu_2, \ldots, \mu_6$ be processor's speed (i.e. service rate of cores in each system). Let $\forall\ t \geq 0\ \{X_1^t\}, \{X_2^t\}, \ldots, \{X_6^t\}$ be the Markov chain, and $\Pi_1, \Pi_2, \ldots, \Pi_6$ the steady-state distribution for each stable multi-core Pstate processor, then*

$$\mu_1 \leq \mu_2, \ldots, \leq \mu_6 \quad \Rightarrow \quad \Pi_6 \leq_{st} \Pi_5, \ldots, \leq_{st} \Pi_1. \tag{3}$$

Proof. The proof is directly derived from Theorem 1 and Lemma 1.

In our systems, we have the same arrivals rate λ in every system. Then for all the states "x", $\lambda_x^{(S_1)} = \lambda_x^{(S_2)} =, \ldots, = \lambda_x^{(S_6)} = \lambda$. Also from the assumption, we have for all state x, $\mu_x^{(S_1)} \leq \mu_x^{(S_2)} \leq, \ldots, \leq \mu_x^{(S_6)}$ as a consequence of $\mu_x^{(S_i)} = \mu_i$, then all conditions are satisfied to apply the theorem cited above. Therefore, we get $\forall\ t \geq 0$, $\{X_6^t\} \leq_{st} \{X_5^t\} \leq_{st}, \ldots, \leq_{st} \{X_1^t\}$. Hence, from Lemma 1 we conclude that $\Pi_6 \leq_{st} \Pi_5, \ldots, \leq_{st} \Pi_1$. □

Property 1. Let $\mathbb{E}[X_1], \mathbb{E}[X_2], \ldots, \mathbb{E}[X_6]$ (resp. $\overline{T}_1, \overline{T}_2, \ldots, \overline{T}_6$) be the mean number of jobs (resp. the mean response time) for stable multi-core Pstate processors $S_1, S_2, \ldots, S_6$ that uses one Pstate, then

$$\Pi_6 \leq_{st} \Pi_5, \ldots, \leq_{st} \Pi_1 \quad \Rightarrow \quad \begin{cases} \mathbb{E}[X_1] \geq \mathbb{E}[X_2], \ldots, \geq \mathbb{E}[X_6], \\ \\ \overline{T}_1 \geq \overline{T}_2, \ldots, \geq \overline{T}_6. \end{cases} \tag{4}$$

Proof. The systems are supposed to be stable, therefore, the steady-state distribution exists. Then

$$\Pi_6 \leq_{st} \Pi_5, \ldots, \leq_{st} \Pi_1 \Rightarrow \sum_{x \geq 0} x\Pi_1(x) \geq \sum_{x \geq 0} x\Pi_2(x) \geq, \ldots, \geq \sum_{x \geq 0} x\Pi_6(x), \tag{5}$$

and using Eq. 2, we get

$$\Pi_6 \leq_{st} \Pi_5, \ldots, \leq_{st} \Pi_1 \Rightarrow \mathbb{E}[X_1] \geq \mathbb{E}[X_2], \ldots, \geq \mathbb{E}[X_6]. \tag{6}$$

Mean response time is obtained using Little's law for the jobs in the system (see Eq. 2).

Corollary 2. *The result of this proposition is the conjunction of Corollary 1 and Property 1. Under stability condition of systems, we state that a higher speed of the Pstate processors (so higher Pstates) implies a lesser mean number of jobs and response time.*
If $\mu_1 \leq \mu_2, \ldots, \leq \mu_6$, then

$$\mathbb{E}[X_1] \geq \mathbb{E}[X_2], \ldots, \geq \mathbb{E}[X_6] \quad and \quad \overline{T_1} \geq \overline{T_2}, \ldots, \geq \overline{T_6}. \tag{7}$$

No proof is needed since Eq. 7 is the result of the conjunction of Eq. 3 and Eq. 4.

3.2   Power and Energy Consumption

Let p_i (*resp.* $p_{i,Id}$) for $i \in \{1, \ldots, 6\}$ be the Pstate power (resp. Pstate idle power) used by a core in a system S_i (see Table 1), and let $0 \leq \alpha < 1$ such that

$$p_{i,Id} = \alpha p_i. \tag{8}$$

Lemma 2. *The mean power consumption of a stable mono Pstate system PW_i is the sum the mean power of the servers in activity $PW_i^{(a)}$ and the idle power $PW_i^{(Id)}$:*

$$PW_i = PW_i^{(a)} + PW_i^{(Id)} = p_{i,Id}C + \frac{(p_i - p_{i,Id})\lambda}{\mu_i}. \tag{9}$$

Proof. Let $\Pi_i(x)$ be the steady-state distribution of the stable Markov chain $\{X_i^t, \ t \geq 0\}$, then

$$\begin{cases} PW_i^{(a)} = \sum_{x=0}^{\infty} \Pi_i(x) \left[p_i * min\{x, C\} \right], \\ PW_i^{(Id)} = \sum_{x=0}^{\infty} \Pi_i(x) \left[(C - min\{x, C\}) * p_{i,Id} \right]. \end{cases} \tag{10}$$

$\sum_{x=0}^{\infty} \Pi_i(x) \left[min\{x, C\} \right]$ represents the mean number of servers in activity of the system S_i. The mean number of jobs in service corresponds to the mean number of servers in activity since a job is served by one server at each time. We will use Little's law for servers in activity, which states that the mean number of jobs in service is the mean service time $\frac{1}{\mu_i}$ (since servers are homogeneous) times the mean arrivals rate λ. Then we have $\sum_{x=0}^{\infty} \Pi_i(x) \left[min\{x, C\} \right] = \frac{\lambda}{\mu_i}$. After simplifications, Eq. 10 becomes:

$$PW_i^{(a)} = \frac{\lambda p_i}{\mu_i} \quad and \quad PW_i^{(Id)} = \left(C - \frac{\lambda}{\mu_i} \right) p_{i,Id}. \tag{11}$$

By summing $PW_i^{(a)}$ and $PW_i^{(Id)}$ we get the expression of the mean power consumption to complete the proof.   □

Notice that the model we study here is a stable a M/M/C queue, and by definition the stability condition of the system is $\lambda < C\mu_i$ so $(C - \frac{\lambda}{\mu_i}) > 0$ then $PW_i^{(Id)} > 0$, therefore, $PW_i > 0$. □

Lemma 3. *Let $E_i^{(Job)}$ be the energy consumption per job in a stable system S_i,*

$$E_i^{(Job)} = \frac{PW_i}{\lambda}. \tag{12}$$

Proof. The energy consumption of a device is the power consumed on a period of time. Therefore, we expressed the energy consumption of a stable system as the mean power consumption PW_i times the mean resident time of a task $\overline{T_i}$. Let Eng_i be this energy

$$Eng_i = PW_i * \overline{T_i}. \tag{13}$$

Eng_i is the energy consumption of the system when considering all jobs. Hence, in order to obtain the energy consumption per one job,

$$E_i^{(Job)} = \frac{Eng_i}{\mathbb{E}[X_i]}. \tag{14}$$

Using Little's law for the mean number of jobs in a steady-state system ($\mathbb{E}[X_i] = \lambda\overline{T_i}$), then by substitution we get Eq. 12, the proof is complete. □

Lemma 4. *Let $E_i^{(Job)}$ (resp. $E_j^{(Job)}$) be the energy consumption per job of a stable system S_i (resp. S_j). Also let PW_i (resp. PW_j) be the corresponding mean power consumption. In below, we present a sufficient and necessary condition for the comparison of the mean power and the energy per job consumption:*

$$\frac{p_i}{p_j} \le \frac{\mu_i g(\mu_j)}{\mu_j g(\mu_i)} \iff \begin{cases} PW_i \le PW_j \\ E_i^{(Job)} \le E_j^{(Job)} \end{cases} \tag{15}$$

where

$$g(\mu_i) = \alpha\mu_i C + (1 - \alpha)\lambda. \tag{16}$$

Proof. By the substitution of $p_{i,Id} = \alpha p_i$, we get

$$PW_i = \left(\frac{\alpha\mu_i C + (1 - \alpha)\lambda}{\mu_i}\right)p_i. \tag{17}$$

then

$$E_i^{(Job)} \le E_j^{(Job)} \Leftrightarrow PW_i \le PW_j \Leftrightarrow \frac{p_i}{p_j} \le \frac{(\alpha\mu_j C + (1 - \alpha)\lambda)\mu_i}{(\alpha\mu_i C + (1 - \alpha)\lambda)\mu_j} \tag{18}$$

and using Eq. 16 the proof is complete. □

Lemma 5. *In this lemma we present a sufficient condition for the comparison of the mean power and the energy per job consumption in two stable systems S_i and S_j (with $i \leq j$):*

$$\mu_i \leq \mu_j \quad and \quad \frac{p_i}{p_j} \leq \frac{\mu_i}{\mu_j} \quad \Longrightarrow \quad \begin{cases} PW_i \leq PW_j \\ E_i^{(Job)} \leq E_j^{(Job)}. \end{cases} \tag{19}$$

Proof. Equation 15 in Lemma 4 presents a sufficient and necessary condition. Therefore:

$$\frac{p_i}{p_j} \leq \frac{\alpha \mu_i \mu_j C + (1-\alpha)\lambda \mu_i}{\alpha \mu_i \mu_j C + (1-\alpha)\lambda \mu_j} \quad \Rightarrow \quad \begin{cases} PW_i \leq PW_j \\ E_i^{(Job)} \leq E_j^{(Job)}. \end{cases} \tag{20}$$

Hence, we only have to verify that

$$\frac{\mu_i}{\mu_j} \leq \frac{\alpha \mu_i \mu_j C + (1-\alpha)\lambda \mu_i}{\alpha \mu_i \mu_j C + (1-\alpha)\lambda \mu_j}. \tag{21}$$

Equation 21 can be expressed as

$$\alpha \mu_i^2 \mu_j C + (1-\alpha)\lambda \mu_i \mu_j \leq \alpha \mu_i \mu_j^2 C + (1-\alpha)\lambda \mu_i \mu_j. \tag{22}$$

After simplification of terms we obtain $\mu_i \leq \mu_j$, which is the assumption stated in the presentation of this lemma. Hence, from Eq. 20 and 21 we obtain Eq. 19, and the proof is complete. □

3.3   Numerical Comparison of the Six Pstates Configurations

We consider here the six systems $S_1, \ldots, S_6$, each system has $C = 20$ cores. The speed of the servers (i.e. cores) depends on the Pstate performed. Also, we suppose that each task requires one core's instruction $b = 1$, therefore, $\mu_i = f_i$. Let $\alpha = 0.25$ which provides a power gain of 75% when the processor is in idle Pstate (see Eq. 8).

Performance, Power and Energy per Job: In Fig. 1 we observe that, (a) by increasing the task's arrivals load, the system fills up more and the waiting time in the queue increases. (b) In stable cases of each system, Pstate6 (contrary to Pstate1) presents the best results in terms of performance i.e. mean number of tasks and waiting time in the network, but requires the highest power consumption (see Fig. 2). These numerical results are clearly matching with our analytical results (Corollary 2, Lemma 4 and Lemma 5).

Condition Verification for Power and Energy Comparison: In Lemma 5 we only need the core's speed and core's power consumption to make a comparison of the mean power consumption and the energy per job consumption between two single-Pstate processors. This comparison is sufficient and not necessary, which means that when $\mu_i \leq \mu_j$ and $\frac{p_i}{p_j} \leq \frac{\mu_i}{\mu_j}$ is verified, then Pstate "i" consumes less mean power and energy per job consumption than the system with Pstate "j". Otherwise, we don't dispose of much information to compare the two systems. In that case, we can use Lemma 4, which, additionally, includes traffic arrivals rate and other parameters. For instance, we observed that $(i = 1, j = 2)$ and $(i = 5, j = 6)$ does not verify the assumption of Lemma 5. Hence, using Lemma 4 we obtained that: (a) for the case of $(i = 1, j = 2)$, $PW_1 \leq PW_2$ (and $E_1^{(job)} \leq E_2^{(job)}$) for all values of λ that makes both systems stable. (b) for the case of $(i = 5, j = 6)$, when $\lambda \leq 34$ we have $PW_5 \leq PW_6$ (and $E_5^{(Job)} \leq E_6^{(Job)}$). Otherwise, $PW_6 < PW_5$. That explains the crossover behavior between the red and the green curve in the right figure of Fig. 2.

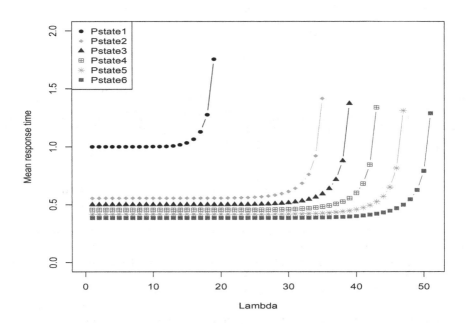

Fig. 1. Mean response time. Parameters: $C = 20$ servers, server's rate (i.e. core's speed) in each Pstate is depicted in Table 1.

4   A Multi-core Processor with Two Pstates

In this model we consider a birth-death process with two Pstates "i" and "j" with $i \leq j$. Servers speed rate depends on the number of tasks in the system.

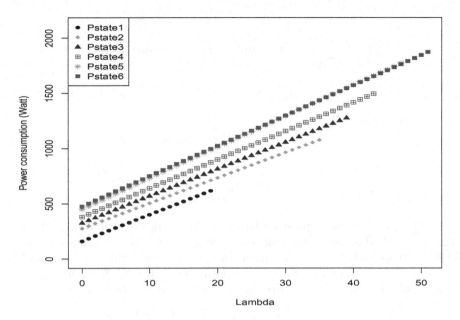

Fig. 2. Mean power consumption. Parameters are the same as in Fig. 1.

It means that if the number of tasks is below a certain threshold "th", then servers (cores) speed corresponds to μ_i, otherwise it is supposed that the queue is well loaded, therefore, servers speed rate are fixed to μ_j. We denote such a system $S(i, j, th)$. Stability condition of this model, that verifies the existence of a steady-state distribution is $\lambda < C\mu_j$.

4.1 Closed Form for the Steady-State Distribution

The processor contains C cores. Therefore, when x tasks are in the system, $min\{C, x\}$ tasks are in service and $(x - min\{C, x\})^+$ are queued. Under the classical assumptions we mention in Sect. 2, $\{X^t_{i,j,th}, \ t \geq 0\}$ is a Markov chain, and the transitions are as follows:

$$
\begin{aligned}
(x) &\rightarrow (x + 1) \ with \ rate \ \lambda, \\
(x) &\rightarrow (\max\{0, x - 1\}) \ with \ rate \ min\{x, C\} \cdot \mu(x).
\end{aligned}
\tag{23}
$$

and service rate $\mu(x)$ is $\mu(x) = \mu_i = \frac{f_i}{b}$ if $x \leq th$ and $\mu(x) = \mu_j = \frac{f_j}{b}$ if $x > th$. Note that, b is the number of instructions per task, and f_i is the frequency (GHz) given in Table 1.

Theorem 2. *The transition diagram of this Markov chain is a birth-death process. Then, under stability condition (i.e. $\lambda < C\mu_j$), we can derive $\Pi_{i,j,th}(x)$ the steady-state probability of the two Pstates processor $S(i, j, th)$.*

Let

$$\Psi(x) = \prod_{k=1}^{x} min\{k, C\}\mu(k), \quad and \quad R = max\{C, th\},$$

then

$$\Pi_{i,j,th}(x) = \begin{cases} \frac{\lambda^x}{\Psi(x)}\Pi_{i,j,th}(0) & \forall\ 0 < x \le R, \\[2mm] \lambda^x \left[(C\mu_j)^{x-R}\Psi(R) \right]^{-1} \Pi_{i,j,th}(0)\ \forall\ x > R, \end{cases} \tag{24}$$

where

$$\Pi_{i,j,th}(0) = \left[1 + \frac{\lambda^{R+1}}{(C\mu_j - \lambda)\Psi(R)} + \sum_{x=1}^{R} \frac{\lambda^x}{\Psi(x)} \right]^{-1}. \tag{25}$$

Proof. The Markov chain we consider is a birth-death process. We set "R" the state at which the system maintains it's service rate. This simplifies the calculation of the steady state distribution. Hence, we have divided the calculation of $\Pi_{i,j,th}(x)$ into two parts. States from 1 to "R" and states from "R" to $+\infty$. From classical equations in a birth-death process, we deduce that:

$$\Pi_{i,j,th}(x) = \begin{cases} \frac{\lambda}{min\{x,C\}\mu(x)}\Pi_{i,j,th}(x-1) & \forall\ 0 < x \le R, \\[2mm] \left(\frac{\lambda}{C\mu_j}\right)^{x-R}\Pi_{i,j,th}(R) & \forall\ x > R. \end{cases} \tag{26}$$

After simple substitutions in Eq. 26 (i.e. to write $\Pi_{i,j,th}(x)$ as a function of $\Pi_{i,j,th}(0)$ for all $x \in [1, +\infty[$), we get Eq. 24. Finally, as the system is supposed to be stable, then Eq. 25 is obtained after the normalization of probabilities ($\sum_{x=0}^{+\infty} \Pi_{i,j,th}(x) = 1$). □

4.2 Mean Number of Jobs and Response Time

Here we seek to compare the mean number of jobs and mean response time between two stable systems $S(i, j, th1)$ with $i \le j$ and $S(k, l, th2)$ with $k \le l$. Note that, to calculate the mean number of jobs $\mathbb{E}[X_{i,j,th}]$ and mean response time $\overline{T}_{i,j,th}$, we use the same approach as in Eq. 2.

Corollary 3. *We consider the two stable systems described above, then for any thresholds th1 and th2, we have:*

$$j \le k \quad \Rightarrow \quad \begin{cases} \mathbb{E}[X_{k,l,th2}] \le \mathbb{E}[X_{i,j,th1}] \\[2mm] \overline{T}_{k,l,th2} \le \overline{T}_{i,j,th1}. \end{cases} \tag{27}$$

Proof. Let $\{X_{i,j,th1}^t,\ t \ge 0\}$ (resp. $\{X_{k,l,th2}^t,\ t \ge 0\}$) the Markov chain of the stable system $S(i, j, th1)$ (resp. $S(k, l, th2)$). Also, let $\Pi_{i,j,th1}$ and $\Pi_{k,l,th2}$ be the steady-state distributions of both systems. Let $\lambda_x^{S(i,j,th1)}$, $\lambda_x^{S(k,l,th2)}$ (resp. $\mu_x^{S(i,j,th1)}$, $\mu_x^{S(k,l,th2)}$) be the arrivals rate (resp. service rate) generated at the state "x" in the system $S(i, j, th1)$ and $S(k, l, th2)$.

- We first derive the "st" comparison between the two systems. The arrivals rate λ is supposed the same. Then for all the states "x", $\lambda_x^{S(i,j,th1)} = \lambda_x^{S(k,l,th2)} = \lambda$. Also, we have $j \leq k$ from the assumption, and $i \leq j$ and $k \leq l$ from the definition of systems, then $i \leq j \leq k \leq l$. It means that, for any values of "th1" and "th2", services rates in $S(k,l,th2)$ are always greater than the ones in $S(i,j,th1)$, so $\mu_x^{S(k,l,th2)} \geq \mu_x^{S(i,j,th1)}$. From Theorem 1 we deduce that

$$\{X_{k,l,th2}^t, \ t \geq 0\} \leq_{st} \{X_{i,j,th1}^t, \ t \geq 0\}. \tag{28}$$

- Both systems are supposed stable, then using Lemma 1, we have

$$\{X_{k,l,th2}^t, \ t \geq 0\} \leq_{st} \{X_{i,j,th1}^t, \ t \geq 0\} \Rightarrow \Pi_{k,l,th2} \leq_{st} \Pi_{i,j,th1}. \tag{29}$$

- Finally, the same approach of property 1 is used to get

$$\Pi_{k,l,th2} \leq_{st} \Pi_{i,j,th1} \ \Rightarrow \ \begin{cases} \mathbb{E}[X_{k,l,th2}] \leq \mathbb{E}[X_{i,j,th1}] \\ \\ \overline{T}_{k,l,th2} \leq \overline{T}_{i,j,th1}. \end{cases} \tag{30}$$

By combining the three last equations we obtain Eq. 27, that concludes the proof. $\square$

Corollary 4. *In this corollary, we use another assumption, that concerns two stable systems using the same Pstates and different thresholds $S(i,j,th1)$ and $S(i,j,th2)$. We have:*

$$th1 \leq th2 \ \Rightarrow \ \begin{cases} \mathbb{E}[X_{i,j,th1}] \leq \mathbb{E}[X_{i,j,th2}] \\ \\ \overline{T}_{i,j,th1} \leq \overline{T}_{i,j,th2}. \end{cases} \tag{31}$$

Proof. The approach of the proof is similar to the one in Corollary 3.

- We have $th1 \leq th2 \ \Rightarrow \ \mu_x^{S(i,j,th1)} \geq \mu_x^{S(i,j,th2)}$. It means that system $S(i,j,th1)$ will activate the higher Pstate "j" earlier than system $S(i,j,th2)$. Therefore, from Theorem 1 we deduce that

$$\{X_{i,j,th1}^t, \ t \geq 0\} \leq_{st} \{X_{i,j,th2}^t, \ t \geq 0\}. \tag{32}$$

- Using Lemma 1, and Property 1 for two Pstates systems. We obtain Eq. 31, and the proof is complete. $\square$

4.3 Power and Energy Consumption

Let p_i (resp. p_j) be the power consumption corresponding to Pstate i (resp. Pstate j), with $i \leq j$. The power consumed by the processor depends on the speed of its cores, which is a function of the number of tasks assigned to it. Let (x) be the number of tasks in the processor, then the power consumed by each core, when hosting x tasks, is p_i if $x \leq th$ and p_j if $x > th$.

Lemma 6. *Let $PW_{i,j,th}$ be the mean power consumption of the system $S(i,j,th)$,*

$$PW_{i,j,th} = p_i(1-\alpha)\left[p_i \sum_{x=0}^{th} min\{x,C\}\Pi_{i,j,th}(x) + p_j \sum_{x=th+1}^{+\infty} min\{x,C\}\Pi_{i,j,th}(x)\right]$$
$$+\alpha C\left[p_j + (p_i - p_j)\sum_{x=0}^{th}\Pi_{i,j,th}(x)\right].$$
$$(33)$$

Proof. Let $PW_{i,j,th}^{(a)}$ (resp. $PW_{i,j,th}^{(Id)}$) be the mean power consumption of the servers in activity state (resp. Idle state). Hence, as in Lemma 2, we have

$$PW_{i,j,th} = PW_{i,j,th}^{(a)} + PW_{i,j,th}^{(Id)},\qquad(34)$$

where

$$\begin{cases} PW_{i,j,th}^{(a)} = p_i \sum_{x=0}^{th} min\{x,C\}\Pi_{i,j,th}(x) + p_j \sum_{x=th+1}^{+\infty} min\{x,C\}\Pi_{i,j,th}(x), \\ PW_{i,j,th}^{(Id)} = p_{i,Id} \sum_{x=0}^{th}(C - min\{x,C\})\Pi_{i,j,th}(x) \\ +p_{j,Id}\sum_{x=th+1}^{+\infty}(C - min\{x,C\})\Pi_{i,j,th}(x). \end{cases}$$
$$(35)$$

After simple simplifications: using Eq. 8 in Eq. 35, and by substituting $PW_{i,j,th}^{(a)}$ and $PW_{i,j,th}^{(Id)}$ in Eq. 34, we obtain Eq. 33 and the proof is complete.   □

Note that Lemma 3, for energy consumption per job, still hold for this model. Hence, $E_{i,j,th}^{(Job)} = \frac{PW_{i,j,th}}{\lambda}$.

4.4   Optimization Under Energy and Response Time Constraints

- From Corollary 4, we conclude that when comparing two stable systems $S(i,j,th1)$ and $S(i,j,th2)$ with $th1 \leq th2$ then $\overline{T}_{i,j,th1} \leq \overline{T}_{i,j,th2}$. Therefore, the mean response time (and mean number of jobs) is an increasing function of the threshold. Hence, to minimize the mean response time in a two-Pstates system, $th = 1$ is the optimal threshold to use.
- The "st" comparison does not hold for the mean power consumption (resp. energy per job) function. The function is not monotone with the threshold.
- So in order to optimize the mean power consumption (resp. energy consumption), we shall consider an exhaustive algorithm to obtain for the best threshold.

In the following, we suggest merging the mean response time $\overline{T}_{i,j,th}$ and the energy per job consumption $E_{i,j,th}^{(job)}$ in a single function to minimize. Let $c1$ (resp. $c2$) be the cost of the mean response time (resp. energy per job consumption). Let Θ be the total cost to minimize.

$$\Theta = \overline{T}_{i,j,th} * c1 + E_{i,j,th}^{(job)} * c2 \qquad(36)$$

To analyze Eq. 36, we propose the algorithm below. This algorithm, under stability constraint, generates the best threshold (in the range $[1,\ldots,THMAX]$)

to use for each couple of Pstates (i, j). Given the input parameters: number of servers C, services rate (given in Table 1), an upper bound $THMAX$ for the thresholds, arrival rate of tasks λ, and the rewards costs $c1$ and $c2$.

Algorithm 1: Purchasing the best threshold for each two-Pstates system

---

    Input   : Number of servers C, arrivals rate λ, rewards cost $c1$, $c2$, and a value of THMAX.

    Output: Threshold that minimizes cost function Θ for each couple of Pstates (i, j)

1 **for** (i, j) *with* $1 \leq i \leq j \leq 6$ **do**
2   **if** $\lambda < C\mu_j$ **then**   // The system is stable
3   initiate "th^*";
4   **for** $th \leftarrow 1$ **to** $THMAX$ **do**
5   1) Calculate the steady-state distribution (Equation 24 and 25) for the system $S(i, j, th)$;
6   2) Derive the mean response time $\overline{T}_{i,j,th}$ (Equation 2) and energy per job consumption $E_{i,j,th}^{(job)}$;
7   3) Calculate the cost function (Equation 36) ;
8   4) Update "th^*" if the current cost function is lower than the cost function for the previous iteration.
9   **end**
10   Print "The best threshold for the couple of Pstates (i,j) is "th^*" ;
11   **else**
12   Print "The system is not stable for the couple of Pstates (i,j)";
13   **end**
14 **end**

4.5 Numerical Results

We now investigate the influence of cost values $c1$ and $c2$ on Pstates (i, j) and their optimal thresholds. We considered the following parameters: arrivals rate $\lambda = 20$, number of servers $C = 20$, and the maximal threshold to purchase is $THMAX = 100$. Services rate are inspired from Table 1 with $b = 1$. Cost values ($c1$ and $c2$) are taken randomly in order to reflect the energy or performance constraints. We present here two experiments:

- In the first experiment (Fig. 3), we seek to minimize the cost function under the energy per job constraints. Therefore, we consider higher $c2$ costs and lower $c1$ costs. In particular, $c1 = 1$ and $c2 = 50$.
- In the second experiment (Fig. 4), we focus on response time constraints. Therefore, we consider higher $c1$ costs (and lower $c2$ costs). So $c1 = 200$ and $c2 = 1$.

Note that, the following results are derived from Algorithm 1. In the figures below, white areas are due to "$i \leq j$" which is an assumption of a two-Pstates (i, j) system. Colored square areas represents the values of the objective function (Eq. 36). Blue integer in colored areas is the value of the optimal threshold we obtained (in the range $[1, \ldots, THMAX]$) that minimizes the objective function for the corresponding couple of Pstates (i,j). Case of $i = j$ the threshold has no meaning since the system remains in the same Pstate, that explains the "-" entry in the figure.

Optimization Under Energy per Job Constraints. Recall that in this experiment $c1 = 1$ and $c2 = 50$. We observe in Fig. 3, that:

- Energy per job consumption increases with the value of Pstate "i". Therefore, the best couples of Pstates minimizing the objective function are in the green range $(i = 1, j \in \{2, \ldots, 6\})$. In order to reduce the energy consumption, the system opts for low Pstates.
- Optimal thresholds (blue integers in the colored squares) for the energy per job, achieves the highest levels (up to $th = THMAX = 100$). Notice that, under high energy costs, it is consistent that the system switches "lately" to the higher Pstate "j" which consumes more. The system remains mostly all the time in Pstate "i". That explains the unnoticed modifications of cost function when changing the Pstate "j".

Optimization Under Performance Constraints: To evaluate the cost function under response time requirement, we considered a second experiment: $c1 = 200$ and $c2 = 1$. We observe in Fig. 4, that:

- The system opts for high Pstates to reduce the response time component in the cost function.
- The optimal thresholds corresponds to a very small values. It shows that the Pstate processor switches quickly its cores to the higher Pstate "j", i.e. after having 1, 2, 3, 4, or 5 jobs in the system.
- When increasing $c1$ cost (above 500), we observe that the best threshold for all configurations is $th = 1$. Energy per job cost is irrelevant with regard to the response time's cost. Therefore, the system behaves as there is no constraints on the energy per job. In that case, we have proved (as a consequence of Corollary 4) that response time function increases with the threshold, which explains the best threshold $th = 1$.

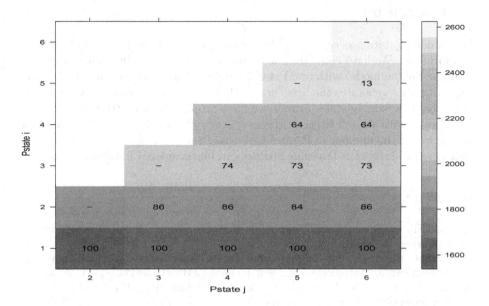

Fig. 3. $c1 = 1$, $c2 = 50$: objective function and optimal threshold for all pairs (i, j). For instance, the optimal threshold for the entry $i = 2$, $j = 3$ is $th = 86$ with a total cost of $\Theta = 1833.38$.

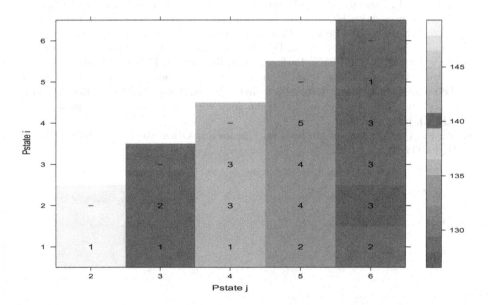

Fig. 4. $c1 = 200$, $c2 = 1$: objective function and optimal threshold for all pairs (i, j). For instance, the optimal threshold for the entry $i = 2$, $j = 3$ is $th = 2$ with a total cost of $\Theta = 140.6$.

5   Conclusion

We model a Pstate processor as a birth-death process in order to provide analytical formulas. We have compared, analytically and numerically, different configurations for the model with one Pstate. We have also proposed an optimization algorithm that generates the best threshold to use in the model with two Pstates. Note that, the steady-state formula we have proposed can be trivially extended to a model with $N > 2$ Pstates. However, the optimization algorithm may not be efficient if the number of Pstates to be used (and $THMAX$) is very high. In that case, a Markovian Decision Process can be considered in future work.

References

1. CIGREF: Du Green IT au Green by IT, Exemples d'applications dans les Grandes Entreprises. Technical report (2017)
2. Peyman, M.: A state of the art survey on DVFS techniques in cloud computing environment. J. Multidiscip. Eng. Sci. Technol. (JMEST) **3**, 4740–4743 (2016)
3. Guerreiro, J., Ilic, N., Roma, N., Tomás, P.: DVFS-aware application classification to improve GPGPUs energy efficiency. Parallel Comput. **83**, 93–117 (2019)
4. Basmadjian, R., Niedermeier, F., de Meer, H.: Modelling performance and power consumption of utilisation-based DVFS using M/M/1 queues. In: Proceedings of the Seventh International Conference on Future Energy Systems (e-Energy '16), New York (2016)
5. Kuehn, P.J., Mashaly, M.: DVFS-power management and performance engineering of data center server clusters. In: 15th Annual Conference on Wireless On-demand Network Systems and Services (WONS) (2019)
6. Gandhi, A., Harchol-Balter, M., Das, R., Lefurgy, C.: Optimal power allocation in server farms. SIGMETRICS Perform. Eval. Rev. **37**(1), 157–168 (2009)
7. Advanced Micro Devices (AMD) Inc.: power and cooling in the data center - Addressing today's and tomorrow's challenges with the AMD Opteron TM processor and AMD PowerNow TM technology with Optimized Power Management (OPM), pp. 1–8 (2005)
8. Stoyan, D.: Comparison Methods for Queues and Other Stochastic Models. Wiley, Berlin (1983)

Performance Models of NFV-Based Hybrid Systems for Delay-Sensitive Services

Mitsuki Sato[1], Kohei Kawamura[1], Ken'ichi Kawanishi[2(✉)],
and Tuan Phung-Duc[3]

[1] Graduate School of Science and Technology, Gunma University, Kiryu, Japan
{t211d037,t201d025}@gunma-u.ac.jp
[2] Department of Informatics, Gunma University, Maebashi, Japan
kawanisi@gunma-u.ac.jp
[3] Department of Policy and Planning Sciences, University of Tsukuba,
Tsukuba, Japan
tuan@sk.tsukuba.ac.jp

Abstract. NFV (Network Functions Virtualization) technology widely deployed in recent cloud computing is essential to provide network services. NFV is useful for saving operational cost because VNFs (Virtualized Network Functions) are created dynamically by scaling in/out virtual machines. On the other hand, utilizing existing computing resources, which are provided by legacy equipments and thus do not possess adaptive property of allocating virtual computing resources, is of importance to contribute for maintaining the required performance. In this paper, we propose queueing models of hybrid systems that utilize the NFV technology as well as legacy network equipments. Examples of such systems include 5G networks systems. We focus on a scenario of providing delay-sensitive real-time services, and in particular evaluate the delay performance.

Keywords: Cloud computing · Network functions virtualization · Markov chains · Delay-sensitive services · Delay performance

1 Introduction

Recently, NFV (Network Functions Virtualization) technology is commonly used for cloud computing on which various IT (Information Technology) services are provided. In the NFV environments, VNFs (Virtualized Network Functions) play a key role. VNFs realize network services such as firewalls and routings in a software-based manner, and are run on several virtual machines. Accordingly, VNF instances are created dynamically by scaling in/out virtual machines. Therefore, NFV enables us to adjust required computing resources such as CPU and memory, and thus is useful for service providers not only to meet on-demand requests but to reduce operational cost. It is also advantageous to save energy.

Supported in part by JSPS KAKENHI Grant Numbers 20K04980 and 21K11765.

P. Ballarini et al. (Eds.): EPEW 2021/ASMTA 2021, LNCS 13104, pp. 181–196, 2021.
https://doi.org/10.1007/978-3-030-91825-5_11

While pure NFV-based systems allow service providers to build network services platforms efficiently, it would be favorable to utilize widely deployed existing legacy network equipments. In fact, we can find such a scenario in typical cellular networks, where network service providers are forced to provide new advanced services efficiently, at the same time keeping in mind making effective use of existing legacy equipments. Using legacy equipments is not effective in saving energy, but improves user performance. Actually, VNF instances may not promptly response to service requests due to some setup time during which service requests cannot receive services. This is not the case for legacy equipments because they are always powered on and ready for providing services.

In this paper, we study the performance of such hybrid systems by simple analytical models. In this subject, prior work [9] assumes that the service rate provided by virtual machines is equivalent to that of legacy servers while [10] assumes that these service rates are different but jobs serving in a virtual machine may be transferred to a legacy server. We assume the service rate of the virtual machines is not the same to that of legacy server and a job is served by the same server until completion. In most cases, we expect that the virtual machines, which are based on advanced technology, provide higher service rates than the legacy servers. To the best of our knowledge, our scenario has not been investigated in the literature. We believe that our scenario in this paper would be more practical. We also assume that hybrid systems provide time-constrained services and thus jobs have to be completed by service-specific deadlines. A typical example of the time-constrained services is real-time video delivery, because the QoS (Quality of Service), and more importantly QoE (Quality of Experience) of video delivery services are delay-sensitive. We summarize our goals of this paper as follows:

- Using the matrix-analytic approach, we evaluate performance of the hybrid systems in steady state. We assume inter-arrival times of jobs, service times of legacy servers and virtual machines, setup times of virtual machines, and deadline times of jobs are distributed exponentially and independently each other. These assumptions allow us to formulate the system dynamics by continuous-time Markov chains (CTMC). Using the Markov chains, we discuss the effect of the service rate of virtual machines on waiting time performance of jobs.
- For the system model with exponential deadline times, we analyze the waiting time distribution of jobs. It should be noted that we cannot use the distributional Little's law in the model. Therefore, we resort to an absorbing Markov chain to evaluate waiting time distribution as in [8]. Using the waiting time distribution, we evaluate the expectation of the waiting time conditional to the jobs that receive their services or abandon the queue.
- We also consider the case where the deadline is deterministic, not random, for all jobs. In this case, we propose an approximate but analytically tractable model to evaluate the stationary performance. It is clear that the deterministic deadline does not possess the so-called Markov property. As a heuristic approach for this challenging issue, we propose a Markov chain model using the result in [3]. We validate the approach by computer simulation.

The rest of the paper is organized as follows. In Sect. 2, we introduce two system models to describe the dynamics of the hybrid systems. In Sect. 3, we provide performance measures of the system models. In Sect. 4, we discuss the system performance by illustrating numerical examples of the waiting time performance. Finally, we conclude the paper in Sect. 5.

2   System Models

In this section, we introduce two system models described by CTMC for the hybrid systems. Our models are basically variants of multiserver queueing models with finite capacity. One of the distinctive properties of the models is the setup policy. In [1], multiserver queues with *staggered* setup policy are investigated. The feature of the staggered setup policy is that the number of servers in setup in a time is limited to one. In our system models, however, we assume that the several servers are allowed to setup simultaneously, not limited to one.

We assume the hybrid system has ℓ legacy servers, and the maximum number of virtual machines that can be scaled out is limited to v. The legacy servers are always powered on, do not require setup, and are available upon job arrivals if the legacy servers are idle. On the other hand, the virtual machines are ready to serve jobs, provided that setup is completed. We assume the system capacity is limited to $K(\geq \ell + v)$. It should be noted that the capacity includes jobs in service by legacy servers or virtual machines as well as jobs waiting for setup or service completion. If the system is full upon job arrivals, then the jobs are blocked and immediately lost.

Suppose that a job arrives when the system is empty, i.e., there is no legacy server in active, i.e., processing a job, and all virtual machines are powered off. Since legacy servers is always powered on, the job is assigned to one of the idle legacy servers and the job's service immediately begins. The jobs are served according to the FCFS discipline. If there is no idle legacy server upon job's arrival, then one of the virtual machines is powered on and setup of the virtual machine is initiated. Since the setup needs some time to complete, the job must wait in queue during the setup time, and the service of the job begins after the virtual machine completes the setup. We describe the policy of job assignment when there are some virtual machines in setup.

1. Once the setup of the virtual machine completes and the service begins, then job is dedicated to the virtual machine until the job's service ends.
2. If one of the legacy servers or virtual machines completes the service before the setup completion of the virtual machine initiated by a job, then the job is assigned to the legacy server or the virtual machine that completes the service. In this case, there are two subcases:
   (a) If the number of jobs waiting for service is larger than the number of virtual machines in setup, then the virtual machine continues to setup.
   (b) If not, then one of the virtual machines in setup is powered off.

Remark 1. The first policy of job assignment implies that jobs continue to receive services from virtual machines, not transferred to legacy servers, once jobs are assigned to virtual machines. In contrast to [9], this policy is unfavorable in terms of energy saving. But we believe that such policy is practical.

Remark 2. The second policy, which is favorable to save energy, is also adopted in our models as in [9]. As a result, the number of virtual machines in setup is always given by the minimum of the number of jobs waiting in queue and the number of virtual machines not serving jobs.

If there are no waiting jobs when the virtual machine completes the service, then the virtual machine is immediately powered off. If not, the virtual machine begins the service of the job waiting for service without setup according to the FCFS discipline.

We assume inter-arrival times of jobs are exponentially distributed with mean $1/\lambda$. We also assume that the service times of legacy servers are exponentially distributed with mean $1/\mu$. The service and setup times of virtual machines are also exponentially distributed with mean $1/\nu$ and $1/\alpha$, respectively. Since we assume that jobs may abandon the queue before receiving their services, we specify the probability distribution of the time to abandon the queue. We analyze two cases. First, we assume that the abandonment time is exponentially distributed with mean $1/\theta$. Second, we consider the deterministic deadline $\tau = 1/\theta$. The first case allows us to construct CTMC. However, the description of the second case by CTMC is hard and hence the exact analysis of the deterministic deadline is challenging. To overcome the difficulty, we propose an approximate and analytically tractable model.

2.1 Markov Model with Random Reneging

First, we consider the system model where the abandonment time is exponentially distributed. We formulate the system by CTMC. We call it Markov model with random reneging.

Let us consider a two-dimensional CTMC $\{X(t) = (C(t), N(t)); t \geq 0\}$ on state space $\mathcal{S} = \{(i,j) \mid i = 0, 1, \ldots, v, j = 0, 1, \ldots, K - i\}$, where $C(t)$ denotes the number of active virtual machines (processing a job) at time t, and $N(t)$ denotes the sum of the number of jobs being served by legacy servers and waiting jobs at time t. The infinitesimal generator Q of $\{X(t); t \geq 0\}$ has the block tridiagonal form given by

$$
Q = \begin{bmatrix}
Q_{0,0} & Q_{0,1} & O & \cdots & & O \\
Q_{1,0} & Q_{1,1} & Q_{1,2} & \ddots & & \vdots \\
O & Q_{2,1} & \ddots & & \ddots & O \\
\vdots & \ddots & \ddots & & \ddots & Q_{K-1,K} \\
O & \cdots & & O & Q_{K,K-1} & Q_{K,K}
\end{bmatrix},
$$

where O denotes a zero matrix with an appropriate dimension. The block matrices $Q_{j,j+1}$ $(0 \leq j < \ell)$, $Q_{j,j}$ $(0 \leq j \leq \ell)$ and $Q_{j,j-1}$ $(0 < j \leq \ell)$ are $(v+1) \times (v+1)$ matrices given by

$$Q_{j,j+1} = \lambda I, \quad Q_{j,j} = \begin{bmatrix} -q_{0,j} & 0 & \cdots\cdots & 0 \\ \nu & -q_{1,j} & \ddots & \vdots \\ \vdots & 2\nu & \ddots & \ddots & \vdots \\ \vdots & & \ddots & \ddots & 0 \\ 0 & \cdots & \cdots & v\nu & -q_{v,j} \end{bmatrix}, \quad Q_{j,j-1} = j\mu I,$$

where I is the identity matrix with an appropriate dimension. The exit rate $q_{i,j}$ of state $(i,j) \in S$ for $0 \leq i \leq v$ and $0 \leq j \leq \ell$ is given by

$$q_{i,j} = \lambda + j\mu + i\nu.$$

The matrices $Q_{j,j+1}$ $(\ell \leq j < \ell+v)$, $Q_{j,j}$ $(\ell < j \leq \ell+v)$ and $Q_{j,j-1}$ $(\ell < j \leq \ell+v)$ are $(v+1) \times (v+1)$ matrices given by

$$Q_{j,j+1} = \lambda I, \quad Q_{j,j} = -\mathrm{diag}\{q_{0,j}, q_{1,j}, \ldots, q_{v,j}\},$$

$$Q_{j,j-1} = \{\ell\mu + (j-\ell)\theta\}I + \nu\mathrm{diag}\{0, 1, \ldots, v\} + \begin{bmatrix} 0 & p_{0,j}\alpha & 0 & \cdots & 0 \\ 0 & \ddots & p_{1,j}\alpha & \ddots & \vdots \\ \vdots & \ddots & \ddots & \ddots & 0 \\ \vdots & & \ddots & \ddots & p_{v-1,j}\alpha \\ 0 & \cdots & \cdots & 0 & 0 \end{bmatrix},$$

where $p_{i,j} = \min\{j - \ell, v - i\}$. The exit rate $q_{i,j}$ of state $(i,j) \in S$ for $0 \leq i \leq v$ and $\ell < j \leq \ell+v$ is given by

$$q_{i,j} = \lambda + \ell\mu + (j-\ell)\theta + i\nu + p_{i,j}\alpha.$$

The matrices $Q_{j,j+1}(\ell+v \leq j < K)$, $Q_{j,j}(\ell+v < j \leq K)$ and $Q_{j,j-1}(\ell+v < j \leq K)$ are $(K-j+1) \times (K-j)$, $(K-j+1) \times (K-j+1)$ and $(K-j+1) \times (K-j+2)$ matrices respectively, and are explicitly given by

$$Q_{j,j+1} = \begin{bmatrix} \lambda & 0 & \cdots & 0 \\ 0 & \lambda & \ddots & \vdots \\ \vdots & \ddots & \ddots & 0 \\ 0 & \cdots & 0 & \lambda \\ & & \mathbf{0} \end{bmatrix}, \quad Q_{j,j} = -\mathrm{diag}\{q_{0,j}, q_{1,j}, \ldots, q_{K-j,j}\},$$

$$Q_{j,j-1} = \left[\{\ell\mu + (j-\ell)\theta\}I \; \mathbf{0}^\top\right] + \left[\nu\mathrm{diag}\{0, 1, \ldots, v\} \; \mathbf{0}^\top\right]$$
$$+ \begin{bmatrix} 0 & p_{0,j}\alpha & 0 & \cdots & 0 \\ \vdots & \ddots & p_{1,j}\alpha & \ddots & \vdots \\ \vdots & & \ddots & \ddots & 0 \\ 0 & \cdots & \cdots & 0 & p_{K-j,j}\alpha \end{bmatrix},$$

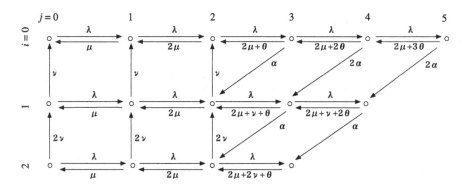

Fig. 1. The state transition diagram of Markov chain with random reneging.

where $\mathbf{0}$ is row vector of zeros with an appropriate dimension and $\mathbf{0}^{\top}$ is its transpose. The exit rate $q_{i,j}$ of state $(i,j) \in \mathcal{S}$ for $0 \leq i \leq K-j$ and $\ell+v < j \leq K$ is given by

$$q_{i,j} = \begin{cases} \lambda + \ell\mu + (j-\ell)\theta + iv + p_{i,j}\alpha, & j < K, \\ \ell\mu + (j-\ell)\theta + iv + p_{i,j}\alpha, & j = K. \end{cases}$$

In Fig. 1, we show an example of the state transition diagram with $\ell = 2, v = 2$ and $K = 5$. For state $(i,j) \in \mathcal{S}$, there are i active virtual machines, $\min\{\ell, j\}$ active legacy servers, and there are $[j-\ell]^{+}$ jobs waiting for service, where $[x]^{+} = \max\{x, 0\}$. In addition, for the state $(i,j) \in \mathcal{S}$ with $j > \ell$, there are $\min\{v - i, j - \ell\}$ virtual machines in setup.

It is clear that $\{X(t); t \geq 0\}$ is an irreducible Markov chain. Since $\mathcal{S}$ is finite, $\{X(t); t \geq 0\}$ is positive recurrent and thus the Markov chain $\{X(t); t \geq 0\}$ has the unique stationary distribution $\boldsymbol{\pi}$ defined by

$$\boldsymbol{\pi}Q = \mathbf{0}, \quad \boldsymbol{\pi}e = 1,$$

where e is column vector of ones with an appropriate dimension. Similar to Q, the stationary distribution $\boldsymbol{\pi}$ is also block partitioned as

$$\boldsymbol{\pi} = (\boldsymbol{\pi}_0, \boldsymbol{\pi}_1, \ldots, \boldsymbol{\pi}_K),$$
$$\boldsymbol{\pi}_j = (\pi_{0,j}, \pi_{1,j}, \ldots, \pi_{\min(K-j,v),j}), \quad j = 0, 1, \ldots, K,$$

where $\pi_{i,j}$ is the stationary distribution of state $(i,j) \in \mathcal{S}$ and is equal to the limiting distribution, i.e.,

$$\pi_{i,j} = \lim_{t \to \infty} \mathrm{P}(C(t) = i, N(t) = j), \quad (i,j) \in \mathcal{S}.$$

Remark 3. If $\ell = 0$, then the system has no legacy servers. In this case, the system model substantially reduces to that in [7,8], in which we can obtain the stationary distribution efficiently by exploiting the special structure of the block matrices and the local balance equation. Since we cannot apply the specially tailored approach when $\ell > 0$, we obtain the stationary distribution by using standard method, e.g., the approach in [4].

2.2 Virtual Waiting Time Process

Next, we analyze the stationary waiting time of a job in the Markov model with random reneging. To this end, we consider the virtual waiting time of an accepted job. Let V denote the virtual waiting. The actual waiting time W of an accepted job is given by $W = \min\{V, Z\}$, where Z is a random variable exponentially distributed with mean $1/\theta$. Since V and Z are independent, we have

$$P(W > t) = e^{-\theta t} P(V > t), \quad t \geq 0.$$

Therefore, we are lead to consider $P(V > t)$. Let P_B denote the probability that a job that sees K jobs in the system upon arrival. By PASTA (Poisson arrivals see time averages) property, P_B is given by

$$P_B = \sum_{i=0}^{v} \pi_{i,K-i}.$$

Then, the conditional probability that an arriving job who is accepted by the system and sees state $(i,j) \in \mathcal{S}$ upon arrival is given by

$$\bar{\pi}_{i,j} = \begin{cases} \dfrac{\pi_{i,j}}{1 - P_B}, & i = 0,1,\ldots,v, \ j = 0,1,\ldots,K-i-1, \\ 0, & i = 0,1,\ldots,v, \ j = K-i. \end{cases}$$

Using $\bar{\pi}_{i,j}$, we obtain

$$P(V > t) = \sum_{(i,j)\in\mathcal{S}_q} \bar{\pi}_{i,j} P(V > t \mid C(0) = i, N(0) = j), \quad t \geq 0.$$

where $\mathcal{S}_q = \{(i,j) \in \mathcal{S} \mid j = \ell, \ell+1, \ldots, K-i\}$. We can obtain $P(V > t \mid C(0) = i, N(0) = j)$ by considering an absorbing Markov chain as in [8]. Let $\widehat{Q}$ denote the infinitesimal generator of the absorbing Markov chain. Then it can be written to be

$$\widehat{Q} = \begin{bmatrix} 0 & 0 \\ s & S \end{bmatrix},$$

where S is the sub-generator describing the state transitions of transient states and $s = -Se$. We can construct S to have the block structure and explicitly obtain S using the system parameters. We omit the detail because it is almost the same as [8]. Using an appropriate vector β comprising $\bar{\pi}_{i,j}$ such that $\beta e = \sum_{(i,j)\in\mathcal{S}_q} \bar{\pi}_{i,j}$, which gives the probability that an accepted job is delayed, the stationary distribution of the virtual waiting time is described by the phase-type distribution with representation (β, S) and hence $P(V > t) = \beta \exp[St]e$ for $t \geq 0$. Therefore, we have

$$P(W > t) = e^{-\theta t}\beta \exp[St]e = \beta \exp[(S - \theta I)t]e, \quad t \geq 0.$$

It implies that the distribution of W is also of phase-type but with different representation $(\boldsymbol{\beta}, \boldsymbol{S} - \theta \boldsymbol{I})$. Let $\mathbb{E}[W]$ and $\mathbb{E}[W^2]$ denote the first and second moments of W, respectively. Using the property of the phase-type distribution, we have

$$\mathbb{E}[W] = \boldsymbol{\beta}[\theta \boldsymbol{I} - \boldsymbol{S}]^{-1}\boldsymbol{e}, \quad \mathbb{E}[W^2] = 2\boldsymbol{\beta}[\theta \boldsymbol{I} - \boldsymbol{S}]^{-2}\boldsymbol{e}.$$

It should be noted that we can readily obtain $\mathbb{E}[W]$ by Little's law even though the system model does not allow us to apply the distributional Little's law. However, it is not for $\mathbb{E}[W^2]$. Therefore, the absorbing Markov chain is useful to evaluate not only the distribution of W but higher moments of W. In addition, it helps us evaluate fine-grained performance measures of the system model. For example, let $\mathbb{E}[W \mid \mathrm{Ab}]$ denote the conditional expectation of waiting time given that the jobs abandon the queue. It is shown in [8] that $\mathbb{E}[W \mid \mathrm{Ab}]$ can be written in terms of $\mathbb{E}[W]$ and $\mathbb{E}[W^2]$ as

$$\mathbb{E}[W \mid \mathrm{Ab}] = \frac{\mathbb{E}[W^2]}{2\mathbb{E}[W]} = \frac{1 + c_W^2}{2}\mathbb{E}[W],$$

where c_W is the coefficient of variation of the waiting time.

2.3   Markov Model with Deterministic Reneging

In contrast to exponential random abandonment times, the stochastic process of the system model with deterministic reneging time is hard to describe. This is because the system dynamics cannot be determined completely by only specifying the number of jobs and the active virtual machines, if the abandonment time is constant. It is possible to recover the Markov property by adding the waiting time of each job (actually the waiting time of jobs in head of the queue is enough if the deadline is common to all jobs), but the analysis is complicated. Therefore, we propose an approximate but tractable Markov chain model. We call it Markov model with deterministic reneging.

Our approach is basically due to the idea in [6], which follows the result in [3]. The idea of [3] is to consider a probability density function $p_k(\cdot)$ of the waiting time of the *oldest* job in the system but not yet served, under the condition that there are k jobs waiting, in steady state. It should be noted that the model in [3] assumes homogeneous servers, where service rates are identical to all servers. In contrast, the model in this paper has heterogeneous servers, i.e., legacy servers and virtual machines can serve jobs simultaneously but with different service rates. We propose an approximate Markov model that has $p_k(\cdot)$ given by

$$p_k(x) = \frac{x^{k-1}e^{-\ell\mu x}}{\displaystyle\int_0^\tau t^{k-1}e^{-\ell\mu t}dt}, \quad x \in [0, \tau].$$

Our proposal implies that that $p_k(\cdot)$ of the dynamics of the system model is approximated as if only legacy servers can serve jobs. Then, we assume that the rate r_k that a waiting job leaves the queue under the condition that there are k jobs waiting is given by $r_k = p_k(\tau)$, i.e.,

$$r_k = \frac{\tau^{k-1}e^{-\ell\mu\tau}}{\displaystyle\int_0^\tau t^{k-1}e^{-\ell\mu t}dt}, \quad k = 1, 2, \ldots, K - \ell. \tag{1}$$

We construct the Markov model with deterministic reneging by replacing the abandonment rate $k\theta$ of the Markov model with random reneging with r_k.

Remark 4. Since r_k's in (1) corresponds to assuming the situation where the services of the system model are provided by only legacy servers, we expect that our proposal well approximates the actual system model, if setup time of virtual machines is large, or equivalently small value of α. This is because that virtual machines are substantially not available and almost all jobs are served by only legacy servers in such a situation even though virtual machines are created. We validate the approximate approach by computer simulation.

We describe the approximate Markov model based on (1). Let us consider a Markov chain $\{\tilde{X}(t); t \geq 0\}$ on the same state space $\mathcal{S}$ of $\{X(t); t \geq 0\}$. Let $\tilde{Q}$ denote the infinitesimal generator of $\{\tilde{X}(t); t \geq 0\}$. Then, similar to the case of Q, we can arrange $\tilde{Q}$ to have the tridiagonal block structure given by

$$\tilde{Q} = \begin{bmatrix} \tilde{Q}_{0,0} & \tilde{Q}_{0,1} & O & \cdots & & O \\ \tilde{Q}_{1,0} & \tilde{Q}_{1,1} & \tilde{Q}_{1,2} & \ddots & & \vdots \\ O & \tilde{Q}_{2,1} & \ddots & \ddots & & O \\ \vdots & \ddots & \ddots & \ddots & \tilde{Q}_{K-1,K} \\ O & \cdots & O & \tilde{Q}_{K,K-1} & \tilde{Q}_{K,K} \end{bmatrix}.$$

The block matrices $\tilde{Q}_{j,j}$ $(0 \leq j \leq K)$, $\tilde{Q}_{j,j+1}$ $(0 \leq j < K)$ and $\tilde{Q}_{j,j-1}$ $(0 < j \leq K)$ have the same dimensions corresponding to $Q_{j,j}$ $(0 \leq j \leq K)$, $Q_{j,j+1}$ $(0 \leq j < K)$ and $Q_{j,j-1}$ $(0 < j \leq K)$, respectively.

For $\tilde{Q}_{j,j+1}$ $(0 \leq j < K)$, we have $\tilde{Q}_{j,j+1} = Q_{j,j+1}$. For $\tilde{Q}_{j,j}$ $(0 \leq j \leq \ell)$ and $\tilde{Q}_{j,j-1}$ $(0 < j \leq \ell)$, we have $\tilde{Q}_{j,j} = Q_{j,j}$ and $\tilde{Q}_{j,j-1} = Q_{j,j-1}$, respectively. The other block matrices are almost the same to those of $\{X(t); t \geq 0\}$. For $\ell < j \leq \ell + v$, we have

$$Q_{j,j-1} = \{\ell\mu + r_{j-\ell}\}I + v\text{diag}\{0, 1, \ldots, v\} + \begin{bmatrix} 0 & p_{0,j}\alpha & 0 & \cdots & 0 \\ 0 & \ddots & p_{1,j}\alpha & \ddots & \vdots \\ \vdots & \ddots & \ddots & \ddots & 0 \\ \vdots & & \ddots & \ddots & p_{v-1,j}\alpha \\ 0 & \cdots & & 0 & 0 \end{bmatrix},$$

and for $\ell + v < j \leq K$,

$$\tilde{Q}_{j,j-1} = \left[\{\ell\mu + r_{j-\ell}\}I\ 0^\top\right]$$

$$+ \left[\nu\mathrm{diag}\{0,1,\ldots,v\}\ 0^\top\right] + \begin{bmatrix} 0 & p_{0,j}\alpha & 0 & \cdots & 0 \\ \vdots & \ddots & p_{1,j}\alpha & \ddots & \vdots \\ \vdots & & \ddots & \ddots & 0 \\ 0 & \cdots & & 0 & p_{K-j,j}\alpha \end{bmatrix}.$$

For $\ell < j \leq K$, we have

$$\tilde{Q}_{j,j} = -\mathrm{diag}\{\tilde{q}_{0,j}, \tilde{q}_{1,j}, \ldots, \tilde{q}_{\min\{K-j,v\},j}\},$$

where the exit rate $\tilde{q}_{i,j}$ of state $(i,j) \in \mathcal{S}$ for $0 \leq i \leq v$ and $\ell < j \leq K$ is given by

$$\tilde{q}_{i,j} = \begin{cases} \lambda + \ell\mu + r_{j-\ell} + iv + p_{i,j}\alpha, & j < K, \\ \ell\mu + r_{j-\ell} + iv + p_{i,j}\alpha, & j = K. \end{cases}$$

Similar to $\{X(t); t \geq 0\}$, we can see that $\{\tilde{X}(t); t \geq 0\}$ is also an irreducible finite Markov chain. Therefore, there exists the unique stationary distribution $\tilde{\pi}$ of $\{\tilde{X}(t); t \geq 0\}$ defined by

$$\tilde{\pi}\tilde{Q} = 0, \quad \tilde{\pi}e = 1.$$

3   Performance Measures

In this section, we show some performance measures obtained by the stationary distribution. At first, we show the performance measures explicitly only for the Markov chain model with random reneging.

Let $\mathbb{E}[W]$ denote the expectation of the waiting time of jobs in queue accepted by the Markov model with random reneging. By Little's law, we have

$$\mathbb{E}[W] = \frac{\mathbb{E}[Q]}{\lambda(1 - P_B)},$$

where $\mathbb{E}[Q]$ denotes the expectation of the number of jobs waiting. For the Markov model with random reneging, we have

$$\mathbb{E}[Q] = \sum_{j=\ell+1}^{K} (j-\ell)\pi_j e = \sum_{j=\ell+1}^{K} (j-\ell) \sum_{i=0}^{\min\{K-j,v\}} \pi_{i,j}.$$

It should be equal to the one obtained by the absorbing Markov chain, i.e., $\mathbb{E}[W] = \beta[\theta I - S]^{-1}e$. In the numerical examples shown later, we show the results computed by the absorbing Markov chain.

Let P_D denote the probability that an accepted job abandons the queue before receiving service. It is defined as the ratio of the intensity of abandonments to the intensity of arrival of jobs who are not blocked. Then, P_D of the Markov model with random reneging is given by

$$P_D = \frac{\sum_{j=\ell+1}^{K}(j-\ell)\theta\pi_j e}{\lambda(1-P_B)}.$$

Then, for the Markov model with random reneging, we can readily obtain

$$P_D = \theta\mathbb{E}[W].$$

Note that the above relation holds if the abandonment time is exponentially distributed [2,11]. Therefore, we can find P_D through $\mathbb{E}[W]$ for the Markov model with random reneging by using the proportional relationship.

The other performance measure of interest and feasible to compute by the Markov model with random reneging is the conditional expectation of waiting time given that the jobs receive services. If we denote it by $\mathbb{E}[W \mid \mathrm{Sr}]$, by law of total probability, we have

$$\mathbb{E}[W] = \mathbb{E}[W \mid \mathrm{Ab}]P_D + \mathbb{E}[W \mid \mathrm{Sr}](1-P_D).$$

Using the relation $P_D = \theta\mathbb{E}[W]$ and results in [8], we have

$$\mathbb{E}[W \mid \mathrm{Sr}] = \frac{\mathbb{E}[W]}{1-\theta\mathbb{E}[W]}\left(1 - \theta\mathbb{E}[W]\frac{1+c_W^2}{2}\right).$$

Therefore, we can evaluate $\mathbb{E}[W \mid \mathrm{Sr}]$ in terms of the first and second moments of W, which in turn can be computed by the distribution of W for Markov model with random reneging.

Next, we describe performance measures for the model with constant deadline time. Let $\tilde{P}_B$ denote the probability that an arriving job is blocked and lost in the Markov model with deterministic reneging. Since jobs arrive according to Poisson processes, we can apply the PASTA property to obtain

$$\tilde{P}_B = \sum_{i=0}^{v} \tilde{\pi}_{i,K-i}.$$

Let $\mathbb{E}[\tilde{W}]$ denote the expectation of the waiting time of accepted job in queue for the Markov model with deterministic reneging. Thanks to the Little's law, we have

$$\mathbb{E}[\tilde{W}] = \frac{\mathbb{E}[\tilde{Q}]}{\lambda(1-\tilde{P}_B)},$$

where $\mathbb{E}[\tilde{Q}]$ is the mean number of jobs waiting in queue or the Markov model with deterministic reneging, and defined by

$$\mathbb{E}[\tilde{Q}] = \sum_{j=\ell+1}^{K} (j-\ell) \sum_{i=0}^{\min\{K-j,v\}} \tilde{\pi}_{i,j}.$$

Therefore, we can evaluate $\mathbb{E}[\tilde{W}]$ in terms of $\mathbb{E}[\tilde{Q}]$ that can be computed by the stationary distribution of the Markov model with deterministic reneging.

4 Numerical Examples

In this section, we show numerical examples of delay performance of both Markov models with random reneging and deterministic reneging. For the deterministic reneging, we also provide results of computer simulation (Monte Carlo simulation) to validate the Markov model with deterministic reneging. In computer simulation of the deterministic reneging, the inter-arrival times, service times of legacy servers and virtual machines, and setup times are all sampled from exponential distributions, but the deadline times are constant. We repeated 100 computer simulations to collect experiment results and obtain the sample mean of delay. For each experiment, we obtain results by running simulation until 10,000,000 jobs are accepted. This implies that at least 10,000,000 jobs are generated because some of jobs may be blocked and lost upon arrival. As a result, we obtain the length of 95% confidence interval less than 10^{-3} for the sample mean of delay in our parameter setting.

4.1 Random Reneging Model

First, we show numerical examples of the Markov model with random reneging. We set default values of system parameters $K = 30, \ell = 10, v = 10$. The arrival rate λ is varied within the interval $(0, 60]$. The default values of other parameters are set as $\mu = 1, \theta = 0.01$.

Figure 2 shows $\mathbb{E}[W]$ versus λ with several values of parameters α and ν. As pointed out in [9], we observe that the impacts of λ on $\mathbb{E}[W]$ is divided into four phases for small values of α, i.e., $\alpha = 0.01, 0.05, 0.1$. Starting from zero, $\mathbb{E}[W]$ rapidly increases, and then decreases as increasing λ. Finally, $\mathbb{E}[W]$ approaches to a bound specified by system capacity K. Such a complicated behavior of $\mathbb{E}[W]$ is due to the long setup time of virtual machines. In fact, we do not clearly observe the behavior when $\alpha = 1$.

With regard to the impacts of ν on $\mathbb{E}[W]$, we observe that asymptotic values of $\mathbb{E}[W]$ for large λ are placed in the reverse order of the values of ν. This is intuitively acceptable because virtual machines with high service rates contribute to reduce the waiting time of jobs. Though it may seem counterintuitive, we can find some range of λ, where $\mathbb{E}[W]$ with $\alpha = 0.01$ decreases as decreasing ν.

In Fig. 3, we show $\mathbb{E}[W \mid \mathrm{Ab}]$ and $\mathbb{E}[W \mid \mathrm{Sr}]$ versus λ with $\alpha = 0.01$ and several values of ν. We observe that $\mathbb{E}[W \mid \mathrm{Ab}]$ and $\mathbb{E}[W \mid \mathrm{Sr}]$ almost synchronize. Note here that $\mathbb{E}[W \mid \mathrm{Sr}]$ is dominant for $\mathbb{E}[W]$. In fact, $\mathbb{E}[W]$ is bounded above by about one in this parameter setting, P_D is bounded above by $P_D = \theta \mathbb{E}[W] \approx 0.01$ and thus $\mathbb{E}[W] \approx \mathbb{E}[W \mid \mathrm{Sr}]$ in this parameter setting. Therefore, this does not lead us to obtain any information about $\mathbb{E}[W \mid \mathrm{Ab}]$. Thanks to the absorbing Markov chains, however, we find the behavior of $\mathbb{E}[W \mid \mathrm{Ab}]$.

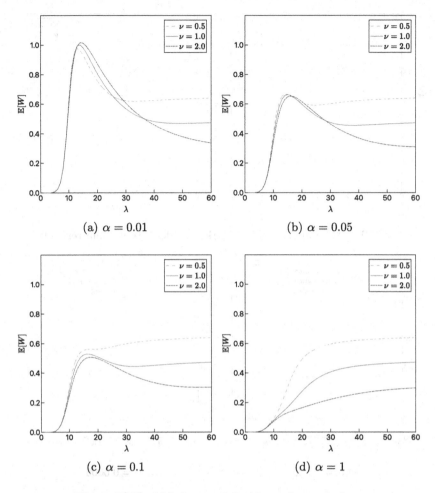

(a) $\alpha = 0.01$

(b) $\alpha = 0.05$

(c) $\alpha = 0.1$

(d) $\alpha = 1$

Fig. 2. $\mathbb{E}[W]$ of Markov model with random reneging.

4.2 Deterministic Reneging Model

Next, we discuss the performance measures of the Markov model with determin-
istic reneging. In Fig. 4, we show $\mathbb{E}[\tilde{W}]$ against the arrival rate λ of the Markov
model with random reneging for $\alpha = 0.005, 0.0005$. In this case, we set other
parameters as follows: system parameters are $K = 200, \ell = 100, v = 40$, service
rates of legacy servers and virtual machines are $\mu = 1$ and $\nu = 1.5$, and the
constant deadline time is specified by $1/\theta$ with $\theta = 1.2$.

We observe that the Markov model with deterministic reneging almost agrees
with computer simulation results. Since the Markov model with deterministic
reneging assumes (1), which is just a simple heuristic approach, the agreement
is somewhat surprising. It should be noted that the distribution of the time to
abandon the queue drastically affects the mean waiting time. Indeed, we observe

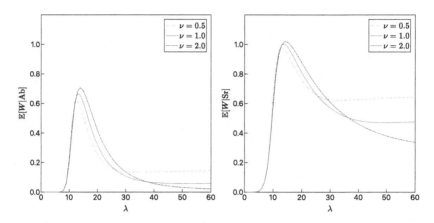

Fig. 3. $\mathbb{E}[W \mid \text{Ab}]$ and $\mathbb{E}[W \mid \text{Sr}]$ of Markov model with random reneging.

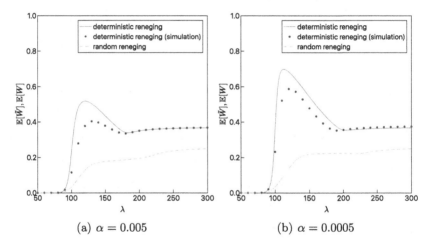

(a) $\alpha = 0.005$ (b) $\alpha = 0.0005$

Fig. 4. $\mathbb{E}[\tilde{W}]$ of Markov model with deterministic reneging. For comparison, $\mathbb{E}[\tilde{W}]$ by computer simulation and $\mathbb{E}[W]$ of Markov model with random reneging are shown.

that the deterministic reneging induces long waiting time compared with the random reneging. Therefore, we are not allowed to use the Markov model with random reneging as a performance model for the constant deadline time. Since the hybrid systems with deterministic deadline is more practical but the exact analysis seems challenging, the heuristic approach in this paper is useful enough to estimate the mean waiting time without relying on time-consuming computer simulation.

5   Concluding Remarks

In this paper, we have considered NFV-based hybrid systems. In addition to VNF instances, the hybrid systems utilize legacy network equipments for the purpose of effective use of them. One of the salient features of the hybrid systems is that VNF instances are created dynamically by scaling in/out virtual machines. While legacy servers are always powered on and thus can immediately provide services, VNF instances cannot begin services until virtual machines are ready for services after setup completion. We have assumed that the hybrid systems process time-constrained jobs, such as real-time video delivery. We have considered two cases where the time to abandon the queue is exponentially distributed, and deterministic. We have analyzed the former case using the Markov model formulated by QBD processes. We have also obtained waiting time distribution allowing us to evaluate conditional expectation of waiting time. For the latter, we have proposed an analytically tractable Markov model based on a heuristic approach and validated the Markov model by computer simulation.

Since we have restricted to the validation of the heuristic approach only in terms of the mean waiting time in this paper, we need more extensive simulation experiments under practical parameter settings to enhance reliability of the approach. While we have resorted to a heuristic approach to construct the Markov model with deterministic deadline, it would be more reliable if the hybrid systems are analyzed by rigorous approach such as [5], where delayed offloading in mobile networks are investigated. These issues are left for future work.

References

1. Artalejo, J.R., Economou, A., Lopez-Herrero, M.J.: Analysis of a multiserver queue with setup times. Queueing Syst. **52**, 53–76 (2005)
2. Baccelli, F., Hebuterne, G.: On queues with impatient customers. In: Kylstra, F.J. (ed.) Performance'81, pp. 159–179 (1981)
3. Barrer, D.Y.: Queuing with impatient customers and ordered service. Oper. Res. **5**(5), 650–656 (1957)
4. Gaver, D., Jacobs, P., Latouche, P.: Finite birth-and-death models in randomly changing environments. Adv. Appl. Probab. **16**, 715–731 (1984)
5. Kawanishi, K., Takine, T.: The M/PH/1+D queue with Markov-renewal service interruptions and its application to delayed mobile data offloading. Perform. Eval. **134**, 102002 (2019)
6. Lee, K., Lee, J., Yi, Y., Rhee, I., Chong, S.: Mobile data offloading: how much can WiFi deliver? IEEE/ACM Trans. Netw. **21**(2), 536–550 (2013)
7. Phung-Duc, T., Kawanishi, K.: Energy-aware data centers with s-staggered setup and abandonment. In: Wittevrongel, S., Phung-Duc, T. (eds.) ASMTA 2016. LNCS, vol. 9845, pp. 269–283. Springer, Cham (2016). https://doi.org/10.1007/978-3-319-43904-4_19
8. Phung-Duc, T., Kawanishi, K.: Delay performance of data-center queue with setup policy and abandonment. Ann. Oper. Res. **293**(1), 269–293 (2019). https://doi.org/10.1007/s10479-019-03268-1
9. Phung-Duc, T., Ren, Y., Chen, J.-C., Yu, Z.-W.: Design and analysis of deadline and budget constrained autoscaling (DBCA) algorithm for 5G mobile networks. In: Proceedings of IEEE CloudCom 2016, Luxembourg, Luxembourg, 12–15 December (2016). https://doi.org/10.1109/CloudCom.2016.0030

10. Ren, Y., Phung-Duc, T., Liu, Y.-K., Chen, J.-C., Lin, Y.-H.: ASA: adaptive VNF scaling algorithm for 5G mobile networks. In: Proceedings of IEEE CloudNet 2018, Tokyo, Japan, 22–24 October (2018). https://doi.org/10.1109/CloudNet. 2018.8549542
11. Zohar, E., Mandelbaum, A., Shimkin, N.: Adaptive behavior of impatient customers in tele-queues: theory and empirical support. Manag. Sci. **48**(4), 566–583 (2002)

A Stochastic SVIR Model with Imperfect Vaccine and External Source of Infection

Maria Gamboa[1](✉)[ID], Martín López-García[2][ID],
and Maria Jesus Lopez-Herrero[1][ID]

[1] Faculty of Statistical Studies, Complutense University of Madrid, Madrid, Spain
{mgamboa,lherrero}@ucm.es
[2] School of Mathematics, University of Leeds, Leeds, UK
M.LopezGarcia@leeds.ac.uk

Abstract. A stochastic SIR (Susceptible - Infected - Recovered) type model, with external source of infection, is considered for the spread of a disease in a finite population of constant size. Our interest is in studying this process in the situation where some individuals have been vaccinated prior to the start of the epidemic, but where the efficacy of the vaccine to prevent infection is not perfect. The evolution of the epidemic is represented by an absorbing three-dimensional continuous-time Markov chain. We focus on analysing the time for a threshold number of individuals to become infected, and carry out a global sensitivity analysis for the impact of varying model parameters on the summary statistic of interest.

Keywords: Stochastic epidemic model · Markov chain · Time to absorption · Imperfect vaccine

1 Introduction

Infectious diseases have been a serious threat to society throughout history. Plague, cholera and smallpox are examples of epidemics in the past that killed many people. This is a problem that we still suffer today, with emerging diseases such as Ebola, SARS and COVID-19 that continue to claim lives every day.

Understanding epidemic processes is vitally important to forecast the incidence of a disease and to establish mitigation strategies, and mathematical modelling has proven to be a robust tool in this area. Deterministic models have been widely used due to their mathematical tractability [1,2], and are especially relevant when considering large populations or when stochastic effects can be neglected. On the other hand, when considering small populations or if extinction events play a relevant role, stochastic models need to be considered instead

Supported by the Government of Spain, Department of Science, Innovation and Universities; European Commission project: PGC2018-097704-B-I00 and Banco Santander and Complutense University of Madrid, Pre-doctoral Contract: CT 42/18-CT43/18.

P. Ballarini et al. (Eds.): EPEW 2021/ASMTA 2021, LNCS 13104, pp. 197–209, 2021.
https://doi.org/10.1007/978-3-030-91825-5_12

of classic ones due to the influence on the impact of the disease of random differences in infectiousness and susceptibility among individuals, while these random effects tend to cancel out each other as population size increases [3,4].

The Kermack and McKendrick model [5] has probably been the most influential in representing the spread of an epidemic in the last decades. It is a compartmental deterministic model that classifies individuals according to their "state" with respect to the disease over time: susceptible (S), infected (I) and recovered (R). This SIR model is appropriate for describing a disease for which individuals develop permanent immunity after infection. The SIR model, and a number of different variations, has been widely analysed both for homogeneous [6,7] and heterogeneous populations [8]. In these systems, of particular interest can be specific summary statistics that characterize an outbreak, such as the size of the outbreak [9], its length [10,11] or the reproduction number [12].

Vaccination is an effective preventive measure to limit or avoid an outbreak, where the presence of a high percentage of vaccinated individuals in a given population can prevent transmission, reducing the size and impact of epidemic outbreaks, or the probability of these outbreaks happening at all. A number of mathematical models have considered vaccinated individuals as an extra compartment in the model [13], and some studies have added vaccination strategies into these mathematical models [14–17]. In some cases, vaccines do not provide permanent immunity, and boosters are required [18]. In other occasions, a vaccine might not be fully effective in preventing disease [19], and a proportion of vaccinated individuals might still be partially susceptible against infection. In this situation of an *imperfect* vaccine, the population runs the risk of losing or not achieving herd immunity [20].

In the literature we can find examples of studies assuming either fully protective [21] or imperfect [22,23] vaccines. In [24,25], authors quantify disease transmission in a stochastic SIS model with external source of infection and imperfect vaccine and study preventive measures surrounding vaccination. Under the assumption of imperfect vaccine, authors in [26] study the stationary distribution of the system for a closed population in a stochastic SVIR-type model. On the other hand, in [27] the time to extinction is studied for a non-linear incidence rate model.

In this paper, we consider a SVIR model with imperfect vaccine and external source of infection for a finite homogeneous population of fixed size. Our interest is in analysing the time until a threshold number of individuals get infected, as a way of quantifying the timescales for disease spread. We do this by representing the epidemic process in terms of a multidimensional continuous-time Markov chain (CTMC), and studying a time to absorption in this process. We show how a particular organization of states in this CTMC leads to the study of a level-dependent quasi birth-and-death process (LD-QBD) [28], and propose an efficient scheme to analyse the summary statistic of interest. Our methodology is based on the analysis of Laplace-Stieltjes transforms and the implementation of first-step arguments, adapting techniques in [24,25].

This paper is organized as follows. In Sect. 2 we introduce the SVIR stochastic model with imperfect vaccine and external source of infection. In Sect. 3 we define the summary statistic of interest, and provide an efficient algorithm to compute any of its moments. In Sect. 4 we illustrate our methodology by carrying out a global sensitivity analysis on model parameters. Finally, we present our conclusions in Sect. 5, and discuss possible future lines of research.

2   Model Description

We model the spread of an infectious disease across a population of constant size N, where a percentage of individuals are vaccinated at time $t = 0$ as a prophylactic device to control disease spread. We assume that vaccine is not perfect so that vaccinated individuals can get the infection with probability $h \in (0,1)$, which we refer to as the vaccine failure probability. Vaccine protection lasts for at least the length of an outbreak, hence further vaccination during the outbreak is not considered. We consider SIR-type dynamics, so that infected individuals become recovered after their infectious period, and denote the recovery rate by γ. Transmission can occur through direct contact, with rate β, or due to an external source of infection, with rate ξ.

We represent this epidemic process in terms of a three-dimensional continuous-time Markov chain (CTMC) $\mathcal{X} = \{(V(t), S(t), I(t)) : t \geq 0\}$, where $V(t)$, $S(t)$ and $I(t)$ represent the number of vaccinated, susceptible and infected individuals in the population at time $t \geq 0$. Given that the population size remains constant, it is clear that $R(t) = N - V(t) - S(t) - I(t)$ represents the number of recovered individual at time t. If one assumes that there are no recovered individuals at the beginning of the epidemic process, the initial state is given by $(V(0), S(0), I(0)) = (v_0, s_0, N - v_0 - s_0)$, for some $v_0, s_0 \geq 0$, with $v_0 + s_0 \leq N$. The state space of the Markov chain is then given by

$$S = \{(v, s, i) : 0 \leq v \leq v_0, 0 \leq s \leq s_0, 0 \leq v + s + i \leq N\}, \tag{1}$$

which is finite and contains $(v_0 + 1)(s_0 + 1)(N + 1 - \frac{s_0 + v_0}{2})$ states, with a unique absorbing state $(0, 0, 0)$.

We assume that recoveries and contacts between individuals happen independently of each other, with exponentially distributed inter-event times. The evolution of the epidemic process over time is represented by transitions between states in S, where the possible events/transitions are outlined in Table 1. In particular, given the current state $(v, s, i) \in S$, possible events are:

(E_1) A susceptible individual gets infected, which occurs with rate

$$\lambda_{s,i} = s \left(\frac{\beta i}{N} + \xi \right).$$

(E_2) Considering imperfect vaccination with vaccine failure probability h, a vaccinated individual can still become infected at rate

$$\eta_{v,i} = vh \left(\frac{\beta i}{N} + \xi \right).$$

(E_3) An infectious individual recovers with rate

$$\gamma_i = \gamma i.$$

Table 1. Possible events and their transition rates.

Effective outgoing event	Transition	Rate
Infection of susceptible individual	$(v, s, i) \rightarrow (v, s-1, i+1)$	$\lambda_{s,i}$
Infection of vaccinated individual	$(v, s, i) \rightarrow (v-1, s, i+1)$	$\eta_{v,i}$
Recovery	$(v, s, i) \rightarrow (v, s, i-1)$	γ_i

Times spent at each state $(v, s, i) \in \mathcal{S}$ are independent and exponentially distributed random variables, with rate $q_{v,s,i} = \lambda_{s,i} + \eta_{v,i} + \gamma_i$. The dynamics of $\mathcal{X}$ is determined by its infinitesimal generator, $\mathbf{Q}$, which one can efficiently construct by organising first the space of states $\mathcal{S}$ in terms of *levels* and *sub-levels*. In particular, for a particular initial state $(v_0, s_0, N - s_0 - v_0)$,

$$\mathcal{S} = \cup_{v=0}^{v_0} l(v),$$
$$l(v) = \cup_{s=0}^{s_0} l(v, s), \quad 0 \leq v \leq v_0,$$
$$l(v, s) = \{(v, s, i) \in \mathcal{S} : \ 0 \leq i \leq N - v - s\}, \quad 0 \leq s \leq s_0, \ 0 \leq v \leq v_0.$$

We note that the number of states in each sub-level is $\#l(v, s) = N - v - s + 1$, while the number of states in each level is $\#l(v) = (s_0 + 1)(N - v + 1) - \frac{s_0(s_0+1)}{2}$.

By ordering states within each sub-level as

$$(v, s, 0) \prec (v, s, 1) \prec \cdots \prec (v, s, N - v - s),$$

and ordering then states by sub-levels and levels, the infinitesimal generator of $\mathcal{X}$, $\mathbf{Q}$, is given by

$$\mathbf{Q} = \begin{pmatrix} \mathbf{Q}_{0,0} & & & & \\ \mathbf{Q}_{1,0} & \mathbf{Q}_{1,1} & & & \\ & \mathbf{Q}_{2,1} & \mathbf{Q}_{2,2} & & \\ & & \ddots & \ddots & \\ & & & \mathbf{Q}_{v_0,v_0-1} & \mathbf{Q}_{v_0,v_0} \end{pmatrix},$$

with $v_0, s_0 \geq 0$ and $v_0 + s_0 \leq N$.

We note that sub-matrices $\mathbf{Q}_{v,v^*}$ are of dimensions $\#l(v) \times \#l(v^*)$. Sub-matrices $\mathbf{Q}_{v,v}$, for $0 \leq v \leq v_0$, contain rates corresponding to transitions between states within the level $l(v)$. These events, according to the definition of levels and Table 1, correspond to susceptible individuals becoming infected, or infected individuals recovering. On the other hand, sub-matrices $\mathbf{Q}_{v,v-1}$, for $1 \leq v \leq v_0$, correspond to transitions from states in level $l(v)$ to states in level $l(v-1)$, which occur due to vaccinated individuals becoming infected. More specifically, sub-matrices $\mathbf{Q}_{v,v^*}$ are described as follows:

$$\mathbf{Q}_{v,v-1} = \begin{pmatrix} \mathbf{A}_{v,v-1}(0,0) & & & \\ & \mathbf{A}_{v,v-1}(1,1) & & \\ & & \ddots & \\ & & & \mathbf{A}_{v,v-1}(s_0,s_0) \end{pmatrix}, \quad 1 \le v \le v_0,$$

$$\mathbf{Q}_{v,v} = \begin{pmatrix} \mathbf{A}_{v,v}(0,0) & & & \\ \mathbf{A}_{v,v}(1,0) & \mathbf{A}_{v,v}(1,1) & & \\ & \mathbf{A}_{v,v}(2,1) & \mathbf{A}_{v,v}(2,2) & \\ & & \ddots & \ddots \\ & & \mathbf{A}_{v,v}(s_0,s_0-1) & \mathbf{A}_{v,v}(s_0,s_0) \end{pmatrix}, \quad 0 \le v \le v_0.$$

Sub-matrices $\mathbf{A}_{v,v-1}(s,s)$, for $1 \le v \le v_0$, $0 \le s \le s_0$, have dimensions $(N - v - s + 1) \times (N - v - s + 2)$, and contain the transition rates from states in sub-level $l(v,s)$ to states in sub-level $l(v-1,s)$. These transitions represent infections of vaccinated individuals. Sub-matrices $\mathbf{A}_{v,v}(s,s)$ contain the transition rates from states in sub-level $l(v,s)$ to states within the same sub-level, and correspond to recoveries of infected individuals. Sub-matrices $\mathbf{A}_{v,v}(s,s-1)$ contain transition rates from states in sub-level $l(v,s)$ to states in sub-level $l(v,s-1)$, corresponding to infections of susceptible individuals. In particular, these sub-matrices are defined as follows:

- For $0 \le v \le v_0$, $0 \le s \le s_0$, $\mathbf{A}_{v,v}(s,s)$ is a matrix of dimensions $(N - v - s + 1) \times (N - v - s + 1)$, with

$$\mathbf{A}_{v,v}(s,s) = \begin{pmatrix} -q_{v,s,0} & & & \\ \gamma & -q_{v,s,1} & & \\ & 2\gamma & -q_{v,s,2} & \\ & & \ddots & \ddots \\ & & & (N-v-s)\gamma & -q_{v,s,N-v-s} \end{pmatrix}.$$

- For $1 \le v \le v_0$, $0 \le s \le s_0$, $\mathbf{A}_{v,v-1}(s,s)$ is a matrix of dimensions $(N - v - s + 1) \times (N - v - s + 2)$, with

$$\mathbf{A}_{v,v-1}(s,s) = \begin{pmatrix} 0 & \eta_{v,0} & & \\ & 0 & \eta_{v,1} & \\ & & \ddots & \ddots \\ & & & 0 & \eta_{v,N-v-s} \end{pmatrix}.$$

202 M. Gamboa et al.

- For $0 \leq v \leq v_0$, $1 \leq s \leq s_0$, $\mathbf{A}_{v,v}(s, s-1)$ is a matrix of dimensions $(N - v - s + 1) \times (N - v - s + 2)$, with

$$
\mathbf{A}_{v,v}(s, s-1) = \begin{pmatrix} 0 & \lambda_{s,0} & & & \\ & 0 & \lambda_{s,1} & & \\ & & \ddots & & \ddots \\ & & & 0 & \lambda_{s,N-v-s} \end{pmatrix}.
$$

3 Time Until M Individuals Get Infected

In this section, we analyse the speed of transmission by focusing on the time that it takes for a threshold number M of individuals to get infected, $W(M)$. $W(M)$ is a non-negative continuous random variable that denotes the time elapsed until a total of M individuals become infected. In order to analyse this summary statistic, we redefine the CTMC as $\mathcal{X}^* = \{(J(t), S(t), I(t)) : t \geq 0\}$ where $S(t)$ and $I(t)$ denote the number of susceptible and infected individuals respectively at time t, and $J(t) = S(t) + V(t)$ represents the sum of vaccinated and susceptible individuals at time t. For an initial state (j_0, s_0, i_0) and a threshold value M of interest, with $1 \leq M \leq N$, $W(M)$ can be defined as

$$
W_{j_0,s_0,i_0}(M) = inf\{t \geq 0 : J(t) = N - M \mid (J(0), S(0), I(0)) = (j_0, s_0, i_0)\}.
$$

To analyse this random variable, one can study the evolution of the Markov chain $\mathcal{X}^*$ in the set of states $\mathcal{S}^* = \{(j, s, i) : N - M \leq j \leq j_0, max(0, j + s_0 - j_0) \leq s \leq s_0, max(0, N - M - j + 1) \leq i \leq N - j\}$, and where trivially $W_{j_0,s_0,i_0}(M) \equiv 0$ if $M \leq N - j_0$. Then, the variable $W_{j_0,s_0,i_0}(M)$ can be studied as first-passage time to the set of absorbing states $\mathcal{S}_M^* = \{(N - M, s, i) \in \mathcal{S}^*\}$ of the Markov chain $\mathcal{X}^*$.

For any initial state (j_0, s_0, i_0), and threshold value of interest $1 \leq M \leq N$, it is clear that $\mathbb{P}(W_{j_0,s_0,i_0}(M) < +\infty) = 1$, since the external source of infection ensures that all individuals will eventually become infected. On the other hand, the definition of $W_{j_0,s_0,i_0}(M)$ for the initial state of interest (j_0, s_0, i_0) can be extended to any other state $(j, s, i) \in \mathcal{S}^*$, and the random variable of interest $W_{j_0,s_0,i_0}(M)$ can be studied by analysing as well the auxiliary ones $W_{j,s,i}(M)$, $(j, s, i) \in \mathcal{S}^*$. In particular, we can introduce the Laplace-Stieltjes transforms for any $(j, s, i) \in \mathcal{S}^*$ as $\phi_{j,s,i}(z) = E\left[e^{-zW_{j,s,i}}\right]$, $z \in \mathbb{C}$, with $Re(z) \geq 0$, and where we omit M from notation from now on.

The Laplace-Stieltjes transforms $\phi_{j,s,i}(z)$ satisfy a set of linear equations, which is obtained via first-step arguments by conditioning on the possible transitions out of state $(j, s, i) \in \mathcal{S}^*$. In particular,

$$\phi_{j,s,i}(z) = (1 - \delta_{i,0})\frac{\gamma_i}{z + q_{j-s,s,i}}\phi_{j,s,i-1}(z)$$

$$+(1 - \delta_{s,0})\frac{\lambda_{s,i}}{z + q_{j-s,s,i}}\phi_{j-1,s-1,i+1}(z)$$

$$+\frac{\eta_{j-s,i}}{z + q_{j-s,s,i}}\phi_{j-1,s,i+1}(z), \tag{2}$$

where $\delta_{i,j}$ represents the Kronecker's delta function, defined as 1 when $i = j$, and 0 otherwise. This system of equations has boundary conditions $\phi_{N-M,s,i}(z) = 1$ for those states at which the number M of infections is reached. We can also note that, by definition, $\phi_{j,s,i}(0) = 1$, for any $(j, s, i) \in \mathcal{S}^*$.

These Laplace-Stieltjes transforms could be computed, at any point $z \in \mathbb{C}$, by solving system (2). Furthermore, with the help of numerical methods for Laplace transforms inversion, it is possible to calculate the probability distribution function of $W(M)$ [29, 30]. Although the numerical inversion is indeed possible, it is many times computationally not feasible. However, our interest instead here is in computing different order moments of these variables. In particular, moments can be computed from direct differentiation of the transform, as

$$m_{j,s,i}^k = E\left[W_{j,s,i}^k\right] = (-1)^k\frac{d^k\phi_{j,s,i}(z)}{dz^k}\bigg|_{z=0}, \quad k \geq 1. \tag{3}$$

Thus, by differentiating Eq. (2) with respect to z k times ($k \geq 1$) and evaluating at $z = 0$, we obtain the equations involving the moments as

$$q_{j,s,i}m_{j,s,i}^k = km_{j,s,i}^{k-1} + \lambda_{s,i}m_{j-1,s-1,i+1}^k + \eta_{j-s,i}m_{j-1,s,i+1}^k + \gamma_i m_{j,s,i-1}^k, \tag{4}$$

with boundary conditions $m_{j,s,i}^0 = 1$, $m_{N-M,s,i}^k = 0$ for any $k \geq 1$.

The loop-free structure of the transition rates of the CMTC $\mathcal{X}^*$ allows one to compute moments in a recursive way from the system above, for increasing values of $k \geq 1$ and taking into account that moments of order 0 are trivially equal to 1. Algorithm 1 outlines how to carry out this computation in an efficient and ordered way, which is based on Theorem 1 below. Proof of Theorem 1 is omitted here for the sake of brevity, since it consists of a recursive solution scheme directly based on Eq. (4).

Algorithm 1. Computation of the kth-order moments of the random variable $W_{j_0,s_0,i_0}(M)$, for $1 \le k \le k_{max}$ for some maximum desired order k_{max}

Input: $j_0, s_0, i_0, N, M, \beta, \xi, \gamma$ and $kmax$.

Step 1: Set $j = N - M$

    Step 1a: Set $s = max(0, j + s_0 - j_0)$

    Step 1b: Set $k = 0$ and set $m^0_{N-M,s,i} = 1$ for $max(0, N-M-j+1) \le i \le N-j$.

    Step 1c: Set $k = k+1$, set $m^k_{N-M,s,i} = 0$ for $max(0, N-M-j+1) \le i \le N-j$.

    Step 1d: If $k < kmax$, go to Step 1c.

    Step 1e: Set $s = s + 1$. If $s \le s_0$, go to Step 1b.

Step 2: Set $j = N - M + 1$.

    Step 2a: Set $s = max(0, j + s_0 - j_0)$.

    Step 2b: Set $k = 0$ and set $m^0_{j,s,i} = 1$ for $max(0, N - M - j + 1) \le i \le N - j$.

    Step 2c: Set $k = 1$ and set $m^k_{j,s,i}$ for $max(0, N - M - j + 1) \le i \le N - j$, from (6).

    Step 2d: Set $k = k+1$ and compute $m^k_{j,s,i}$ for $max(0, N-M-j+1) \le i \le N-j$, from (7)-(8).

    Step 2e: If $k < kmax$, go to Step 2d.

    Step 2f: If $s < s_0$, set $s = s + 1$ and go to Step 2b.

Step 3: If $j = j_0$, stop.

Step 4: Set $j = j + 1$.

    Step 4a: Set $s = max(0, j + s_0 - j_0)$.

    Step 4b: Set $k = 0$ and set $m^0_{j,s,i} = 1$ from $max(0, N - M - j + 1) \le i \le N - j$.

    Step 4c: Set $k = k+1$ and compute $m^k_{j,s,i}$ for $max(0, N-M-j+1) \le i \le N-j$, from (7)-(8).

    Step 4d: If $k < kmax$, go to Step 4c.

    Step 4e: If $s < s_0$, set $s = s + 1$ and go to Step 4b.

Step 5: If $j < j_0$, go to Step 4. If $j = j_0$, stop.

Output: $m^k_{j_0,s_0,i_0}$, for $0 \le k \le kmax$.

Theorem 1. *Given a number of initial vaccinated and susceptible individuals $v_0 \ge 0$ and $s_0 \ge 0$, with $0 \le v_0 + s_0 \le N$ and an integer k, $k \ge 0$, and $1 \le M \le N$, the central moments of order k of the variable $W_{j_0,s_0,i_0}(M)$, are obtained from the following expressions for all $(j, s, i) \in \mathcal{S}^*$:*

$$m^0_{j,s,i} = 1, \quad m^k_{N-M,s,i} = 0, \; for \; k \ge 1, \tag{5}$$

$$m^1_{N-M+1,s,i} = \sum_{r=0}^{i} \frac{i! \frac{\gamma^{i-r}}{r!}}{\prod_{l=r}^{i} q_{N-M-s+1,s,l}}, \tag{6}$$

$$m^k_{j,s,i} = \sum_{r=0}^{i} \frac{i! \frac{\gamma^{i-r}}{r!} T^k_{j,s,r}}{\prod_{l=r}^{i} q_{j-s,s,l}} \; for \; k \ge 1 \tag{7}$$

with

$$T^k_{j,s,i} = k m^{k-1}_{j,s,i} + (1 - \delta_{s,0})\lambda_{s,i} m^k_{j-1,s-1,i+1} + (1 - \delta_{j,s})\eta_{j-s,i} m^k_{j-1,s,i+1}. \tag{8}$$

4   Results

In this section, we illustrate our analysis in Sect. 3 by carrying out a global sensitivity analysis on model parameters for the summary statistic of interest. We set the recovery rate $\gamma = 1.0$ in all the numerical experiments, so that the time unit is taken as the expected time that an infected individual takes to recover. We consider a population of $N = 100$ individuals here, and assume that 50% of them are partially protected against the infection through the vaccine, so that the initial state is $(v_0, s_0, i_0) = (50, 49, 1)$.

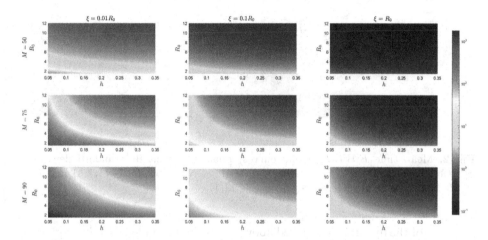

Fig. 1. Mean time $E[W(M)]$ until M individuals get infected, for different values of M, R_0, h and ξ. Initial state $(v_0, s_0, i_0) = (50, 49, 1)$.

In Fig. 1, we plot $\overline{W}_M = E[W(M)]$ for different values of the Basic Reproduction Number, $R_0 = \beta/\gamma$, ξ, h and M. The average time to reach a total of M infections increases with increasing values of M, as one would expect. On the other hand, $\overline{W}_M$ decreases with the external source of infection rate, ξ, since these external infections can contribute towards reaching the threshold M. An interplay can be observed between the value of the reproduction number R_0 and the vaccine failure probability h, so that large values of $\overline{W}_M$ can be due to small transmission rates (small R_0) or to small probability of vaccine failure, h. We note that the value of M, together with the proportion of individuals initially vaccinated, are directly relevant to understand the dynamics in Fig. 1. The relevance of h is observed to be smaller for $M = 50$, since in this situation the outbreak can reach 50 infections just by those infections suffered by susceptible individuals in this system. On the other hand, increasing values of M require infections to happen among the vaccinated sub-population, and thus small values of the vaccine failure probability lead to significantly increased times $\overline{W}_M$ to reach M infections. We also note that, for small values of ξ (e.g.; $\xi = 0.01R_0$), the mean time $\overline{W}_M$ to reach M infections can span several orders of magnitude for different values of the parameters (M, R_0, h). This can be explained by the fact that, if the external source of infection is small and the outbreak was to finish without the level M of infections being reached, one would need to wait until a subsequent outbreak to occur in the remaining susceptible/vaccinated population, which would take long under small values of ξ. Larger values of ξ lead to "overlapping" outbreaks, where external infections can constantly occur, facilitating smaller values of the mean time $\overline{W}_M$.

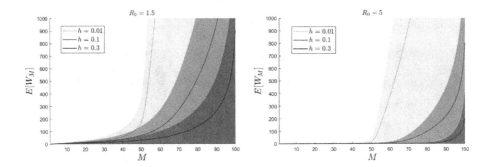

Fig. 2. Mean time $E[W(M)]$ (solid curves) plus and minus its standard deviation $\sigma[W(M)]$ (shaded area) versus M, for $\xi = 0.01R_0$, $N = 100$, $R_0 \in \{1.5, 5\}$ and $h \in \{0.01, 0.1, 0.3\}$. Initial state $(v_0, s_0, i_0) = (50, 49, 1)$.

Some of the dynamics described above can be better understood by exploring Fig. 2, that shows the expected time elapsed until M infections have been reached as a function of M, for a relatively small value of $\xi = 0.01R_0$ and for several values of h and two different values of R_0. Shaded areas are obtained by considering $E[W(M)] \pm \sigma[W(M)]$. As expected, increasing values of R_0 or decreasing values of h lead to increasing times to reach M infections. On the other hand, vaccines with higher probability of failure lead to situations where less time is needed in order to reach M infections, and in consequence the expansion of the disease is faster. This behaviour reveals the importance of the vaccine effectiveness. Particularly interesting is the asymptotic behaviour of the curves, where the time to reach M infections can significantly increase when approaching particular values of M in some situations. This is directly related to the vaccine failure probability h, and the initial number of susceptible and vaccinated individuals $(s_0, v_0) = (49, 50)$. In particular, and when focusing for example on $R_0 = 1.5$ and $h = 0.01$, the small vaccine failure probability means that infections in order to reach the threshold value M are likely to occur among susceptible individuals, and unlikely to happen among vaccinated ones. Since 50% of the population is vaccinated, and we start with 1 infected individual, up to $M = 50$ individuals can become infected in relatively short periods of time (given that $R_0 = 1.5$) by infections happening in the susceptible pool. However, as soon as M exceeds the value 50 infections among the vaccinated pool are required to happen for this threshold to be reached. These infections would be rare ($h = 0.01$), leading to significant increases in the expected time $E[W(M)]$. These behaviours nicely illustrate the protection that a nearly perfect vaccination confers to the pool of vaccinated individuals, where a relatively fast outbreak (due to $R_0 = 1.5 > 1$) would decelerate when approaching $M = s_0 + i_0$. For significantly small values of h (e.g.; $h = 0.01$), the dynamics described above are relatively similar regardless of considering $R_0 = 5$ instead of $R_0 = 1.5$, although increasing values of R_0 facilitate an *overshoot* effect, as can be observed when comparing the two plots in Fig. 2. For relatively larger vaccine failure probabilities (e.g.; $h = 0.1$ or $h = 0.3$), these asymptotic behaviours can be partially compensated by increasing values of R_0, where some infections in the vaccinated pool can be achieved due to the large value of R_0, facilitating the attainment of the threshold number of infections M.

Numerical experiments show that the expected value of $W(M)$ presents an increasing behaviour, as a function of M. Moreover, when we increase the vaccination coverage v_0 and keep fixed the remaining model parameters, the mean time to achieve a number of M infections also increases. This is in accordance to intuition because when an outbreak starts with a big proportion of vaccine protected individuals, infections are becoming less likely and the time to infect M individuals is larger in comparison with outbreaks starting with a lesser number of vaccinated individuals.

Computational times are very high and complexity increases when considering populations larger than 1000 individuals. For instance, when $N = 1000$ individuals the state space $\mathcal{S}^*$ contains around 4.16×10^7 states, while for a population of 10000 individuals the number of states increases to 4.16×10^{10}. The elapsed time to compute $E[W(M)]$ takes around 4 min when $N = 1000$ and it lasts more than 5 h when $N = 10000$, in a personal computer with 8 GB of RAM, M1 memory Chip with GPU of 7 Kernels.

5   Conclusions

In this paper, we have considered a stochastic SVIR model with imperfect vaccine and external source of infection. We have represented this in terms of a multidimensional continuous-time Markov chain, and have showed that by appropriately ordering its space of states in terms of levels and sub-levels, this leads to the study of a LD-QBD. Our interest was in analysing the speed at which the epidemic occurs, by studying the time to reach a threshold number M of infections in the population. By means of first-step arguments, we have obtained a system of linear equations which can be solved efficiently and recursively, as outlined in Algorithm 1. In our results in Sect. 4, we have illustrated our methodology by carrying out a wide sensitivity analysis on model parameters, where an interplay can be observed between the reproduction number R_0, the threshold of interest M, the vaccine failure probability h, the external source of infection rate ξ, and the initial number of vaccinated individuals v_0. Our techniques can in principle be applied in order to study other summary statistics of potential interest in this system, such as the exact reproduction number [24,31] or the time until the end of the outbreak [8]. This remains the aim of future work.

References

1. Heathcote, H.: Asymptotic behavior in a deterministic epidemic model. Bull. Math. Biol. **35**(5), 607–14 (1973). https://doi.org/10.1016/S0092-8240(73)80057-6
2. Bartlett, M.S.: Deterministic and stochastic models for recurrent epidemics. In: Contributions to Biology and Problems of Health. University of California Press California (2020). https://doi.org/10.1525/9780520350717-007
3. Andersson, H., Britton, T.: Stochastic Epidemic Models and Their Statistical Analysis. Lecture Notes in Statistics, Springer, New York (2000). https://doi.org/10.1007/978-1-4612-1158-7
4. Allen, L.J.S.: A primer on stochastic epidemic models: formulation, numerical simulation, and analysis. Infect. Dis. Model. **2**(2), 128–142 (2017). https://doi.org/10.1016/j.idm.2017.03.001
5. Kermack, W.O., McKendrick, A.G.: A contribution to the mathematical theory of epidemics. Proc. R. Soc. Lond. Ser. A **115**, 700–721 (1927). https://doi.org/10.1098/rspa.1927.0118
6. Bailey, N.T.: A simple stochastic epidemic. Biometrika **37**(3–4), 193–202 (1950). https://doi.org/10.2307/2333107
7. Whittle, P.: The outcome of a stochastic epidemic–a note on Bailey's paper. Biometrika **42**(1–2), 116–122 (1955). https://doi.org/10.2307/2333427
8. López-García, M.: Stochastic descriptors in an SIR epidemic model for heterogeneous individuals in small networks. Math. Biosci. **271**, 42–61 (2016). https://doi.org/10.1016/j.mbs.2015.10.010
9. Artalejo, J.R., Economou, A., Lopez-Herrero, M.J.: Stochastic epidemic models with random environment: quasi-stationarity, extinction and final size. J. Math. Biol. **67**(4), 799–831 (2013). https://doi.org/10.1007/s00285-012-0570-5
10. Almaraz, E., Gómez-Corral, A.: On SIR-models with Markov-modulated events: length of an outbreak, total size of the epidemic and number of secondary infections. Discret. Contin. Dyn. Syst.-B **23**(6), 2153 (2018). https://doi.org/10.3934/dcdsb.2018229
11. Gamboa, M., Lopez-Herrero, M.J.: On the number of periodic inspections during outbreaks of discrete-time stochastic SIS epidemic models. Mathematics **6**(8), 128 (2018). https://doi.org/10.3390/math6080128
12. Jacquez, J.A., O'Neill, P.: Reproduction numbers and thresholds in stochastic epidemic models I. Homogeneous populations. Math. Biosci. **107**(2), 161–186 (1991). https://doi.org/10.1016/0025-5564(91)90003-2
13. Nguyen, C., Carlson, J.M.: Optimizing real-time vaccine allocation in a stochastic SIR model. PloS One **11**(4), e0152950 (2016). https://doi.org/10.1371/journal.pone.0152950
14. Ball, F., Sirl, D.: Evaluation of vaccination strategies for SIR epidemics on random networks incorporating household structure. J. Math. Biol. **76**(1), 483–530 (2018). https://doi.org/10.1007/s00285-017-1139-0
15. Arino, J., McCluskey, C.C., van den Driessche, P.: Global results for an epidemic model with vaccination that exhibits backward bifurcation. SIAM J. Appl. Math. **64**(1), 260–276 (2003). https://doi.org/10.1137/S0036139902413829
16. Kribs-Zaleta, C.M., Martcheva, M.: Vaccination strategies and backward bifurcation in an age-since-infection structured model. Math. Biosci. **177**, 317–332 (2002). https://doi.org/10.1016/S0025-5564(01)00099-2
17. Ball, F., Lyne, O.: Optimal vaccination schemes for epidemics among a population of households, with application to variola minor in Brazil. Stat. Methods Med. Res. **15**(5), 481–497 (2006). https://doi.org/10.1177/0962280206071643

18. Charania, N.A., Moghadas, S.M.: Modelling the effects of booster dose vaccination schedules and recommendations for public health immunization programs: the case of Haemophilus influenzae serotype b. BMC Public Health **17**(1), 1–8 (2017). https://doi.org/10.1186/s12889-017-4714-9

19. Gandon, S., Mackinnon, M.J., Nee, S., Read, A.F.: Imperfect vaccines and the evolution of pathogen virulence. Nature **414**(6865), 751–756 (2001). https://doi.org/10.1038/414751a

20. Magpantay, F.M., Riolo, M.A., De Celles, M.D., King, A.A., Rohani, P.: Epidemiological consequences of imperfect vaccines for immunizing infections. SIAM J. Appl. Math. **74**(6), 1810–1830 (2014). https://doi.org/10.1137/140956695

21. Iannelli, M., Martcheva, M., Li, X.Z.: Strain replacement in an epidemic model with super-infection and perfect vaccination. Math. Biosci. **195**, 23–46 (2005). https://doi.org/10.1016/j.mbs.2005.01.004

22. Demicheli, V., Rivetti, A., Debalini, M.G., Di Pietrantonj, C.: Vaccines for measles, mumps and rubella in children. Cochrane Database Syst. Rev. 2, CD004407 (2012). https://doi.org/10.1002/ebch.1948

23. Ball, F., O'Neill, P.D., Pike, J.: Stochastic epidemic models in structured populations featuring dynamic vaccination and isolation. J. Appl. Probab. **44**(3), 571–585 (2007). https://doi.org/10.1239/jap/1189717530

24. Gamboa, M., Lopez-Herrero, M.J.: Measuring infection transmission in a stochastic SIV model with infection reintroduction and imperfect vaccine. Acta Biotheoretica **68**(4), 395–420 (2020). https://doi.org/10.1007/s10441-019-09373-9

25. Gamboa, M., Lopez-Herrero, M.J.: The effect of setting a warning vaccination level on a stochastic SIVS model with imperfect vaccine. Mathematics **8**(7), 1136 (2020). https://doi.org/10.3390/math8071136

26. Kiouach, D., Boulaasair, L.: Stationary distribution and dynamic behaviour of a stochastic SIVR epidemic model with imperfect vaccine. J. Appl. Math. (2018). https://doi.org/10.1155/2018/1291402

27. El Koufi, A., Adnani, J., Bennar, A., Yousfi, N.: Analysis of a stochastic SIR model with vaccination and nonlinear incidence rate. Int. J. Diff. Equ. (2019). https://doi.org/10.1155/2019/9275051

28. Gómez-Corral, A., López-García, M., Lopez-Herrero, M.J., Taipe, D.: On first-passage times and sojourn times in finite QBD processes and their applications in epidemics. Mathematics **8**(10), 1718 (2020). https://doi.org/10.3390/math8101718

29. Gloub, G.H., Van Loan, C.F.: Matrix computations, 3rd edn. Johns Hopkins University Press (1996). https://doi.org/10.2307/3619868

30. Cohen, A.M.: Numerical Methods for Laplace Transforms Inversion. Springer, Boston (2007). https://doi.org/10.1007/978-0-387-68855-8

31. Artalejo, J.R., Lopez-Herrero, M.J.: On the exact measure of the disease spread in stochastic epidemic models. Bull. Math. Biol. **75**, 1031–1050 (2013). https://doi.org/10.1007/s11538-013-9836-3

Analysis of Single Bacterium Dynamics in a Stochastic Model of Toxin-Producing Bacteria

Jamie Paterson[1], Martín López-García[1(✉)], Joseph Gillard[2],
Thomas R. Laws[2], Grant Lythe[1], and Carmen Molina-París[1,3]

[1] School of Mathematics, University of Leeds, Leeds, UK
m.lopezgarcia@leeds.ac.uk
[2] CBR Division, Defence Science and Technology Laboratory, Salisbury, UK
[3] T-6 Theoretical Biology and Biophysics, Theoretical Division,
Los Alamos National Laboratory, Los Alamos, NM, USA

Abstract. We stochastically model two bacterial populations which can produce toxins. We propose to analyse this biological system by following the dynamics of a single bacterium during its lifetime, as well as its progeny. We study the lifespan of a single bacterium, the number of divisions that this bacterium undergoes, and the number of toxin molecules that it produces during its lifetime. We also compute the mean number of bacteria in the genealogy of the original bacterium and the number of toxin molecules produced by its genealogy. We illustrate the applicability of our methods by considering the bacteria *Bacillus anthracis* and antibiotic treatment, making use of *in vitro* experimental data. We quantify, for the first time, bacterial toxin production by exploiting an *in vitro* assay for the *A16R* strain, and make use of the resulting parameterised model to illustrate our techniques.

Keywords: Bacteria · Toxins · Stochastic model · Continuous time · Markov chain · Single cell · Antibiotic

1 Introduction

Mathematical modelling has proven to be a robust approach to analyse biological systems of relevance in infection and immunity at different scales, such as the molecular [24], intra-cellular [6], within-host [7] and population (or epidemic) levels [4]. While deterministic models are usually more amenable for mathematical analysis [1], stochastic methods are generally better suited for characterising biological systems involving few individuals [23] or cells [7], or when extinction events play a crucial role [5]. Markov processes, either in discrete or continuous time, have been used in such instances given their mathematical convenience [2]. While non-Markovian dynamics are typically more difficult to analyse [8,14], the Markovian or memoryless property usually allows for mathematical tractability and efficient numerical implementation [12].

© The Author(s) 2021
P. Ballarini et al. (Eds.): EPEW 2021/ASMTA 2021, LNCS 13104, pp. 210–225, 2021.
https://doi.org/10.1007/978-3-030-91825-5_13

When considering a population of cells in an immune response, or bacteria during an infection, competition for resources is usually represented in terms of logistic growth models [1]. On the other hand, when individuals behave independently (*e.g.*, they do not compete for common resources), the theory of branching processes [18] has been widely applied to follow these populations (of cells or bacteria) over time. Multi-type branching processes [20] allow one to consider different types of bacteria, which might represent different phenotypes [9] or different spatial locations (*e.g.*, tissues or organs) within the body during an infection [7]. The complexity of these processes, and their mathematical tractability, typically depends on the number of compartments considered, and the number of potential events that can occur in the system (*e.g.*, division or death of bacteria, or bacterial movement across compartments) [26].

Novel technological developments have recently allowed for single cells to be precisely followed, together with their progeny [15,17,19,27]. This motivates the idea of mathematically tracking single individuals in these stochastic systems, and to quantify summary statistics related to the lifetime of a single individual (or bacterium in our case), and its progeny or genealogy. Analysing the dynamics of the system by tracking a single individual has already been proposed in related areas such as population dynamics [13] and, more recently, when analysing the stochastic journey of T lymphocytes in lymph nodes and blood [16].

Bacterial systems have been widely studied with stochastic methods in the past [6,7], yet less attention has been paid to the study of toxin-producing bacteria. The production of toxins over time can be especially relevant for certain kinds of bacteria for which the secreted toxins can cause suppression of the host's immune system, and are a key component of pathogenesis *in vivo* [3]. In this work, we illustrate our single cell approach in a stochastic model of two types of toxin-producing bacteria. In particular, we focus on computing the expected lifespan of a single bacterium in this system, as well as the number of toxin molecules secreted and the number of divisions undergone during its lifetime. We also compute two summary statistics that are directly related to the progeny of a single bacterium: the number of bacteria within its genealogy and the number of toxin molecules produced by its genealogy. We illustrate our results by focusing on the bacterium *Bacillus anthracis* and its anthrax toxins. For the *A16R B. anthracis* strain we quantify for the first time the rate of protective antigen (PA) production making use of published data from an *in vitro* experimental assay [28]. The resulting parametrized mathematical model serves to illustrate our techniques and allows us to consider antibiotic treatment.

The structure of the manuscript is as follows: in Sect. 2 we introduce the mathematical model. The single bacterium model is discussed in Sect. 3. A number of summary statistics of interest related to a single bacterium and its progeny are analytically studied in Sect. 3. Model calibration for the *A16R B. anthracis* strain is carried out in Sect. 4 using data from an *in vitro* experimental assay, and the parameterised model is used in this section to illustrate our methods. Concluding remarks are provided in Sect. 5.

2   The Mathematical Model

Our interest is in modelling a system with two toxin-producing bacterial populations (see Fig. 1). Type-i bacteria, $i \in \{1,2\}$, can divide with rate λ_i, produce toxins with rate γ_i, die with rate μ_i, or become type-j bacteria, $j \in \{1,2\}$ $j \neq i$, with rate ν_{ij}. We propose a stochastic model of these events as a continuous time Markov chain (CTMC) $\mathcal{X} = \{(B_1(t), B_2(t), T(t)) : t \geq 0\}$, where $B_i(t)$ denotes the number of type-i bacteria at time $t \geq 0$, $i \in \{1,2\}$, and $T(t)$ represents the number of toxin molecules at time $t \geq 0$. We assume that bacteria and toxins behave independently of each other, and that toxins are degraded at rate ξ. The space of states of $\mathcal{X}$ is given by $\mathcal{S} = \mathbb{N}_0^3$, where we denote $\mathbb{N}_0 = \mathbb{N} \cup \{0\}$, and the possible one-step transitions between states in $\mathcal{X}$ are depicted in Fig. 1.

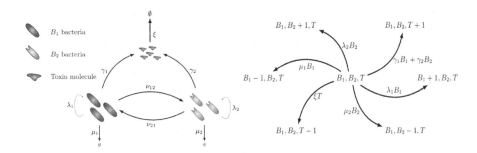

Fig. 1. Left. Diagram showing the dynamics of the two toxin-producing bacterial populations. **Right.** Allowed transitions between states in $\mathcal{X}$ and their rates.

Since each bacterium behaves independently, one can analyse the dynamics of a single bacterium without explicitly modelling the dynamics of the rest of the population. In Sect. 3, we propose a method which allows us to analyse the dynamics of a single bacterium and its progeny. In particular, and by means of first step arguments, we compute the lifespan of a single bacterium, the number of divisions that this bacterium undergoes, and the number of toxin molecules that it produces during its lifetime. We also compute the mean number of cells within the genealogy of the original bacterium and the number of toxin molecules produced by this progeny. We note that a particular advantage of this single bacterium approach is that it can be implemented regardless of the complexity of the model, i.e., regardless of the number of compartments in the model, two compartments in our model (see Fig. 1), or the number of events governing the toxin and bacterial dynamics across compartments, as long as the dynamics of each bacterium is independent of the rest of the population.

3   Dynamics of a Single Bacterium and Its Progeny

Our interest in this section is in following a single bacterium of type i during its lifespan, instead of focusing on the population CTMC $\mathcal{X}$. In particular, we consider a single bacterium (either of type-1 or type-2) at time $t = 0$, and follow its dynamics during its lifetime by studying the continuous time Markov chain $\mathcal{Y} = \{Y(t) : t > 0\}$, where $Y(t)$ represents the *"state"* of the bacterium at time $t \geq 0$. By state, we mean that the bacterium can be of type-1, type-2 or dead at any given time. Thus, $\mathcal{Y}$ is defined on the state space $S = \{B_1, B_2, \emptyset\}$, where B_i here represents the bacterium being of type-i at any given instant, and $\emptyset$ indicates the bacterium is dead. If the bacterium is of type-i at a given instant, meaning that $\mathcal{Y}$ is in state B_i, production of a toxin molecule does not change its state, and $\mathcal{Y}$ remains in B_i. If a division occurs, we randomly choose one of the daughter cells and consider it to be our bacterium of interest, which remains in state B_i.

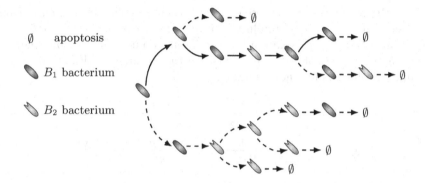

Fig. 2. Example of a stochastic realisation of the population process, starting with one type-1 bacterium. Solid arrows indicate the single bacterium being tracked in process $\mathcal{Y}$. In this realisation, the stochastic process $\mathcal{Y}$ visits states $B_1 \rightarrow B_1 \rightarrow B_1 \rightarrow B_2 \rightarrow B_1 \rightarrow B_1 \rightarrow \emptyset$. Consecutive visits to the same state are due to bacterial division. Toxin production is not explicitly depicted here but can occur during the process.

Figure 2 shows one realisation of the population dynamics for our biological system. The state of the stochastic process $\mathcal{Y}$ only depends on tracking the original bacterium throughout its lifetime, which is depicted via solid arrows. When a division occurs, a daughter is randomly chosen to represent the tracked bacterium of interest. In the following sections we investigate a number of stochastic descriptors or summary statistics that relate to the single bacterium, as well as its genealogy.

3.1 Lifespan of a Bacterium

For an initial bacterium of type i, $i \in \{1, 2\}$, we define its lifespan as the random variable, $T_i = \inf\{t \geq 0 : Y(t) = \emptyset | Y(0) = B_i\}$. We consider the Laplace-Stieltjes transform of T_i given by

$$\phi_i(s) = \mathbb{E}[e^{-sT_i}], \quad \mathrm{Re}(s) \geq 0,$$

which one can compute with first step arguments. This leads to the following equations

$$\phi_1(s) = \frac{\lambda_1}{\Delta_1 + s}\phi_1(s) + \frac{\mu_1}{\Delta_1 + s} + \frac{\gamma_1}{\Delta_1 + s}\phi_1(s) + \frac{\nu_{12}}{\Delta_1 + s}\phi_2(s),$$

$$\phi_2(s) = \frac{\lambda_2}{\Delta_2 + s}\phi_2(s) + \frac{\mu_2}{\Delta_2 + s} + \frac{\gamma_2}{\Delta_2 + s}\phi_2(s) + \frac{\nu_{21}}{\Delta_2 + s}\phi_1(s),$$

with $\Delta_i = \lambda_i + \mu_i + \gamma_i + \nu_{ij}$, $j \in \{1, 2\}$, $j \neq i$. These equations simplify to

$$(\mu_1 + \nu_{12} + s)\,\phi_1(s) = \nu_{12}\phi_2(s) + \mu_1,$$

$$(\mu_2 + \nu_{21} + s)\,\phi_2(s) = \nu_{21}\phi_1(s) + \mu_2.$$

Interestingly, we can see that these equations do not depend on the parameters λ_i (division rate) or γ_i (toxin production rate). This is consistent with our expectations, since division and toxin production events do not affect the lifespan of a bacterium, as can be noticed from inspecting the dynamics in Fig. 1 and Fig. 2. We can find solutions for these equations as follows

$$\phi_1(s) = a^{-1}(s)\frac{1}{\mu_1 + \nu_{12} + s}\left(\frac{\nu_{12}\mu_2}{\mu_2 + \nu_{21} + s} + \mu_1\right),$$

$$\phi_2(s) = a^{-1}(s)\frac{1}{\mu_2 + \nu_{21} + s}\left(\frac{\nu_{21}\mu_1}{\mu_1 + \nu_{12} + s} + \mu_2\right),$$

with $a(s) = 1 - \frac{\nu_{12}\nu_{21}}{(\mu_1 + \nu_{12} + s)(\mu_2 + \nu_{21} + s)}$. We also note that these expressions would simplify for particular scenarios of the bacterial system. For instance, if the change from type-1 to type-2 bacterium was irreversible so that $\nu_{21} = 0$, one obtains

$$\phi_1(s) = \frac{1}{\mu_1 + \nu_{12} + s}\left(\frac{\nu_{12}\mu_2}{\mu_2 + s} + \mu_1\right),$$

$$\phi_2(s) = \frac{\mu_2}{\mu_2 + s},$$

where we note that in this case $T_2 \sim \mathrm{Exp}(\mu_2)$. This is an interesting and important case to consider since the bacterial conversion with rate ν_{12} and reversion rate $\nu_{21} = 0$ represents the irreversible antibiotic treatment we study and analyse in Sect. 4

One can use the Laplace-Stieltjes transform to compute any order moment of T_i by direct differentiation. For example, the average lifetime of a type-i bacterium is given by

$$\mathbb{E}[T_1] = a^{-1}(0)\frac{1}{\mu_1 + \nu_{12}}\left(\frac{\nu_{12}}{\mu_2 + \nu_{21}} + 1\right),$$

$$\mathbb{E}[T_2] = a^{-1}(0)\frac{1}{\mu_2 + \nu_{21}}\left(\frac{\nu_{21}}{\mu_1 + \nu_{12}} + 1\right).$$

The Laplace-Stieltjes transform allows one to find higher order moments. One such example is the second order moment of the lifespan of a bacterium starting in state 1 when $\nu_{21} = 0$, which is given by

$$\mathbb{E}[T_1^2] = \frac{2}{\mu_1 + \nu_{12}} \left(\frac{1}{\mu_1 + \nu_{12}} + \frac{\nu_{12}}{\mu_2(\mu_1 + \nu_{12})} + \frac{\nu_{12}}{\mu_2^2} \right).$$

3.2   Number of Toxin Molecules Produced by a Bacterium in Its Lifetime

We denote by ω_i the random variable that describes the number of toxin molecules produced by the tracked bacterium during its lifetime, if this bacterium is initially of type i, $i \in \{1, 2\}$. We consider its probability generating function defined as follows

$$\psi_i(z) = \mathbb{E}[z^{\omega_i}],$$

for $|z| \leq 1$. By means of a first step argument, one can show that

$$(\mu_1 + \gamma_1(1 - z) + \nu_{12})\psi_1(z) = \nu_{12}\psi_2(z) + \mu_1,$$
$$(\mu_2 + \gamma_2(1 - z) + \nu_{21})\psi_2(z) = \nu_{21}\psi_1(z) + \mu_2.$$

The equations above have the following solutions

$$\psi_1(z) = b^{-1}(z) \frac{1}{\mu_1 + \gamma_1(1 - z) + \nu_{12}} \left(\frac{\nu_{12}\mu_2}{\mu_2 + \nu_{21} + \gamma_2(1 - z)} + \mu_1 \right),$$
$$\psi_2(z) = b^{-1}(z) \frac{1}{\mu_2 + \gamma_2(1 - z) + \nu_{21}} \left(\frac{\nu_{21}\mu_1}{\mu_1 + \nu_{12} + \gamma_1(1 - z)} + \mu_2 \right),$$

with $b(z) = 1 - \frac{\nu_{12}\nu_{21}}{(\mu_1 + \gamma_1(1-z) + \nu_{12})(\mu_2 + \gamma_2(1-z) + \nu_{21})}$. Once again, the particular case where $\nu_{21} = 0$ leads to simplified solutions, given by

$$\psi_1(z) = \frac{1}{\mu_1 + \gamma_1(1 - z) + \nu_{12}} \left(\frac{\mu_2}{\mu_2 + \gamma_2(1 - z)} \nu_{12} + \mu_1 \right),$$
$$\psi_2(z) = \frac{\mu_2}{\mu_2 + \gamma_2(1 - z)}.$$

We note that in this case $\omega_2 \sim \text{Geo}(\frac{\mu_2}{\mu_2 + \gamma_2})$. If $\nu_{21} = 0$, it is also possible to obtain the probability mass function of ω_1, which for $n = 0, 1, 2, \ldots$, it can be written as follows

$$\mathbb{P}(\omega_1 = n) = \gamma_1^n \left(\frac{\mu_2}{\gamma_2 + \mu_2} \nu_{12} + \mu_1 \right) (\gamma_1 + \nu_{12} + \mu_1)^{-(n+1)}$$
$$+ \nu_{12} \frac{\mu_2}{\gamma_2 + \mu_2} \sum_{k=0}^{n-1} \gamma_1^k \left(\frac{\gamma_2}{\gamma_2 + \mu_2} \right)^{n-k} (\gamma_1 + \nu_{12} + \mu_1)^{-(k+1)},$$

where the sum above is equal to 0 when $n = 0$. The mean number of toxin molecules produced by a single bacterium can be computed from direct differentiation of $\psi_i(z)$ with respect to z. One can show that

$$\mathbb{E}[\omega_1] = b^{-1}(1) \frac{1}{\mu_1 + \nu_{12}} \left(\frac{\gamma_2 \nu_{12}}{\mu_2 + \nu_{21}} + \gamma_1 \right),$$

$$\mathbb{E}[\omega_2] = b^{-1}(1) \frac{1}{\mu_2 + \nu_{21}} \left(\frac{\gamma_1 \nu_{21}}{\mu_1 + \nu_{12}} + \gamma_2 \right).$$

3.3   Number of Division Events in the Lifespan of a Bacterium

Let us consider now the number of times that the tracked bacterium divides during its lifetime, D_i, if this bacterium is originally of type i, $i \in \{1, 2\}$. We can define its probability generating function as $\Phi_i(z) = \mathbb{E}[z^{D_i}]$ for $|z| \leq 1$. $\Phi_i(z)$ satisfies the following equations:

$$\Delta_1 \Phi_1(z) = \lambda_1 z \Phi_1(z) + \mu_1 + \gamma_1 \Phi_1(z) + \nu_{12} \Phi_2(z),$$
$$\Delta_2 \Phi_2(z) = \lambda_2 z \Phi_2(z) + \mu_2 + \gamma_2 \Phi_2(z) + \nu_{21} \Phi_1(z).$$

These equations have solutions

$$\Phi_1(z) = c^{-1}(z) \frac{1}{\mu_1 + \nu_{12} + \lambda_1(1-z)} \left(\frac{\mu_2 \nu_{12}}{\mu_2 + \nu_{21} + \lambda_2(1-z)} + \mu_1 \right),$$

$$\Phi_2(z) = c^{-1}(z) \frac{1}{\mu_2 + \nu_{21} + \lambda_2(1-z)} \left(\frac{\mu_1 \nu_{21}}{\mu_1 + \nu_{12} + \lambda_1(1-z)} + \mu_2 \right),$$

with $c(z) = 1 - \frac{\nu_{12} \nu_{21}}{(\mu_1 + \nu_{12} + \lambda_1(1-z))(\mu_2 + \nu_{21} + \lambda_2(1-z))}$. We note that these expressions, as one would expect, do not depend on the toxin production rate, γ_i. The desired average number of divisions is then given by

$$\mathbb{E}[D_1] = c^{-1}(1) \frac{1}{\mu_1 + \nu_{12}} \left(\frac{\lambda_2 \nu_{12}}{\mu_2 + \nu_{21}} + \lambda_1 \right),$$

$$\mathbb{E}[D_2] = c^{-1}(1) \frac{1}{\mu_2 + \nu_{21}} \left(\frac{\lambda_1 \nu_{21}}{\mu_1 + \nu_{12}} + \lambda_2 \right).$$

Once again, particular scenarios might lead to simplified expressions. If one sets $\nu_{21} = 0$, this yields

$$\mathbb{E}[D_1] = \frac{1}{\mu_1 + \nu_{12}} \left(\frac{\lambda_2 \nu_{12}}{\mu_2} + \lambda_1 \right),$$

$$\mathbb{E}[D_2] = \frac{\lambda_2}{\mu_2}.$$

This choice implies $D_2 \sim \text{Geo}(\frac{\mu_2}{\mu_2 + \lambda_2})$.

3.4   Number of Bacteria in the Genealogy of a Bacterium

We focus now on the random variable describing the number of bacteria in the genealogy of the original bacterium (see Fig. 2). We denote this number as G_i, with i indicating the original bacterium type. We restrict ourselves in what follows to computing the expectation value, $\hat{G}_i = E[G_i]$. If G_i denotes the number of bacteria in the progeny (not including the original bacterium itself, so that $G_1 = 15$ in the particular realization depicted in Fig. 2), then its expectation satisfies

$$\hat{G}_1(\mu_1 + \nu_{12} - \lambda_1) = 2\lambda_1 + \nu_{12}(\hat{G}_2 + 1),$$
$$\hat{G}_2(\mu_2 + \nu_{21} - \lambda_2) = 2\lambda_2 + \nu_{21}(\hat{G}_1 + 1).$$

These quantities will be positive and finite only if $\mu_1 + \nu_{12} > \lambda_1$ and $\mu_2 + \nu_{21} > \lambda_2$, which become conditions for the number of cells in the genealogy to be finite. Solutions are given by

$$\hat{G}_1 = g^{-1}\frac{1}{\mu_1 + \nu_{12} - \lambda_1}\left(2\lambda_1 + \nu_{12}\frac{\lambda_2 + 2\nu_{21} + \mu_2}{\mu_2 + \nu_{21} - \lambda_2}\right),$$
$$\hat{G}_2 = g^{-1}\frac{1}{\mu_2 + \nu_{21} - \lambda_2}\left(2\lambda_2 + \nu_{21}\frac{\lambda_1 + 2\nu_{12} + \mu_1}{\mu_1 + \nu_{12} - \lambda_1}\right),$$

with $g = 1 - \frac{\nu_{12}\nu_{21}}{(\mu_1 + \nu_{12} - \lambda_1)(\mu_2 + \nu_{21} - \lambda_2)}$. In order for these averages to be positive, we also require $g > 0$. This leads to a third condition; namely, we have: $\frac{\nu_{21} + \mu_2 - \lambda_2}{\nu_{21}} > \frac{\nu_{12}}{\nu_{12} + \mu_1 - \lambda_1}$. For the specific case when $\nu_{21} = 0$, one obtains

$$\hat{G}_1 = \frac{1}{\mu_1 + \nu_{12} - \lambda_1}\left(2\lambda_1 + \nu_{12}\frac{\lambda_2 + \mu_2}{\mu_2 - \lambda_2}\right),$$
$$\hat{G}_2 = \frac{2\lambda_2}{\mu_2 - \lambda_2}.$$

3.5   Number of Toxin Molecules Produced by the Genealogy of a Bacterium

Our interest is to mathematically describe a system of toxin-producing bacteria, thus, we now compute the number of toxin molecules produced by the progeny of the original bacterium. We then introduce, Ω_i, the number of toxin molecules produced by the genealogy of an initial type-i bacterium, including any toxins produced by this bacterium. We denote its expectation value by $\hat{\Omega}_i = \mathbb{E}[\Omega_i]$, for $i \in \{1, 2\}$. We note that the number of toxin molecules produced by the genealogy of the single bacterium will be finite if and only if the number of bacteria within the genealogy is finite, so that the conditions on the model parameters described in the previous section are needed in what follows. The expected values, $\hat{\Omega}_1$ and $\hat{\Omega}_2$, satisfy

$$(\mu_1 + \nu_{12} - \lambda_1)\hat{\Omega}_1 = \gamma_1 + \nu_{12}\hat{\Omega}_2,$$
$$(\mu_2 + \nu_{21} - \lambda_2)\hat{\Omega}_2 = \gamma_2 + \nu_{21}\hat{\Omega}_1,$$

with solutions

$$\hat{\Omega}_1 = g^{-1}\frac{1}{\mu_1 + \nu_{12} - \lambda_1}\left(\gamma_1 + \nu_{12}\frac{\gamma_2}{\mu_2 + \nu_{21} - \lambda_2}\right),$$

$$\hat{\Omega}_2 = g^{-1}\frac{1}{\mu_2 + \nu_{21} - \lambda_2}\left(\gamma_2 + \nu_{21}\frac{\gamma_1}{\mu_1 + \nu_{12} - \lambda_1}\right).$$

When $\nu_{21} = 0$ the equations simplify to

$$\hat{\Omega}_1 = \frac{1}{\mu_1 + \nu_{12} - \lambda_1}\left(\gamma_1 + \nu_{12}\frac{\gamma_2}{\mu_2 - \lambda_2}\right),$$

$$\hat{\Omega}_2 = \frac{\gamma_2}{\mu_2 - \lambda_2}.$$

We note that there exist links between the expected number of toxin molecules produced by the genealogy and the expected number of bacteria in this genealogy. For instance, when $\nu_{21} = 0$ the average number of bacteria in the genealogy of an original type-2 bacterium, including this original bacterium, is $\hat{G}_2 + 1 = \frac{2\lambda_2}{\mu_2 - \lambda_2} + 1 = \frac{\mu_2 + \lambda_2}{\mu_2 - \lambda_2}$ (see Sect. 3.4). It is clear that, in this case, the genealogy is formed by type-2 bacteria only since $\nu_{21} = 0$. Each of these type-2 bacteria will produce, on average, $\frac{\gamma_2}{\lambda_2 + \mu_2}$ toxins (from a geometric distribution) before they decide their fate (division or death). Thus, the mean number of toxin molecules produced by the genealogy is $\frac{\mu_2 + \lambda_2}{\mu_2 - \lambda_2} \times \frac{\gamma_2}{\lambda_2 + \mu_2} = \frac{\gamma_2}{\mu_2 - \lambda_2} = \hat{\Omega}_2$, as computed above.

4   Results

We now make use of the previous results to analyse the behaviour of *Bacillus anthracis* bacteria, which causes anthrax infection, in the presence of antibiotic treatment. We consider that non-treated fully vegetative *Bacillus anthracis* bacteria form the B_1 compartment in Fig. 1, while the second compartment, B_2, represents bacteria affected by the antibiotic. *B. anthracis* produces three anthrax exotoxin components [22]: protective antigen (PA), lethal factor (LF) and edema factor (EF). The effectiveness of the anthrax toxins in infecting cells and causing symptoms is mainly due to the protective antigen (PA) capsule [21], with which the other toxin components can form complexes [22]. Therefore, we focus here on the production of PA when implementing our methods. We consider an antibiotic treatment, such as Ciprofloxacin, that inhibits bacterial division and triggers cellular death, so that we shall assume $\mu_2 \geq \mu_1$ and $\lambda_2 = 0$. It is to be expected that the production rate of toxin molecules by antibiotic-treated cells would be at most equal to non-treated cells, and thus, we consider $\gamma_2 \leq \gamma_1$. Bacteria become treated at some rate ν_{12}, and we set $\nu_{21} = 0$ to indicate that the process is irreversible. In Sect. 4.1 we leverage data from an *in vitro* assay for the A16R strain of *B. anthracis* [28] to inform our choice of parameters $(\lambda_1, \mu_1, \gamma_1)$. On the other hand, a global sensitivity analysis of model parameters $(\nu_{12}, \mu_2, \gamma_2)$ allows us in Sect. 4.2 to study the impact of treatment on the summary statistics introduced and analysed in Sect. 3, illustrating the applicability of our techniques.

4.1 Parameter Calibration

In Ref. [28] the authors examine the growth of the A16R *B. anthracis* strain by measuring the viable count of colony forming units (CFU) per mL in the assay for the following time points: $t \in \{4\,\text{h}, 8\,\text{h}, 12\,\text{h}, 16\,\text{h}, 20\,\text{h}\}$. They also develop a sandwich ELISA and cytotoxicity-based method to quantify the concentration of PA every two hours during the experiment, from $t = 4\,\text{h}$ to $t = 26\,\text{h}$. In order to exploit this data set, and to estimate representative values for λ_1, μ_1 and γ_1, we consider its corresponding deterministic model (for the first compartment of non-treated bacteria)

$$\frac{dB}{dt} = (\lambda_1 - \mu_1)B, \quad \frac{dT}{dt} = \gamma_1 B - \xi T,$$

where $B(t)$ is the concentration (in units [CFU/mL]) of bacteria at time t, and $T(t)$ the concentration of PA (in units of [ng/mL]). Results from Ref. [28, Figure 1] support bacterial exponential growth during the first 12 h of the experiment. The bacterial population reaches a carrying capacity after this point, indicating that there exists competition for resources. Thus, since our interest (see Fig. 1) is the analysis of non-competing bacteria, we focus here on the first period of the experiment: $t \in [4\,\text{h}, 12\,\text{h}]$. In particular, we set $\lambda_1 = 0.8\,\text{h}^{-1}$ from Ref. [10], and use bacterial counts from Ref. [28, Figure 1] and toxin concentration measurements from Ref. [28, Figure 4] to estimate the bacterial death rate, μ_1, and the toxin production rate, γ_1. Since the dynamics of the toxin population is likely to be dominated by the production of toxins from an exponentially growing bacterial population, we neglect toxin degradation and set $\xi = 0$ in what follows. We acknowledge that this might lead to underestimating the rate γ_1. Yet, the rate ξ has no effect on any of the summary statistics analysed in Sect. 3.

Parameters μ_1 and γ_1 are estimated making use of the *curve_fit* function from the *scipy.optimize* package in *Python*, which is based on a non-linear least squares method. This leads to point estimates $\mu_1 = 0.43\,\text{h}^{-1}$ and $\gamma_1 = 4.63 \times 10^{-6}$ ng CFU^{-1}h^{-1}. A comparison between model predictions and observed measurements is provided in Fig. 3. Finally, in order to use our estimate for γ_1 in the stochastic model from Fig. 1, one needs to convert units (from mass in ng to number of molecules). To do this, we note that PA has a relative molecular mass of 83 kD [11, 25]. This means that 7.2×10^9 PA molecules have an approximate weight of 1ng, so that $\gamma_1 = 3.34 \times 10^4$ molecules CFU^{-1}h^{-1}.

4.2 Summary Statistics

We now perform a global sensitivity analysis on a subset of the model parameters for the summary statistics of interest introduced in Sect. 3. We consider the stochastic model of Fig. 1 with baseline parameter values: $\mu_1 = 0.43$ h^{-1}, $\lambda_1 = 0.8$ h^{-1} and $\gamma_1 = 3.34 \times 10^4$ molecules CFU^{-1} h^{-1}, according to the calibration carried out in the previous section. To analyse the role of antibiotic treatment (B_1

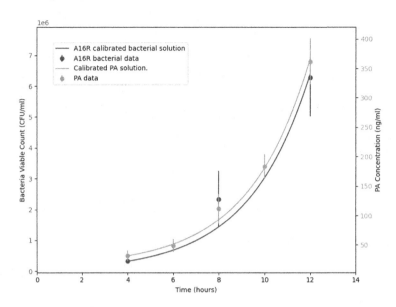

Fig. 3. Mathematical model predictions compared to experimental observations from Ref. [28].

represents non-treated bacteria and B_2 antibiotic-treated bacteria, respectively), we explore parameter regimes with $\nu_{12} > 0$, $\nu_{21} = \lambda_2 = 0$, $\mu_2 \geq \mu_1$ and $\gamma_2 \leq \gamma_1$.

In Fig. 4 we look at summary statistics directly related to the lifetime of a single bacterium. We assume at time $t = 0$ we start with one non-treated bacterium. We first carry out a sensitivity analysis for parameters μ_2, ν_{12} and γ_2. This allows one to analyse the impact of treatment on the tracked bacterium during its lifespan. On the other hand, even when we have a baseline value for μ_1, we vary this parameter when considering the number of divisions undergone by the tracked bacterium, for illustrative purposes. The top-left plot in Fig. 4 shows the impact of treatment on the mean lifespan of the bacterium, $\mathbb{E}[T_1]$, which varies between 1 and 3 h for the parameter values considered. Increasing antibiotic efficiency (in terms of larger values of μ_2 and ν_{12}) leads to shorter lifespans. We note that if one assumes $\mu_2 = \mu_1 = 0.43$ h^{-1}, no effect of treatment on the lifespan is expected, and the value of ν_{12} (which is directly related to the rate at which antibiotic can affect bacteria, as well as the concentration of antibiotic present in the system) becomes irrelevant. Finally, increasing values of μ_2 make the value of ν_{12} increasingly relevant, as one would expect.

The top-right plot of Fig. 4 shows the expected number of divisions undergone by the bacterium, $\mathbb{E}[D_1]$, for a range of μ_1 and ν_{12} values. We note here that since $\lambda_2 = 0$, μ_2 has no effect on D_1. Thus, we vary μ_1 instead. As one would expect, increasing values of ν_{12} and μ_1 lead to fewer bacterial divisions. We indicate that in order for the bacterial population to grow as a function of time, each bacterium (on average) needs to undergo more than one division events. We

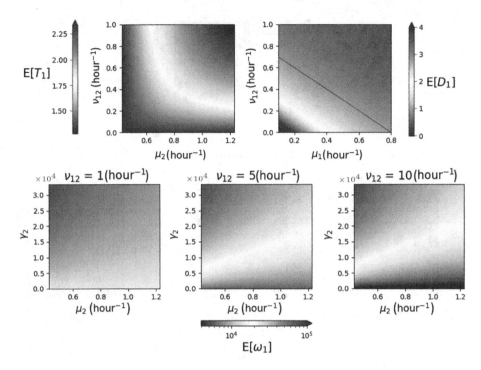

Fig. 4. Top-left. Expected lifespan [*hours*] of a bacterium. **Top-right.** Expected number of divisions during the lifetime of a bacterium. **Bottom.** Expected number of toxin molecules produced by a bacterium during its lifetime for different values of $\nu_{12} \in \{1, 5, 10\}$ (left to right). Units for γ_2 are molecules CFU^{-1} h^{-1}.

highlight the value $\mathbb{E}[D_1] = 1$ by a green line in Fig. 4, which is achieved when $\nu_{12} + \mu_1 = \lambda_1$. The bottom row in Fig. 4 shows the effect of varying ν_{12}, μ_2 and γ_2 on the expected number of toxin molecules produced by a bacterium during its lifetime, $\mathbb{E}[\omega_1]$. Increasing values of μ_2 and ν_{12} can have a significant effect on the number of toxin molecules produced. The values $\gamma_2 = \gamma_1 = 3.34 \times 10^4$ molecules CFU^{-1} h^{-1} and $\mu_2 = \mu_1 = 0.43$ h^{-1} represent no treatment effect for the tracked bacterium, and for these choices the value of ν_{12} has no effect on $\mathbb{E}[\omega_1]$. On the other hand, decreasing values of γ_2 have a significant effect on the predicted number of PA molecules produced, especially for increasing values of ν_{12}.

In Fig. 5 we present summary statistics of relevance to the genealogy of a B_1 bacterium. The top plot of Fig. 5 shows the effect that parameters ν_{12} and μ_1 have on the mean number of cells in the genealogy of a single bacterium, $1 + \hat{G}_1$. We note that, in this plot, the white area corresponds to parameter combinations for which the mean number of cells in the genealogy is not finite. This happens when $\lambda_1 \geq \mu_1 + \nu_{12}$. Values of $\mu_1 + \nu_{12}$ larger but close to λ_1 lead to increasing the mean number of cells in the genealogy, as one would expect. On the other hand, the bottom row of Fig. 5 shows the effect on the number

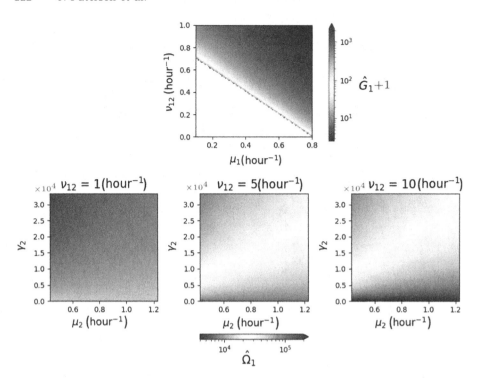

Fig. 5. Top. Mean number of bacteria in the genealogy of a single bacterium. **Bottom.** Mean number of toxin molecules secreted by the genealogy of a single bacterium for different values of $\nu_{12} \in \{1, 5, 10\}$ (from left to right). Units for γ_2 are molecules $\mathrm{CFU}^{-1}\,\mathrm{h}^{-1}$.

of toxin molecules secreted by the genealogy of a single bacterium for varying values of μ_2 and γ_2. We investigate these parameter values for three different choices of $\nu_{12} \in \{1, 5, 10\}$. It is clear that γ_2 has a large impact on the expected value, $\hat{\Omega}_1$, which mimics the similar effect that γ_2 has on its single bacterium counterpart, $\mathbb{E}[\omega_1]$ (see Fig. 4). Figure 4 and Fig. 5 show the significance of ν_{12} on the expected number of toxin molecules produced. Interestingly, as ν_{12} becomes much larger than λ_1, we observe that $\mathbb{E}[\omega_1]$ approaches $\hat{\Omega}_1$, since in this case $1 + \hat{G}_1 \approx 1$ represents the single bacterium of interest.

5   Conclusions

We have defined and analysed a two-compartment stochastic model for toxin-producing bacteria. Our focus has been a number of summary statistics that relate to the lifetime of a single bacterium (tracked over time) and its progeny. In particular, we have studied the lifespan of the bacterium, the number of divisions undergone and the number of toxin molecules produced during its lifetime, as well as the number of cells in its genealogy, and the number of toxin molecules produced by this progeny. We illustrated in Sect. 4 our methods by focusing on the growth of *B. anthracis* bacteria in the presence of antibiotic treatment. To the best of our knowledge, this is the first approach to quantify the PA production rate in this system. We acknowledge that our estimate for this rate might be an underestimate, given that we neglected PA degradation.

We point out that, although the model considered in Fig. 1 is relatively simple, consisting only of two bacterial compartments, our single bacterium approach can be applied to any network *topology* of compartments, as long as the bacteria behave independently, so that the dynamics of a single bacterium can be effectively followed. Implementing our techniques in more complex systems, such as those representing *in vivo* infection and bacterial dissemination between different organs, remains the aim of future work. We also indicate that, while we have analysed probability generating functions and Laplace-Stieltjes transforms in Sect. 3, we have focused in practice, for simplicity and brevity, on computing the first order moments for the summary statistics of interest. However, this approach can be easily generalised to compute higher order moments or probability mass functions.

Acknowledgements. Jamie Paterson is supported by an EPSRC CASE studentship, project reference 2274495, in partnership with the Defence Science and Technology Laboratory (Dstl). This manuscript has been reviewed at Los Alamos National Laboratory, reference number LA-UR-21-26048.

References

1. Allen, L.J.: An Introduction to Mathematical Biology. Upper Saddle River, Hoboken (2007)
2. Allen, L.J.: An Introduction to Stochastic Processes with Applications to Biology. CRC Press, Boco Raton (2010)
3. Banks, D.J., Barnajian, M., Maldonado-Arocho, F.J., Sanchez, A.M., Bradley, K.A.: Anthrax toxin receptor 2 mediates Bacillus anthracis killing of macrophages following spore challenge. Cell. Microbiol. **7**(8), 1173–1185 (2005)
4. Britton, T.: Stochastic epidemic models: a survey. Math. Biosci. **225**(1), 24–35 (2010)
5. Brockwell, P.J.: The extinction time of a general birth and death process with catastrophes. J. Appl. Probab. **23**, 851–858 (1986)
6. Carruthers, J., López-García, M., Gillard, J.J., Laws, T.R., Lythe, G., Molina-París, C.: A novel stochastic multi-scale model of Francisella tularensis infection to predict risk of infection in a laboratory. Front. Microbiol. **9**, 1165 (2018)
7. Carruthers, J., et al.: Stochastic dynamics of Francisella tularensis infection and replication. PLoS Comput. Biol. **16**(6), e1007752 (2020)
8. Castro, M., López-García, M., Lythe, G., Molina-París, C.: First passage events in biological systems with non-exponential inter-event times. Sci. Rep. **8**(1), 1–16 (2018)
9. Choi, P.J., Cai, L., Frieda, K., Xie, X.S.: A stochastic single-molecule event triggers phenotype switching of a bacterial cell. Science **322**(5900), 442–446 (2008)

10. Day, J., Friedman, A., Schlesinger, L.S.: Modeling the host response to inhalation anthrax. J. Theor. Biol. **276**(1), 199–208 (2011)
11. Fabre, L., et al.: Structure of anthrax lethal toxin prepare complex suggests a pathway for efficient cell entry. J. Gen. Physiol. **148**(4), 313–324 (2016)
12. Gillespie, D.T.: Exact stochastic simulation of coupled chemical reactions. J. Phys. Chem. **81**(25), 2340–2361 (1977)
13. Gómez-Corral, A., López-García, M.: Lifetime and reproduction of a marked individual in a two-species competition process. Appl. Math. Comput. **264**, 223–245 (2015)
14. Gómez-Corral, A., López-García, M.: On SIR epidemic models with generally distributed infectious periods: number of secondary cases and probability of infection. Int. J. Biomath. **10**(02), 1750024 (2017)
15. Herzenberg, L.A., Parks, D., Sahaf, B., Perez, O., Roederer, M., Herzenberg, L.A.: The history and future of the fluorescence activated cell sorter and flow cytometry: a view from Stanford. Clin. Chem. **48**(10), 1819–1827 (2002)
16. de la Higuera, L., López-García, M., Castro, M., Abourashchi, N., Lythe, G., Molina-París, C.: Fate of a naive T cell: a stochastic journey. Front. Immunol. **10**, 194 (2019)
17. Johnson, J., Nowicki, M.O., Lee, C.H., Chiocca, E.A., Viapiano, M.S., Lawler, S.E., Lannutti, J.J.: Quantitative analysis of complex glioma cell migration on electrospun polycaprolactone using time-lapse microscopy. Tissue Eng. Part C Methods **15**(4), 531–540 (2009)
18. Kimmel, M., Axelrod, D.E.: Branching Processes in Biology (2002)
19. Krutzik, P.O., Nolan, G.P.: Fluorescent cell barcoding in flow cytometry allows high-throughput drug screening and signaling profiling. Nat. Methods **3**(5), 361–368 (2006)
20. Kyprianou, A.E., Palau, S.: Extinction properties of multi-type continuous-state branching processes. Stochast. Processes Appl. **128**(10), 3466–3489 (2018)
21. Leppla, S.H.: Anthrax toxin. In: Aktories, K., Just, I. (eds.) Bacterial Protein Toxins. HEP, vol. 145, pp. 445–472. Springer, Heidelberg (2000). https://doi.org/10.1007/978-3-662-05971-5_19
22. Liu, S., Moayeri, M., Leppla, S.H.: Anthrax lethal and edema toxins in anthrax pathogenesis. Trends Microbiol. **22**(6), 317–325 (2014)
23. López-García, M.: Stochastic descriptors in an SIR epidemic model for heterogeneous individuals in small networks. Math. Biosci. **271**, 42–61 (2016)
24. López-García, M., Nowicka, M., Bendtsen, C., Lythe, G., Ponnambalam, S., Molina-París, C.: Quantifying the phosphorylation timescales of receptor-ligand complexes: a Markovian matrix-analytic approach. Open Biol. **8**(9), 180126 (2018)
25. Petosa, C., Collier, R.J., Klimpel, K.R., Leppla, S.H., Liddington, R.C.: Crystal structure of the anthrax toxin protective antigen. Nature **385**(6619), 833–838 (1997)
26. Thakur, A., Rescigno, A., Schafer, D.: On the stochastic theory of compartments: II. Multi-compartment systems. Bull. Math. Biol. **35**(1), 263–271 (1973)
27. Westera, L., et al.: Closing the gap between T-cell life span estimates from stable isotope-labeling studies in mice and humans. Blood J. Am. Soc. Hematol. **122**(13), 2205–2212 (2013)
28. Zai, X., et al.: Quantitative determination of lethal toxin proteins in culture supernatant of human live anthrax vaccine bacillus anthracis A16R. Toxins **8**(3), 56 (2016)

EM Based Parameter Estimation for Markov Modulated Fluid Arrival Processes

Salah Al-Deen Almousa[1] , Gábor Horváth[1] , and Miklós Telek[1,2](✉)

[1] Department of Networked Systems and Services, Budapest University of
Technology and Economics, Budapest, Hungary
`almousa@hit.bme.hu, ghorvath@hit.bme.hu, telek@hit.bme.hu`
[2] MTA-BME Information Systems Research Group, ELKH, Budapest, Hungary

Abstract. Markov modulated discrete arrival processes have a wide literature, including parameter estimation methods based on expectation-maximization (EM). In this paper, we investigate the adaptation of these EM based methods to Markov modulated fluid arrival processes (MMFAP), and conclude that only some parameters of MMFAPs can be approximated this way.

Keywords: Markov modulated fluid arrival processes ·
Expectation-maximization method · Parameter estimation

1 Introduction

Markovian queueing systems with discrete customers are widely used in stochastic modeling. Markovian Arrival Process (MAP, [11]), that are able to characterize a wide class of point processes, play an important role in these systems. The properties of MAPs have been studied exhaustively, using queueing models involving MAPs nowadays has become common, queueing networks with MAP traffic have also been investigated. Several methods exist to create a MAP approximating real, empirical data. Some of them aim to match statistical quantities like marginal moments, joint moments, auto-correlation, etc. [9,15] An other approach to create MAPs from measurement data is based on likelihood maximization, which is often performed by Expectation-Maximization (EM) [12]. Several EM-based fitting methods have been published for MAPs, based on randomization [5], based on special structures [7,14], methods that support batch arrivals [4] and those that are able to work with group data [13].

In many systems the workload can be represented easier with continuous, fluid-like models rather than discrete demands [1]. Basic Markovian fluid models have been introduced and analyzed in [2,10], later on several model variants

This work is partially supported by the OTKA K-123914 project and the Artificial Intelligence National Laboratory Programme of Hungary.

P. Ballarini et al. (Eds.): EPEW 2021/ASMTA 2021, LNCS 13104, pp. 226–242, 2021.
https://doi.org/10.1007/978-3-030-91825-5_14

appeared and were investigated. Despite of their practical relevance, the "ecosystem" around fluid models is far less complete than in the discrete case. In particular, fitting methods for Markov modulated fluid arrival processes (MMFAP) are available only to some very restricted cases like on-off models, motivated by telecommunication applications. To the best of our knowledge, fitting methods for the general class of MMFAPs based on likelihood maximization has not been investigated in the past. At first glance adapting the methods available for MAPs might seem feasible, since fluid models can be treated as a limit of a discrete model generating infinitesimally small fluid drops. In fact, adapting the algorithms for MAPs to MMFAPs is not straight forward at all, fitting MMFAPs is a qualitatively different problem.

The rest of the paper is organized as follows. Section 2 introduces the mathematical model and the parameter estimation problem. The next section discusses the applicability of the EM method for MMFAPs. Some implementation details associated with the EM method for MMFAPs are provided in Sect. 4. Finally, Sect. 5 provides numerical experiments about the properties of the proposed method and Sect. 6 concludes the paper.

2 Problem Definition

2.1 Fluid Arrival Process

The fluid arrival process $\{\mathcal{Z}(t) = \{\mathcal{J}(t), \mathcal{X}(t)\}, t > 0\}$ consists of a background continuous time Markov chain $\{\mathcal{J}(t), t > 0\}$ which modulates the arrival process of the fluid $\{\mathcal{X}(t), t > 0\}$. When the Markov chain stays in state i for a Δ long interval a normal distributed amount of fluid arrives with mean $r_i \Delta$ and variance $\sigma_i^2 \Delta$, that is, when $\mathcal{J}(\tau) = i, \forall \tau \in (t, t + \Delta)$

$$\frac{d}{dx} Pr(\mathcal{X}(t + \Delta) - \mathcal{X}(t) < x) = \mathcal{N}(r_i \Delta, \sigma_i^2 \Delta, x), \tag{1}$$

where $\mathcal{N}(\mu, \sigma^2, x) = \frac{1}{\sqrt{2\pi\sigma^2}} e^{-\frac{(x-\mu)^2}{2\sigma^2}}$ is the Gaussian density function. We note that our proposed analysis approach allows negative fluid rates as well. Since the normal distribution has infinite support also in case of strictly positive fluid rates Sect. 4.2 discusses a numerical approach to handle negative fluid samples.

The generator matrix and the initial probability vector of the N-state background continuous time Markov chain (CTMC) are $\mathbf{Q}$ and $\underline{\alpha}$, and the diagonal matrix of the fluid rates and the fluid variances are given by matrix $\mathbf{R}$ with $\mathbf{R}_{i,j} = \delta_{i,j} r_i$, and matrix $\boldsymbol{\Sigma}$ with $\boldsymbol{\Sigma}_{i,j} = \delta_{i,j}\sigma_i^2$, where $\delta_{i,j}$ denotes the Kronecker delta.

Assuming $\mathcal{X}(0) = 0$, the amount of fluid arriving in the $(0, t)$ interval is $\mathcal{X}(t)$, with density matrix defined by

$$[\mathbf{N}(t, x)]_{i,j} = \frac{\partial}{\partial x} \Pr\left(\mathcal{X}(t) < x, \mathcal{J}(t) = j | \mathcal{J}(0) = i\right) \tag{2}$$

The double sided Laplace transform of this quantity regarding the amount of fluid arrived can be expressed as [6]

$$\mathbf{N}^*(t,v) = \int_{x=-\infty}^{\infty} e^{-xv}\mathbf{N}(t,x)dx = e^{(\mathbf{Q}-v\mathbf{R}-v^2\mathbf{\Sigma}/2)t}. \tag{3}$$

The stationary distribution of the CTMC is denoted by vector $\underline{\pi}$, which is the solution of $\underline{\pi}\mathbf{Q} = \underline{0}, \underline{\pi}\mathbb{1} = 1$. In this work, we are interested in the stationary fluid arrival process and assume that the initial probability vector of the background CTMC is $\underline{\alpha} = \underline{\pi}$.

2.2   Measurement Data to Fit

We assume that the data to fit is given by a series of pairs $\mathcal{D} = \{(t_k, x_k); k = 1, \ldots, K\}$, where t_k is the time since the last measurement instant and x_k is the amount of fluid arrived since the last measurement instant (which can be negative as well). That is, the measurement instances are $T_k = \sum_{\ell=1}^{k} t_\ell$ for $k \in \{1, \ldots, K\}$.

The likelihood of the data is defined as

$$\mathcal{L} = \underline{\alpha} \prod_{k=1}^{K} \mathbf{N}(t_k, x_k)\mathbb{1}. \tag{4}$$

Our goal is to find $\mathbf{Q}$, $\mathbf{R}$ and $\mathbf{S}$ which maximize the likelihood.

3   The EM Algorithm

The EM algorithm is based on the observation that the likelihood would be easier to maximize when certain unobserved, hidden variables were known. In our case the hidden variables are related to the trajectory of the hidden Markov chain, specifically

- $J_n^{(k)}$ is the nth state visited by the Markov chain in the kth measurement interval,
- $\theta_n^{(k)}$ is the sojourn time of the nth sojourn of the Markov chain (which is in state $J_n^{(k)}$) in the kth measurement interval,
- $f_n^{(k)}$ is the fluid accumulated during the nth sojourn in the kth measurement interval,
- $n^{(k)}$ is the number of sojourns in the kth measurement interval.

Based on these hidden variables the logarithm of the likelihood is computed in the next section.

3.1   Log-Likelihood as a Function of the Hidden Variables

With the hidden variables defined above, the likelihood $\mathcal{L}$ can be expressed as

$$
\mathcal{L} = \prod_{k=1}^{K} e^{-\theta_1^{(k)} q_{J_1^{(k)}}} \mathcal{N}\left(\theta_1^{(k)} r_{J_1^{(k)}}, \theta_1^{(k)} \sigma^2_{J_1^{(k)}}, f_1^{(k)}\right) q_{J_1^{(k)} J_2^{(k)}}
$$

$$
\cdot e^{-\theta_2^{(k)} q_{J_2^{(k)}}} \mathcal{N}\left(\theta_2^{(k)} r_{J_2^{(k)}}, \theta_2^{(k)} \sigma^2_{J_2^{(k)}}, f_2^{(k)}\right) q_{J_2^{(k)} J_3^{(k)}} \cdots
$$

$$
\cdot e^{-\theta_{n^{(k)}}^{(k)} q_{J_{n^{(k)}}^{(k)}}} \mathcal{N}\left(\theta_{n^{(k)}}^{(k)} r_{J_{n^{(k)}}^{(k)}}, \theta_{n^{(k)}}^{(k)} \sigma^2_{J_{n^{(k)}}^{(k)}}, f_{n^{(k)}}^{(k)}\right)
$$

$$
= \prod_{k=1}^{K} \prod_{n=1}^{n^{(k)}-1} e^{-\theta_n^{(k)} q_{J_n^{(k)}}} \mathcal{N}\left(\theta_n^{(k)} r_{J_n^{(k)}}, \theta_n^{(k)} \sigma^2_{J_n^{(k)}}, f_n^{(k)}\right) q_{J_n^{(k)} J_{n+1}^{(k)}}
$$

$$
\cdot e^{-\theta_{n^{(k)}}^{(k)} q_{J_{n^{(k)}}^{(k)}}} \mathcal{N}\left(\theta_{n^{(k)}}^{(k)} r_{J_{n^{(k)}}^{(k)}}, \theta_{n^{(k)}}^{(k)} \sigma^2_{J_{n^{(k)}}^{(k)}}, f_{n^{(k)}}^{(k)}\right),
$$

where $\mathcal{N}(\mu, \sigma^2, x)$ is the Gaussian density function and $q_i = \sum_{j, j \neq i} q_{ij}$ is the departure rate of state i of the CTMC. Using $\log \mathcal{N}(\mu, \sigma^2, x) = -\frac{c}{2} - \frac{\log \sigma^2}{2} - \frac{(x-\mu)^2}{2\sigma^2}$ with $c = \log 2\pi$ we have

$$
\log\left(\mathcal{N}\left(\theta_n^{(k)} r_{J_n^{(k)}}, \theta_n^{(k)} \sigma^2_{J_n^{(k)}}, f_n^{(k)}\right)\right)
$$

$$
= -\frac{c}{2} - \frac{\log(\theta_n^{(k)} \sigma^2_{J_n^{(k)}})}{2} - \frac{(f_n^{(k)} - \theta_n^{(k)} r_{J_n^{(k)}})^2}{2\theta_n^{(k)} \sigma^2_{J_n^{(k)}}}
$$

$$
= -\frac{c}{2} - \frac{\log(\theta_n^{(k)}) + \log(\sigma^2_{J_n^{(k)}})}{2} - \frac{f_n^{(k)^2} - 2f_n^{(k)} \theta_n^{(k)} r_{J_n^{(k)}} + \theta_n^{(k)^2} r^2_{J_n^{(k)}}}{2\theta_n^{(k)} \sigma^2_{J_n^{(k)}}}
$$

$$
= -\frac{c}{2} - \frac{\log \theta_n^{(k)}}{2} - \frac{\log(\sigma^2_{J_n^{(k)}})}{2} - \frac{f_n^{(k)^2}}{2\theta_n^{(k)} \sigma^2_{J_n^{(k)}}} + \frac{f_n^{(k)} r_{J_n^{(k)}}}{\sigma^2_{J_n^{(k)}}} - \frac{\theta_n^{(k)} r^2_{J_n^{(k)}}}{2\sigma^2_{J_n^{(k)}}},
$$

and the log-likelihood is

$$
\log \mathcal{L} = \sum_{k=1}^{K} \sum_{n=1}^{n^{(k)}-1} -\theta_n^{(k)} q_{J_n^{(k)}} + \log\left(\mathcal{N}\left(\theta_n^{(k)} r_{J_n^{(k)}}, \theta_n^{(k)} \sigma_{J_n^{(k)}}^2, f_n^{(k)}\right)\right) + \log q_{J_n^{(k)} J_{n+1}^{(k)}}
$$

$$
- \theta_{n^{(k)}}^{(k)} q_{J_{n^{(k)}}^{(k)}} + \log\left(\mathcal{N}\left(\theta_{n^{(k)}}^{(k)} r_{J_{n^{(k)}}^{(k)}}, \theta_{n^{(k)}}^{(k)} \sigma_{J_{n^{(k)}}^{(k)}}^2, f_{n^{(k)}}^{(k)}\right)\right)
$$

$$
= \sum_{k=1}^{K} \sum_{n=1}^{n^{(k)}-1} -\theta_n^{(k)} q_{J_n^{(k)}} - \frac{c}{2} - \frac{\log \theta_n^{(k)}}{2} - \frac{\log(\sigma_{J_n^{(k)}}^2)}{2} - \frac{f_n^{(k)\,2}}{2\theta_n^{(k)} \sigma_{J_n^{(k)}}^2} + \frac{f_n^{(k)} r_{J_n^{(k)}}}{\sigma_{J_n^{(k)}}^2}
$$

$$
- \frac{\theta_n^{(k)} r_{J_n^{(k)}}^2}{2\sigma_{J_n^{(k)}}^2} + \log q_{J_n^{(k)} J_{n+1}^{(k)}} - \theta_{n^{(k)}}^{(k)} q_{J_{n^{(k)}}^{(k)}} - \frac{c}{2} - \frac{\log \theta_{n^{(k)}}^{(k)}}{2} - \frac{\log(\sigma_{J_{n^{(k)}}^{(k)}}^2)}{2}
$$

$$
- \frac{f_{n^{(k)}}^{(k)\,2}}{2\theta_{n^{(k)}}^{(k)} \sigma_{J_{n^{(k)}}^{(k)}}^2} + \frac{f_{n^{(k)}}^{(k)} r_{J_{n^{(k)}}^{(k)}}}{\sigma_{J_{n^{(k)}}^{(k)}}^2} - \frac{\theta_{n^{(k)}}^{(k)} r_{J_{n^{(k)}}^{(k)}}^2}{2\sigma_{J_{n^{(k)}}^{(k)}}^2}.
$$

Observe that knowing each individual hidden variable is not necessary to express the log-likelihood. It is enough to introduce the following aggregated measures to fully characterize interval k:

- $\Theta_i^{(k)} = \sum_{n=1}^{n^{(k)}} \theta_n^{(k)} \mathcal{I}_{\left\{J_n^{(k)}=i\right\}}$ is the total time spent in state i,
- $F_i^{(k)} = \sum_{n=1}^{n^{(k)}} f_n^{(k)} \mathcal{I}_{\left\{J_n^{(k)}=i\right\}}$ is the total amount of fluid arriving during a visit in state i,
- $M_i^{(k)} = \sum_{n=1}^{n^{(k)}} \mathcal{I}_{\left\{J_n^{(k)}=i\right\}}$ the number of visits to state i,
- $M_{i,j}^{(k)} = \sum_{n=1}^{n^{(k)}-1} \mathcal{I}_{\left\{J_n^{(k)}=i, J_{n+1}^{(k)}=j\right\}}$ the number of state transitions from state i to state j, additionally
- $L\Theta_i^{(k)} = \sum_{n=1}^{n^{(k)}} \log \theta_n^{(k)} \mathcal{I}_{\left\{J_n^{(k)}=i\right\}}$ is the total time spent in state i,
- $F\Theta_i^{(k)} = \sum_{n=1}^{n^{(k)}} \frac{f_n^{(k)2}}{\theta_n^{(k)}} \mathcal{I}_{\left\{J_n^{(k)}=i\right\}}$ is the total amount of fluid arriving during a visit in state i.

With these aggregate measures the log-likelihood simplifies to

$$
\log \mathcal{L} = \sum_{k=1}^{K} \sum_{n=1}^{n^{(k)}-1} -\theta_n^{(k)} q_{J_n^{(k)}} - \frac{c}{2} - \frac{\log \theta_n^{(k)}}{2} - \frac{\log(\sigma_{J_n^{(k)}}^2)}{2} - \frac{f_n^{(k)\,2}}{2\theta_n^{(k)} \sigma_{J_n^{(k)}}^2} + \frac{f_n^{(k)} r_{J_n^{(k)}}}{\sigma_{J_n^{(k)}}^2}
$$

$$
- \frac{\theta_n^{(k)} r_{J_n^{(k)}}^2}{2\sigma_{J_n^{(k)}}^2} + \log q_{J_n^{(k)} J_{n+1}^{(k)}} - \theta_{n^{(k)}}^{(k)} q_{J_{n^{(k)}}^{(k)}} - \frac{c}{2} - \frac{\log \theta_{n^{(k)}}^{(k)}}{2} - \frac{\log(\sigma_{J_{n^{(k)}}^{(k)}}^2)}{2}
$$

$$
- \frac{f_{n^{(k)}}^{(k)\,2}}{2\theta_{n^{(k)}}^{(k)} \sigma_{J_{n^{(k)}}^{(k)}}^2} + \frac{f_{n^{(k)}}^{(k)} r_{J_{n^{(k)}}^{(k)}}}{\sigma_{J_{n^{(k)}}^{(k)}}^2} - \frac{\theta_{n^{(k)}}^{(k)} r_{J_{n^{(k)}}^{(k)}}^2}{2\sigma_{J_{n^{(k)}}^{(k)}}^2}
$$

$$
= \sum_{k=1}^{K} -\frac{cn^{(k)}}{2} + \sum_{i} \left(-\Theta_i^{(k)} \left(q_i + \frac{r_i^2}{2\sigma_i^2} \right) + F_i^{(k)} \frac{r_i}{\sigma_i^2} - \frac{L\Theta_i^{(k)}}{2} - \frac{F\Theta_i^{(k)}}{2\sigma_i^2} \right.
$$

$$
\left. - M_i^{(k)} \frac{\log \sigma_i^2}{2} + \sum_{j,\,j \neq i} M_{i,j}^{(k)} \log q_{i,j} \right).
$$

3.2 The Maximization Step of the EM Method

The maximization step of the EM method aims to find the optimal value of the model parameters based on the hidden variables. They are obtained from the partial derivatives of the log-likelihood as detailed in Appendix A.

Summarizing the results, the model parameter value which maximizes the log-likelihood based on the hidden variables are

$$
q_{i,j} = \frac{\sum_{k=1}^{K} M_{i,j}^{(k)}}{\sum_{k=1}^{K} \Theta_i^{(k)}}, r_i = \frac{\sum_{k=1}^{K} F_i^{(k)}}{\sum_{k=1}^{K} \Theta_i^{(k)}}, \text{ and } \sigma_i^2 = \frac{\sum_{k=1}^{K} \Theta_i^{(k)} r_i^2 - 2F_i^{(k)} r_i + F\Theta_i^{(k)}}{\sum_{k=1}^{K} M_i^{(k)}}.
$$

That is, $\sum_{k=1}^{K} \Theta_i^{(k)}$, and $\sum_{k=1}^{K} M_{i,j}^{(k)}$ are needed for computing the optimal $q_{i,j}$ parameters and additionally, $\sum_{k=1}^{K} F_i^{(k)}$, $\sum_{k=1}^{K} M_i$ and $\sum_{k=1}^{K} F\Theta_i^{(k)}$ are needed for the optimal r_i and σ_i^2 parameters.

3.3 The Expectation Step of the EM Method

In the expectation step of the EM method the expected values of the hidden variables has to be evaluated based on the samples. Appendix B provides the analysis of those expectations, resulting $E(F_i^{(k)}) = r_i E(\Theta_i^{(k)})$ and $\mathrm{E}\left(F\Theta_i^{(k)} \right) = \mathrm{E}\left(M_i^{(k)} \right) \sigma_i^2 + \mathrm{E}\left(\Theta_i^{(k)} \right) r_i^2$, from which the zth iteration of the EM method updates the fluid rate and variance parameters as

$$
r_i(z+1) = \frac{\sum_{k=1}^{K} \mathrm{E}\left(F_i^{(k)} \right)}{\sum_{k=1}^{K} \mathrm{E}\left(\Theta_i^{(k)} \right)} = r_i(z) \tag{5}
$$

and

$$\begin{aligned}
\sigma_i^2(z+1) &= \frac{\sum_{k=1}^K \mathrm{E}\left(\Theta_i^{(k)} r_i^2(z) - 2F_i^{(k)} r_i(z) + F\Theta_i^{(k)}\right)}{\sum_{k=1}^K \mathrm{E}\left(M_i^{(k)}\right)} \\
&= \frac{\sum_{k=1}^K \mathrm{E}\left(\Theta_i^{(k)} r_i^2(z)\right) - 2\mathrm{E}\left(F_i^{(k)}\right) r_i(z) + \mathrm{E}\left(F\Theta_i^{(k)}\right)}{\sum_{k=1}^K \mathrm{E}\left(M_i^{(k)}\right)} \\
&= \frac{\sum_{k=1}^K \mathrm{E}\left(M_i^{(k)}\right) \sigma_i^2(z)}{\sum_{k=1}^K \mathrm{E}\left(M_i^{(k)}\right)} = \sigma_i^2(z).
\end{aligned} \tag{6}$$

Consequently, the fluid rate and variance parameters remain untouched by the EM method.

Remark 1. This result is in line with the results obtained for discrete arrival processes in [13] considering the special features of the fluid model. That is, we consider the MMPP arrival process, since there is no state transition at the fluid drop arrival, and fluid drops are assumed to be infinitesimal, hence for a finite amount of time there is an unbounded number of fluid drop arrivals. Using these features, equations (21) and (23) of [13] take the form

$$\mathrm{E}\left(Z_i^{[k]}\right) = \sum_{l=0}^{x_k} \int_0^{t_k} [f_k(l,\tau)]_i [b_k(x_k - l, t_k - \tau)]_i d\tau$$

$$\mathrm{E}\left(Y_{ii}^{[k]}\right) = \lambda_{ii} \sum_{l=0}^{x_k-1} \int_0^{t_k} [f_k(l,\tau)]_i [b_k(x_k - l, t_k - \tau)]_i d\tau.$$

Assuming x_k is large, the update of λ_{ii} in the zth step of the iteration is ((12) of [13])

$$\begin{aligned}
\lambda_{ii}(z+1) &= \frac{\sum_{k=1}^K \mathrm{E}\left(Y_{ii}^{[k]}\right)}{\sum_{k=1}^K \mathrm{E}\left(Z_i^{[k]}\right)} \\
&= \lambda_{ii}(z) \frac{\sum_{k=1}^K \sum_{l=0}^{x_k-1} \int_0^{t_k} [f_k(l,\tau)]_i [b_k(x_k - l, t_k - \tau)]_i d\tau}{\sum_{k=1}^K \sum_{l=0}^{x_k} \int_0^{t_k} [f_k(l,\tau)]_i [b_k(x_k - l, t_k - \tau)]_i d\tau} \approx \lambda_{ii}(z).
\end{aligned}$$

The transition rate parameters are updated by the EM method as

$$q_{i,j} = \frac{\sum_{k=1}^K \mathrm{E}\left(M_{i,j}^{(k)}\right)}{\sum_{k=1}^K \mathrm{E}\left(\Theta_i^{(k)}\right)}. \tag{7}$$

The computation of $\mathrm{E}\left(M_{i,j}^{(k)}\right)$ and $\mathrm{E}\left(\Theta_i^{(k)}\right)$ are detailed in Appendix C and the results are summarized in (22) and (23).

4 Implementation Details

The implementation of the EM based parameter estimation method contains some intricate elements which influence the computational complexity and the accuracy of the computations. This section summarizes our proposal for those elements.

4.1 Structural Restrictions of MMFAP Models

In case of many discrete Markov modulated arrival processes (e.g. MAP, BMAP) the representation is not unique, and starting form a given representation of an arrival process infinitely many different, but stochastically equivalent representations of the same process can be generated with similarity transformation. Based on past experience it is known that optimizing non-unique representations should be avoided, since most computational effort of the optimizers is wasted on going back and forth between almost equivalent representations having significantly different parameters. The usual solution to address this issue is to apply some structural restrictions (e.g. the Jordan representation of some of the matrices), which can make the representation unique [15].

In this work, we also apply a structural restriction to make the optimization of MMFAPs more efficient (by making the path to the optimum more straight): We restrict matrix $\mathbf{R}$ to be diagonal such that the diagonal elements of $\mathbf{R}$ are non-decreasing, which makes the representation of an MMFAP unique except for the ordering of states with identical fluid arrival rates.

4.2 Computation of the Double Sided Inverse Laplace Transform

A crucial step of the algorithm both in terms of execution speed and numerical accuracy is that to compute the numerical inverse Laplace transformation (NILT) of the expression in (3). There are many efficient numerical inverse transformation methods for single sided functions [8]. However, in our case the function we have is double sided (as Gauss distributions can be negative, too), and numerical inverse transformation of double sided Laplace transforms are rather limited.

If $f(t)$ is the density of a positive random variable then $\int_{-\infty}^{\infty} e^{-st} f(t) dt = \int_{0}^{\infty} e^{-st} f(t) dt$ and the single and double sided Laplace transforms of $f(t)$ are identical. If $f(t)$ is the density of a random variable which is positive with a high probability then $\int_{-\infty}^{\infty} e^{-st} f(t) dt \approx \int_{0}^{\infty} e^{-st} f(t) dt$. Based on this approximation one can apply single sided numerical inverse Laplace transformation for density functions of dominantly positive random variables.

For a general MMFAP the non-negativity of the fluid increase samples in $T = \{(t_k, x_k); k = 1, \ldots, K\}$ can not be assumed. To make the single sided numerical inverse Laplace transformation appropriately accurate also in this case we apply the following model transformation

$$\mathcal{L}_{\mathbf{Q,R,S}}(\{(t_k, x_k); k = 1, \ldots, K\}) = \mathcal{L}_{\mathbf{Q,R}+c\mathbf{I,S}}(\{(t_k, x_k + ct_k); k = 1, \ldots, K\}),$$

where $\mathcal{L}_{\mathbf{Q},\mathbf{R},\mathbf{S}}(\{(t_1,x_1);(t_2,x_2);\ldots;(t_K,x_K)\}) = \underline{\alpha} \prod_{k=1}^{K} \mathbf{N}(t_k,x_k)\mathbb{1}$ is the likelihood of the samples when $\mathbf{N}(t_k,x_k)$ is computed with $\mathbf{Q},\mathbf{R},\mathbf{S}$ according to (3) and c is an appropriate constant. If c is large enough, the relative difference of the fluid increase samples reduces and the likelihood function gets less sensitive to the changes of the model parameters. If c is small, fluid increase samples might become close to zero and the single sided numerical inverse Laplace transformation might cause numerical issues.

4.3 Reducing Computational Cost for Equidistant Measurement Intervals

For computing the likelihood function, the numerical inverse Laplace transformation of matrix $\mathbf{N}^*(t,v)$ needs to be performed once for each data point, i.e. K times, which might be computationally expensive.

In the special case when the samples are from identical time intervals, that is $t_1 = \ldots = t_K = \bar{t}$, we apply the following approximate approach to reduce the computational complexity to M ($M << K$) numerical inverse Laplace transformation of matrix $\mathbf{N}^*(t,v)$.

- Let $x_{min} = \min(x_1,\ldots,x_K)$, $x_{max} = \max(x_1,\ldots,x_K)$ and $\Delta = (x_{max} - x_{min})/M$.
- Compute $\mathbf{N}(\bar{t}, x_{min} + (m-0.5)\Delta)$ for $m = 1,\ldots,M$ by NILT of $\mathbf{N}^*(t,v)$.
- For $x \in (x_{min} + (m-1)\Delta, x_{min} + m\Delta)$, apply $\mathbf{N}(\bar{t},x) \approx \mathbf{N}(\bar{t}, x_{min} + (m-0.5)\Delta)$.

This way the $[x_{min}, x_{max}]$ range is divided to M equidistant intervals and the ILT is performed once for each. The higher the parameter M, the higher the accuracy, but the slower the procedure.

4.4 Computation of $E(\Theta_i^{(k)})$ and $E(M_{i,j}^{(k)})$

Let us introduce the forward and backward likelihood vectors for the beginning and the end of the kth observation period

$$\hat{\underline{f}}_k = \underline{\alpha}\left(\prod_{\ell=1}^{k-1} \mathbf{N}(t_\ell,x_\ell)\right) = \underline{\alpha} \prod_{\ell=1}^{k-1} \mathrm{ILT}_{v\to x_\ell}\mathbf{N}^*(t_\ell,v), \tag{8}$$

$$\hat{\underline{b}}_k = \left(\prod_{\ell=k+1}^{K} \mathbf{N}(t_\ell,x_\ell)\right)\mathbb{1} = \prod_{\ell=k+1}^{K} \mathrm{ILT}_{v\to x_\ell}\mathbf{N}^*(t_\ell,v)\mathbb{1}. \tag{9}$$

and the forward and backward likelihood vectors for an internal point in the kth observation period as

$$\underline{f}_k(t,x) = \underline{\alpha}\left(\prod_{\ell=1}^{k-1} \mathbf{N}(t_\ell,x_\ell)\right)\mathbf{N}(t,x), \tag{10}$$

$$\underline{b}_k(t,x) = \mathbf{N}(t,x)\left(\prod_{\ell=k+1}^{K} \mathbf{N}(t_\ell,x_\ell)\right)\mathbb{1}. \tag{11}$$

We note that, using $\hat{f}_k$ and $\hat{b}_k$, the likelihood can be expressed as

$$\mathcal{L} = \underline{\alpha} \cdot \underline{b}_1(t_1, x_1) = \underline{f}_{k-1}(t_{k-1}, x_{k-1}) \cdot \underline{b}_k(t_k, x_k) = \underline{f}_K(t_K, x_K)\mathbb{1}$$
$$= \underline{\alpha} \cdot \hat{\underline{b}}_0 = \hat{\underline{f}}_\ell \cdot \hat{\underline{b}}_{\ell-1} = \hat{\underline{f}}_{K+1}\mathbb{1},$$

for any $k = 2, \ldots, K - 1$ and $\ell = 1, \ldots, K$.

To compute the expected value of $\Theta_i^{(k)}$, the integrals of the forward and backward likelihood vectors have to be evaluated. The special form of the integrals allows for simplifications as

$$E(\Theta_i^{(k)}) = \int_{x=0}^{x_k} \int_{t=0}^{t_k} [\underline{f}_k(t, x)]_i \cdot [\underline{b}_k(t_k - t, x_k - x)]_i \, dt \, dx$$

$$= \hat{\underline{f}}_k \left(\int_{x=0}^{x_k} \int_{t=0}^{t_k} \mathbf{N}(t, x)\underline{e}_i \cdot \underline{e}_i^T \mathbf{N}(t_k - t, x_k - x) \, dt \, dx \right) \hat{\underline{b}}_k$$

$$= \hat{\underline{f}}_k \, \mathrm{ILT}_{v \to x_k} \left(\int_{t=0}^{t_k} \mathbf{N}^*(t, v)\underline{e}_i \cdot \underline{e}_i^T \mathbf{N}^*(t_k - t, v) \, dt \right) \hat{\underline{b}}_k$$

$$= \hat{\underline{f}}_k \, \mathrm{ILT}_{v \to x_k} \left(\int_{t=0}^{t_k} e^{(\mathbf{Q} - v\mathbf{R} - v^2\mathbf{\Sigma}/2)t} \underline{e}_i \cdot \underline{e}_i^T e^{(\mathbf{Q} - v\mathbf{R} - v^2\mathbf{\Sigma}/2)(t_k - t)} \, dt \right) \hat{\underline{b}}_k$$

$$= \hat{\underline{f}}_k \, \mathrm{ILT}_{v \to x_k} \left([\mathbf{0}\ \mathbf{I}] \, e^{\begin{bmatrix} \mathbf{Q} - v\mathbf{R} - v^2\mathbf{\Sigma}/2 & \underline{e}_i \cdot \underline{e}_i^T \\ \mathbf{0} & \mathbf{Q} - v\mathbf{R} - v^2\mathbf{\Sigma}/2 \end{bmatrix} t_k} \begin{bmatrix} \mathbf{I} \\ \mathbf{0} \end{bmatrix} \right) \hat{\underline{b}}_k.$$

That is, the convolution integral is replaced by the evaluation of a matrix exponential of double size [16]. In a similar manner, the expected value of $M_{i,j}^{(k)}$ is

$$E(M_{i,j}^{(k)}) = \int_{x=0}^{x_k} \int_{t=0}^{t_k} [\underline{f}_k(t, x)]_i \cdot q_{ij} \cdot [\underline{b}_k(t_k - t, x_k - x)]_j \, dt \, dx$$

$$= q_{ij} \, \hat{\underline{f}}_k \, \mathrm{ILT}_{v \to x_k} \left([\mathbf{0}\ \mathbf{I}] \, e^{\begin{bmatrix} \mathbf{Q} - v\mathbf{R} - v^2\mathbf{\Sigma}/2 & \underline{e}_i \cdot \underline{e}_j^T \\ \mathbf{0} & \mathbf{Q} - v\mathbf{R} - v^2\mathbf{\Sigma}/2 \end{bmatrix} t_k} \begin{bmatrix} \mathbf{I} \\ \mathbf{0} \end{bmatrix} \right) \hat{\underline{b}}_k.$$

We note that these expressions give an interpretation for the zth iteration of the EM method for $q_{i,j}$

$$q_{i,j}(z+1) = \frac{\sum_{k=1}^{K} \text{E}\left(M_{i,j}^{(k)}\right)}{\sum_{k=1}^{K} \text{E}\left(\Theta_i^{(k)}\right)} =$$

$$q_{i,j}(z) \frac{\left(\underline{\hat{f}}_k \; \text{ILT}_{v \to x_k}\left(\left[\mathbf{0} \; \mathbf{I}\right] e^{\begin{bmatrix} \mathbf{Q} - v\mathbf{R} - v^2\Sigma/2 & \underline{e_i} \cdot \underline{e_j}^T \\ 0 & \mathbf{Q} - v\mathbf{R} - v^2\Sigma/2 \end{bmatrix} t_k} \begin{bmatrix} \mathbf{I} \\ \mathbf{0} \end{bmatrix}\right) \underline{\hat{b}}_k\right)}{\left(\underline{\hat{f}}_k \; \text{ILT}_{v \to x_k}\left(\left[\mathbf{0} \; \mathbf{I}\right] e^{\begin{bmatrix} \mathbf{Q} - v\mathbf{R} - v^2\Sigma/2 & \underline{e_j} \cdot \underline{e_i}^T \\ 0 & \mathbf{Q} - v\mathbf{R} - v^2\Sigma/2 \end{bmatrix} t_k} \begin{bmatrix} \mathbf{I} \\ \mathbf{0} \end{bmatrix}\right) \underline{\hat{b}}_k\right)},$$

that is, $q_{i,j}(z+1)$ is the product of $q_{i,j}(z)$ and an actual guess dependent value.

4.5   Computation of $\underline{\hat{f}}_k$ and $\underline{\hat{b}}_k$

The computation of $\underline{\hat{f}}_k$ and $\underline{\hat{b}}_k$ follows a similar pattern and contains the same difficulties, except that $\underline{\hat{f}}_k$ is computed from $k = 0$ onward and $\underline{\hat{b}}_k$ is computed from $k = K$ downward. The main implementation issue with the computation of $\underline{\hat{f}}_k$ and $\underline{\hat{b}}_k$, is to avoid under-/overflow during the computation. We adopted the under-/overflow avoiding method proposed in [3].

5   Numerical Examples

5.1   MMFAP Simulator

For the numerical evaluation of the proposed method we developed a simulator which generates the required number (K) of traffic samples based on matrices $\mathbf{Q}$, $\mathbf{R}$ and $\mathbf{S}$. In each step of the simulation, the program samples the next state transition of the Markov chain and checks if it occurs before or after the next measurement instance. In the first case it samples the accumulated fluid until the next state transition and performs the state transition, in the second case it samples the accumulated fluid until the next measurement instance and maintains the state of the Markov chain.

To reduce the computational time of the fitting procedure by applying the approximate approach introduced in Sect. 4.3, the simulator generates the data samples such that $t_1 = \ldots = t_K = 1$.

For the MMFAP with

$$\mathbf{Q}_{slow} = \begin{bmatrix} -0.8 & 0.5 & 0.3 \\ 0.6 & -0.7 & 0.1 \\ 0.2 & 0.3 & -0.5 \end{bmatrix}, \; \mathbf{R} = \begin{bmatrix} 2 & 0 & 0 \\ 0 & 4 & 0 \\ 0 & 0 & 8 \end{bmatrix}, \; \mathbf{S} = \begin{bmatrix} 0.01 & 0 & 0 \\ 0 & 0.02 & 0 \\ 0 & 0 & 0.04 \end{bmatrix} \quad (12)$$

the histogram of the samples is presented in Fig. 1a. The histogram indicates that the Markov chain is "slow" in this case, i.e., it stays in a single state (e.g. state i) during the measurement interval of length 1 with high probability and accumulates $\mathcal{N}(r_i, \sigma_i^2)$ distributed amount of fluid during this interval. That is the explanation of the peaks at around $r_1 = 2$, $r_2 = 4$ and $r_3 = 8$. It is also visible that the transitions between state 1 and 2 are faster than the transitions to and from state 3 and consequently, the histogram indicates fluid samples in the $x \in (2, 4)$ interval. These samples might come from measurement intervals starting from state 1 with $r_1 = 2$ and moving to state to with $r_2 = 4$, or vice versa.

To indicate the effect of the "speed" of the Markov chain on the histogram of the generated samples, Fig. 1b depicts the histogram when the Markov chain is "fast", namely 10 times faster, $\mathbf{Q}_{fast} = 10\mathbf{Q}_{slow}$. The "fast" Markov chain experiences state transitions during the measurement interval with very high probability and the amount of fluid accumulated during the interval gets to be less dependent on the state of the Markov chain at the beginning of the measurement interval.

When the variance is low, as it is in this example, one can easily predict the values of the $\mathbf{R}$ matrix with the "slow" Markov chain, while for the "fast" Markov chain the values of the $\mathbf{R}$ matrix is not possible to guess based on the histogram. Still, the minimal and the maximal sample values allows to estimate the minimal and the maximal fluid rates of matrix $\mathbf{R}$.

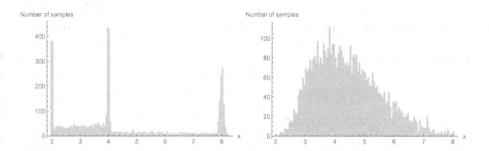

Fig. 1. Histogram of 5000 generated samples with $\mathbf{Q}_{slow}$, $\mathbf{R}$ and $\mathbf{S}$ and $\mathbf{Q}_{fast}$, $\mathbf{R}$ and $\mathbf{S}$ defined in (12).

5.2 Approximating Q with the EM Method

Based on 300 samples of the MMFAP with

$$\bar{\mathbf{Q}} = \begin{bmatrix} -2 & 2 \\ 4 & -4 \end{bmatrix}, \quad \bar{\mathbf{R}} = \begin{bmatrix} 4 & 0 \\ 0 & 8 \end{bmatrix}, \quad \bar{\mathbf{S}} = \begin{bmatrix} 0.01 & 0 \\ 0 & 0.02 \end{bmatrix}, \tag{13}$$

we approximate the MMFAP starting from $\bar{\mathbf{Q}}_0 = \begin{bmatrix} -1 & 1 \\ 2.5 & -2.5 \end{bmatrix}$, $\bar{\mathbf{R}}_0 = \bar{\mathbf{R}}$, $\bar{\mathbf{S}}_0 = \bar{\mathbf{S}}$ with the EM method. The evolution of the log-likelihood value and the transition rates of the Markov chain are depicted in Fig. 2a and 2b respectively, where the dotted horizontal lines refer to the MMFAP according to (13), which was used to generate the samples. The figure indicates that the obtained transition rates provide a bit higher log-likelihood than the ones in (13). Additionally, the figures report convergence after ~ 25 iterations of the EM method.

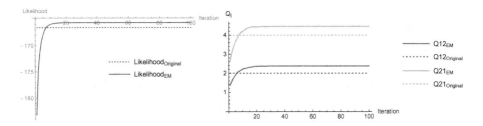

Fig. 2. Behaviour of the EM method based on 300 samples generated from $\bar{\mathbf{Q}}$, $\bar{\mathbf{R}}$ and $\bar{\mathbf{S}}$ in (13) with initial guess $\bar{\mathbf{Q}}_0$, $\bar{\mathbf{R}}_0$ and $\bar{\mathbf{S}}_0$. According to (5) and (6), $\bar{\mathbf{R}}_0$ and $\bar{\mathbf{S}}_0$ remained unchanged during the EM iterations.

Similarly, we evaluated the EM based approximation of the MMFAP defined in (12) based on 1000 samples starting from $\mathbf{Q}_0 = \begin{bmatrix} -0.8 & 0.5 & 0.3 \\ 0.6 & -0.7 & 0.1 \\ 0.2 & 0.3 & -0.5 \end{bmatrix}$, $\mathbf{R}_0 = \mathbf{R}$, $\mathbf{S}_0 = \mathbf{S}$.

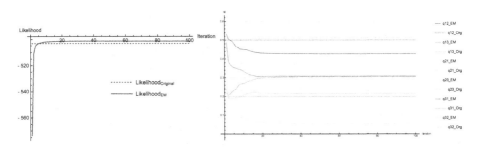

Fig. 3. Behaviour of the EM method based on 1000 samples generated from $\mathbf{Q}$, $\mathbf{R}$ and $\mathbf{S}$ in (12) with initial guess $\mathbf{Q}_0$, $\mathbf{R}_0$ and $\mathbf{S}_0$.

The evolution of the log-likelihood value and the transition rates of the Markov chain along the EM iterations are depicted in Fig. 3a and 3b respectively. Figure 3a indicates that similar to the 2×2 example in Fig. 2a the likelihood value increased above the one obtained from original MMFAP. At the same time, the transition rate values in Fig. 3b differ more significantly from ones of the original MMFAP than in Fig. 2b, which might be a consequence of a looser relation between the transition rates and the likelihood value in higher dimensions.

6   Conclusions

The EM method is commonly applied for parameter estimation of Markov modulated models. In this paper we consider the fitting of MMFAP and recognize that the EM method is not applicable for optimizing the fluid rate and variance parameters. As a result, we investigated the properties of the EM based MMFAP method for fitting the parameters of the governing Markov chain via numerical experiments.

Appendix

A   Maximizing the Model Parameters

Assuming $q_i = \sum_{j, j \neq i} q_{i,j}$, the derivatives of $\log \mathcal{L}$ are as follows:

$$
\frac{\partial}{\partial q_{i,j}} \log \mathcal{L} = \frac{\partial}{\partial q_{i,j}} \sum_{k=1}^{K} \left(-\Theta_i^{(k)} q_{i,j} + M_{i,j}^{(k)} \log q_{i,j} \right)
$$

$$
= \sum_{k=1}^{K} \left(-\Theta_i^{(k)} + M_{i,j}^{(k)} \frac{1}{q_{i,j}} \right),
$$

$$
\frac{\partial}{\partial r_i} \log \mathcal{L} = \frac{\partial}{\partial r_i} \sum_{k=1}^{K} \left(-\Theta_i^{(k)} \frac{r_i^2}{2\sigma_i^2} + F_i^{(k)} \frac{r_i}{\sigma_i^2} \right)
$$

$$
= \sum_{k=1}^{K} \left(-\Theta_i^{(k)} \frac{r_i}{\sigma_i^2} + F_i^{(k)} \frac{1}{\sigma_i^2} \right),
$$

$$
\frac{\partial}{\partial \sigma_i^2} \log \mathcal{L} = \frac{\partial}{\partial \sigma_i^2} \sum_{k=1}^{K} \left(-\Theta_i^{(k)} \frac{r_i^2}{2\sigma_i^2} + F_i^{(k)} \frac{r_i}{\sigma_i^2} - \frac{1}{2\sigma_i^2} F \Theta_i^{(k)} - M_i^{(k)} \frac{\log \sigma_i^2}{2} \right)
$$

$$
= \frac{\partial}{\partial \sigma_i^2} \sum_{k=1}^{K} \left(\left(-\Theta_i^{(k)} r_i^2 + 2 F_i^{(k)} r_i - F \Theta_i^{(k)} \right) \frac{1}{2\sigma_i^2} - M_i^{(k)} \frac{\log \sigma_i^2}{2} \right)
$$

$$
= \sum_{k=1}^{K} \left(\left(-\Theta_i^{(k)} r_i^2 + 2 F_i^{(k)} r_i - F \Theta_i^{(k)} \right) \frac{-1}{2(\sigma_i^2)^2} - M_i^{(k)} \frac{1}{2\sigma_i^2} \right).
$$

The optimal parameter values are obtained where the derivative is zero:

$$0 = \sum_{k=1}^{K} \Theta_i^{(k)} - M_{i,j}^{(k)}/q_{i,j} \longrightarrow q_{i,j} = \frac{\sum_{k=1}^{K} M_{i,j}^{(k)}}{\sum_{k=1}^{K} \Theta_i^{(k)}}. \tag{14}$$

$$0 = \sum_{k=1}^{K} \frac{\Theta_i^{(k)} r_i - F_i^{(k)}}{\sigma_i^2} \longrightarrow r_i = \frac{\sum_{k=1}^{K} F_i^{(k)}}{\sum_{k=1}^{K} \Theta_i^{(k)}}. \tag{15}$$

$$0 = \sum_{k=1}^{K} \left(\Theta_i^{(k)} r_i^2 - 2F_i^{(k)} r_i + F\Theta_i^{(k)} \right) \frac{1}{\sigma_i^2} - M_i^{(k)}$$

$$\longrightarrow \sigma_i^2 = \frac{\sum_{k=1}^{K} \Theta_i^{(k)} r_i^2 - 2F_i^{(k)} r_i + F\Theta_i^{(k)}}{\sum_{k=1}^{K} M_i^{(k)}}. \tag{16}$$

B   Expected Values of the Hidden Parameters

For $E(F_i^{(k)})$ we have

$$\mathrm{E}\left(F_i^{(k)}\right) = E_{\Theta_i^{(k)}} E_{F_i^{(k)}|\Theta_i^{(k)}} (F_i^{(k)}) = E_{\Theta_i^{(k)}} r_i \Theta_i^{(k)} = r_i E(\Theta_i^{(k)}). \tag{17}$$

For $F\Theta_i^{(k)} = \sum_{n=1}^{n^{(k)}} \frac{f_n^{(k)2}}{\theta_n^{(k)}} \mathcal{I}_{\left\{ J_n^{(k)} = i \right\}}$, we have

$$\mathrm{E}\left(F\Theta_i^{(k)}\right) = \mathrm{E}\left(\sum_{n=1}^{n^{(k)}} \frac{f_n^{(k)2}}{\theta_n^{(k)}} \mathcal{I}_{\left\{ J_n^{(k)} = i \right\}} \right)$$

$$= \mathrm{E}_{\left\{ n^{(k)}, \theta_1^{(k)}, \dots, \theta_{n^{(k)}}^{(k)} \right\}} \left(\sum_{n=1}^{n^{(k)}} \mathrm{E}_{\{f_n^{(k)}|\theta_n^{(k)}\}} \left(\frac{f_n^{(k)2}}{\theta_n^{(k)}} \mathcal{I}_{\left\{ J_n^{(k)} = i \right\}} \right) \right). \tag{18}$$

Let $\mathcal{X}(\mu, \sigma^2)$ denote a normal distributed random variable with mean μ and variance σ^2. Its second moment is $\mathrm{E}\left(\mathcal{X}^2(\mu, \sigma^2)\right) = \sigma^2 + \mu^2$. When $\theta_n^{(k)} = x$ then $f_n^{(k)}$ is $\mathcal{X}(xr_i, x\sigma_i^2)$ distributed and $\mathrm{E}\left(f_n^{(k)2}\right) = \mathrm{E}\left(\mathcal{X}^2(xr_i, x\sigma_i^2)\right) = x\sigma_i^2 + x^2 r_i^2$, that is

$$\mathrm{E}_{\{f_n^{(k)}|\theta_n^{(k)}\}} \left(\frac{f_n^{(k)2}}{\theta_n^{(k)}} \mathcal{I}_{\left\{ J_n^{(k)} = i \right\}} \right) = \frac{\mathrm{E}_{\{f_n^{(k)}|\theta_n^{(k)}\}} \left(f_n^{(k)2} \right)}{\theta_n^{(k)}} \mathcal{I}_{\left\{ J_n^{(k)} = i \right\}}$$

$$= \frac{\theta_n^{(k)} \sigma_i^2 + \theta_n^{(k)2} r_i^2}{\theta_n^{(k)}} \mathcal{I}_{\left\{ J_n^{(k)} = i \right\}} = \left(\sigma_i^2 + \theta_n^{(k)} r_i^2 \right) \mathcal{I}_{\left\{ J_n^{(k)} = i \right\}}. \tag{19}$$

Substituting (19) into (18) results

$$
\begin{aligned}
\mathrm{E}\left(F\Theta_i^{(k)}\right) &= \mathrm{E}\left(\sum_{n=1}^{n^{(k)}} \frac{f_n^{(k)\,2}}{\theta_n^{(k)}} \, \mathcal{I}_{\left\{J_n^{(k)}=i\right\}}\right) \\
&= \mathrm{E}_{\{n^{(k)},\theta_1^{(k)},\ldots,\theta_{n^{(k)}}^{(k)}\}}\left(\sum_{n=1}^{n^{(k)}} \mathrm{E}_{\{f_n^{(k)}|\theta_n^{(k)}\}}\left(\frac{f_n^{(k)\,2}}{\theta_n^{(k)}} \, \mathcal{I}_{\left\{J_n^{(k)}=i\right\}}\right)\right) \\
&= \mathrm{E}_{\{n^{(k)},\theta_1^{(k)},\ldots,\theta_{n^{(k)}}^{(k)}\}}\left(\sum_{n=1}^{n^{(k)}} \left(\sigma_i^2 + \theta_n^{(k)}r_i^2\right) \mathcal{I}_{\left\{J_n^{(k)}=i\right\}}\right) \\
&= \mathrm{E}\left(M_i^{(k)}\right)\sigma_i^2 + \mathrm{E}\left(\Theta_i^{(k)}\right)r_i^2.
\end{aligned}
$$

C Numerical Computation of the Expected Value of the Hidden Parameters

In the E-step we compute the expected value of the hidden parameters for given $\underline{\alpha}$, $\mathbf{Q}$, $\mathbf{R}$, $\mathbf{S}$ and observed data (t_k, x_k) for $k = 1, \ldots, K$. For the expected values of $\Theta_i^{(k)}$ we have

$$
\mathrm{E}\left(\Theta_i^{(k)}|t_k, x_k\right) = \mathrm{E}\left(\sum_{n=1}^{n^{(k)}} \theta_n^{(k)} \, \mathcal{I}_{\left\{J_n^{(k)}=i\right\}}\Big|t_k, x_k\right) = \mathrm{E}\left(\int_{t=0}^{t_k} \mathcal{I}_{\{\mathcal{J}(t)=i|x_k\}}dt\right)
\tag{20}
$$

$$
\begin{aligned}
&= \int_{t=0}^{t_k} \mathrm{E}\left(\mathcal{I}_{\{\mathcal{J}(t)=i|x_k\}}\right)dt = \int_{t=0}^{t_k} \Pr\left(\mathcal{J}(t)=i|x_k\right)dt \\
&= \sum_k \sum_\ell \int_{t=0}^{t_k} \Pr\left(\mathcal{J}(0)=k, \mathcal{J}(t)=i, \mathcal{J}(t_k)=\ell|x_k\right)dt \\
&= \sum_k \sum_\ell \int_{t=0}^{t_k} \Pr\left(\mathcal{J}(0)=k\right) \\
&\qquad \int_{x=0}^{x_k} \lim_{\Delta\to 0}\frac{1}{\Delta}\Pr\left(x \le \mathcal{X}(t) < x+\Delta, \mathcal{J}(t)=i|\mathcal{J}(0)=k, \mathcal{X}(0)=0\right) \\
&\qquad \lim_{\Delta\to 0}\frac{1}{\Delta}\Pr\left(x_k \le \mathcal{X}(t_k) < x_k+\Delta, \mathcal{J}(t)=\ell|\mathcal{J}(t)=i, \mathcal{X}(t)=x\right)dx\,dt \\
&= \alpha_k \int_{t=0}^{t_k}\int_{x=0}^{x_k} \mathbf{N}(t,x)\underline{e}_i\underline{e}_i^T\mathbf{N}(t_k-t, x_k-x)\mathbb{1}\,dx\,dt
\end{aligned}
\tag{21}
$$

where the jth element of vector $\underline{\alpha}_k$ is $\Pr\left(\mathcal{J}(0)=j\right)$ and $\underline{e}_i$ is the ith unit column vector.

According to (21), (10) and (11), the expected value of $\Theta_i^{(k)}$ is

$$
E(\Theta_i^{(k)}) = \int_{x=0}^{x_k}\int_{t=0}^{t_k} [\underline{f}_k(t,x)]_i \cdot [\underline{b}_k(t_k-t, x_k-x)]_i \, dt\,dx.
\tag{22}
$$

In a similar manner, the expected value of $M_{i,j}^{(k)}$ is

$$E(M_{i,j}^{(k)}) = \int_{x=0}^{x_k} \int_{t=0}^{t_k} [\underline{f}_k(t,x)]_i \cdot q_{i,j} \cdot [\underline{b}_k(t_k - t, x_k - x)]_j \, dt \, dx. \qquad (23)$$

References

1. Anick, D., Mitra, D., Sondhi, M.M.: Stochastic theory of a data-handling system with multiple sources. Bell Syst. Tech. J. **61**(8), 1871–1894 (1982)
2. Asmussen, S.: Stationary distributions for fluid flow models with or without Brownian noise. Commun. Stat. Stoch. Models **11**(1), 21–49 (1995)
3. Bražėnas, M., Horváth, G., Telek, M.: Parallel algorithms for fitting Markov arrival processes. Perform. Eval. **123–124**, 50–67 (2018)
4. Breuer, L.: An EM algorithm for batch Markovian arrival processes and its comparison to a simpler estimation procedure. Ann. Oper. Res. **112**(1), 123–138 (2002)
5. Buchholz, P.: An EM-algorithm for MAP fitting from real traffic data. In: Kemper, P., Sanders, W.H. (eds.) TOOLS 2003. LNCS, vol. 2794, pp. 218–236. Springer, Heidelberg (2003). https://doi.org/10.1007/978-3-540-45232-4_14
6. Horváth, G., Rácz, S., Telek, M.: Analysis of second-order Markov reward models. In: The International Conference on Dependable Systems and Networks, DSN/PDS 2004, Florence, Italy, pp. 845–854. IEEE CS Press, June 2004
7. Horváth, G., Okamura, H.: A fast EM algorithm for fitting marked Markovian arrival processes with a new special structure. In: Balsamo, M.S., Knottenbelt, W.J., Marin, A. (eds.) EPEW 2013. LNCS, vol. 8168, pp. 119–133. Springer, Heidelberg (2013). https://doi.org/10.1007/978-3-642-40725-3_10
8. Horváth, I., Horváth, G., Almousa, S.A.D., Telek, M.: Numerical inverse Laplace transformation using concentrated matrix exponential distributions. Perform. Eval. (2019)
9. Mitchell, K., van de Liefvoort, A.: Approximation models of feed-forward G/G/1/N queueing networks with correlated arrivals. Perform. Eval. **51**(2), 137–152 (2003)
10. Mitra, D.: Stochastic theory of a fluid model of producers and consumers coupled by a buffer. Adv. Appl. Probab. **20**(3), 646–676 (1988)
11. Neuts, M.F.: A versatile Markovian point process. J. Appl. Probab. **16**, 764–779 (1979)
12. Okamura, H., Dohi, T.: Faster maximum likelihood estimation algorithms for Markovian arrival processes. In: 2009 Sixth International Conference on the Quantitative Evaluation of Systems, pp. 73–82. IEEE (2009)
13. Okamura, H., Dohi, T., Trivedi, K.: Markovian arrival process parameter estimation with group data. IEEE/ACM Trans. Netw. **17**, 1326–1339 (2009)
14. Rydén, T.: An EM algorithm for estimation in Markov-modulated Poisson processes. Comput. Stat. Data Anal. **21**(4), 431–447 (1996)
15. Telek, M., Horváth, G.: A minimal representation of Markov arrival processes and a moments matching method. Perform. Eval. **64**(9–12), 1153–1168 (2007)
16. Van Loan, C.: Computing integrals involving the matrix exponential. IEEE Trans. Autom. Control **23**(3), 395–404 (1978)

Reinforcement Learning with Model-Based Approaches for Dynamic Resource Allocation in a Tandem Queue

Thomas Tournaire[1,2]($^{\boxtimes}$), Jeanne Barthelemy[2], Hind Castel-Taleb[2], and Emmanuel Hyon[3,4]

[1] Nokia Bell Labs France, Nozay, France
thomas.tournaire@nokia.com
[2] Samovar, Telecom SudParis, Paris, France
[3] LIP6 UMR 7606, Sorbonne Universités CNRS, Paris, France
[4] Université Paris Nanterre, Nanterre, France

Abstract. We consider three-tier network architecture modeled with two physical nodes in tandem where an autonomous agent controls the number of active resources on each node. We analyse the learning of auto-scaling strategies in order to optimise both performance and energy consumption of the whole system. We compare several model-based reinforcement learning with model-free Q-learning algorithm. The relevance of these algorithms is to faster update Q-value function with an additional planning phase allowed by approximated model of the dynamics of the environment. Secondly, we consider the same tandem queue scenario with MMPP (Markov modulated Poisson process) for arrivals. In this context, the arrival rate is varying over time and this information is hidden to the agent. Our goal is to assess the robustness of such model-based reinforcement learning algorithms in this particular scenario.

Keywords: Model-based reinforcement learning · Tandem queues · Energy saving · Cloud · QoS guarantee

1 Introduction

Resource auto-scaling [2] technique is a very efficient solution in data center owners for adapting resource provisioning to a variable service demand, by setting up activation and deactivation of Virtual Machines (VMs) according to the workload [6]. However, finding the policy that tailors resources to demand is a crucial point that requires accurate assessment of both the energy expended and the performance of the system. Unfortunately, these two measures are inversely proportional, which motivates researchers to evaluate them simultaneously via a unique global cost function.

Among others, Markov Decision Process (MDP) has been widely used to model resource management, it performs actions (activation or deactivation) on

© Springer Nature Switzerland AG 2021
P. Ballarini et al. (Eds.): EPEW 2021/ASMTA 2021, LNCS 13104, pp. 243–263, 2021.
https://doi.org/10.1007/978-3-030-91825-5_15

the resources according to the queue state. Different algorithms exist to find the optimal policy, but they require the knowledge of the underlying transition probabilities [23]. On the other hand, Reinforcement Learning (RL) techniques learn the optimal policy without requiring the knowledge of the statistics of the system. A model-free reinforcement learning techniques with Q-Learning for autonomic resource allocation in the cloud was proposed by Dutreilh et al. in [9]. They focus on a single node queuing system where the agent has to control the number of active resources to satisfy Quality of Service (QoS) constraints. Moreover, [13] proposes Q-learning to derive auto-scaling policies in the cloud and consider again a single node. Both show that while RL holds great promise for learning adaptive control policies, it still suffers from slow convergence and detrimental random exploration. In this context, model-based reinforcement learning techniques [18] can decrease exploration steps by learning a model of the environment, allowing the agent to update faster the Q-Value by a supplementary planning phase.

In this paper, we focus on N-tier architectures [12] which are the main software application for client-server architectures. We model such network architecture with queues in tandem and consequently consider a three-tier system. The underlying system is represented by a Semi Markov Decision Process in continuous time with a discounted criteria (a similar SMDP for average case appears in [16]) but for our RL purpose the agent does not know it precisely.

It exists in the literature several works studying the structure of the optimal policy in such systems. We refer to [23] for solving a single node system with hysteresis policies and showing improvement while considering the policy structure in the MDP algorithms. Also in [16] is shown that under some assumptions the optimal policy in a tandem queue system is a bang-bang control policy with monotonicity properties. However, the aim of this work is not to focus on the structural form of the policy but on the relevance of reinforcement learning algorithms to find the optimal policy. Indeed, Reinforcement learning is a promising axis in the discipline of queuing systems and many recent papers present RL solutions for these systems [10]. Our goal is to adopt practitioners model-based RL approaches [18], especially Dyna architectures [21] to assess their relevance, if ever, or to improve them for dynamic resource allocation in queuing network systems. The key contributions of this paper are as follows:

1. We propose two queuing systems for modelling three-tier architectures: tandem queueing model with Poisson arrivals, and then with (MMPP) Markov modulated Poisson process;
2. We integrate literature considerations regarding experimental comparison of reinforcement learning algorithms;
3. We implement several versions of Dyna architecture and experimentally show that model-based reinforcement learning techniques can outperform classical model-free algorithms such as Q-Learning, on cloud auto-scaling applications;
4. We study the robustness of RL algorithms on a partially observable system with Markov modulated Poisson process arrivals where arrival rate is varying

over time and hidden to the agent. Nevertheless, we show that model-based methods suffer more than Q-learning in changing environments.

The remainder of this paper is organised as follows. Section 2 reviews the related works. In Sect. 3, we describe the tandem queueing model associated with the three-tier architecture and the Markov decision process. We also detail the cost function used to express the reward in terms of performance and energy consumption. In Sect. 4, we describe model-based reinforcement learning algorithms and discuss about their parameterisation. Section 5 presents the Markov modulated Poisson process model where the agent is in partial observation mode as the variation of arrival rate is hidden. Section 6 shows experimental results and comparison between assessed algorithms. Finally, achieved results are discussed in the conclusion and comments about further research issues are given.

2   Related Works

We are faced with several challenges: energy consumption reduction, optimal management in the cloud to satisfy quality of services, queuing model representation for cloud architectures and reinforcement learning solution for cloud resource allocation.

2.1   Energy and Performance Management in the Cloud

Energy consumption increase is one of the many challenges facing large-scale computing. When the resource utilisation is too low, some of them can be turned off to save energy, without sacrificing performance. In a data center, the power consumption can be divided into static and dynamic parts. The static parts are the base costs of running the data center when being idle and the dynamic costs depend on the current usage. In [15], a power-aware model was defined to estimate the dynamic part of energy cost for a virtual machine (VM) of a given size, this model keeps the philosophy of the pay as you go model but based on energy consumption. Two main approaches of physical server resource management have been proposed to improve the energy efficiency: shutdown or switching on servers or VMs which is referred as dynamic power management [6], and scaling of the CPU performance referred as Dynamic Voltage and Frequency Scaling [14]. Shutdown strategies (considered here) are often combined with consolidation algorithms that gather the load on a few servers to favour the shutdown of the others. Hence, managing energy by switching on or switching off VM is an intuitive and fairly widespread way to save energy. However, as quoted in [6], coarse techniques of shutdown are most often not the best solution to achieve energy reduction. Indeed, shutdown policies suffer from energy and time losses when switching off and switching on take longer than the idle period.

2.2 Control Management for Queueing Models and Markov Decision Process

In order to represent the problems with activations and deactivations of virtual machines, server farm models have been proposed [1,17]. Usually, these server farm models are modeled with multi-server queueing systems [4]. Although multi-server queues allow a fairly fine representation of the dynamicity due to virtualisation, these models do not address issues related to the internal network of the cloud since all VM are considered as parallel resources. Then, these models are appropriate for studying simple nodes of several servers but does not extend to more complex model as N-tier and we must consider more complex queueing systems.

If queuing models allow us to easily compute performance metrics, the decision making for switching on or switching off the VM requires an additional step which remains a key point. The computation of the optimal actions has led to a large field of researches and methods. Dynamic control and especially Markov decision Processes appear to be the most direct method. Hence, numerous works have been devoted to compute optimal policy in multi-server queue models with MDP (see [23] and references therein). However, Markov Decision Process framework requires a perfect knowledge of the model (queuing statistics such as arrival or service distributions etc.). Unhappily, these values are not always known in practice and Reinforcement Learning techniques should be applied to dynamic resource allocation to overcome this lack of information.

2.3 Three-Tier Software Architecture

A three-tier architecture [12] is a software architecture where the application is decomposed into three logical tiers: the presentation layer or user interface, the application layer where data is processed and the data storage layer. This architecture is widely used in client-server applications such as a web applications. The main benefit of three-tier architecture is the local decomposition for each tier on their own infrastructure. Each tier treats a specific task and can be managed independently (scaling, updates, etc.). Currently multi-tier applications are subject to modernisation, using cloud-native technologies such as containers and can be instantiated in well-known cloud providers such as Amazon Web Scaling [3]. Such architectures can be modeled by queuing models especially networks of queues (see [24]). This makes our tandem multi-queue networks particularly relevant for modeling a three-tier architecture.

2.4 Reinforcement Learning for Resource Allocation

A very complete survey about reinforcement learning techniques for auto-scaling in the cloud can be found in [10]. Many works for RL application on cloud resource allocation problems are presented and different taxonomies about the resolution techniques, the criteria to optimize as well as the type of problem are given (ours is a scaling problem). The vast majority of RL techniques are model-free: most often Q-learning and Deep Q-learning techniques.

For scaling problems, the seminal model-free proposals [9], considers model-free Q-Learning algorithm, where an agent has to control the number of resources to optimise, a cost function taking into account virtual machine costs and SLA. Most of the following works focus mainly Q-learning algorithm on a single node but have different optimisation criteria or state space: the work [11] considers the response time and the average resource utilization and [25] proposes an approach based on inhomogeneous VM. At last [13] assesses the applicability of such approaches. Moreover, as far as we know, it does not exist any work about reinforcement learning for scaling problems with several nodes in tandem.

In [10] only two methods are classified as model-based techniques and these two works ought to estimate the transition probabilities and thereafter solve the problem with MDP algorithms. Hence, due to the requirement of having a complete model of the environment in Model-based methods, [5] estimates the probability distribution of the transition between states and the planning of the policy is done in an offline mode. However, as quoted in [18], the model based framework is larger than the simple models presented in [10]. This, coupled with the very few amount of works in literature about model-based reinforcement learning for cloud applications call for further research in this topic and for comparison of model-free algorithms.

3 Tandem Queue Model

This section presents the behaviour of our three-tier architecture. We describe the queuing system as well as the transition probabilities and the costs. We strongly insist that we describe here how the system works and not as it is seen by the controller agent. The agent's observation and its knowledge of the system are described later.

3.1 Model Description

We model the 3-tier software architecture by two nodes (or multi-server stations) in tandem, where one node acts for one tier: application tier and data tier. Each node is represented by a multi server queue (or a buffer, where requests wait for a service) and servers (or Virtual Machines: VMs) which can be activated or deactivated by a controller. We assume that each node has a finite capacity, let B_1 (resp. B_2) the capacity of node 1 (respectively node 2), where the capacity of the node represents the maximum number of requests either waiting for a service or in service. Each VM is represented by a server and we define by K_1 (respectively K_2) the maximum number of usable VMs in node 1 (respectively node 2) knowing that we must have at least one machine activated. All virtual machines (VMs) in a given node are homogeneous, and the service rates can be modelled by an exponential distribution with rate μ_i for node i ($i = 1, 2$) (Fig. 1).

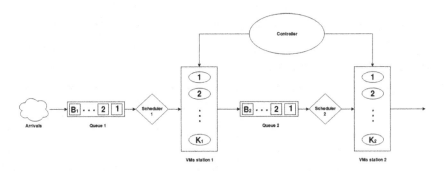

Fig. 1. Tandem queue representation of three-tier architecture

We suppose that requests arrive in the system only in node 1 and arrivals follow a Poisson process with parameter λ. When a request arrives in node 1 and finds B_1 customers in node 1, then it is lost. Otherwise, it waits in the queue until a server becomes free, then after being served in node 1 the request enters in node 2 unless the second queue is full in which case the request is lost. Once the request finishes its service in node 2, it leaves the system.

At each transition epoch, a single controller manages the number of activated VM in each tier and can decide to turn on or turn off virtual machines or to do nothing. Only one VM can be deactivated or activated in each station each time. The only actions that the controller can trigger are then activation or deactivation on each node.

3.2 Semi Markov Decision Process Description

System's Dynamics. The system state includes the current number of requests in node 1, denoted by m_1, in node 2, denoted by m_2, as well as the number of active servers in node 1, denoted by k_1, and in node 2, denoted by k_2. Thus, the state space is defined by $\mathcal{S}$ with $s = (m1, m2, k1, k2) \in \mathcal{S}$ such that $0 \leq m1 \leq B1$, $0 \leq m2 \leq B2$, $1 \leq k_1 \leq K_1$, and $1 \leq k_2 \leq K_2$.

We denote by a_i the action available in node i (with $i \in \{1, 2\}$). It can take three values: **1** if we activate one VM; -1 if we deactivate one VM; and **0** if the number is left unchanged. Any action taken by the controller is the couple of actions in each of the nodes. Hence, the action space is $\mathcal{A}$ where $a = (a_1, a_2) \in \mathcal{A}$ with $a_i \in \{-1, 0, 1\}$.

We describe now the transitions. We consider here that the controller can observe the system just after any change in the state and reacts. The actions are instantaneous. We describe now the effects of the controller's action. Its knowledge of the system as well as the way it decides which action to perform are described in Sect. 4. After an action, the system evolves until the next transition occurs. We define the effect of the action in node i by $N(k_i + a_i) = \min\{\max\{1, k_i + a_i\}, K_i\}$.

So, at state $s = (m1, m2, k1, k2)$, after triggering action $a = (a_1, a_2)$, the possible transitions are:

In case of an arrival in queue 1, we move, with rate λ, in:

$$s_1' = \big(\min(m_1 + 1, B1), m_2, N(k_1 + a_1), N(k_2 + a_2)\big).$$

In case of departure from queue 1 and entry in queue 2, we move, with rate $\mu_1 \min(m_1, N(k_1 + a_1))$ in:

$$s_2' = \big(\max(m_1 - 1, 0), \min(m_2 + 1, B_2), N(k_1 + a_1), N(k_2 + a_2)\big).$$

In case of departure from queue 2, we move, with rate $\mu_2 \min(m_2, N(k_2 + a_2))$, in:

$$s_3' = \big(m_1, \max(m_2 - 1, 0), N(k_1 + a_1), N(k_2 + a_2)\big).$$

We define the transition rate of state-action pair (s, a) by $\Lambda(s, a)$ such that:

$$\Lambda(s, a) = \lambda + \mu_1 \min\{m_1, N(k_1 + a_1)\} + \mu_2 \min\{m_2, N(k_2 + a_2)\}.$$

System's Costs. We approach the problem with a cost-aware model. We introduce here several costs that allow us to manage the trade-off between quality of services (QoS) and energy consumption. There are *instantaneous costs* that are charged only once where C_A denotes the activation cost of a VM, C_D its deactivation cost and C_R the cost of rejecting a request. There are *accumulated costs* that accumulate over time: C_S denote the cost per time unit of using a VM and C_H the cost per time unit of holding a request in the system. In [22] is presented a way to give values to these costs so that they have a real meaning regarding cloud infrastructure. Formally, after triggering action $a = (a_1, a_2)$ in state $s = (m_1, m_2, k_1, k_2)$, the accumulated cost c_i at node i equals:

$$c_i(s, a) = N_i(k_i + a_i) \cdot C_S + m_i \cdot C_H$$

and the instantaneous cost at node i is equal to:

$$h_i(s, a) = C_A \cdot \mathbb{1}_{\{a_i = 1\}} + C_D \cdot \mathbb{1}_{\{a_i = -1\}} + \frac{\lambda}{\Lambda(s, a) + \gamma} C_R \mathbb{1}_{\{m_i = B_i\}} \text{ if } i = 1;$$

$$h_i(s, a) = C_A \cdot \mathbb{1}_{\{a_i = 1\}} + C_D \cdot \mathbb{1}_{\{a_i = -1\}} + \frac{\min\{m_1, N(k_1 + a_1)\} \cdot \mu_1}{\Lambda(s, a) + \gamma} C_R \mathbb{1}_{\{m_i = B_i\}}$$

if $i = 2$.

Terms in front of the reject cost C_R comes from the probability that the event is an arrival and the buffer is full.

Objective Function. We consider here a continuous time discounted model. The discount factor is denoted by γ. We define a Markov Deterministic stationary policy π as a mapping from state space to action space $\pi : \mathcal{S} \to \mathcal{A}$. This mapping defines the action to be performed in a given state s. From discounted model in [19], we define the stage cost function $\mathcal{R}$ as:

$$\mathcal{R}(s, a) = \frac{1}{\Lambda(s, a) + \gamma} [c_1(s, a) + c_2(s, a)] + h_1(s, a) + h_2(s, a).$$

We search the best policy π to minimise the expected discounted expected reward, then the objective function is:

$$V^*(s) = \min_\pi \mathrm{E}^\pi \left[\sum_{k=0}^\infty \exp^{-\gamma t_k} \mathcal{R}(s_k, \pi(s_k)) \mid s_0 = s \right],$$ (1)

with t_k the epoch and s_k the state of the kth transition.

Uniformisation. Most reinforcement learning models and more precisely model-based applications deal with discrete time scenario. In order to handle this control model with an RL approach we will uniformise the continuous time model to obtain a discrete time model on which we can use standard methods of R.L. some of which are already implemented in R.L. libraries.

The way to proceed (see details in Chap. 11 of [19]) is first to define a constant which is finite and larger or equal to the maximum transition rate. Here we take $\tilde{\Lambda} = \lambda + K_1 \cdot \mu_1 + K_2 \cdot \mu_2$. Then, for each state-action couple (s, a) we add a transition associated with a pseudo event so that the process remains in the same state s. In (s, a), this transition has rate $\tilde{\Lambda} - (\Lambda(s, a))$ thus the transition rate is constant with rate $\tilde{\Lambda}$ regardless (s, a). The transition probabilities in the uniformised model are denoted by $\tilde{p}(s'|(s, a))$ and are given by:

$$\tilde{\Lambda} \times \tilde{p}(s'|s, a) = \begin{cases} \lambda & \text{if } s' = s'_1 \\ \mu_1 \min\{m_1, N(k_1 + a_1)\} & \text{if } s' = s'_2 \\ \mu_2 \min\{m_2, N(k_2 + a_2)\} & \text{if } s' = s'_3 \\ (\tilde{\Lambda} - \Lambda(s, a)) & \text{when } s' = s \\ 0 & \text{otherwise}. \end{cases}$$

The states s'_2 and s'_3 are defined in the system's dynamic part above. It exists only few states for which we naturally jump into the same state (*i.e.* $s' = s$). These states are such that $m_1 = B_1$ and $N(k_1 + a_1) = a_1$ and $N(k_2 + a_2) = a_2$ and the event is an arrival. In such cases the transition probability toward $s' = s$ is equal to $\tilde{\Lambda} - (\Lambda(s, a) - \lambda)/\tilde{\Lambda}$, see [19].

The stage costs also need to be modified. Now they are given in the discounted model by:

$$\tilde{\mathcal{R}}(s, a) = \frac{\Lambda(s, a) + \gamma}{\tilde{\Lambda} + \gamma} \mathcal{R}(s, a).$$

Then the Bellman Equation of such a model is:

$$V^*(s) = \min_{a \in \mathcal{A}} \left(\tilde{\mathcal{R}}(s, a) + \frac{\tilde{\Lambda}}{\tilde{\Lambda} + \gamma} \sum_{s'} \tilde{p}(s'|(s, a)) V^*(s') \right).$$

Note that this equation is never solved in R.L. methods but we solve it in our numerical experiments to assess the precision of the algorithms. Indeed its solution is the theoretical optimal value.

4   Model-Based Reinforcement Learning

This section is devoted to the reinforcement learning elements. We first recall the basics of reinforcement learning as presented in [21]. Then, we detail theoretically and practically some model-based algorithms. Most of them are discussed in [18]. In this work, we want to compare several tabular value-based model-based RL algorithms with state of the art model-free algorithms, such as tabular Q-Learning, in the tandem queue scenario. We would like to know if we can benefit from learning the model underlying the environment and if it is worth increasing complexity to speed up convergence.

4.1   Reinforcement Learning

Basics of Reinforcement Learning. A general presentation of Reinforcement Learning can be found in [21]. This dynamic learning is related to the control of dynamic systems. It involves an agent which should take actions in an environment to maximize some cumulative reward. The environment is a view of the stochastic system. The theoretical model behind this framework is that of Markov decision process [19]. It allows to describe the interactions between the agent and its environment in terms of states, actions, and rewards.

The state space $\mathcal{S}$ represents the environment knowing that this one is the view of the system by the agent. The action space $\mathcal{A}$ represents the set of controls available to the agent. At each step, the agent interacts with the environment by the mean of an action, it receives a reward. We denote by $\mathcal{R}(s, a)$ the immediate reward received after taking action a in state s. The environment behaves stochastically from step to step and $p(s'|(s, a))$ denotes the probability to move in state s' at the next step after taking action a in state s. However the agent does not know neither the transition probabilities nor the reward function.

The goal of the agent is to maximise the total discounted reward received during an episode (or conversely to minimise the cost). For this, the agent has to determine which actions will bring the greatest long-term benefit and then determine the best policy π. The return of an episode of length T is defined as the expected sum of the discounted rewards applying a policy π:

$$V^\pi(s) = \mathrm{E}^\pi \Big(\sum_{k=0}^{T} \gamma^t \mathcal{R}(s_k, \pi(s_k)) \mid s_0 = s \Big).$$

When the optimisation criterion is an infinite horizon criterion then the episodes are repeated many times to approximate the infinite horizon expected cost. We also define the action-value function (also called Q-function) which represents the reward received applying action a and then applying policy π:

$$Q_\pi(s, a) = \tilde{\mathcal{R}}(s, a) + \gamma \sum_{s'} p(s'|(s, a)) V^\pi(s').$$

We have $V_\pi(s) = \arg\max_a Q_\pi(s, a)$ and $\pi^* = \arg\max_\pi V_\pi$.

Model-Free and Model-Based Definition. A model $\mathcal{M}$ mimics the dynamics behaviour of the environment. Given a state s and an action a, it predicts the next state s' and the reward $\mathcal{R}(s, a)$. Basically, we have $\mathcal{M} = \{p, \mathcal{R}\}$. In reinforcement learning scenarios, such model is unknown and a learning is needed to overcome this lack of information by updating online the policy or the value function. Thus two frameworks exist: model-free learning in which agent updates its policy or value function directly from experience; and model-based learning in which agent tries to approximate the model $\mathcal{M}$. So agent can do **planning** and update its value function or policy. The RL agent performs planning by sampling trajectories with its model using past experiences and updating its value function again with Q-learning update. A major benefit of model-based techniques is that it requires limited amount of experience and will achieve a better policy with fewer interactions.

The Reinforcement Learning Model. We are dealing with a RL model therefore the uniformised model of Sect. 3 is not known by the controller. Indeed, the controller does not have any information about queuing statistics (arrival rate, etc.), therefore does not know the dynamics of the system. The environment has state space $\mathcal{S}$ and action space $\mathcal{A}$ similarly as the SMDP model. The discount factor equals $\tilde{\Lambda}/(\tilde{\Lambda} + \gamma)$. When the agent interacts with the environment, the SMDP model is simulated. It will behave by returning a state, sampled according to the uniformised transition probabilities $\tilde{p}$, and return the uniformised costs $\tilde{\mathcal{R}}(s, a)$. The cost and the new state are the only information that the controller will discover.

We consider in this work countable discrete state space where the agent can evaluate value function, policy, transition and reward matrices with tabular forms. We want to minimise the expected discounted cumulative costs.

4.2 Advanced Model-Based Algorithms

We present now the model-based framework (see details in [18]). The generic model-based reinforcement learning algorithm is given in pseudo-code in Algorithm 1 and its different steps are detailed below.

Algorithms Architectures. For this purpose we implemented several versions of the Dyna-Q architecture [21]. The generic Dyna-Q algorithm behaves as follow: the RL agent interacts with the real system and directly updates from experiences its value function Q (model-free update). In the same stage, it also learns a model $\mathcal{M}$ of the dynamics of the environment. Its model $\mathcal{M}$ acts as a predictor to know what will be the next state s' and reward r when it chooses action a in state s. Then the agent can do a planning phase by simulating trajectories with $\mathcal{M}$ to additionally update its Q value function or policy.

What Model to Learn. The difference between assessed algorithms relies in the model choice. We describe two different model-based process: *buffer-oriented* and *model-oriented* which we refer respectively as Dyna-Q-buffer and Dyna-Q-model.

In the offline planning phase, the first process will sample trajectories with past experiences stored in a buffer, thus calling agent's memory. It will plan with $\mathcal{M}(s,a)$ to predict new state s' and reward r from past experiences. We can consider this method as model-based method assuming the environment is deterministic. After each transition $s_t, a_t \rightarrow r_{t+1}, s_{t+1}$, the model maintains in its memory for the couple s_t, a_t the prediction of the values of r_{t+1}, s_{t+1}. This prediction is a deterministic mapping from the couple s_t, a_t. Thus, if the model is called for a state-action pair (s,a) that has been observed before, it will returns the last-observed next state and next reward as its prediction.

The second process aims to learn a model of the dynamics of the environment, i.e. transition probabilities $\bar{p}$ and reward function $\bar{\mathcal{R}}$. The agent will be able to generate trajectories with the learned model: $s', r \leftarrow \bar{p}(s,a), \bar{\mathcal{R}}(s,a)$ and estimate the return to update its value function. In *model-oriented* algorithms, we consider tabular representation of transition matrix $\tilde{p}$ and reward matrix $\mathcal{R}$ as a model for the agent. Updating the model means to approximate $\tilde{p}$ and $\mathcal{R}$ by counting occurrences in tuples (s,a) and (s,a,s'):

$$\bar{p}(s'|(a,s)) = \frac{\#(s,a,s')}{\#(s,a)}$$

Buffer Replay for Planning. After interaction with the environment, RL agents can store their past experiences in a buffer and can reuse these experiments to do offline planning. The algorithms randomly samples only from state-action pairs that have previously been experienced. So the model is never queried with a pair about which it has no information. Prioritised sweeping [21] is a heuristic that rearranges the buffer so agent can plan with efficient tuples. It sorts the replay buffer with TD-error criteria. The algorithm moves tuples with highest TD-error at the beginning of the replay buffer so the agent can re-update its Q-value function on these uncertain tuples when planning. This process is integrated in two previous algorithms and we denote: Dyna-Q-buffer-prior and Dyna-Q-model-prior. Finally, Dyna-Q-+ [21] is an extended version of Dyna-Q algorithm in which the agent keeps track of time steps elapsed since the last trial of a pair and encourages with extra reward exploration of long-time unseen pairs. Again, we extend our two main algorithms with this extended version: we will refer to Dyna-Q-buffer-plus and Dyna-Q-model-plus.

Algorithms Parameterisation. In this part, we describe the different elements on which we can operate in the model-based algorithms. Indeed, many factors affect the quality of learning: the frequency of model update and the different planning parameters (frequency, state space browsing).

When to Update Model. Knowing when to update the model is a complex task. Indeed, updating the model too often will increase complexity and, during the early phase of learning, agent will plan using a very poor precision approximated model.

Algorithm 1: Model-based reinforcement learning algorithms

Input: V_0, Q_0, ν: frequency model update, β: frequency planning, η: planning
depth, ζ: planning breadth

Output: π^*, Q^*

Data: System dynamics unknown

```
/* Loop until end of episodes                                    */
```

1   **for** $e \in EPISODE$ **do**

2   Select state $s_0 \in \mathcal{S}$ `// Initial state`

3   **for** $i \in ITERATION$ **do**

4   Take action $a \in \mathcal{A}$ with epsilon-greedy policy   `// Action selection`

5   Observe s' and reward $r(s,a)$ and collect tuple $< s, a, r, s' >$
        `// Collecting tuples data`

6   Model-free update
$$Q(s,a) = Q(s,a) + \alpha\big[r(s,a) + \gamma\max'_a Q(s',a') - Q(s,a)\big]$$

7   **if** $i \% \nu = 0$ **then**

8   Update transitions $T(s,a,s')$ and rewards $R(s,a)$
            `// Update/Learn model `$\tilde{M}$

9   **if** $i \% \beta = 0$ **then**

10   **for** $b \in 1, .., \zeta$ **do**

11   Select (s,a) from replay buffer

12   **for** $p \in 1, ..., \eta$ **do**

13                 $s', r \leftarrow \tilde{T}(s,a,s'), \tilde{R}(s,a)$ `// With model-oriented`
                `methods`

14                 $s', r \leftarrow \text{Buffer}(s,a)$ `// With buffer-oriented methods`

15   Q learning update $Q(s,a)$

Integration Planning and Learning. In [21], Dyna-Q algorithm does planning at every iteration. When the agent performs an action in the real system, it stores the experience tuple in the buffer and do planning steps just after. This has proven to outperform classical model-free techniques such as Q-learning on simple environments with small state space such as games (mazes, etc.). However, this can be too costly in network systems where the state space includes many metrics such as number of packets, number of resources, etc. This raises the question of how frequently an agent should do planning.

Planning Depth and Breadth. An autonomous agent can do planning with a model to imagine consequences of a policy or actions. Based on a model of the dynamics of the environment, he can memorise a given state s, then imagine an action $a = \pi(s)$ and evaluate its consequences with the model $s', r = \mathcal{M}(s,a)$. Now, an agent can perform two imaginary evaluations. Either it starts from a state s and evaluates an episode (depth), or it browses the state space in large and evaluates some actions in multiple independent states s (breadth). Several parameterisation were assessed in this work but only one will be displayed in the results for ease of comparison between model-free and dyna-based algorithms.

5   Robustness of Model-Based RL

In this section we want to assess robustness of reinforcement learning algorithms and especially model-based algorithms. For this purpose, we consider variations in the requests arrival rate with Markov Modulated Poisson Process system. However, these variations remain ignored by the agent which expects a constant intensity. We study how the algorithms react to sudden and large increases of packet arrivals and if they can quickly adapt their policy to overcome bursts traffic. This is a key element in network in which statistics are often not very precise. This is all the more important for model-based methods, since it was quoted in [10] that offline policies might no be adequate as soon as there exist changes in the dynamics of the environment. We think that this lack of flexibility comes from the offline learning of the policy in model-based methods. The offline learning is done from a predefined model which does not allow to adapt to changes in the dynamics of the environment.

5.1   Markov Modulated Poisson Process

We modify now the arrival process. We want to integrate a variability feature and we consider a Markov modulated Poisson process (MMPP) where arrival rates vary over time. The process switches between different Poisson process which differ by their intensity indicated by their arrival rate λ_j. We assume we have J phases corresponding to a specific arrival rate λ_j. The switch between the phases follows a continuous time Markov chain. It is represented by a birth and death process [20] with rates $\mathcal{Q}$. We move from phase j to phase $j+1$ with rate $q_{j,j+1}$ and from phase j to phase $j-1$ with rate $q_{j,j-1}$. Usually [20], $q_{j,j+1}$ and $q_{j,j-1}$ for all j are much smaller than arrival rates $\{\lambda_j\}_j$. We extend the previous state $s = (m_1, m_2, k_1, k_2)$ to a new state representation integrating the phase:

$$s = (m_1, m_2, k_1, k_2, j)$$

with λ_j the current arrival rate in phase j.

In our model, we assume that $J = 2$. The first phase is considered as *normal* and the second phase called *burst* is a phase with a very high intensity. We denote by $q_{1,2}$ the transition rate from phase 1 to phase 2 and by $q_{2,1}$ the transition rate from phase 2 to phase 1.

MMPP System's Dynamics. We need to integrate new phase transitions in the whole environment dynamics. Under one phase j, we have the dynamics detailed in Sect. 3.2 for arrival rate λ_j. The system can have two additional events corresponding to a change of its phase, *i.e.* the current arrival rate changes. Thus:

We move from $s = (j, m_1, m_2, N(k_1 + a_1), N(k_2 + a_2))$ with rate $q_{j,j+1}$ to $s'_4 = (m_1, m_2, N(k_1 + a_1), N(k_2 + a_2), j + 1)$, from $s = (j, m_1, m_2, N(k_1 + a_1), N(k_2 + a_2))$ with rate $q_{j,j-1}$ to $s'_5 = (m_1, m_2, N(k_1 + a_1), N(k_2 + a_2), j - 1)$.

We define the new transition rate by state by $\Lambda(s, a)$ such that:

$$\Lambda(s,a) = q_{j,j+1} + q_{j,j-1} + \lambda_j + \mu_1 \min\{m_1, N(k_1+a_1)\} + \mu_2 \min\{m_2, N(k_2+a_2)\}$$

The cost function remains the same since variation in the arrival rate does not induce additional costs.

MMPP Uniformisation. For the MMPP uniformisation process we take $\tilde{\Lambda} = \max_j q_{j,j+1} + \max_j q_{j,j-1} + \max_j \lambda_j + K_1 \cdot \mu_1 + K_2 \cdot \mu_2$. Then, for each state-action couple (s,a) we add a transition associated with a pseudo event so that the process remains in the same state s. In (s,a), this transition has rate $\tilde{\Lambda} - \Lambda(s,a)$ such that the transition rate of the whole point process is constant with rate $\tilde{\Lambda}$ regardless (s,a). The transition probabilities in the uniformised model are denoted by $\tilde{p}(s'|(s,a))$ and satisfy:

$$\tilde{\Lambda} \times \tilde{p}(s'|s,a) = \begin{cases} \lambda_j & \text{if } s' = s_1' \\ \mu_1 \cdot \min\{m_1, N(k_1+a_1)\} & \text{if } s' = s_2' \\ \mu_2 \cdot \min\{m_2, N(k_2+a_2)\} & \text{if } s' = s_3' \\ q_{j,j+1} & \text{if } s' = s_4' \\ q_{j,j-1} & \text{if } s' = s_5' \\ \left(\tilde{\Lambda} - \Lambda(s,a)\right) & \text{when } s' = s \\ 0 & \text{otherwise} \end{cases}.$$

Similarly as before, the states for which it exists a natural jump into the same state are those already described for which the event is an arrival. In such a case the transition probability toward $s' = s$ is equal to $\left(\tilde{\Lambda} - (\Lambda(s,a) - \lambda)\right)/\tilde{\Lambda}$.

5.2 Partially Observable System

As already said, we consider a model with variation of arrivals, thus the system behaves as the SMDP described just above and the system state is described by $s = (m_1, m_2, k_1, k_2, j)$. However, we assume that the learning agent ignores these variations. Hence, the system can be only observed partially and the environment state is described by $o = (m_1, m_2, k_1, k_2)$. In this way, the two system states $(m_1, m_2, k_1, k_2, 1)$ and $(m_1, m_2, k_1, k_2, 2)$ translate in the same environment state (m_1, m_2, k_1, k_2). The goal is to assess and compare the robustness of reinforcement learning algorithms in the context where the agent does not have knowledge of an explanatory variable.

6 Experimental Results

6.1 Comparison Criteria Between Algorithms

Directly adapting methods to compare algorithms used by practitioners is proving to have little relevance and utility. We adopt the guidelines proposed in the guide for rigorous comparisons of reinforcement learning algorithms [8]. Comparisons are divided into two main parts: comparison while learning and comparison after learning on test environment.

Learning Curves Comparison. First, we compare different algorithms during the learning phase. Instead of only comparing the final performances of the RL methods after t timesteps in test environment, we can compare performances along learning. Indeed, performance measures while learning are represented by a learning curve. This reveal differences in speed of convergence and can provide more robust comparisons.

Test Policy Comparison. RL algorithms should also be assessed offline. The algorithm performance after t iterations is measured as the average of the returns over N evaluation episodes conducted independently after training. The evaluation is done implementing the policy π returned by the RL method at step t.

Comparison Criteria Used. Based on literature and experimental considerations, we devised our own comparison criteria, mainly for the learning comparison between algorithms. We express two comparisons regarding learning curves. The first one is the mean reward obtained after each learning episodes e. The second one is the value function V for a given state s which is retrieved at the end of each episode by:

$$V(s) = max_a Q(s, a)$$

The mean reward during an episode e is the average discounted reward obtained over the learning episode. These two curves indicates how fast the convergence is and how quickly RL algorithms can adapt the policy in environment with changes such as Sect. 5.1. After learning, we evaluate the learned policy of each algorithm in a test environment. We evaluate all policies in Monte Carlo simulations (starting from a state s_0) for 50 episodes of 10000 iterations. We finally take the mean discounted reward obtained overall and can compare the goodness of the algorithms. We also compare our results with the optimal value $V(s_0)$ obtained by MDP policy computed with Value Iteration and which serves as a baseline.

6.2   Environments and Simulation Parameters

Gym Environments. OpenAI Gym [7] is a toolkit for developing and comparing reinforcement learning algorithms. We developed a python simulator under a Gym environment to simulate the tandem queue model. Two environments were implemented: **Environment 1** represents the tandem queue model and **Environment 2** represents the model with MMPP. The goal is to compare model-based and model-free algorithms on both environments.

Parameters. We ran multiple simulations on both environments. We first describe cloud simulations parameters that are common to all experiments. We focus on one main cloud scenario: $B_1 = B_2 = 10$, $K_1 = K_2 = 3$ and costs: $\{C_a = 1, C_d = 1, C_s = 2, C_h = 2, C_r = 10\}$ and on a larger scale scenario $B_1 = B_2 = 40$, $K_1 = K_2 = 5$. Learning phase runs for 450 episodes of length 10000 iterations. One iteration corresponds to a transition from state s with action a to state s' with reward r. The agent runs epsilon-greedy policy over the whole learning process, with initial epsilon set to $\epsilon = 1$ and epsilon decay $\epsilon_{dec} = 0.99$. We decrease epsilon value after each episode. The discount rate is set to $\gamma = 0.9$. Despite our discussion about the algorithms parameterisation, in this paper experiments are performed with only one set of parameters. This includes frequency update of the model (for *model* algorithms, frequency of planning phases and the depth of the planning phase (number of planning iterations).

6.3   Results

First, we display experimental results for the environment 1 with the normal scenario. In this case, we consider the two families of model-based algorithms: *model*-oriented or *buffer*-oriented. Next, we leave aside *model*-oriented algorithms and we provide a comparison on the large scale cloud scenario to confirm the preliminary results related to *buffer*-oriented methods. At last, we provide experimental results on the MMPP environment.

Moreover, during the first environment study, we seek to consider two quantitative elements: the quantity of interactions as well as the execution time. Indeed, these are two important elements for a machine learning analyst. It must be noticed that, due to the offline updates, the complexities of the model-based algorithms are greater than these of model-free algorithms. Thus model-based algorithms have a larger running time for the same number of interactions with the environment. Henceforth, we provide two comparisons: one where the number of interactions are the same between model-based algorithms and Q-learning and a second one where the running times are the same. In this second case, this means that Q-learning does additional iterations and requires more interactions.

First Environment Experiments

Learning Curves Comparison. We display in Fig. 2 the mean reward obtained during learning episodes for all algorithms and observe a gain from dyna-Q-buffer and dyna-Q-buffer-plus methods compared to Q-learning. We first see from Fig. 2 that *model*-oriented algorithms suffers compared to *buffer*-oriented algorithms. Indeed, we observe poor performances and a very high execution time for these methods. In Fig. 3 is displayed value function of a given state s over learning episodes. On the left hand side is given the value function for a same number of online iterations between Q-learning and dyna-sampling methods and on the right hand side is given the value function where Q-learning has more online iterations to compensate the lack of offline update. Again, we see that dyna-Q-buffer and dyna-Q-buffer-plus converge much faster than Q-learning. Unfortunately, we obtain poor performances with prioritised sweeping heuristic and this requires more work to fine-tune this method.

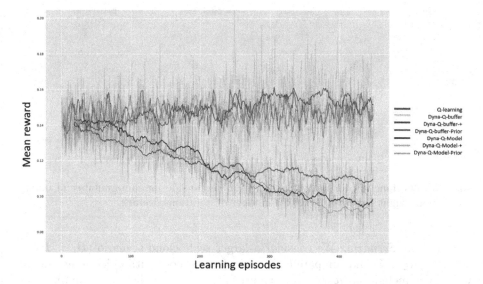

Fig. 2. Mean reward over learning episodes in tandem queue environment

Policies Comparison in Test Environments. We also compare algorithms by evaluating their learned policies in test environment. We provide to them the same amount of learning time (Fig. 3 right hand side). We run the obtained policy in a test environment for an evaluation which takes the average discounted reward (return) obtained starting from a given state s. We can see that *buffer*-oriented methods have converged while Q-learning has not. We do not display here *model*-oriented algorithms with model approximation of $\tilde{p}$ since we have already shown in learning curves their bad efficiency, due to poor offline planning update with $\tilde{p}, \mathcal{R}$. We observe in Table 1 that model-based *buffer*-oriented methods outperform Q-learning and are very close to the MDP policy performances.

Table 1. Mean cost over test episodes for policy evaluation after fixed period of learning

Algorithms policy	Mean cost Env 1	Mean cost Env 1 larger scale	Mean cost Env 2
Random policy	3,37	3.47	3,58
MDP policy	2,25	2.73	Unknown
Q-learning	2,45	3.17	2,63
Dyna-Q-buffer	2,26	2,84	3,05
Dyna-Q-buffer-plus	2,26	2,87	2,92
Dyna-Q-buffer-prior	2,6	3,2	3,11

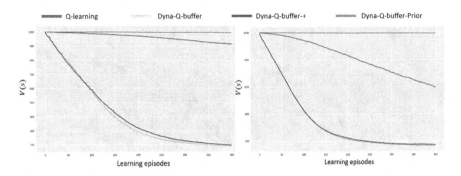

Fig. 3. Value function for a given state s (left hand-side for same number of online interactions, right hand-side for same number of iterations overall)

Larger Scale Scenario. We consider a larger scale cloud scenario ($B_1 = B_2 = 40, K_1 = K_2 = 5$) and compare *buffer*-oriented methods with Q-learning. Again, we do not include *model*-oriented algorithms for this higher scale comparison, neither *buffer*-oriented prioritised sweeping algorithm. We provide learning performances in Fig. 4 and show policy evaluation in Table 1. We demonstrate again that *buffer*-oriented techniques have better performances than Q-learning for a same learning time.

Concluding Remarks. First we show in the environment 1 that model-oriented algorithms denoted *model* are not efficient due to an extreme increase of complexity and planning with low precision of the approximation of transitions. We also exhibit that *buffer*-oriented techniques provide better performances than model-free Q-learning, by accelerating the convergence. This appears in the learning curves and on policy evaluation in test environments.

MMPP Environment Experiments

Environment Characteristics. We consider a scenario where burst arrivals can appear. The goal is to assess robustness of RL algorithms to sudden changes in the system. Arrival rates λ_j can take two values: 5, 20. We define transition rate between phases by: $q_{0,1} = 1$ and $q_{1,0} = 10$. This leads to low chances to have burst phase and when it happens to remain in this phase a very short amount of time. We compare model-based *buffer*-oriented algorithms with Q-learning on learning phase and test phase with evaluation of learned policies, on the low scale cloud scenario ($B_1 = B_2 = 10, K_1 = K_2 = 3$).

Policy Evaluation. We show that model-based techniques suffers from variation in queuing statistics. We can see in Table 1 that mean cost measured after simulations is better in model-free technique compared to model-based ones for a same learning time. One explanation is that model-based RL agent keeps updating its value function based on an outdated model of the world. Therefore, they

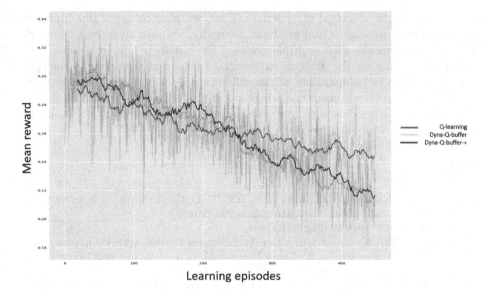

Fig. 4. Average reward over learning episodes in tandem queue environment with larger scale

do not integrate new dynamics change and blend model-free update with model-based update resulting in poor performances. These findings confirm theoretical assumptions about model-based techniques for varying environment. The MDP policy computation is still in progress for further comparison about mean reward.

Concluding Remarks. We illustrate that model-based RL algorithms are not robust to environment changes, especially varying queuing statistics such as arrival rate. We show on policy's evaluation in test environments that Q-learning performs better than model-based algorithms after a same learning time. Thus we can conclude that model-based methods are not robust (in line with [18]) to changes in statistics. This is due to the fact that the agent updates its value function with an outdated model for a certain time therefore fading the model-free update.

7 Conclusion

In this paper we have shown that Dyna-buffer-based algorithms were outperforming model-free Q-learning algorithm in the tandem queue environment, thus demonstrating the sample-efficiency and faster convergence of model-based algorithms. Yet, model-oriented algorithms are suffering of high complexity and dynamics poor approximation. We underline that the approach presented here can be applied to more general network models and with a large set of distributions for arrivals or services and even directly by using traffic traces of cloud platforms. We expect these results would be still valid in these cases.

This work raises several perspectives. First, we plan to apply the approach presented here on a real cloud platform. Secondly, we want to extend this work to network of queues with multiple nodes. This is namely very relevant in the field of 5G slicing for the dynamic allocation of resources according to the QoS (Quality of service) constraints of the different classes of slices. We also want to consider larger-scale scenarios where we might have to approximate policy, value function or models with functional forms such as neural networks, linear approximation, Gaussian processes, etc.

References

1. Adan, I.J., Kulkarni, V.G., Van Wijk, A.C.C.: Optimal control of a server farm. INFOR: Inf. Syst. Oper. Res. **51**(4), 241–252 (2013)
2. Amazon: AWS Auto Scaling (2018)
3. Amazon: AWS Serverless Multi-Tier Architectures (2019). AWS Whitepaper
4. Artalejo, J.R., Economou, A., Lopez-Herrero, M.J.: Analysis of a multiserver queue with setup times. Queueing Syst. **51**(1–2), 53–76 (2005)
5. Barrett, E., Howley, E., Duggan, J.: A learning architecture for scheduling workflow applications in the cloud. In: European Conference on Web Services, pp. 83–90 (2011)
6. Benoit, A., Lefèvre, L., Orgerie, A.-C., Rais, I.: Reducing the energy consumption of large scale computing systems through combined shutdown policies with multiple constraints. Int. J. High Perform. Comput. Appl. **32**(1), 176–188 (2018)
7. Brockman, G., et al.: OpenAI GYM. arXiv:1606.01540 (2016)
8. Colas, C., Sigaud, O., Oudeyer, P.Y.: A Hitchhiker's guide to statistical comparisons of reinforcement learning algorithms. In: RML@ICLR (2019)
9. Dutreilh, X., et al.: Using reinforcement learning for autonomic resource allocation in clouds: towards a fully automated workflow. In: ICAS (2011)
10. Gari, Y., Monge, D.A., Pacini, E., Mateos, C., Garino, C.G.: Reinforcement learning-based application autoscaling in the cloud: a survey. Eng. Appl. Artif. Intell. **102** (2021)
11. Horovitz, S., Arian, Y.: Efficient cloud auto-scaling with SLA objective using Q-learning. In: IEEE FiCloud (2018)
12. IBM: IBM Cloud Learn Hub (2020)
13. Jin, Y., et al.: Testing a Q-learning approach for derivation of scaling policies in cloud-based applications. In: ICIN, 1–3 (2018)
14. Krzywda, J., Ali-Eldin, A., Carlson, T. E., Ostberg, P.-O., Elmroth, E.: Power-performance tradeoffs in data center servers: DVFS, CPU Pinning, horizontal and vertical scaling. Future Gener. Comput. Syst. **81**, 114–128 (2018)
15. Kurpicz, M., Orgerie, A.-C., Sobe, A.: How much does a VM cost? Energy-proportional accounting in VM-based environments. In: PDP, pp. 651–658 (2016)
16. Liu, Z., Deng, W., Chen, G.: Analysis of the optimal resource allocation for a Tandem Queueing system. Math. Probl. Eng. **2017**, 1–10 (2017)
17. Mitrani, I.: Managing performance and power consumption in a server farm. Ann. Oper. Res. **202**, 121–134 (2013)
18. Moerland, T., Broekens, J., Jonker, C.: Model-based Reinforcement Learning: A Survey. arXiv:2006.16712v3 (2021)
19. Puterman, M.: Markov Decision Processes: Discrete Stochastic Dynamic Programming. Wiley, Hoboken (1994)

20. Scott, S., Smyth, P.: The Markov modulated poisson process and Markov Poisson cascade with applications to Web traffic modeling. Bayesian Stat. **7** (2003)
21. Sutton, R.S., Barto, A.G.: Reinforcement Learning: An Introduction. The MIT Press, Cambridge (2015)
22. Tournaire, T., Castel-Taleb, H., Hyon, E., Hoche, T.: Generating optimal thresholds in a hysteresis queue: a cloud application. In: IEEE Mascots (2019)
23. Tournaire, T., Castel-Taleb, H., Hyon, E.: Optimal control policies for resource allocation in the cloud: comparison between markov decision process and heuristic approaches. CoRR, abs/2104.14879 (2021)
24. Urgaonkar, B., Pacificiy, G., Shenoy, M.S.P., Tantawi, A.: An analytical model for Multi-Tier internet services and its applications. ACM SIGMETRICS Perform. Eval. Rev. **33**, 291–302 (2005)
25. Wei, Y., Kudenko, D., Liu, S., Pan, L., Wu, L., Meng, X.: A reinforcement learning based auto-scaling approach for SaaS providers in dynamic cloud environment. Math. Probl. Eng. **2019**, 1–11 (2019)

Performance Analysis of Production Lines Through Statistical Model Checking

Paolo Ballarini[1]([✉]) and András Horváth[2]

[1] MICS, CentraleSupélec, Université Paris-Saclay, Gif-sur-Yvette, France
paolo.ballarini@centralesupelec.fr
[2] Dipartimento di Informatica, Università di Torino, Turin, Italy
horvath@di.unito.it

Abstract. Design and maintenance of reliable manufacturing systems calls for the development of formal models that allow for performance analysis. We consider the class of manufacturing systems such that the production of a workpiece consists of a sequence of manufacturing stages performed by fault-prone, repairable, workstations, equipped with finite-sized input buffers. We name this kind of systems production lines. Relying on an expressive property specification formalism, namely the hybrid automata specification language, we introduce a framework that allows for 1) the automatic generation of stochastic Petri nets models of arbitrary sized production lines and 2) the generation of a number of sophisticated performance indicators (in terms of hybrid automata) for analysing the dynamics of a production line. We validate our approach by presenting a number of experiments executed by means of the statistical model checker Cosmos.

Keywords: Manufacturing systems · Hybrid automata specifications · Statistical model checking

1 Introduction

The design of modern industrial production systems is strongly affected by product quality and delivery reliability requirements. The ability to guarantee that products are issued within given time deadlines and that they match given quality standards are essential factors throughout the design and maintenance of a manufacturing system. In this paper we consider *synchronous* production lines, namely production systems consisting of a sequence of unreliable machines separated by finite-size buffers (where pieces are temporally stored throughout the intermediate phases of manufacturing) and whose dynamics is of a strongly synchronous nature. As shown in previous works (e.g., [5,6,12–14]), this kind of production systems are suitable to be modelled through discrete-time Markov chains (DTMCs). Relying on the statistical model checking (SMC) framework Cosmos [15], in this paper we introduce an integrated framework for formal modeling and performance analysis of production lines. Our framework allows

© Springer Nature Switzerland AG 2021
P. Ballarini et al. (Eds.): EPEW 2021/ASMTA 2021, LNCS 13104, pp. 264–281, 2021.
https://doi.org/10.1007/978-3-030-91825-5_16

for the generation of a stochastic Petri net representation of a production line of arbitrary size (i.e. arbitrary number of production machines and buffers' size) and, at the same, to introduce a number of sophisticated key performance indicators (KPIs) formally defined through a property language which uses hybrid automata as a formalism to specify a performance indicators (namely the hybrid automata specification language (HASL) [9]).

The paper is organized as follows: Sect. 2 describes the production lines class of systems. Section 3 gives an overview of the HASL model checking framework, including the stochastic Petri net formalism it refers to. Section 4 shows the main contribution of the paper, namely the Petri net encoding of production lines and the definition of a few production lines related KPIs in HASL terms. In Sect. 5 we present experiments demonstrating the proposed framework. Conclusions are drawn in Sect. 7.

2   Production Line Systems

A linear production system (or production line for short) is a type of manufacturing system in which parts visit a number of workstations (called also machines) in a fixed order (see Fig. 1) and following a single path (no branching). Machines process one part per time unit. In many industries there are production processes that can be described by such a simple model, for example food industry, automotive industry, paper industry, or semiconductor manufacturing.

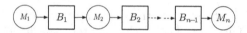

Fig. 1. A production line with n machines.

In this paper we assume two-state machines: a machine can be either up (or UP or U for short) or broken (or $DOWN$ or D for short). The state of a machine changes in a random fashion. In each time unit (called also slot) an operational machine can break with probability p_i, $1 \leq i \leq n$, where n is the number of machines. A broken down machine gets instead repaired with probability r_i, $1 \leq i \leq n$. Standalone availability of a given machine M_i, i.e., the probability that M_i is up assuming that it is not affected by the rest of the system, can be calculated simply as $r_i/(r_i + p_i)$.

Between two adjacent machines there is a buffer where parts can be stored. Accordingly, the number of buffers is $n - 1$. The capacity of the ith buffer, i.e., the number of parts it can hold, is denoted by n_i with $1 \leq i \leq n - 1$.

We refer to the ith machine by M_i and to the ith buffer by B_i. As for terminology, B_i is called the upstream buffer of M_{i+1} and the downstream buffer of M_i. The first machine does not have an upstream buffer and we assume that parts are always available to the first machine (it never gets starved). The last machine does not have a downstream buffer and we assume that it can always output processed parts (i.e., it never gets blocked because of its full downstream buffer).

A machine is operational if it is UP, its upstream buffer is not empty (i.e., it is not starved) and its downstream buffer is not full (i.e., it is not blocked). An important and natural assumption to keep in mind is that only operational machines can break down. In other words, a machine that is UP but starved and/or blocked cannot break down, it remains in its UP state.

Let us turn our attention to the dynamics of the model. As already mentioned, we assume that things happen in slots, i.e., we have a discrete time model. We can think of the update of the state of the model in a time unit as a two-phase procedure: first, the states of all the machines are determined (probabilistically) then the buffer occupancies are changed accordingly. If machine M_i is operational and does not break down (which happens with probability $1 - p_i$) then it takes one part from its upstream buffer and puts one part in its downstream buffer. If it breaks down then it does not move any part. If it is UP but either starved or blocked or both then it remains UP with probability 1 and does not move any part. If machine M_i is $DOWN$ then it gets repaired with probability r_i and in the same slot it can move a part from its upstream buffer to its downstream buffer. If it remains $DOWN$ (this happens with probability $1 - r_i$) then it does not move any part.

The state of the model is given by the state of the machines and the number of parts in the buffers, i.e. by a $(2n - 1)$-tuple of the form

$$(m_1, b_1, m_2, b_2, \ldots b_{n-1}, m_n)$$

where $m_i \in \{U, D\}$ is the state of the i-th machine and $b_i \in \{0, \ldots n_i\}$ is the number of parts in buffer B_i. The stochastic process at hand is a discrete time Markov chain (DTMC).

In order to give an example for a transition of the DTMC, consider a line with three machines being in the state $(U, 4, D, 0, U)$ with buffer sizes $n_1 = n_2 = 10$. Accordingly, the last machine is starved and cannot change state. The first machine either breaks down or remains up and the second either gets repaired or remains down. This means that there are four possible transitions with associated probabilities as follow

$$(U, 4, D, 0, U) \xrightarrow{(1-p_1)\cdot(1-r_2)} (U, 5, D, 0, U), (U, 4, D, 0, U) \xrightarrow{p_1 \cdot r_2} (D, 3, U, 1, U)$$

$$(U, 4, D, 0, U) \xrightarrow{(1-p_1)\cdot r_2} (U, 4, U, 1, U), (U, 4, D, 0, U) \xrightarrow{p_1 \cdot (1-r_2)} (D, 4, D, 0, U)$$

Figure 2 shows the complete state-transition graph of the DTMC for a 2 machines line and assuming $n_1 = 4$ as the size of the single buffer (for the sake of space we denote $\bar{p} = 1 - p$). Assuming that the initial state is $(U, 0, U)$ the number of reachable states is 13. Note that the initial state is transient meaning that the system cannot reach it once it is left. The state space is irregular close to the boundaries (empty or full buffer). For example, having the buffer full is possible only with M_1 up and M_2 down.

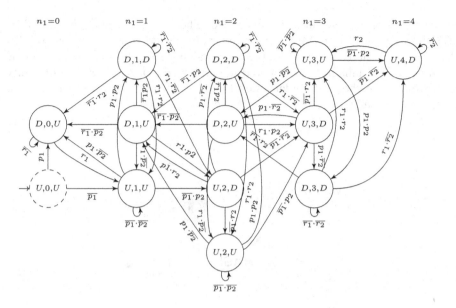

$n_1=0$ $\qquad$ $n_1=1$ $\qquad$ $n_1=2$ $\qquad$ $n_1=3$ $\qquad$ $n_1=4$

Fig. 2. The DTMC for the MBM system with buffer size $n_1 = 4$ consists of 13 states.

3   HASL Model Checking

In order to develop our framework for formal modelling and performance analysis of production lines (see Sect. 4) we rely on the HASL [9] statistical model checking (SMC) platform and the Cosmos tool [10,15]. SMC approaches [3] spread out as of early 2000 as an alternative to classical *numerical* probabilistic model checking approaches [7,16] which, requiring the storage of a model's state-space, do not scale up w.r.t. the model's dimension. The overall idea with SMC is to obtain an estimate of some performance indicator (i.e. property) by sampling of (finite length) paths issued by stochastic simulation of the model: as a consequence there's no need to build, hence to store, the model's state space, therefore SMC can be applied even to infinite-state systems. The property to estimate is formally expressed either as a temporal logic formula (associated to a given grammar) or through an automata formalism (e.g. timed-automata, hybrid automata).

HASL-SMC Scheme. The HASL model checker (Fig. 3) takes 3 inputs: a discrete-state stochastic process, belonging to a very generic class which we simply name Discrete Event Stochastic Process (DESP), denoted $\mathcal{M}$, a linear hybrid automaton (LHA), denoted $\mathcal{A}$, and a target expression denoted Z. The model checker iteratively simulates paths from the product process $\mathcal{M} \times \mathcal{A}$ (see below for details about $\mathcal{M} \times \mathcal{A}$ semantics) ending the simulation of a path as soon as either the path reaches an *accepting location* of $\mathcal{A}$ (in which case the values stored in $\mathcal{A}$'s variables are used for estimating Z), or the simulation halts (because a deadlock state of $\mathcal{M} \times \mathcal{A}$ is reached), in which case the path, as well as the values of $\mathcal{A}$'s variables, are discarded. Therefore $\mathcal{A}$ can be seen as a *filter* of paths σ sampled from $\mathcal{M}$: the accepted σ are used to produce the output, i.e. the confidence interval of the sample mean value $\overline{Z}$, which means that the true mean value of Z falls in the estimated interval is $[\overline{Z} - \delta, \overline{Z} + \delta]$ with probability ϵ, where ϵ is the chosen confidence level and δ the chosen width of the interval (both ϵ and δ being auxiliary inputs of the model checker).

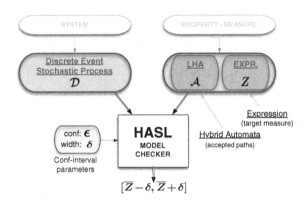

Fig. 3. HASL-SMC schema: sampled paths are filtered by a LHA and the accepted ones used for a confidence interval estimate of the target measure.

Discrete Event Stochastic Processes and Paths. HASL-SMC refers to a generic class of stochastic processes, named DESP, which subsumes, but is not limited to, Markov chains. A DESP is a discrete-state stochastic process with no specific restriction imposed on the type of probability distributions used to model the delay with which an event occurs. Therefore the time at which an event occurs can be described by any form of probability distribution, including, e.g., discrete Dirac distributions. For the sake of simplicity we omit here the formal definition of DESP [9]. Intuitively a DESP is characterised by a set of states S, a set of events E (describing the possible kind of state transitions), plus a number of auxiliary functions capturing the relevant stochastic aspects of the model (e.g. the probability distribution associated to each event). The notion of path of a DESP is relevant for understanding the semantics of temporal logic properties. For $\mathcal{M}$ a DESP model we denote $Path(\mathcal{M})$ the set of possible paths of $\mathcal{M}$ where a path $\sigma \in Path(\mathcal{M})$ is a (possibly infinite) sequence $\sigma = s_0 \xrightarrow{t_0} s_1 \xrightarrow{t_1} s_2 \xrightarrow{t_2} \dots \xrightarrow{t_{n-1}} s_n \dots$ with $t_i \in \mathbb{R}_{>0}$ being the sojourn-time in state $s_i \in S$. Alternatively we may adopt the following enriched notation $\sigma = s_0 \xrightarrow[e_{1j}]{0.25} s_1 \xrightarrow[e_{2j}]{0.5} s_2 \xrightarrow[e_{3j}]{0.15} s_3 \xrightarrow[e_{4j}]{1} \dots$ where e_{ij} indicates that event e_j occurred on the i-th transition of the path. Notice that trajectories of DESP are *càdlàg* (i.e. step) functions of time (e.g. blue plot in Fig. 5). It can be shown that a DESP model $\mathcal{M}$ induces a probability space over the set of events $2^{Path(\mathcal{M})}$, where the probability of a set of paths $E \in 2^{Path(\mathcal{M})}$ is given by the probability of their common finite prefix [8].

Petri Nets Representation of DESP. For practical reasons HASL-SMC (and the Cosmos tool) uses Petri nets as a high-level formalism for representing DESP models. Specifically HASL uses the non-Markovian generalisation of Generalised Stochastic Petri Nets (GSPN) [4] as the modelling language for DESPs. For the sake of space here we only give a very short account of the GSPN formalism that we use for characterising the models of production lines (see Sect. 4): we take for granted the basics about Petri nets (i.e. the notion of token game, transition enabling/firing, input/output/inhibitor arcs). We refer the reader to [9] for an exhaustive treatment.

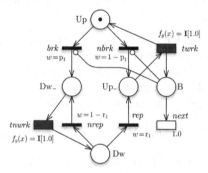

Fig. 4. Example of non-Markovian extension of a GSPN model.

A GSPN model is a bipartite graph consisting of two classes of nodes: *places* (circles), which may contain tokens, and *transitions* (bars), which describe how tokens move between places. The state of a GSPN is given by the *marking* (number of tokens in each one) of its places. Transitions of a GSPN can be either *timed* (denoted by thick bars) or *immediate* (denoted by filled-in thin bars). Immediate transitions represent activities that do not consume time: typically they are used either to disambiguate between non-deterministic choices (through *priorities*) or to model a probabilistic choice (by means of *weights*). Stochastic timed transitions are used to model time consuming events. With non-Markovian GSPN (NMGSPN) we distinguish between Exponentially distributed (thick, empty bars) and Generally distributed (thick, filled-in bars) timed transitions. In order to disambiguate concurrent occurrence of timed transitions (which may exist in case e.g. of Dirac distributions) a *weight* value is associated even to generally distributed timed transitions (the weight is used to probabilistically choose one out of several concurrently occurring timed transition).

Figure 4 shows an example of NMGSPN representing the dynamic of a production machine that can undergo faults and repairs and that is equipped with an output buffer. Mutually exclusive places Up and Dw represent the current operational (Up) or broken (Dw) state of the machine. Place B is the output buffer. The 2 pairs of immediate transitions brk (weight), $nbrk$ (weight $w = p_1$, resp. $w = 1 - p_1$) and rep, $nrep$ (weight $w = r_1$, resp. $w = 1 - r_1$) are used to model probabilistic choices representing the probability that the machine breaks down, when operational (i.e. p_1), and that it gets repaired, when broken (i.e. r_1). Three time consuming activities are modelled through timed transitions: $next$, which is assumed to follow an Exponential distribution of rate 1.0 (Exponential transitions are depicted by thick empty rectangles) and $tnwork$ and $twork$, which are assumed to follow a Deterministic (Dirac) distribution of duration 1.0 (transitions following a General distribution are depicted by thick filled-in rectangles). Furthermore notice that, because of the inhibitor arc, transitions brk and $nbrk$ are enabled only when place B is empty. Finally observe that the priority value of each transition is, by default, equal to 1, hence no priority is considered in this example as it will be the case for all models discussed in this paper.

Hybrid Automaton as Property Specification. The driving principle of HASL model checking is that of employing a LHA as a *path selector* through synchronisation with the considered model $\mathcal{M}$. A HASL-LHA $\mathcal{A}$ has access to certain elements of model $\mathcal{M}$, namely the events and the state-variables of $\mathcal{M}$. Formally a LHA for HASL is defined as an n-tuple:

$$\mathcal{A} = \langle E, L, \Lambda, Init, Final, X, flow, \rightarrow \rangle$$

where: E is a finite alphabet of events (the transitions of the GSPN representation of of $\mathcal{M}$ that drive the synchronisation); L is a finite set of locations; $\Lambda : L \rightarrow Prop$, a location labelling function (*Prop* being the set of atomic proposition built on top of variables X); $Init$ is a subset of L called the initial locations; $Final$ is a subset of L called the final locations; $X = (x_1, ... x_n)$ a n-tuple of data variables; $flow : L \mapsto Ind^n$ is a function which gives for each location the rate at which variable x_i evolves (where the rate for variable x_i is given by an *indicator function* that depends on the state of the model $\mathcal{M}_\theta$); $\rightarrow \subseteq L \times ((\mathsf{Const} \times 2^E) \uplus (\mathsf{IConst} \times \{\sharp\})) \times \mathsf{Up} \times L$, a set of edges, where the notation $l \xrightarrow{\gamma, E', U} l'$ means that $(l, \gamma, E', U, l') \in \rightarrow$, with Const the set of constraints, whose elements are boolean combinations of inequalities of the form $\sum_{1 \leq i \leq n} \alpha_i x_i + c \prec 0$ where α_i and c are constants, $\prec \in \{=, <, >, \leq, \geq\}$, whereas IConst is the set of left-closed constraints. Selection of a model's trajectories with an automaton $\mathcal{A}$ is achieved through *synchronization* of $\mathcal{M}$ with $\mathcal{A}$, i.e. by letting $\mathcal{A}$ synchronises its transitions with the transitions of the trajectory σ being sampled. To this aim, an LHA for HASL admits two kinds of transitions: *synchronizing* transitions (associated with a subset $E \subseteq \Sigma$ of event names, with ALL denoting Σ), which may be traversed when an event (in E) is observed on σ (for example a reaction occurs), and *autonomous* transitions (denoted by $\sharp$) which are traversed autonomously (and have priority over synchronised transitions), on given conditions, typically to update relevant statistics or to terminate (accept) the analysis of σ.

Synchronisation of Model and LHA. The synchronisation of $\mathcal{M}$ with $\mathcal{A}$ boils down to the characterisation of the so-called product process $\mathcal{M} \times \mathcal{A}$ whose semantics we describe intuitively here (referring the reader to [9] for its formal characterisation). For $\mathcal{M}$ a DESP with state space S and $\mathcal{A}$ a LHA with locations L and variables X the states of $\mathcal{M} \times \mathcal{A}$ are triples (s, l, ν) where $s \in S$ is the current state of $\mathcal{M}$, $l \in L$ is the current location of $\mathcal{A}$ and $\nu : X \to \mathbb{R}^{|X|}$ is the current value of the variables of $\mathcal{A}$. The semantics of $\mathcal{M} \times \mathcal{A}$ naturally yields a stochastic simulation procedure which is implemented by the HASL model checker. The paths of $\mathcal{M} \times \mathcal{A}$ sampled by the HASL simulator are composed of two kinds of transitions: *synchronising transitions*, that correspond to a simultaneous occurrence of a transition in $\mathcal{M}$ and one in $\mathcal{A}$, as opposed to, *autonomous transitions*, that correspond to the occurrence of a transition in $\mathcal{A}$ without any correspondence in $\mathcal{M}$. Therefore a path of $\mathcal{M} \times \mathcal{A}$ can be seen as the result of the synchronisation of a path σ of $\mathcal{M}$ with $\mathcal{A}$ (or conversely we may say that a path of $\mathcal{M} \times \mathcal{A}$ always admits a projection over $\mathcal{M}$). Given σ a path of $\mathcal{M}$ we denote $\sigma \times \mathcal{A}$ the corresponding path of $\mathcal{M} \times \mathcal{A}$. For example if $\sigma : s \xrightarrow[e_1]{t_1} s_1 \xrightarrow[e_2]{t_2} s_2 \dots$ is a path of $\mathcal{M}$ such that after sojourning for t_1 in its origin state s, switches to s_1 through occurrence of event e_1, and then, at t_2, switches to s_2, through event e_2 (and so on), then the corresponding path $\sigma \times \mathcal{A}$ in the product process may be $\sigma \times \mathcal{A} : (s, l, \nu) \xrightarrow[e_1]{t_1} (s_1, l_1, \nu_1) \xrightarrow[\sharp]{t_1^*} (s_1, l_2, \nu_2) \xrightarrow[e_2]{t_2} (s_2, l_3, \nu_3) \dots$ where, the sequence of transitions e_1 and e_2 observed on σ is interleaved with an autonomous transition (denoted $\sharp$) in the product process: i.e. from state (s_1, l_1, ν_1) the product process jumps to state (s_1, l_2, ν_2) (notice that state of $\mathcal{M}$ does not change) before continuing mimicking σ. The semantics of the product process is detailed in Example 1.

Example 1. Figure 5 depicts a toy GSPN model (left) of a simple, faulty, production machine together with a toy LHA (right) for assessing properties of it. The GSPN represents a simple machine that iteratively switches between a working state (place Up) and a broken state (place $Down$) and that may produce workpieces to be placed in an output buffer B. When Up the machine may produce, with a delay given by transition wrk a new piece and place in buffer B, or it can change its state to $Down$, with a delay given by transition $nwrk$. When $Down$ the machine gets repaired and returns to Up with a delay given by transitions rep. For simplicity all transitions in this example are assumed to be timed and exponentially distributed (with rate λ_1, λ_2, λ_3 and λ_4 respectively). The two locations LHA (top right) refers to the GSPN model (top left) with $L = \{l_0, l_1\}$ and l_0 the initial location, l_1 the final location. The LHA variables are $X = \{t, x_1, n_2\}$ with t a clock variable, x_1 a real valued variable (for measuring the average number of pieces in buffer B) and n_2 an integer variable (for counting the number of occurrences of the brk events). Variables with non-null flow are clock t, whose flow is inherently constant and equal to 1, and x_1, whose flow is given by B, i.e. the marking of place B. Therefore (while in l_0) x_1 measure the integral of the marking of B (based on σ_B the projection w.r.t. to B of the synchronising trajectory σ). The synchronisation with a path σ issued of the GSPN model is

272   P. Ballarini and A. Horváth

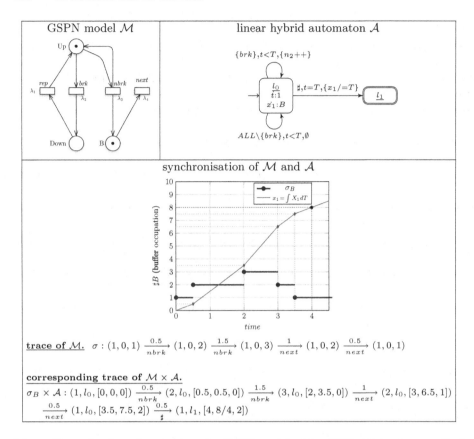

Fig. 5. Example of synchronisation of a GSPN model with a LHA. (Color figure online)

as follows: at time $t = 0$ the LHA starts in l_0 and stays there up to $t = T$ (T being a constant assumed to value 4). As soon as $t = T$ the synchronisation with σ ends as the *autonomous* transition $l_0 \xrightarrow{\sharp, t=T, \{x_1/=T\}} l_1$ becomes enabled hence is fired (by definition autonomous transitions, which are labelled with $\sharp$, have priority over *synchronised* transitions in HASL). As long as $t < T$ the LHA is in l_0 where it synchronises with the occurrences of the GSPN tranistions: on occurrence of brk transition $l_0 \xrightarrow{\{brk\}, t<T, \{n_2++\}} l_0$ (which is *synchronised* on event set $\{brk\}$) is fired hence increasing the counter n_2, whereas on occurrence of any other GSPN transition the LHA transition $l_0 \xrightarrow{ALL\backslash\{brk\}, t<T, \emptyset} l_0$ (which is *synchronised* on event set $ALL\backslash\{brk\}$, where ALL denotes all transitions of the GSPN) fires without updating any variable. Finally on ending the synchronisation with σ variable x_1 is updated to x_1/T which corresponds to average marking of place B observed over the time interval $[0, T]$. Such a LHA can therefore be used (through iterated synchronisation with a sufficiently large number of trajectories) for estimating the confidence interval of random variables such as the

"average number of pieces in the buffer" as well as the "number of *break* events" observed over time interval $[0, T]$.

HASL Target Expression. The second component of an HASL specification is an expression Z given by grammar (1). Z is associated to a LHA $\mathcal{A}$ and expresses the target measure whose confidence interval should be estimated based on the paths accepted by $\mathcal{A}$.

$$
\begin{aligned}
Z ::= \ & AVG(Y) \mid Z + Z \mid Z \times Z \mid Pdist \\
Pdist ::= \ & PDF(Y, step, start, stop)) \mid CDF(Y, step, start, stop) \mid PROB() \\
Y ::= \ & c \mid Y + Y \mid Y \times Y \mid Y/Y \mid last(y) \mid min(y) \mid max(y) \\
y ::= \ & c \mid x \mid y + y \mid y \times y \mid y/y
\end{aligned} \tag{1}
$$

There are two main types of expressions Z: $AVG(Y)$ (where AVG indicates *mean value of*) and $Pdist$ (indicating a probability distribution or probability value expression). Y represent a random variable built on top of algebraic combination of some *path* operators applied to an LHA variable y, i.e. $last(y)$ (i.e. the last value that y has at the end of an accepted path, $min(y)(resp.\max(y))$, the min (resp. max) value of y along an accepted path. Conversely Z expressions of $Pdist$ type include $PDF(Y, step, start, stop)$, which allows for estimating the PDF of random variable Y computed by discretisation of the support set $[start, stop]$ in $(stop - start)/step$ sub-intervals of size $step$ and similarly $CDF(Y, step, start, stop)$, Finally expression $PROB()$ allows for estimating the probability that a path is accepted, otherwise said $PROB()$ is used to estimating the probability of the paths event set represented by the considered automaton $\mathcal{A}$. For example, referring to the LHA $\mathcal{A}$ of Fig. 5, the HASL specification $(\mathcal{A}, AVG(last(x_1)))$ corresponds to the *mean number of pieces in buffer B within T*, while $(\mathcal{A}, PDF(last(x_1), 0.1, 0, 10))$ corresponds to the *PDF of the mean number of pieces in buffer B within T computed over $[0, 10]$ with discretisation step* 0.1.

4   HASL Based Performance Analysis of Production Lines

4.1   Stochastic Petri Net Encoding of Production Lines

As discussed in Sect. 2, the kind of production lines we consider can be conveniently modelled by a DTMC, i.e., a discrete-time, memoryless stochastic process. On the other hand, Cosmos, which supports DESP models in form of NMGSPNs, allows us to use a wide range of delay distributions and, as a consequence, gives us the possibility to formulate models whose behaviour is not memoryless. As it happens, even if the model we consider is associated with a discrete time memoryless process, and consequently all sojourn times are geometric, the most straightforward encoding of the considered system in Cosmos is by means of a NMGSPN whose timed transitions have deterministic firing times.

Generally speaking deterministic firing times may lead to a non-memoryless process but in the NMGSPN encoding we propose we make sure that the resulting stochastic process is indeed a DTMC. In particular, we let all deterministic transitions share the same firing time that corresponds to one slot of the DTMC (hence the actual interleaving of transition firing is irrelevant). Furthermore we have seen that the dynamics of the production line systems is of a *fully synchronous* nature: the state of the machines as well as that of the buffers are updated synchronously in each time slot. For sake of simplicity, we model such synchronousity by splitting the model dynamics into two phases which can be conveniently be done with NMGSPN: in the first phase the machine states are updated through a probabilistic switch modelled using immediate transitions (i.e., no time is consumed) while in the second the buffer occupation levels are changed accordingly using deterministic transitions whose firing time correspond to one time slot of the corresponding DTMC.

NMGSPN Models of Production Lines. We illustrate the NMGSPN encoding of production lines by means of the 3-machines case given in Fig. 6. For each machine there are four places. Places Up_i and Dw_i are used to represent the fact that machine i is up or down at the beginning of a time slot of the DTMC. Places Up_{i-} and Dw_{i-}, where token sojourn times are 0, indicate instead the state of the machine at the end of the time slot. Places Up_i and Dw_i are connected to places Up_{i-} and Dw_{i-} by immediate transition whose weights are set according to the break down and repair probabilities (p_i and r_i). Deterministic transitions whose firing time corresponds to one time slot are used to update the places B_i that represent buffers. In particular, $tnwrk_i$ and $twrk_i$ are the transitions that represent the fact that machine i does not process or does process a part, respectively. Blocking of machines due to full buffers are modelled by inhibitor arcs with multiplicity corresponding to buffer capacity (nb_i in the figure). Blocking by starvation is modelled instead by arcs that connect place B_i to transitions brk_i and $nbrk_i$.

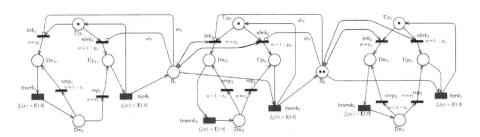

Fig. 6. NMGSPN model of the DTMC for a 3-machines production line (initially machines are assumed to be all UP as buffer B_2 contains 2 pieces and B_1 is empty).

Generator of NMGSPN Models. We developed a generator of NMGSPN models (in the .gspn Cosmos format) of production lines of arbitrary size[1]. For what concerns the dimension of the Petri net model of production lines, in case of N machines we have (corresponding to straightforward adaptation of that depicted in Fig. 6) $4 \cdot N + (N-1)$ places and $6 \cdot N$ transitions ($4 \cdot N$ of which are immediate and $2 \cdot N$ are deterministic).

4.2   HASL Performance Indicators for Production Lines

To asses the dynamics of production lines we identified a number of relevant KPIs including the following ones: $\phi_1 \equiv$ *"the probability distribution of the time the system takes to produce N pieces"*; $\phi_2 \equiv$ *"mean delay to empty buffer B_i after it got full"*; $\phi_{2b} \equiv$ *"probability to empty buffer B_i within T after it got full"*; $\phi_3 \equiv$ *"the mean (or similarly, the probability distribution of the) occupation level of buffer B_i within T"*. Table 1 shows the LHA encoding of the above listed KPIs (automaton $\mathcal{A}_i$ corresponds to ϕ_i). Notice that if $\mathcal{A}_1$, $\mathcal{A}_2$ and $\mathcal{A}_3$ accept all paths

Table 1. Linear hybrid automata for performance indicators of production lines.

ID	LHA	description
$\mathcal{A}_1$	$ALL\backslash\{twrkM\}, x<N, \emptyset$ l_0 $t{:}1$ $\sharp, x{=}N, \emptyset$ $\to l_1$ $\{twrkM\}, x<N, \{x{+}{+}\}$	accepts all paths on observing the production of N-th piece (i.e. transition $tworkM$) **variables:** x: num. produced pieces t: time until N produced pieces
$\mathcal{A}_2$	$ALL, \top, \{x{:}{=}B_i\}$ $ALL, \top, \{x{:}{=}B_i\}$ l_0 $t{:}1$ $\sharp, x{=}n_i, \emptyset$ l_1 $t_1{:}1$ $\sharp, x{=}N, \emptyset$ l_2	accepts all paths such that the occupation of buffer B_i reached N after it was n_i **variables:** x: num. pieces in B_i t: time to N pieces in B_i t_1: time to N pieces in B_i after there were n_i
$\mathcal{A}_{2b}$	$ALL, \top, \{x{:}{=}B_i\}$ $ALL, \top, \{x{:}{=}B_i\}$ l_0 $t{:}1$ $\sharp, x{=}n_i, \emptyset$ l_1 $t_1{:}1$ $\sharp, x{=}N \wedge t_1{=}T, \emptyset$ l_2	accepts only paths such that the occupation of buffer B_i reached N with delay T after it was n_i **variables:** x: num. pieces in B_i t: time to N pieces in B_i t_1: time to N pieces in B_i after there were n_i
$\mathcal{A}_3$	$ALL, t<T, \emptyset$ l_0 $t{:}1$ $x_i{:}B_i$ $\sharp, t{=}T, \{x_i/{=}T\}$ l_1	accepts all paths of duration T **variables:** t: time x_i: integral of occupation of B_i (for $t < T$), then mean occupation of B_i (at $t = T$)

---

[1] available here https://gitlab-research.centralesupelec.fr/2011ballarinp/cosmos_prod uctionlines.

5   Experiments

In this section we report about a selection of the experiments we run on Cosmos based on the HASL encoding of the KPIs described in Sect. 4.2.

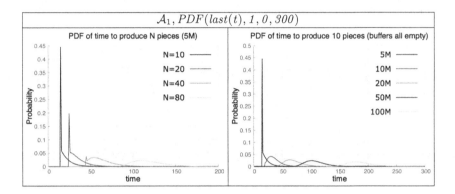

Fig. 7. Measured distribution of probability of delay to produce N pieces.

PDF of the Time to Produce N Pieces. Figure 7 (lhs) shows the probability distribution of the time needed for producing N pieces for a production line with 5 machines. On the rhs, the distribution of the time to produce 10 parts is shown for different number of machines. Note that we use the term PDF (which often stands for probability density function) because in cosmos the keyword to obtain such measures is PDF but, since in our case the model is discrete time, what we obtain is a probability mass function.

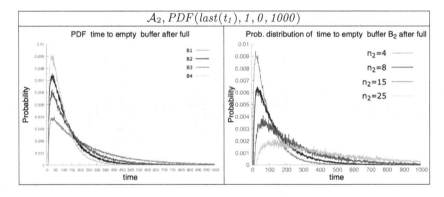

Fig. 8. Measured distribution of probability of delay to empty a buffer after it is full.

PDF of the Time to Empty a Buffer After it is Full. Figure 8 shows the estimated probability distribution for the delay to empty a buffer after it gets full[2]. The obtained plots (lhs) indicate that a downstream buffer is more likely to empty faster than an upstream one, which, intuitively, is sensible since a downstream buffer has fewer (breakable) downstream machines that can slow down its emptying. Curves on rhs show that increasing the size of a buffer (B_2) spreads the probability of the delay for emptying B_2.

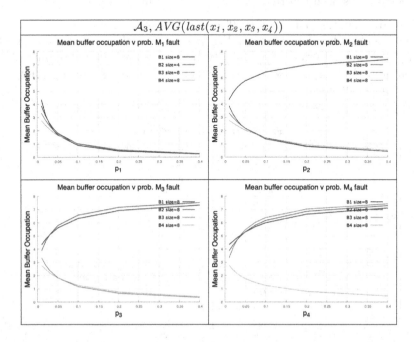

Fig. 9. Mean buffer occupation versus fault probability of machine M_i

Mean Buffer Occupation Versus Machines' Fault Probability. Figure 9 depicts plots referring to the mean occupation level of each buffer obtained by varying the break down probability of a single machine. All results here have been obtained using (a 4 variables extension of) the LHA $\mathcal{A}_3$ (see Table 1)[3]. As one may expect the outcome of these experiments indicate that augmenting the probability that M_i breaks down increases the mean occupation of the upstream buffers while decreasing that of the downstream buffers. More specifically, we observe that beyond a certain value of p_i the closest upstream (downstream) buffer has the highest (smallest) mean occupation, however there exists

---

[2] Results obtained with $\mathcal{A}_2$ on a 5M production line with buffers of equal size 8 an machines of equal fault/repair (0.01/0.1) probability.

[3] Experiments obtained using $\mathcal{A}_3$ using $T = 500$ as time horizon and the 5M model with fault probability $p_j = 0.01$, for all machines M_j $j \neq i$ and repair probabilities $r_1 = 0.1, r_2 = 0.2, r_3 = 0.15, r_4 = 0.18, r_5 = 0.1$ with buffers of equal size $n = 8$ all initially empty.

a crossover point below which the furthers upstream (downstream) buffers have a (slightly) higher occupation of the one which is closest to the M_i. Such an "inversion" phenomenon may be explained as inertia: not breaking "enough" machine M_i has little consequences on its neighbours buffers (i.e. B_{i-1} and B_{i+1}), whereas only when breaking becomes considerably likely the effect on neighbors buffers becomes evident (i.e. B_{i-1} getting more occupied than its upstream buffers and B_{i+1} getting less occupied than its downstream buffers).

Probability Distribution of Buffer Occupation. Figure 10 depicts results of experiments related to those of Fig. 9: we use again $\mathcal{A}_3$ but this time for estimating the distribution of probability of the mean occupation of buffers (B_1, top row, and B_2, bottom row) versus the fault probability of the upstream, resp. downstream, machine.

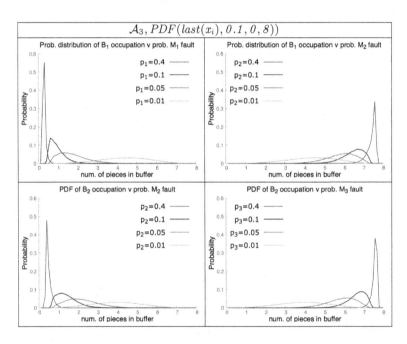

Fig. 10. Probability distribution of mean buffer occupation versus fault probability.

The obtained plots indicate that increasing the probability that a machine breaks down has the effect of concentrating the probability mass towards lower (higher) occupation level for the downstream (upstream) buffer.

6   Discussion

The realm of stochastic simulation based tools for performance analysis of production systems includes commercial products such as the Arena simulation software [2] and the 3D Flexsim simulation tool [1]. Those tools are equipped with custom made modelling frameworks through which models of production lines are obtained by composition of basic building blocks (*modules*) and then analysed by estimation of "classical" performance indices (e.g., total production, average waiting time in queue, maximum waiting time in queue, etc.) through discrete-event simulation engines.

We briefly discuss how the framework we introduce in this paper compares with those tool. From a modelling point of view tools such as Arena and Flexsim are more general as, differently from our framework, they are equipped with all elements, including *branching* (i.e., *fork*) and *join* blocks, that are necessary to model generic production systems. Furthermore, the Arena software allows also for modelling of continuous-state production systems (to consider, for example, production of liquids), hence, differently from our framework, is not limited to discrete-state systems. In principle, our framework could straightforwardly be extended to model also continuous-state systems by including fluid Petri-nets amongst the supported modelling formalisms.

The added value of a model-checking framework like ours lies in the *separation of concerns* between the model construction which is, inherently, kept independent of model analysis. By its very nature model checking, provides the modeller with (besides a modelling language) a *property language* for formally stating performance measures of interest. The expressive power of a formalism like HASL allows us to conceive performance indicators well beyond the "classical ones" that can be measured with standard stochastic simulation tools (e.g., Arena and Flexsim). For example, measures such as, e.g., *distribution of buffer occupation over a certain period* or *distribution of time to empty a buffer once it gets full* (Fig. 8), in our understanding cannot be assessed through classical stochastic simulation tools unless the simulator engines (and possibly the models) are adapted/enriched with the necessary means for measuring the considered indicators. HASL model checking can therefore be viewed as a generic simulation based tool which one can use to assess whatever indicator can be expressed in terms of the HASL formalism, without requiring any adaptation of the simulation engine nor any enrichment of the model. Finally HASL based model checking provides the user with arbitrary statistical guarantees as the output of a KPI assessment is obtained by confidence interval estimation yield with a desired *confidence* and *accuracy* (chosen by the user as inputs of the estimation process, hence affecting the number of simulation runs generated for the estimation).

7 Conclusion

We presented a framework for encoding of DTMC models of production lines in terms of stochastic Petri nets. This allowed us to take advantage of the expressive power of the HASL property specification framework in order to characterise a number of sophisticated key indicators through which, at the aid of the Cosmos tool, we assessed the performances of production lines. The switch to

HASL/Cosmos allowed us to overcome the limited expressiveness we faced in a previous study of the performances of production lines we carried out [11] through CSL model checking using the PRISM tool [17]. Future developments include the development and analysis of furthers relevant performance indicators, as well as the extension of the proposed modelling framework to variants of systems considered here, e.g. closed-loops production lines and/or production systems with branching topologies.

References

1. https://www.flexsim.com/
2. https://www.rockwellautomation.com/en-us/products/software/arena-simulation.html
3. Agha, G., Palmskog, K.: A survey of statistical model checking. ACM Trans. Model. Comput. Simul. **28**(1) (2018)
4. Marsan, M.A., Balbo, G., Conte, G., Donatelli, S., Franceschinis, G.: Modelling with Generalized Stochastic Petri Nets. John Wiley & Sons, Hoboken (1995)
5. Angius, A., Horváth, A., Colledani, M.: Moments of accumulated reward and completion time in Markovian models with application to unreliable manufacturing systems. Perform. Eval. **75**, 69–88 (2014)
6. Angius, A., Colledani, M., Horváth, A., Gershwin, S.B.: Analysis of the lead time distribution in closed loop manufacturing systems. IFAC-PapersOnLine **49**(12), 307–312 (2016)
7. Baier, C., Haverkort, B., Hermanns, H., Katoen, J.-P.: On the logical characterisation of performability properties. In: Montanari, U., Rolim, J.D.P., Welzl, E. (eds.) ICALP 2000. LNCS, vol. 1853, pp. 780–792. Springer, Heidelberg (2000). https://doi.org/10.1007/3-540-45022-X_65
8. Baier, C., Haverkort, B., Hermanns, H., Katoen, J.P.: Model-checking algorithms for continuous-time Markov chains. Softw. Eng. IEEE Trans. **29**, 524–541 (2003)
9. Ballarini, P., Barbot, B., Duflot, M., Haddad, S., Pekergin, N.: Hasl: a new approach for performance evaluation and model checking from concepts to experimentation. Perform. Eval. **90**, 53–77 (2015)
10. Ballarini, P., Djafri, H., Duflot, M., Haddad, S., Pekergin, N.: COSMOS: a statistical model checker for the hybrid automata stochastic logic. In: Proceedings of the 8th International Conference on Quantitative Evaluation of Systems (QEST 2011), pp. 143–144. IEEE Computer Society Press, September 2011
11. Ballarini, P., Horváth, A.: Formal analysis of production line systems byprobabilistic model checking tools. In: Proceedings of the 2021 IEEE Emerging Technology and Factory Automation (ETFA) (2021)
12. Colledani, M., Horvath, A., Angius, A.: Production quality performance in manufacturing systems processing deteriorating products. CIRP Ann. Manuf. Technol. **64**, 431–434 (2015)
13. Colledani, M., Tolio, T.: Integrated quality, production logistics and maintenance analysis of multi-stage asynchronous manufacturing systems with degrading machines. CIRP Ann. Manuf. Technol. **61**(1), 455–458 (2012)
14. Colledani, M., Angius, A., Horvàth, A.: Lead time distribution in unreliable production lines processing perishable products. In: Proceedings of the 2014 IEEE Emerging Technology and Factory Automation (ETFA), pp. 1–8 (2014)
15. Cosmos home page. https://cosmos.lacl.fr/

16. Kwiatkowska, M., Norman, G., Parker, D.: Stochastic model checking. In: Bernardo, M., Hillston, J. (eds.) SFM 2007. LNCS, vol. 4486, pp. 220–270. Springer, Heidelberg (2007). https://doi.org/10.1007/978-3-540-72522-0_6
17. Kwiatkowska, M., Norman, G., Parker, D.: PRISM 4.0: verification of probabilistic real-time systems. In: Gopalakrishnan, G., Qadeer, S. (eds.) CAV 2011. LNCS, vol. 6806, pp. 585–591. Springer, Heidelberg (2011). https://doi.org/10.1007/978-3-642-22110-1_47

Reliability Reference Model for Topology Configuration

Paul Kevin Reeser[(✉)] (iD)

AT&T Labs, Middletown, NJ 07748, USA
preeser@att.com

Abstract. We use a simple 2-tiered reference model consisting of servers and sites to illustrate an approach to topology configuration and optimization, with a focus on addressing geo-redundancy questions like how many sites, and how many servers per site, are required to meet performance and reliability requirements. We first develop a multi-dimensional component *failure mode* reference model, then reduce this model to a one-dimensional service *outage mode* reference model. The key contribution is the *exact* derivation of the outage and restoral rates from the set of 'available' states to the set of 'unavailable' states using an adaptation of the hyper-geometric "balls in urns" distribution with unequally likely combinations. We describe a topology configuration tool for optimizing resources to meet requirements and illustrate effective use of the tool for a hypothetical VoIP call setup protocol message processing application.

Keywords: FMEA · Geo-redundant · Topology · Configuration · Optimization · Tool

1 Introduction

Most availability analyses typically start by characterizing a *failure mode* reference model that captures the underlying hardware (HW) and software (SW) components that constitute an application deployment. From a performance:reliability:cost *optimization* perspective, these models are typically used to determine the minimal topology required to meet the distributed application capacity and availability requirements.

In this paper, we use a simple 2-tiered reference model consisting of servers and sites to illustrate an approach to topology configuration, with the focus on addressing common geo-redundancy questions like how many sites, and how many servers per site, are required to meet the requirements. Even for this simple 2-tiered reference model, the number of states grows exponentially. We demonstrate techniques to reduce the state space of more complex models, and we show how to collapse a state transition diagram into a one-dimensional representation in terms of the amount of available server capacity, where transitions can occur across multiple levels.

At the service level, application outages matter more than individual HW or SW failures. Thus, we next develop a generic outage mode reference model based on the one-dimensional representation of the failure model. The key contribution of this paper is the exact derivation of the outage and restoral rates from the superset of 'available' states to the superset of 'unavailable' states as a function of the number of servers, the number of sites, and the minimum required server capacity level.

© Springer Nature Switzerland AG 2021
P. Ballarini et al. (Eds.): EPEW 2021/ASMTA 2021, LNCS 13104, pp. 282–297, 2021.
https://doi.org/10.1007/978-3-030-91825-5_17

There are many papers analyzing and optimizing the availability of redundant, distributed topologies (c.f., [1–4] and the many references therein), especially in the context of storage systems and virtualized applications. Yet to our knowledge, none have derived the exact general formula for the composite service outage and restoral rates, based on the hyper-geometric distribution with unequally likely combinations.

2   Reference Failure Model

2.1   Notation and Input Parameters

Typical availability analyses start by characterizing a *failure mode* reference model that captures the underlying elements constituting the application deployment. These failure models can vary in their level of detail, from simple block diagrams to sophisticated failure trees. As we show, in most practical cases this detail can be aggregated to reduce model complexity to a one-dimensional state space without losing underlying individual component failure and restoral rates, dependencies, or interactions.

Typical reliability optimization questions that these models need to address include the "how many eggs in one basket" type: How many application processes can run on a single host? How many host servers in one rack? How many racks in one datacenter site? How many sites per region, etc.? For the analysis to follow, we assume a very simple 2-tiered reference model consisting of servers and sites, and we focus on the common geo-redundancy questions: how many sites and how many servers per site? Generalization to more than two tiers is straightforward.

Let M denote the number of geographically diverse sites (e.g., datacenters) and let N denote the number of hosts (servers). For simplicity, we assume that N is an integer multiple of M, and that N identical hosts are spread evenly across M identical sites. Let $J = N/M$ denote the number of hosts per site. Hosts and sites are the HW elements.

We assume that a single identical application instance is running on each host, and the set of J instances at each site make up the resident application function. Instances and resident functions are the SW elements. We assume that hosts and their associated instances are tightly coupled (that is, if a host is down then its associated instance is unavailable, and vice versa). Similarly, we assume that sites and their resident function (set of J instances) are tightly coupled (that is, if a site is down its resident function is unavailable, and vice versa). Let K denote the minimum number of instances required for service to be up (i.e., to have adequate capacity to serve the workload).

Next, let $\left\{\lambda_I^{-1}, \lambda_F^{-1}, \lambda_H^{-1}, \lambda_S^{-1}\right\}$ denote the mean time between failure (MTBF) and let $\left\{\mu_I^{-1}, \mu_F^{-1}, \mu_H^{-1}, \mu_S^{-1}\right\}$ denote the mean time to restore (MTTR) of the {Instance SW, Function SW, Host HW, and Site HW} respectively. Then the typical failure modes and associated effects (capacity impacts) for this canonical reference model are given in Table 1. We include default MTBF/MTTR values in brackets [ ] that will be used for the simple example later in the paper. A SW fault impacting a single instance could be a buffer overflow that leads to an application restart, while a fault impacting an entire resident function could be the corruption of shared data. A HW failure impacting a single host could be a fan failure, while a failure impacting an entire site could be a transfer switch failure following a commercial power outage.

Table 1. Simple failure mode reference model.

Failure Mode	Count	MTBF λ^{-1}		MTTR μ^{-1}		Capacity Impact
Single SW instance	N	λ_I^{-1}	[3 mo]	μ_I^{-1}	[1hr]	1 instance
Resident SW function	M	λ_F^{-1}	[2 yr]	μ_F^{-1}	[6hr]	Up to J instances
Single HW host	N	λ_H^{-1}	[6 mo]	μ_H^{-1}	[2hr]	1 instance
Entire HW site	M	λ_S^{-1}	[2 yr]	μ_S^{-1}	[4hr]	Up to J instances

2.2 Probability State Space

Now that we have developed the failure mode reference model, we next typically develop a state space transition diagram and solve for the stationary distribution. In order to make the analysis tractable, we assume that failure and restoral rates are exponentially distributed, and that the associated stochastic process forms a Markov chain (MC). The first key step in our approach is to *characterize states in terms of the amount of available capacity*. To illustrate for this simple reference model, let the M-tuple $(j_1, ..., j_m, ..., j_M)$ denote the number of instances up at each site $m = 1, ..., M$, where $0 \le j_m \le J$. There are $(J + 1)^M$ total states. A 'level' in the state diagram consists of all states with n total instances up, where $\sum_{m=1}^{M} (j_m) = n$ for every state on level n ($0 \le n \le N$). For all levels where $n \ge K$, the service is up; otherwise, service is down.

Next, we specify the state transitions. In this simple reference model, events can result in 1-level transitions in the case of host/instance failure and restoral, or up to J-level transitions in the case of site/function failure and restoral. Finally, we enumerate and solve the resulting balance equations to determine the state probabilities. The state diagram quickly becomes unwieldy as N and M grow, and the equations become impossible to solve explicitly without the use of tools to find the matrix eigenvalues.

Figure 1 shows the state space and feasible transitions for the small reference model of $N = 6, M = 2$, and $K = 3$. Service is available for unshaded ('up') states and unavailable for shaded ('down') states where $j_1 + j_2 < 3$. Straight transition arrows correspond to single host/instance failure and restoral, while curved arrows correspond to entire site/function failure and restoral. Table 2 (in Sect. 2.3) lists the associated transition rates.

As a prelude to the *outage mode* reference model presented later in this paper, looking closely we see that a service outage can occur from any state other than the level 6 'all up' state (3,3). In general, an outage can occur from any level n state where $n - J < K$. The key contribution of this paper is the exact derivation of the composite outage and restoral rates between the superset of available ('up') states and the superset of unavailable ('down') states as function of the input parameters N, M, and K, using an adaptation of the hyper-geometric "balls in urns" distribution with unequally likely combinations. Knowing these rates is critical when sizing deployments for services with stringent (e.g., FCC reportable [5]) outage and restoral requirements.

2.3 Collapsing Failure Modes

In order to develop a generic framework, the second key step in advancing the state space modeling is *collapsing the failure modes*; that is, combining all (HW and SW) failure and restoral rates causing single instance as well as single site transitions. To this end,

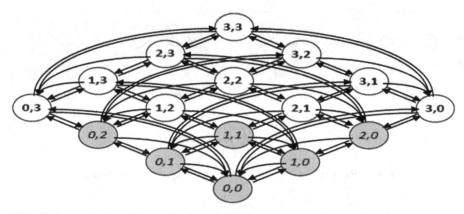

Fig. 1. State transition diagram for the small reference model of $N = 6$, $M = 2$, and $K = 3$.

let $\{A_I, A_F, A_H, A_S\}$ denote the availability $A = \mu/(\lambda + \mu)$ and let $\{\rho_I, \rho_F, \rho_H, \rho_S\}$ denote the utilization (load) $\rho = \lambda/\mu$ of the {instance SW, function SW, host HW, and site HW}, respectively. First, the easy part is combining the failure rates, loads, and availabilities. Let

$$\begin{aligned}
\lambda_N &\equiv host \ (HW + SW) \ failure \ rate = \lambda_I + \lambda_H \\
\lambda_M &\equiv site \ (HW + SW) \ failure \ rate = \lambda_F + \lambda_S
\end{aligned} \tag{1}$$

$$\begin{aligned}
\rho_N &\equiv host \ (HW + SW) \ failure \ load = \rho_I + \rho_H \\
\rho_M &\equiv site \ (HW + SW) \ failure \ load = \rho_F + \rho_S
\end{aligned} \tag{2}$$

$$\begin{aligned}
A_N &\equiv host \ (HW + SW) \ availability = A_I A_H \\
A_M &\equiv site \ (HW + SW) \ availability = A_F A_S
\end{aligned} \tag{3}$$

Now we come to the more interesting part: composite restoral rates. Let

$$\begin{aligned}
\mu_N &\equiv host \ (HW + SW) \ restoral \ rate \\
\mu_M &\equiv site \ (HW + SW) \ restoral \ rate
\end{aligned} \tag{4}$$

The most suitable approach to collapsing restoral rates depends on the particular failure mode interactions and dependencies. Figure 2 shows four different example models, all leading to different values for μ_N and μ_M. First, Model 1 is most appropriate if all failure activity stops when any failure occurs. In this case, it can be shown that

$$\begin{aligned}
\text{Model 1:} \quad \mu_N &= \lambda_N/\rho_N = \lambda_N/(\rho_H + \rho_I) \\
\mu_M &= \lambda_M/\rho_M = \lambda_M/(\rho_S + \rho_F).
\end{aligned} \tag{5}$$

Next, Model 2 is most appropriate if all failure activity stops when all failures occur. In this case, it can be shown that

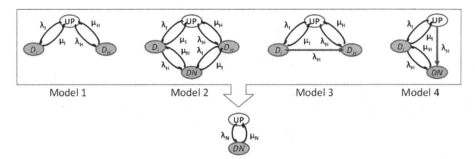

Fig. 2. Examples of different models for collapsing failure modes.

$$\text{Model 2:} \quad \mu_N = \lambda_N / (\rho_N + \rho_I \rho_H) = \lambda_N / (\rho_H + (1 + \rho_H)\rho_I)$$
$$\mu_M = \lambda_M / (\rho_M + \rho_F \rho_S) = \lambda_M / (\rho_S + (1 + \rho_S)\rho_F). \quad (6)$$

Next, Model 3 is most appropriate if all failure activity stops when a select failure occurs (host failure for μ_N or site failure for μ_M). In this case, it can be shown that

$$\text{Model 3:} \quad \mu_N = \lambda_N / \left(\rho_H + (1 + \rho_H)\frac{\lambda_I}{\lambda_H + \mu_F} \right)$$
$$\mu_M = \lambda_M / \left(\rho_S + (1 + \rho_S)\frac{\lambda_F}{\lambda_S + \mu_F} \right). \quad (7)$$

Finally, Model 4 is most appropriate if all failure activity stops when a select failure occurs, and restoral activity is sequential (e.g. host then instance for μ_N or site then resident function for μ_M). In this case, it can be shown that

$$\text{Model 4:} \quad \mu_N = \lambda_N / \left(\rho_H + (1 + \rho_H)\frac{\lambda_N}{\mu_F} \right)$$
$$\mu_M = \lambda_M / \left(\rho_S + (1 + \rho_S)\frac{\lambda_M}{\mu_S} \right). \quad (8)$$

Each model is suitable for different reliability scenarios. The simplicity of Model 1, for instance, makes it a good choice when combining many failure modes (e.g., internal components of a server). Model 2 works well if all element failures and replacements are independent (e.g., PC peripheral devices). Models 3 and 4 are most suitable if failure modes are hierarchical (e.g., user session controlled by application SW running on server HW). Model 4 is most appropriate for our reference failure model, since the instance (or function) sits on top of the underlying host (or site) HW, and recovery involves replacing the HW and restarting the SW in sequence.

While these example state space aggregation models are exact in terms of the *mean* restoral rate, the resulting model may no longer form a MC. For tractability of analysis, we assume that the aggregate restoral rates are still exponentially distributed, and the resulting collapsed model still forms a MC. There is much literature (c.f., [6–9] and references therein) addressing the Markovian implications of collapsing chains.

Now that for λ_N, λ_M, μ_N, and μ_M are defined, Table 2 lists the state transition rates associated with the small reference model of $N = 6$, $M = 2$, and $K = 3$ illustrated in Fig. 1.

Table 2. State transition rates for the small reference model of $N = 6$, $M = 2$, and $K = 3$.

Transition	Parameter Range	Rate
$i,j \rightarrow i-1,j$	$0 < i \leq J, 0 \leq j \leq J$	$i\lambda_N$
$i,j \rightarrow i,j-1$	$0 \leq i \leq J, 0 < j \leq J$	$j\lambda_N$
$i,j \rightarrow 0,j$	$0 < i \leq J, 0 \leq j \leq J$	λ_M
$i,j \rightarrow i,0$	$0 \leq i \leq J, 0 < j \leq J$	λ_M
$i,j \rightarrow i+1,j$	$0 \leq i < J, 0 \leq j \leq J$	$i\mu_N$
$i,j \rightarrow i,j+1$	$0 \leq i \leq J, 0 \leq j < J$	$j\mu_N$
$0,j \rightarrow J,j$	$0 \leq j \leq J$	μ_M
$i,0 \rightarrow i,J$	$0 \leq i \leq J$	μ_M

Additional complexities can easily be incorporated without complicating the analysis. For example, an important implication of network function virtualization (NFV) is the increased importance and added difficulty of fault *detection* and test *coverage*. Separating SW from HW (with possibly different vendors for each) creates additional reliability requirements enforcement challenges, such as how to ensure that different vendors have robust defect instrumentation and detection mechanisms if failures lie within the interaction between SW and HW, and how to ensure that test coverage is adequate. From an analysis standpoint, we can easily include detection and coverage. Let C_x denote the coverage factors and let v_x^{-1} denote the uncovered MTTRs (including detection time) for $x \in \{I, F, H, S\}$. Then replace μ_x by $\mu'_x = C_x\mu_x + (1 - C_x)v_x$.

As another example, consider scheduled maintenance. Single instance or host maintenance could be rolling application or firmware upgrades. Resident function or site maintenance could be shared database upgrades or power backup testing. Let δ_x denote the maintenance rates, let γ_x^{-1} denote the maintenance MTTRs, and let $\pi_x = \delta_x/\gamma_x$ denote the maintenance load for $x \in \{I, F, H, S\}$. Then we can replace λ_x by $\lambda'_x = \lambda_x + \delta_x$, ρ_x by $\rho'_x = \rho_x + \pi_x$, and μ_x by $\mu'_x = \lambda'_x/\rho'_x$.

2.4 Collapsing Failure Levels

Now that we have collapsed failure modes, the last key step in refining our state space representation is to *collapse the failure levels* by combining all states with the same number of available instances (capacity levels) and consolidating capacity level transition rates. Figure 3 illustrates the approach for our small reference failure model of $N = 6$, $M = 2$, and $K = 3$. As can be seen, our state space is reduced to $N + 1$ states, and our individual transitions are consolidated. All transitions due to failure/restoral of a single instance/host result in single-level transitions (—). Some single- and all multi-level transitions (---) are due to failure/restoral of an entire resident function/site. For this analysis, we again assume that the aggregate transition rates are exponentially distributed, and the resulting collapsed model still forms a MC.

As noted previously, the key contribution of this paper is the exact derivation of these composite transition rates, and in particular, the outage rate from the superset of unshaded 'up' states to the superset of shaded 'down' states as a function of N, M, K.

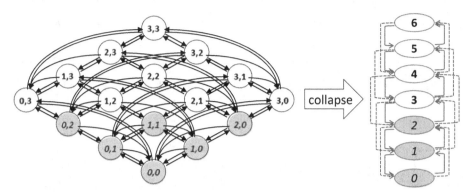

Fig. 3. Collapsing failure levels for the small reference model of $N = 6$, $M = 2$, and $K = 3$.

3 Reference Outage Model

3.1 Notation and Formulation

At the service level, application outages matter more than individual failures. Thus, we next develop a generic *outage mode* reference model (based on the failure modes). Let $n \in [0, N]$ denote the number of instances up and let $m \in [0, M]$ denote the number of sites up. Next, let P_n denote the probability that n instances are up ($0 \le n \le N$), let P_{UP} denote the probability that $\ge K$ instances are up (adequate capacity), and let $P_{DN} = 1 - P_{UP}$ denote the probability that $<K$ instances are up (service outage). Finally, let $F \equiv \lambda_D^{-1}$ denote the mean time between service-affecting outages and let $R \equiv \mu_U^{-1}$ denote the mean time to restore service following an outage.

Using the reference model of $N = 6$, $M = 2$, $K = 3$ as an example, P_6 is the probability that both sites and all instances are available, given by $P_6 = A_M^2 A_N^6$. Next, P_5 is the probability that both sites and 5 out of 6 instances are available, given by $P_5 = A_M^2 \binom{6}{5} A_N^5 (1 - A_N)$. Similarly, $P_4 = A_M^2 \binom{6}{4} A_N^4 (1 - A_N)^2$. Next, P_3 is the probability that 2 sites and 3 of 6 instances are available, plus the probability that 1 site and all instances in the other site are available, given by $P_3 = A_M^2 \binom{6}{3} A_N^3 (1 - A_N)^3 + \binom{2}{1} A_M (1 - A_M) A_N^3$. P_2 and P_1 are similar. Finally, P_0 is the probability that both sites but no instances are available, plus the probability that 1 site is available but no instances in the other site are available, plus the probability that no sites are available, given by $P_0 = A_M^2 (1 - A_N)^6 + \binom{2}{1} A_M (1 - A_M)(1 - A_N)^3 + (1 - A_M)^2$.

Generalizing, the capacity level state probabilities P_n are given by

$$P_n = \sum_{m=n/J}^{M} \binom{M}{m} A_M^m (1 - A_M)^{M-m} \binom{mJ}{n} A_N^n (1 - A_N)^{mJ-n}, \tag{9}$$

where $\lceil x \rceil$ in (9) denotes the smallest integer $\ge x$.

The probability that the service is up P_{UP} and the *ratio F/R* are given by

$$P_{UP} = \sum_{n=K}^{N} P_n \quad and \quad \frac{F}{R} = \frac{\mu_U}{\lambda_D} = \frac{P_{UP}}{1 - P_{UP}}. \tag{10}$$

In preparation for the analysis to follow, we decompose P_n as

$$P_n = \sum_{m=n/J}^{M} P_{n|m} P_M(m), \quad where$$

$$P_{n|m} = \binom{mJ}{n} A_N^n (1 - A_N)^{mJ-n} \quad and \quad P_M(m) = \binom{M}{m} A_M^m (1 - A_M)^{M-m}. \tag{11}$$

3.2   Balls in Urns Formulation

Note that $\frac{F}{R} = \frac{\mu_U}{\lambda_D}$ is expressed as a ratio in (10), thus all that remains is to determine λ_D (the transition rate from the *up* super-state to the *down* super-state). Figure 4 shows relevant transitions from an *up* state to a *down* state. For $K + J \leq n \leq N$, transitions from n to the *down* super-state (*DN*) are not possible. For $K + 1 \leq n \leq K - 1 + J$, transitions from $n \to DN$ can occur if 1 of m sites fails. And for $n = K$, transitions from $K \to DN$ *can* occur if 1 of m sites fails and *do* occur if 1 of K instances fails. Let $m * (n)$ denote the number of sites with at least enough $(n - K + 1)$ instances up, such that its failure leaves $< K$ instances up. We now need to determine $m*$ for each applicable n.

We can now begin to put mathematical structure around the solution. λ_D is given by

$$\lambda_D = \sum_{n=K}^{min(K-1+J,N)} \left[\sum_{m=n/J}^{M} P_{n|m}[m^*(n)] P_M(m) \right] \lambda_M + P_K K \lambda_N, \tag{12}$$

where $m * (n)$ is the number of sites out of m with $> n - K$ instances up. The quantities $P_{n|m}[m^*(n)] P_M(m)$ are the (weighted) combinations of ways to distribute n instances to m sites. The inner sum is across all sites m that could be up $\left(m \geq \lceil \frac{n}{J} \rceil \right)$. The outer sum is across all states n where transition from n to *DN* due to site failure is possible.

The solution is a specialized "balls in urns" problem involving the hyper-geometric distribution. There are N balls (instances) distributed in M urns (sites) *with exactly J balls in each urn*. Of the population of N balls, n are *UP* balls and $N - n$ are *DN* balls. For $M = 2$, there are $\binom{n}{i}\binom{N-n}{J-i} / \binom{N}{J} = \binom{J}{i}\binom{J}{n-i} / \binom{N}{n}$ ways of distributing J instances to site 1 (S1) such that i instances are *UP* and $J - i$ instances are *DN*, with the remaining instances in site 2 (S2). For $M = 3$, there are $\binom{J}{i}\binom{J}{j}\binom{J}{n-i-j} / \binom{N}{n}$ ways of distributing i *UP* instances to S1, j *UP* instances to S2, and $n - i - j$ *UP* instances to site 3 (S3). For general M, there are $\binom{J}{i}\binom{J}{j} \cdots \binom{J}{z}\binom{J}{n-i-j-\ldots-z} / \binom{N}{n}$ ways of appropriately distributing n *UP* instances to M sites.

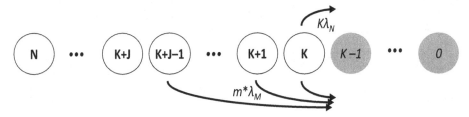

Fig. 4. Feasible transitions from an 'up' state to a 'down' state.

For simplicity (for now), consider the case of $M = 2$ sites. It would seem that

$$\lambda_D = \sum_{n=K}^{\min(K-1+J,N)} P_n \left[\sum_{i=\max(0,n-J)}^{\min(n,J)} \frac{\binom{J}{i}\binom{J}{n-i}}{\binom{N}{n}} \left[I_{i>n-K} + I_{i<K} \right] \right] \lambda_M + P_K K \lambda_N.$$

$$(13)$$

The sum of indicator functions $\left[I_{i>n-K} + I_{i<K} \right]$ in (13) is the number of sites with enough *UP* instances to cause an outage if the site fails.

3.3 Unequal Combinations

The problem with (13) is that the $\binom{J}{i}\binom{J}{n-i} / \binom{N}{n}$ combinations are not all equally likely. If all sites are up, then all *DN* instances must be due to individual failures, thus all combinations are equally likely (and if $n > (M - 1)J$, then all sites are up). And all combinations where every site has >0 *UP* instances are equally likely. However, combinations with 0 *UP* instances in a site could be due to J individual *DN* instances or 1 *DN* site. Hence, we need to condition on m; that is, $P_n = \sum_{m=n/J}^{M} P_{n|m} P_M(m)$.

To illustrate, Fig. 5 shows the 41 feasible combinations for our small reference failure model of $N = 6$, $M = 2$, $K = 3$. Transitions from n to *DN* are possible for $3 \leq n \leq 5$. For each n, there are $\binom{6}{n}$ distributions of n *UP* instances into 2 sites, and $\binom{3}{i}\binom{3}{n-i}$ distributions of i *UP* instances to S1 and $n - i$ *UP* instances to S2, where $n - 3 \leq i \leq 3$. Each row in each matrix represents a unique distribution of instances for that n, where the first 3-tuple of columns corresponds to S1 and the second 3-tuple corresponds to S2.

As can be seen, for $n = 5$ (left) there are 6 distributions of 5 *UP* instances to 2 sites (3 with 2 in S1 and 3 with 3 in S1). Since both sites have *UP* instances, both sites are up. Since $n = 5 > J = 3$, only site failures (not individual instance failures) can result in an outage. Since all combinations are the result of a single instance failure, all combinations are equally likely. Finally, $[I_{i>2} + I_{i<3}] = 1$ for all combinations.

For $n = 4$ (center), there are 15 equally likely distributions of 4 *UP* instances (3 with 1 in S1, 9 with 2 in S1, and 3 with 3 in S1). The main difference is that for the 9 combinations with 2 in S1 (and 2 in S2), $[I_{i>1} + I_{i<3}] = 2$ (i.e., failure of either site results in an outage). For the remaining 6 combinations, $[I_{i>1} + I_{i<3}] = 1$.

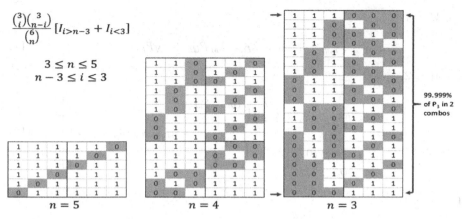

Fig. 5. Feasible combinations for the small reference model of $N = 6, M = 2, K = 3$.

For $n = 3$ (right), the flaw in the 'equally likely' assumption is exposed. There are 20 distributions of 3 *UP* instances in 2 sites (1 with 0 in S1, 9 with 1 in S1, 9 with 2 in S1, 1 with 3 in S1). The 18 combinations with either 1 or 2 *UP* instances in S1 (and vice versa in S2) are the result of single instance failures, and all are equally likely. The 2 combinations with either 0 or 3 in S1 (and vice versa in S2) could result from *either* 3 individual instance failures *or* 1 site failure, so these combinations are more likely. In fact, for the Table 1 defaults, these combinations account for 99.999% of P_3.

To further illustrate, the erroneous "equally likely combinations" formula suggests

$$\lambda_D = \left\{ P_5 \frac{3[1]+3[1]}{6} + P_4 \frac{3[1]+9[2]+3[1]}{15} + P_3 \frac{1[1]+9[2]+9[2]+1[1]}{20} \right\} \lambda_M + P_3 3\lambda_N$$
$$= \{P_5[1.0] + P_4[1.6] + P_3[1.9]\}\lambda_M + P_3 3\lambda_N. \tag{14}$$

For $M = 2$, this scenario of unequal combinations can only happen when $i = 0$ or $n - i = 0$ (that is, when one site has no UP VMs). The result from the correct formula looks like

$$\lambda_D = \left\{ \begin{array}{l} P_5 \frac{3[1]+3[1]}{6} + P_4 \frac{3[1]+9[2]+3[1]}{15} \\ +P_{3|2} \frac{1[1]+9[2]+9[2]+1[1]}{20} P_M(2) + P_{3|1} \frac{1[1]}{1} P_M(1) \end{array} \right\} \lambda_M + P_3 3\lambda_N$$
$$= \{P_5[1.0] + P_4[1.6] + P_{3|2}[1.9]P_M(2) + P_{3|1}[1.0]P_M(1)\}\lambda_M + P_3 3\lambda_N. \tag{15}$$

3.4 Outage Rate

As illustrated in this example, we can account for the fact that not all combinations are equally likely by breaking P_n apart and conditioning on m. The resulting exact formula for λ_D for general M, N, and K is given by

$$\lambda_D = \lambda_M \sum_{n=K}^{\min(K-1+J,N)} \left\{ \sum_{m=n/J}^{M} P_{n|m} \mathcal{F}_{n,m} P_M(m) \right\} + \lambda_N P_K K. \tag{16}$$

For $M = 1$,

$$\mathcal{F}_{n,m} = 1 \quad \text{and} \quad \lambda_D = \lambda_M P_{UP} + \lambda_N P_K K. \tag{17}$$

For $M = 2$,

$$\mathcal{F}_{n,m} = \sum_{\substack{i = \max(0, \\ n - (m-1)J)}}^{\min(n,J)} \left[\frac{\binom{J}{i}\binom{(m-1)J}{n-i}}{\binom{mJ}{n}} [I_{i>n-K} + I_{n-i>n-K}] \right]. \tag{18}$$

For $M = 3$,

$$\mathcal{F}_{n,m} = \sum_{\substack{i = \max(0, \\ n - (m-1)J)}}^{\min(n,J)} \sum_{\substack{j = \max(0, \\ n - (m-2)J - i)}}^{\min(n-i,J)} \left[\frac{\binom{J}{i}\binom{J}{j}\binom{(m-2)J}{n-i-j}}{\binom{mJ}{n}} * \atop [I_{i>n-K} + I_{j>n-K} + I_{i+j<K}] \right]. \tag{19}$$

For $M = 4$,

$$\mathcal{F}_{n,m} = \sum_{\substack{i = \max(0, n- \\ (m-1)J)}}^{\min(n,J)} \sum_{\substack{j = \max(0, n- \\ (m-2)J - i)}}^{\min(n-i,J)} \sum_{\substack{k = \max(0, n- \\ (m-3)J - i - j)}}^{\min(n-i-j,J)} \left[\frac{\binom{J}{i}\binom{J}{j}\binom{J}{k}\binom{(m-3)J}{n-i-j-k}}{\binom{mJ}{n}} * \atop [I_{i>n-K} + I_{j>n-K} + I_{k>n-K} + I_{i+j+k<K}] \right]. \tag{20}$$

We now have the exact formula for the mean time between service outages $F = \lambda_D^{-1}$, and can also compute the mean time to restore service $R = \mu_D^{-1} = F(1 - P_{UP})/P_{UP}$. As we show next, these are powerful tools to facilitate the analysis and optimal sizing of application topologies to meet service performance and reliability requirements.

Although the equation for $\mathcal{F}_{n,m}$ becomes increasingly more awkward to express for increasing M, it is very straightforward to program algorithmically for computation. The combined number of terms evaluated is loosely bounded by 2^N, so computational complexity grows roughly exponentially with N. For a given SW application, the total number of servers N is dictated by capacity needs and remains relatively insensitive to the number of sites M across which those servers are spread.

Furthermore, in many practical applications, the number of deployment sites M tends to remain relatively small due to failover latency, data replication/backup, and cost constraints. For example, Amazon Web Services (AWS) has four domestic US Regions, each with 3–6 Availability Zones [10]. Applications requiring ultra-high availability can deploy in $M = 4$ domestic regions, while latency-sensitive applications can be regionalized and deploy in $M = 3$–6 availability zones per region. Thus, the modest values for M considered here do in fact reflect realistic deployment scenarios.

4 Example Application

4.1 Requirements and Assumptions

As a hypothetical example, consider a Voice over IP (VoIP) call setup message processing application. The goal is to cost-effectively size the application (M sites and N virtual instances) to satisfy the following requirements and assumptions:

- Application (service) availability $A \geq 0.99999$.
- Adequate capacity to process 600 VoIP calls/sec.
- Peak traffic rate $1.5\times$ average traffic rate.
- Mean message processing latency ≤ 30 ms, and 95th percentile ≤ 60 ms.
- Service outages lasting longer than 30 min are reportable.
- Probability of a reportable outage in 1 year $\leq 1\%$.
- An outage occurs if available capacity $<50\%$ (2X over-engineering).
- Local- and geo-redundancy required (minimum 2+ instances at each of 2+ sites).
- Instances implemented as virtual machines (VMs) of the 4-vCPU flavor.

4.2 Capacity and Latency Requirements

First, we consider the latency requirements to determine the required number of instances N. Given that voice call arrivals are reasonably random, and protocol message processing time is reasonably constant, we assume an $M/D/C$ service model, where C is the required number of vCPUs. Let $E(W)$ and $V(W)$ denote the mean and variance of the waiting time W prior to service. Tijms [11] provides the exact, non-trivial solution for the distribution of $W(t)$, together with a recursive computational algorithm. For simplicity, we use the well-known Kingman-Köllerstörmheavy traffic $GI/G/C$ two-moment approximations below (c.f., [12] and references therein) for $E(W)$ and $V(W)$ based on the coefficients of variation C_a^2 and C_s^2 of the arrival process and the service process (where $C_a^2 = 1$ and $C_s^2 = 0$ for the $M/D/C$ system). Then the mean and variance of the waiting time W are given by

$$E(W) \cong T^0 \left(\frac{\rho}{1-\rho} \left[\frac{C_a^2 + C_s^2}{2} \right] \right) = T^0 x, \text{ and} \tag{21}$$

$$V(W) \cong \left(T^0 \right)^2 C_s^2 + \left(T^0 \right)^2 \left\{ \left(\frac{\rho}{1-\rho} \right)^2 \left[\frac{C_a^2 + C_s^2}{2} \right]^2 \left[1 + \frac{4(1-\rho)C_s^2}{\rho(1+C_s^2)} \right] \right\} = \left(T^0 x \right)^2, \tag{22}$$

where T^0 is the no-load message processing (code execution) time and $x = \frac{\rho}{2(1-\rho)}$. This Kingman-Köllerstörmapproximation assumes that W is exponentially distributed with mean $T^0 x$, and latency $T = T^0 + W$ is a shifted exponential. The 95th percentile latency is given approximately by $T^0 + 3E(W) = T^0(1 + 3x)$. Thus, the performance requirements, combined with the capacity requirement of 600 calls/sec, become $T^0 \leq min\left\{ \frac{0.03}{1+x}, \frac{0.06}{1+3x} \right\}$, where $x = \frac{\rho}{2(1-\rho)}$, $\rho = \frac{600T^0}{C}$, and $C = $ number of vCPUs.

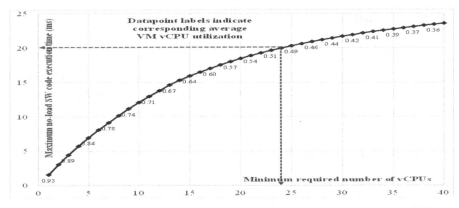

Fig. 6. Relationship between allowable processing time and required number of vCPUs.

This result yields an explicit relationship (observed in [13]) between the maximum allowable processing time T^0 and minimum required number of vCPUs C, as shown in Fig. 6. For $\rho < 2/3$, the mean delay requirement is more constraining, while for $\rho > 2/3$, the 95th percentile requirement is more constraining. Since $\rho \leq 50\%$ is required to ensure adequate capacity in the event of site failure, $T^0 = 20$ ms and $C = 24$. Finally, since SW instances are of the 4-vCPU flavor, $N = 6$ instances are required ($J = K = 3$). Note that this relationship places a requirement on the SW, and if the SW vendor cannot meet this 20 ms execution time target, then more vCPUs will be required.

4.3   Reference Outage Model and Availability Requirement

Next, given the proposed minimum topology $M = 2, N = 6$, and $J = K = 3$ that satisfies the capacity and latency requirements, we can now apply the reference outage model. For the default MTBF/MTTR values in Table 1, the model output parameters, explicit formulae, and resulting values are given in Table 3. As can be seen, based on the assumed MTBFs and MTTRs for this topology, $F = 323567$ h and $R = 67$ min.

Table 3. Model parameters, explicit formulae, and resulting values for $N = 6, M = 2, K = 3$.

Parameter	Formula	Value
P_6	$A_M^2 A_N^6$	9.9340E-01
P_5	$A_M^2 6A_N^5(1-A_N)$	5.4445E-03
P_4	$A_M^2 15A_N^4(1-A_N)^2$	1.2433E-05
P_3	$A_M^2 20A_N^3(1-A_N)^3 + 2A_M(1-A_M)A_N^3$	1.1373E-03
P_2	$A_M^2 15A_N^2(1-A_N)^4 + 2A_M(1-A_M)3A_N^2(1-A_N)$	3.1166E-06
P_1	$A_M^2 6A_N(1-A_N)^5 + 2A_M(1-A_M)3A_N(1-A_N)^2$	2.8468E-09
P_0	$A_M^2(1-A_N)^6 + 2A_M(1-A_M)(1-A_N)^3 + (1-A_M)^2$	3.2550E-07
P_{UP}	$P_6 + P_5 + P_4 + P_3$	0.99999656
P_{DN}	$P_2 + P_1 + P_0$	0.00000344
F	$[P_5 1.0\lambda_M + P_4 1.6\lambda_M + P_3 1.9\lambda_M + P_3 3\lambda_N]^{-1}$	323567 hr
R	FP_{DN}/P_{UP}	1.11 hr

Now, consider the availability requirement and assume (worst case) that all outages occur during peak traffic periods, where the peak-to-average traffic ratio $\sigma = 1.5$. Then (see [13]) $F \geq \sigma \, RA/(1 - A) = 166498$ h. Since $323567 > 166498$, the availability requirement is met, and it appears that the minimum $M = 2, N = 6$ topology is sufficient. However, we must verify that this solution meets the reportable outage requirement.

4.4   Reportable Outage Requirement

Next, consider the service outage requirement $P(no\ outages > 30\ min\ in\ 8760\ h)$

$$= \sum_{n=0}^{\infty} P(no\ outages > 0.5\ hours \mid n\ outages)P(n\ outages\ in\ 8760\ hours)$$

$$= \sum_{n=0}^{\infty} \left[1 - e^{-\frac{\mu}{2}}\right]^n \frac{(8760\lambda)^n e^{-8760\lambda}}{n!} = e^{-8760\lambda} \sum_{n=0}^{\infty} \frac{\left(8760\lambda\left[1-e^{-\frac{\mu}{2}}\right]\right)^n}{n!} \qquad (23)$$

$$= e^{-8760\lambda} e^{8760\lambda[1-e^{-\mu/2}]} = e^{-8760\lambda e^{-\mu/2}} \geq 99\% \ .$$

Then $\lambda e^{-\mu/2} \leq -ln(0.99)/8760 = 871613^{-1}$ and $F \geq 871613 e^{-0.5/1.11} = 556564$ h. Since $323567 < 556564$, the reportable outage requirement is not met.

There are a number of possible options, all easily evaluated using the reference outage model. First, we can model the effect of hardening elements by increasing MTBFs and/or decreasing MTTRs (details omitted). Hardening instance SW (increasing λ_I^{-1} from 3 to 13 months) or the resident function SW (increasing λ_F^{-1} from 2 to 6.4 years) both result in increasing F above 556564 h. Interestingly, decreasing SW MTTRs is not as effective because in this example (where the reportable service outage requirement is most constraining), the solution is more sensitive to failure rates than to restoral rates. Notably, hardening the HW (increasing MTBFs or decreasing MTTRs) does not help, lending analytical support to the trend of using commodity hosts and public cloud sites instead of high-end servers and hardened datacenters.

Next, we can add instances (N) and/or sites (M). Adding a fourth host/instance to each site ($M = 2, N = 8, J = 4$) easily meets the requirement. Surprisingly, adding a third site and redistributing the hosts ($M = 3, N = 6, J = 2$) also easily meets the requirement. The reason is that although site failures are now more frequent with three sites, the much more probable {1 site + 1 instance} duplex failure is no longer an outage mode.

5   Topology Configuration Tool

Given the minimal topology description of M sites, N hosts, $J = N/M$ instances/site, and K instances required for service to be up, and given the basic failure and restoral rates $\{\lambda_I, \lambda_F, \lambda_H, \lambda_S\}$ and $\{\mu_I, \mu_F, \mu_H, \mu_S\}$ of the {instance SW, function SW, host HW, site HW}, we determine the exact formulae for the service availability $A = P_{UP}$, the mean time between service outages $F = \lambda_D^{-1}$, and the mean time to restore service $R = \mu_D^{-1}$. This reference outage model forms the kernel for a *topology configuration and optimization tool*. Instead of inputting M, N, and K, and computing A and F, we want to input *requirements* for availability A and capacity K (and possibly other metrics), and compute the most cost-effective system topology M and N.

Consider the following topology configuration and optimization algorithm.

Inputs:

- MTBFs and MTTRs for {instance, function, host, site}
- Annualized capital and operational expense costs {C_M, O_M, C_N, O_N}
- Required availability A and capacity K
- Required local- and geo-redundancy $J \geq j \geq 1$ and $M \geq m \geq 1$
- Required mean outage and restoral times $F \geq f$ and $R \leq r$, etc.

Objective function:

Minimize {$(C_M + O_M)M + (C_N + O_N)N$} subject to $P_{UP} \geq A$, $J \geq j$, $M \geq m$, $F \geq f$, $R \leq r$, etc.

Given the inputs, the approach is to compute a family of feasible solution pairs {M,N} that are generally in the range {m,N_{max}},..., (M_{max},j). The most cost-optimal topology is then easily determined given the capital and operational expense cost parameters.

Algorithm:

1. Start by setting $M = m$ and $A_N = 1$ (i.e., only site failures can occur).
2. Solve for {P_{UP}, F, R}.
3. If any outputs {P_{UP}, F, R} do not meet their respective requirements (that is, no feasible solution exists for M for *any* N), then increment $M \leftarrow M + 1$ and go to step 2.
4. Set $J = \max(\lceil K/M \rceil, j)$, $N = MJ$, and $A_N = \mu_N/(\lambda_N + \mu_N)$.
5. Solve for {P_{UP}, F, R}. If any outputs {P_{UP}, F, R} do not meet their respective requirements, then increment $N \leftarrow N + M$ and $J \leftarrow J + 1$, and repeat step 5.
6. Record {M, N} as a feasible solution pair.
7. If $J > j$, then increment $M \leftarrow M + 1$ and go to step 4; otherwise, stop.
8. The optimal solution is determined from the resulting set of feasible pairs {M, N}.

6   Conclusions

In this paper, we illustrate an approach to topology configuration and optimization, with a focus on addressing geo-redundancy issues like how many sites, and how many servers per site, are required to meet performance and reliability requirements. We first develop a multi-dimensional component *failure mode* reference model, then show how to reduce this model to a one-dimensional service *outage mode* reference model. We describe a topology configuration tool for optimizing resources to meet requirements and illustrate its effective use for a hypothetical application.

The key contribution of this work is the exact derivation of the composite *service* outage and restoral rates as a function of the number of servers, the number of sites, and the minimum required server capacity level, using a novel adaptation of the hyper-geometric "balls in urns" distribution with unequally likely combinations.

Acknowledgements. We wish to thank D. Hoeflin for his invaluable assistance with the example in §4, and C. R. Johnson for her insightful comments on the manuscript.

References

1. Lai, C., Xie, M., Poh, K., Dai, Y., Yang, P.: A model for availability analysis of distributed software/hardware systems. Inf. Softw. Technol. **44**(6), 343–350 (2002). https://doi.org/10.1016/S0950-5849(02)00007-1

2. Jammal, M., Kanso, A., Heidari, P., Shami, A.: A formal model for the availability analysis of cloud deployed multi-tiered applications. In: Proceedings IEEE International Conference on Cloud Engineering Workshop (IC2EW), Berlin, Germany, pp. 82–87 (2016). https://doi.org/10.1109/IC2EW.2016.21

3. Spinnewyn, B., Mennes, R., Botero, J., Latré, S.: Resilient application placement for geo-distributed cloud networks. J. Netw. Comput. Appl. **85**, 14–31 (2016). https://doi.org/10.1016/j.jnca.2016.12.015

4. Do, T., Kim, Y.: Topology-aware resource-efficient placement for high availability clusters over geo-distributed cloud infrastructure. IEEE Access **7**, 107234–107246 (2019). https://doi.org/10.1109/ACCESS.2019.2932477

5. Network Outage Reporting System (NORS). https://www.fcc.gov/network-outage-reporting-system-nors. Accessed 06 June 2021

6. Burke, C., Rosenblatt, M.: A Markovian function of a Markov chain. Ann. Math. Stat. **29**(4), 1112–1122 (1958)

7. Rosenblatt, M.: Functions of a Markov process that are Markovian. J. Math. Mech. **8**(4), 585–596 (1959)

8. Hachigian, J.: Collapsed Markov chains and the Chapman-Kolmogorov equation. Ann. Math. Stat. **34**(1), 233–237 (1963)

9. Tolver, A.: Introduction to Markov Chains. Lecture Notes, Copenhagen, Denmark (2016)

10. Amazon Web Services (AWS), Global Infrastructure, Regions and Availability Zones. https://aws.amazon.com/about-aws/global-infrastructure/regions_az/?p=ngi&loc=2. Accessed 05 Sept 2021

11. Tijms, H.: Stochastic Modelling and Analysis: A Computational Approach, pp. 333–340, Wiley, Chichester (1986)

12. Kleinrock, L.: Queueing Systems: Computer Applications, vol. 2, pp. 46–50. Wiley, New York (1976)

13. Reeser, P., Johnson, C.: Novel reliability methodology for virtual solutions. In: Proceedings IEEE ComSoc International Communications Quality and Reliability Workshop (CQR), Naples, FL, USA, pp. 1–6 (2019). https://doi.org/10.1109/CQR.2019.8880108

Splittable Routing Games in Ring Topology with Losses

Sami Dallali[4], Clara Fontaine[4,5,6], and Eitan Altman[1,2,3($\boxtimes$)]

[1] INRIA Sophia Antipolis - Méditerranée, Valbonne, France
`Eitan.Altman@inria.fr`
[2] CERI/LIA, University of Avignon, Avignon, France
[3] lincs, 23 Ave d'Italie, 75013 Paris, France
[4] CentraleSupélec, CS, Paris Saclay, Gif sur Yvette, France
`sami.dallali@student-cs.fr`
[5] Cornell University, Engineering Management and EE, Ithaca, USA
`cyf3@cornell.edu`
[6] Centre for Quantum Technologies at the National Univ of Singapore,
Singapore, Singapore

Abstract. We consider a splittable atomic game with lossy links on a ring in which the cost that each player i minimizes is their own loss rate of packets. The costs are therefore non-additive (unlike costs based on delays or tolls) and moreover, there is no flow conservation (total flow entering a link is greater than the flow leaving it). We derive a closed-form for the equilibrium, which allows us to obtain insight on the structure of the equilibrium. We also derive the globally optimal solution and obtain conditions for the equilibrium to coincide with the globally optimal solution.

Keywords: Routing games · Loss probabilities · Ring topology

1 Introduction

We study routing on a ring network in which traffic originates from nodes on the ring and is destined to the center node. Each node has two possible paths: either a direct path from the node to the center node or a two-hop path in which the packet is first relayed to the next node on the ring and then takes the direct link from that next node to the common destination. The traffic originating from a given node is assumed to form an independent Poisson process with some intensity (which we call the demand). Beyond forwarding the traffic that arrives from the previous node, each node has to decide what fraction of its own traffic would be routed on each one of the two possible paths to the destination.

Routing games of this type have been intensively studied both in the road traffic community [13] as well as in the community of telecommunications network [12] under additive costs (such as delays or tolls) and conservation constraints (at each node, the sum of incoming flow equals the sum of outgoing flows) [9]. In this paper, we depart from these assumptions by considering loss networks in which

© Springer Nature Switzerland AG 2021
P. Ballarini et al. (Eds.): EPEW 2021/ASMTA 2021, LNCS 13104, pp. 298–307, 2021.
https://doi.org/10.1007/978-3-030-91825-5_18

losses may occur at all links: there are links with i.i.d. losses (relay links) and collision losses (on direct links between source and the common destination node).

Two levels of system modeling are presented here: a flow level in which routing decisions are taken, and a more detailed packet-level modeling that determines the losses and thus the interference between flows from different sources. The decisions of a node concern only the fraction of packets originated in that node which will be routed on each of the two paths available to traffic from that node. Then the actual packets to be transmitted over each one of the paths are selected according to an i.i.d. Bernoulli thinning process. The decisions are thus the Bernoulli thinning parameters.

The ring topology has received much less attention than the parallel-link topology as well as the load-balancing triangular topology introduced by Prof Kameda and his students. Although in practice, the ring topology may seem to be a toy problem, we do encounter ring networks quite often in practice, mostly in runabouts. Ring topologies can also be found in access to communication networks, both in local area networks – see IEEE 802.5 token ring standard and the metropolitan area network FDDI [4].

Related Work. Previous studies of routing games with circular topology have appeared in citecircle1,burra,chen. [7,8] consider linear costs, and none of these references consider non-conservation of flows. We note also that in [1], there are either bi-directional roads or two rings, one in each of two directions (clockwise and anti-clockwise), and cars have to decide which direction to drive in. In [3] other non-additive cost criteria have been introduced in a context of load-balancing games (triangle topology) where their performance measure is related to blocking probabilities. See also [2] and [5] for other related work.

Focusing on symmetric ring networks, our main contribution is to obtain closed-form expressions with the help of Maple. This includes best response functions, derivatives of the costs that are used to compute the best response, and the symmetric equilibrium. We derive the globally optimal solution as well as the equilibrium solution.

2   The Model

We consider K nodes on a circle, indexed by $0, 1, ..., K - 1$, see Fig. 1.

Each node k is connected to a set N_k containing N players. Each player $(n, k) \in N_k$ has to ship a strictly positive demand ϕ_k to a destination Δ common to all players. Each of the Nk players generates packets following an independent Poisson process. The player decides with what probability to send an arrival that it generates over one of two possible paths to Δ; the packet can use a direct transmission link $D(k)$ or an indirect path consisting of first relaying the packet to node $k + 1$ and only then transmit it to Δ through $D(k+1)$ (note that node indices are modulo K.) Let x_k^n and α_k^n denote the amount and fraction, respectively, of class k flow originating from player (n, k) through the direct path, i.e., through link $D(k)$. We call

$$\alpha = (\alpha_k^n, k = 0, ..., K - 1, n = 1, ..., N)$$

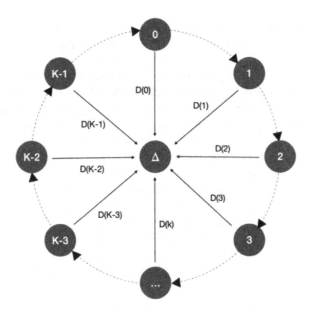

Fig. 1. Ring network topology

the assignment or th action vector. For a given demand vector $\phi = (\phi_k^n, k = 0, ..., K-1, n = 1, ..., N)$, the assignment vector α completely specifies the set of flows

$$\mathbf{x} = (x_k^n, k = 0, ..., K-1, n = 1, ..., N) = (\alpha_k^n \phi_k^n, k = 0, ..., K-1, n = 1, ..., N).$$

With probability α_j^n, a packet originating from player n in N_j takes a direct path to the center, and otherwise it takes an indirect path: it is first relayed to node $j+1$ and then forwarded to the destination through link $D(k+1)$.

Loss Probabilities. We consider two types of losses: (1) i.i.d. losses at the relay: a packet originating from node k is lost if relayed to node $k+1$ with probability q_k. (2) collision losses on the links D_k: whenever an arrival occurs while there is another packet in service then there is a loss. The transmission duration of a packet in link $D(k)$ is exponentially distributed with parameter μ_k.

The total flow sent to the link $D(k)$ consists of the superposition of (1) the Poisson flows that arrive at node k and are transmitted over $D(k)$, and (2) the Poisson flow originated in node $k-1$ consisting of the packets that were not lost in the relay to node k.

Thus the total rate of arrivals to $D(k)$ is

$$R(k) = \sum_{n=1}^{N} \phi_k^n \alpha_k^n + \phi_{k-1}^n (1 - \alpha_{k-1}^n)(1 - q_k)$$

The loss probability of packets at $D(k)$ is

$$P(k) = \frac{R(k)}{R(k) + \mu_k} \tag{1}$$

By the same arguments, the loss probability of packets at $D(k+1)$ is

$$P(k+1) = \frac{R(k+1)}{R(k+1) + \mu_{k+1}} \tag{2}$$

The total rate of losses of packets of player (n, k) is

$$J_k^n(\alpha) = \phi_k^n \Big(\alpha_k^n P(k) + (1 - \alpha_k^n) \big[q_k + (1 - q_k) P(k+1) \big] \Big) \tag{3}$$

In the rest of the paper we assume that ϕ_k^n, μ_k and q_k are constant, independent of k and n.

3   The Globally Optimal Solution and the Equilibrium

3.1   Minimizing Average Loss Probability

Consider a symmetric multi-strategy α for all players, i.e. in which α_k^n are the same for all players (n, k). Without loss of generality, let $\phi = 1/N$. Then the rate of arrival of packets at the links $D(k)$ is $R = 1 - (1 - \alpha)q$. The loss probability on link $D(k)$ is

$$P_l = \frac{R}{R + \mu} = 1 - \frac{\mu}{1 - (1 - \alpha)q + \mu}$$

so the global loss probability is

$$\pi = P_l + (1 - \alpha)q.$$

It is minimized at $\alpha = 1$, which means that all players take direct path to the destination.

3.2   Equilibrium

Assume that player (n, k) deviates and plays b instead of playing α. Let u be the multi-strategy after the deviation. Then

$$R(k, u) = [\alpha + (1 - \alpha)(1 - q)]\frac{N - 1}{N} + [b + (1 - \alpha)(1 - q)]\frac{1}{N}$$

$$= \frac{b - \alpha}{N} + (1 - \alpha)(1 - q)$$

$$R(k+1, u) = [\alpha + (1 - \alpha)(1 - q)]\frac{N - 1}{N} + [\alpha + (1 - b)(1 - q)]\frac{1}{N}$$

$$= \alpha + (1 - \alpha)(1 - q) - (1 - q)\frac{b - \alpha}{N}$$

The loss probability for a player (k, n) is given by

$$J_k^n(u) = \frac{1}{N}\Big(bP(k, u) + (1 - b)\big[q + (1 - q)P(k + 1, u)\big]\Big)$$

where the path loss probabilities $P(k, u)$ and $P(k + 1, u)$ are given in (1)–(2).
To obtain the equilibrium, we:

1. differentiate $J_k^n(u)$ with respect to b, and obtain (using Maple) the expression in Fig. 6. Equating the expression to 0 allows us to obtain the best response action $b = f(a)$ that minimizes the loss probabilities of a player when all others use α.
2. obtain the symmetric equilibrium by computing the fixed point of the mapping $b = f(\alpha)$. This leads to

$$\alpha^* = -\frac{N\mu q - Nq^2 + Nq + q^2 - 2q + 1}{Nq^2 - q^2 + 2q - 2}$$

If $0 < \alpha^* \leq 1$, then it is a symmetric equilibrium. If the fixed point is greater than 1, then the symmetric equilibrium is the policy $\alpha = 1$ for all players. In that case, the equilibrium is globally optimal and only the direct links are used. Thus, the equilibrium coincides with the globally optimal policy for all q large enough.

3.3 Best Response

As already mentioned, we are able with the help of Maple to get an explicit cumbersome expression for the best response. This allows us to obtain a much simpler expression for the equilibrium (as a function of the parameters). We present in Fig. 2 the best response b as a function of q.

3.4 When Is the Globally Optimal Policy $\alpha = 1$ an Equilibrium

Consider the cost $J_k^n(b, 1)$ for some player (n, k) where $(b, 1)$ is the policy where all players use $\alpha = 1$ and the deviating player (k, n) uses b.

Theorem 1. *A necessary and sufficient condition for $\alpha = 1$ to be a symmetric equilibrium is that*

$$qN > \frac{1}{1 + \mu}$$

Proof. A necessary and sufficient condition for the symmetric policy $\alpha = 1$ to be an equilibrium is that the cost for the deviating player be decreasing in b at $b = 1$. This is equivalent to the following first-order condition. The derivative of

$$\Big(-2\,a^2q^2 - 2\,a\mu q^2 + a\,q^3 + 4\,a^2q + 2\,a\mu q - a\,q^2 - 2\,\mu^2q + 3\,\mu q^2 - q^3 - 2\,a^2 - a\,q$$

$$+\,2\,\mu^2 - 6\,\mu q + 3\,q^2$$

$$+\Big(-a^4q^6 - a^3\mu q^6 + 6\,a^4q^5 + 3\,a^3\mu q^5 + 3\,a^3q^6 - 3\,a^2\mu^2q^5 + 3\,a^2\mu q^6 - 13\,a^4q^4 + 4\,a$$

$$-\,22\,a^3\mu q^3 + 51\,a^3q^4 - 18\,a^2\mu^2q^3 + 13\,a^2\mu q^4 + 22\,a^2q^5 + 15\,a\mu^3q^3$$

$$-\,35\,a\mu^2q^4 + 19\,a\mu q^5 + a\,q^6 - \mu^4q^3 + 3\,\mu^3q^4 - 3\,\mu^2q^5 + \mu q^6 - 4\,a^4q^2$$

$$+\,24\,a^3\mu q^2 - 62\,a^3q^3 - a^2\mu^2q^2 + 28\,a^2\mu q^3 - 64\,a^2q^4 - 24\,a\mu^3q^2 + 72\,a\mu^2q^3$$

$$-\,43\,a\mu q^4 - 8\,a\,q^5 + 5\,\mu^4q^2 - 18\,\mu^3q^3 + 21\,\mu^2q^4 - 8\,\mu q^5 - 8\,a^3\mu q + 36\,a^3q^2$$

$$+\,12\,a^2\mu^2q - 62\,a^2\mu q^2 + 94\,a^2q^3 + 12\,a\mu^3q - 59\,a\mu^2q^2 + 37\,a\mu q^3 + 26\,a\,q^4$$

$$-\,8\,\mu^4q + 39\,\mu^3q^2 - 57\,\mu^2q^3 + 26\,\mu q^4 - 8\,a^3q - 4\,a^2\mu^2 + 40\,a^2\mu q - 73\,a^2q^2$$

$$+\,12\,a\mu^2q + 2\,a\mu q^2 - 44\,a\,q^3 + 4\,\mu^4 - 36\,\mu^3q + 75\,\mu^2q^2 - 44\,\mu q^3 - 8\,a^2\mu$$

$$+\,28\,a^2q + 4\,a\mu^2 - 20\,a\mu q + 41\,a\,q^2 + 12\,\mu^3 - 48\,\mu^2q + 41\,\mu q^2 - 4\,a^2 + 8\,a\mu$$

$$-\,20\,a\,q + 12\,\mu^2 - 20\,\mu q + 4\,a + 4\,\mu\Big)^{n} + a + 3\,\mu - 3\,q + 1\Big)\Big/\Big(a\,q^3 - 3\,a\,q^2$$

$$+\,\mu q^2 - q^3 + 4\,a\,q - \mu q + 3\,q^2 - 2\,a - 3\,q + 1\Big)$$

Fig. 2. Expression for the best response function $b = f(\alpha)$

the cost of the deviating player evaluated at $b = 1$ when all other players use $\alpha = 1$ is non-negative. Calculation in Maple yields

$$\left.\frac{dJ_k^n(b,1)}{db}\right|_{b=1} = -\frac{\mu(N\mu q + Nq - 1)}{(1+\mu)^2} \qquad (4)$$

This concludes the proof.

We conclude that if one invests in improving a communication channel thus decreasing the loss probabilities (the parameter q in our case), then as a result we may end up worsening the performance for all the users in the system.

4 Numerical Examples

With the help of Maple, we obtained a simple expression for the equilibrium as a function of the parameters of the system. The following experiments allow us to get an insight on equilibrium behavior.

4.1 The Equilibrium

We depict in Fig. 3 the parameter α^* defining the symmetric equilibrium as a function of the loss rate parameter q. The following parameters are fixed: $N = 1$ and $\mu = 1$. As long as α^* is in the unit interval, it is the equilibrium. This is the case for $q \leq 0.5$. For larger q, the corresponding symmetric equilibrium is $\alpha = 1$.

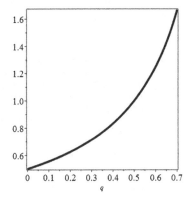

Fig. 3. The symmetric equilibrium as a function of the loss probability parameter q.

4.2 The Best Response

In Fig. 4, we depict the loss probability of a player that plays b when the others play α. b is varied while α is held fixed. The figure thus allows us to identify the best response b. In the case that the best response $b(\alpha)$ to a given α in the unit interval satisfies $b(\alpha) = \alpha$, then α is a symmetric equilibrium.

The following parameters are fixed: $N = 1$ and $\mu = 1$. There are three curves: the first corresponds to $\alpha = 0.5012$ and $q = 0.01$, the second to $q = 0.3$ and $\alpha = 0.55$ and the third to $q = 0.9$ and $\alpha = 1$. In the two first curves, b for which the derivative of the cost is zero is within the unit interval and is thus the best response to α. In the first curve, $\alpha = 0.5012$ is a fixed point of the best response function $b(\alpha)$. This confirms that $\alpha = 0.5012$ is an equilibrium, which can be seen from Fig. 3.

In the third curve, b that minimizes the cost is larger than 1 and the best response is obtained on the boundary $b = 1$ and not on the b for which the derivative w.r.t. b is zero.

4.3 Non-equilibrium of the Globally Optimal Policy $\alpha = 1$

We depict the best response to the globally optimal policy in Fig. 5, for $N = 1$ and $\mu = 1$ being held fixed, as a function of the loss parameter q. For the above parameters, the best response function is given in Fig. 2.

We observe that for all $q < 0.5$, the best response to other players sending their traffic through the direct path (i.e. $\alpha = 1$) is to play $b < 1$. We conclude that for these q, $\alpha = 1$ is not an equilibrium. In contrast, for all $q \geq 0.5$, the best response to $\alpha = 1$ is also $\alpha = 1$ and hence the global optimal policy is an equilibrium policy.

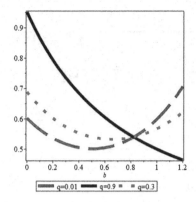

Fig. 4. The symmetric equillibrium as a function of the loss probability parameter q.

4.4 The Derivative of the Cost of a Player

In order to provide an expression for the equilibrium, we had to differentiate the cost of the player that uses b while others use a. Below is long expression obtained by Maple for the derivative w.r.t. b of the cost for the player, i.e. of the loss probability of packets of player (k, n) who plays b when all others play α.

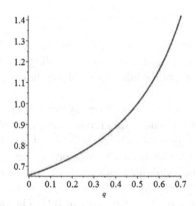

Fig. 5. Sending all the traffic through direct paths is not an equilibrium. The Figure shows the best response of a player when the others use $\alpha = 1$ as a function of the loss parameter q.

$$dJbaN := \frac{(a + (1-a)(1-q))\left(1 - \frac{1}{N}\right) + \frac{b + (1-a)(1-q)}{N}}{(a + (1-a)(1-q))\left(1 - \frac{1}{N}\right) + \frac{b + (1-a)(1-q)}{N} + \mu}$$

$$+ \frac{b}{N\left((a + (1-a)(1-q))\left(1 - \frac{1}{N}\right) + \frac{b + (1-a)(1-q)}{N} + \mu\right)}$$

$$- \frac{b\left((a + (1-a)(1-q))\left(1 - \frac{1}{N}\right) + \frac{b + (1-a)(1-q)}{N}\right)}{\left((a + (1-a)(1-q))\left(1 - \frac{1}{N}\right) + \frac{b + (1-a)(1-q)}{N} + \mu\right)^2 N} - q$$

$$- \frac{(1-q)\left((a + (1-a)(1-q))\left(1 - \frac{1}{N}\right) + \frac{a + (1-b)(1-q)}{N}\right)}{(a + (1-a)(1-q))\left(1 - \frac{1}{N}\right) + \frac{a + (1-b)(1-q)}{N} + \mu} + (1$$

$$- b)\left(\frac{(1-q)(-1+q)}{N\left((a + (1-a)(1-q))\left(1 - \frac{1}{N}\right) + \frac{a + (1-b)(1-q)}{N} + \mu\right)}\right.$$

$$\left. - \frac{(1-q)\left((a + (1-a)(1-q))\left(1 - \frac{1}{N}\right) + \frac{a + (1-b)(1-q)}{N}\right)(-1+q)}{\left((a + (1-a)(1-q))\left(1 - \frac{1}{N}\right) + \frac{a + (1-b)(1-q)}{N} + \mu\right)^2 N}\right)$$

Fig. 6. Expression for the derivative of the cost of the player (k, n) with respect to b.

5   Conclusions

We have seen that for any choice of system parameters, the equilibrium performance improves when increasing the loss probability parameter q on the relay links. The equilibrium loss probability of a player is thus decreasing in q. This is a Braess-type paradox.

Furthermore, for any parameters of the system, if the number N of players at each node is large enough, then the globally optimal policy is an equilibrium; this means that in the regime of a large number of players (called a Wardrop equilibrium), the above type of paradox does not occur.

The original Braess paradox [6] was shown to hold in a framework of a very large number of players (Wardrop equilibrium). Later on it was shown to occur also in the case of any number $N > 1$ players in [11]. The paradox we introduced, known as the Kameda paradox, does not occur in the case of a very large number of players. This was shown for standard delay-type cost functions in [10] for a triangular network topology.

References

1. Altman, E., Estanislao, A., Panda, M.: Routing games on a circle. In: NetGCOOP 2011: International Conference on NETwork Games, COntrol and OPtimization, Telecom SudParis et Université Paris Descartes, Paris, France (October 2011). hal-00644364

2. Altman, E., Kuri, J., El-Azouzi, R.: A routing game in networks with lossy links. In: 7th International Conference on NETwork Games COntrol and OPtimization (NETGCOOP 2014), Trento, Italy (October 2014). hal-01066453

3. Altman, E., El Azouzi, R., Abramov, V.: Non-cooperative routing in loss networks. Perform. Eval. **49**(1–4), 257–272 (2002). RR-4405, INRIA. <inria-00072183>

4. Altman, E., Liu, Z.: Improving the stability characteristics of asynchronous traffic in FDDI token ring. [Research Report] RR-1934, INRIA, pp. 21 (1993)

5. Boukoftane, A., Altman, E., Haddad, M., Oukid, N.: Paradoxes in a multi-criteria routing game. In: Duan, L., Sanjab, A., Li, H., Chen, X., Materassi, D., Elazouzi, R. (eds.) GameNets 2017. LNICST, vol. 212, pp. 165–172. Springer, Cham (2017). https://doi.org/10.1007/978-3-319-67540-4_15

6. Braess, D.: Uber ein Paradox aus der Verkehrsplanung. Unternehmensforschung **12**, 258–268 (1968)

7. Burra, R., Singh, C., Kuri, J., Altman, E.: Routing on a ring network. In: Song, J.B., Li, H., Coupechoux, M. (eds.) Game Theory for Networking Applications. EICC, pp. 25–36. Springer, Cham (2019). https://doi.org/10.1007/978-3-319-93058-9_3. hal02417

8. Chen, X., Doerr, B., Hu, X., Ma, W., van Stee, R., Winzen, C.: The price of anarchy for selfish ring routing is two. In: Goldberg, P.W. (ed.) WINE 2012. LNCS, vol. 7695, pp. 420–433. Springer, Heidelberg (2012). https://doi.org/10.1007/978-3-642-35311-6_31

9. Haddad, M., Altman, E., Gaillard, J.: Sequential routing game on the line: transmit or relay? In: 2012 International Conference on Communications and Information Technology (ICCIT), Hammamet, pp. 297–301 (2012)

10. Kameda, H., Altman, E., Kozawa, T., Hosokawa, Y.: Braess-like paradoxes in distributed computer systems. IEEE Trans. Autom. Control **45**, 1687–1691 (2000)

11. Korilis, Y.A., Lazar, A.A., Orda, A.: Avoiding the Braess paradox in non-cooperative networks. J. Appl. Probab. **36**, 211–222 (1999)

12. Orda, A., Rom, R., Shimkin, N.: Competitive routing in multiuser communication networks. IEEE/ACM Trans. Netw. **1**(5), 510–521 (1993)

13. Patriksson, M.: The Traffic Assignment Problem: Models and Methods. VSPBV, The Netherlands (1994)

Routing of Strategic Passengers in a Transportation Station

Dimitrios Logothetis$^{(\boxtimes)}$ ⓘ and Antonis Economou ⓘ

Department of Mathematics, National and Kapodistrian University of Athens,
Panepistemiopolis, Athens 15784, Greece
{dlogothetis,aeconom}@math.uoa.gr

Abstract. We study the routing decisions of passengers in a transportation station, where various types of facilities arrive with limited seating availability. The passengers' arrivals occur according to a Poisson process, the arriving instants of the transportation facilities form independent renewal processes and the seating availability at the successive visits of the facilities correspond to independent random variables, identically distributed for each facility. We analyze the strategic passenger behavior and derive the equilibrium strategies. We also discuss the associated social welfare optimization problem.

Keywords: Queueing · Strategic customers · Equilibrium strategies · Optimal policies · Social welfare · Transportation station · Parallel queues · Routing · Network

1 Introduction

1.1 Overview and Scope

One can often model the components of production, transportation, service, communication, or distributed computer systems as parallel queues, in which jobs (customers) are processed by various servers. The servers could be production stations, vehicles, office employees or public agents, or communication and computer systems. In many cases, multiple jobs can be processed simultaneously by one of many such service facilities.

In the present paper, we assume for concreteness the framework of a transportation station (hub). In such a station, multiple types of transportation facilities (e.g. buses, trains etc.) are scheduled to arrive and serve the incoming passengers who decide, upon arrival, which facility are going to wait. To this end, we consider a stylized stochastic model for the study of strategic customer behavior in such service systems which can be represented as a network of parallel clearing systems that operate independently. New customers arrive at the system according to a Poisson process and each transportation facility visits the station at the instants of a renewal process, independently of the arrival process and the other facilities' visit processes. Upon arriving at the station, the customers

P. Ballarini et al. (Eds.): EPEW 2021/ASMTA 2021, LNCS 13104, pp. 308–324, 2021.
https://doi.org/10.1007/978-3-030-91825-5_19

decide which of the available transportation facilities (if any), is preferable for them, according to the well-known Wardrop principle (see [30]). Based on their decisions, customers will then start to accumulate at the corresponding waiting room or platform. The model is assumed unobservable, i.e., the customers know the operational and economic parameters of the station, but cannot observe the number of customers who wait at the various platforms.

A key feature of this model is that each transportation facility has limited capacity at its visits at the station. Thus, there is a positive probability for any customer that he/she may not be accommodated by his/her chosen facility. In this case, such customers seek some alternative option elsewhere (we model this as an outside opportunity).

The paper focuses primarily on analyzing the strategic behavior of rational customers regarding their routing decisions and on deriving the customer equilibrium and socially optimal routing strategies. A theoretical comparison between these two cases shows that the equilibrium and socially optimal routing decisions do not coincide, and, in general, customers use the facilities more than what is socially desirable.

1.2 Literature Review

Formulating the behavior of individual customers as a game and identifying equilibrium strategies enables a more meaningful assessment of how various system parameters or control policies affect the effective arrival rate, which represents the demand for the service provided. It is for this reason that the papers on strategic analysis in queueing systems have been proliferated in the last decades. The study of queueing systems under a game theoretic perspective was initiated by Naor [22] who studied the strategic customer behavior and the associated social welfare and profit maximization problems for the observable M/M/1 queue regarding the join-or-balk dilemma. Naor showed that the strategic customer behavior can lead to inefficient system outcomes and the system administrator can rely on monetary transfers to coordinate the system. Edelson and Hildebrand [14] studied the unobservable counterpart of Naor's model. An extensive survey of the main methodological approaches and several results in this broader area can be found in the monographs by Hassin and Haviv [17], Stidham [29] and Hassin [16]. Finally, Patriksson [24] provides a unified account of models and methods for the problem of estimating equilibrium traffic flows in urban areas and a survey of the scope and limitations of present traffic models.

In the present study, we analyze the strategic routing decisions of customers in a network of parallel clearing systems. In each node, all present customers are removed instantaneously at the events of renewal processes which represent the visits of the transportation facilities. This model is related to the stream of research that studies the strategic customer behavior in clearing systems and also extends to the body of research related to strategic behavior in queueing networks.

The literature on clearing systems from a performance evaluation point of view is vast and goes back to Stidham [28]. However, the consideration of customer strategic behavior in such systems is a recent endeavor. Some key references in this thread of research are the papers by Economou and Manou [13], Manou, Economou and Karaesmen [21], Manou, Canbolat and Karaesmen [20], Czerny, Guo and Hassin [12] and Canbolat [10]. However, all these papers focus on the join-or-balk dilemma of the customers, and not on the routing decision which is the object of the present paper.

If one drops the assumption of the instantaneous removal of customers that cannot be accommodated, the appropriate model is a queueing system with batch services of finite size. The analysis of strategic customer behavior in such systems has been only recently initiated in the framework of the M/M/1 queue with batch services, see e.g., Bountali and Economou [6,7] and [8].

Another thread of research that is closely related to the present work started with the seminal paper of Bell and Stidham [5] who studied the routing decisions of customers in a system of several parallel M/G/1 queues, with the same coefficient of variation for the service times. In that paper, the authors determined the equilibrium and socially optimal strategies. Moreover, in the exponential case, they also provided closed-form expressions. They showed that when arriving customers are free to choose which server to join, they over-utilize the low-cost/high-speed servers causing higher total costs in comparison to the socially optimal policy. In general, it is known that the user equilibrium, where the customers try to maximize their expected utility, and the social optimum, do not coincide, both in the contexts of probabilistic routing, and state-dependent routing (see e.g., Whitt [31], and Cohen and Kelly [11]). Other important related works concerning the strategic customer behavior in queueing networks with parallel servers are the papers [1–4,9,15,23,26,27]. However, all these works study the strategic behavior regarding routing in parallel queues that process customers one by one, unlike our work which considers routing in parallel clearing systems.

1.3   Organization of the Paper

The rest of the paper is structured as follows: In Sect. 2, we introduce the model and calculate the key quantities that we will use in the sequel. Sections 3 and 4 include the main analysis for the computation of the equilibrium and socially optimal strategies, and their comparison. Finally, Sect. 5 contains a discussion of the results and points to directions for future study.

2   The Model

2.1   Mathematical Description and Notation

We consider a transportation station modeled as a network of n parallel clearing systems (platforms), operating independently. The system i corresponds to the platform of a transportation facility that visits the station according to a renewal process $\{M_i(t)\}$. We denote the corresponding generic inter-visit time by X_i, its distribution function by $F_{X_i}(t)$, its mean and second moment by $\bar{x}_i$

and $\bar{x}_{i(2)}$, respectively, and its density (in case of an absolute continuous r.v.) by $f_{X_i}(t)$. Each transportation facility has its own capacity which is revealed when it reaches the station. We assume that the successive capacities of the transportation facility of type i form a sequence of i.i.d. random variables. We denote a generic random variable representing the capacity at a visit of facility i by C_i and its probability mass function by $(g_i(j) : j \in \{1, 2, ..., m_i\})$.

Customers arrive at the transportation station according to a Poisson process with rate λ and are strategic in the sense that they decide which of the available transportation facilities they might use. After choosing a facility of type i, a customer moves directly to the corresponding facility's waiting room/platform which has infinite waiting space. He/she waits there for the transportation facility to arrive along with other customers who have made the same decision. The service reward is r_i and will be received only in the case where the next arriving facility has the capacity to accommodate him/her. Moreover, the waiting cost is c per time unit. The customer has also the option to balk from the station which corresponds to zero cost and a diminished service value which is denoted by v. This value can be set to zero for simplicity.

Each time that a transport facility of type i visits the station, having capacity C_i, it will accommodate at most C_i customers. Moreover, we assume that those customers who cannot be accommodated will not wait for the next facility, but they will abandon the station. This is a realistic assumption when the cost of waiting a whole inter-visit time is prohibitive. Thus, the number of customers served by a facility i equals the minimum of its capacity and the number of present customers at its arrival instant. After each visit of a facility, its platform is left empty.

Finally, we assume that the boarding discipline into the various transportation facilities is the First-Come-First-Served (FCFS). Later, we will see that the results remain valid for the Last-Come-First-Served (LCFS) and the Random-Order (RO) disciplines as well.

2.2 Information Structure - Strategies

We assume that arriving customers have complete knowledge of the operational and economic parameters of the model. However, they do not receive any information about the actual state of the station, i.e., the model is unobservable. This situation can be modeled as a symmetric game among the customers. In this case, a customer's strategy is a vector $\mathbf{q} = (q_0, q_1, ..., ..., q_n)$, where $q_i \in [0, 1], i \geq 1$ is the probability of choosing facility i and q_0 is the probability of choosing the balking/outside option. These probabilities sum to 1, i.e. $\sum_{i=0}^{n} q_i = 1$.

2.3 Customer's Utility Function

A key quantity for the study of strategic customer behavior in this model is the expected net benefit of a tagged customer who decides to wait for a given facility i, given that the population of customers follows a strategy $\mathbf{q}$. A moment of reflection reveals that this quantity depends only on q_i, so it will be denoted by $G_i(q_i)$. Let

- $P_i(q_i)$ be the conditional probability that the tagged customer will be served by the next facility of type i given that the customers who choose the i facility will join with probability q_i, and
- E_i be the expectation of his/her waiting time till the visit of the next facility of type i.

The expected utility of the tagged customer who chooses to wait for facility's i next visit, is

$$G_i(q_i) = r_i P_i(q_i) - cE_i. \tag{1}$$

Using simple probabilistic arguments, $G_i(q_i)$ can be found in a more explicit form.

Proposition 1 (Expected utility). *The expected utility of a tagged customer who chooses facility i given that all other customers choose facility i with probability q_i, is equal to*

$$G_i(q_i) = r_i \sum_{j=1}^{m_i} g_i(j) \int_0^\infty p_j(\lambda q_i a) \frac{1 - F_{X_i}(a)}{\bar{x}_i} da - c \frac{\bar{x}_{i(2)}}{2\bar{x}_i}, \quad i = 1, 2, \ldots, n, \tag{2}$$

where

$$p_j(s) = \sum_{i=0}^{j-1} e^{-s} \frac{s^i}{i!} = \frac{1}{(j-1)!} \int_s^\infty u^{j-1} e^{-u} du = \frac{\Gamma(j, s)}{(j-1)!}, \tag{3}$$

(with $\Gamma(j, s)$ being the upper incomplete gamma function). For the balking/outside option, $i = 0$, we set $G_0(q_0) = v = 0$. Also, we note that $p_j(0) = 1$

Proof. For the computation of the boarding probabilities, $P_i(q_i)$, we consider a given renewal cycle and condition on the capacity of facility i being $C_i = j$. We denote this quantity by $P_i(q_i, j)$. When the boarding discipline is the $FCFS$, a tagged customer who chooses this facility will find a seat only if the capacity at its next visit suffices to accommodate him/her. In other words, if the capacity is j, then, he/she will be served if and only if at most $j - 1$ other customers have arrived during the age of the corresponding renewal process. But, it is well-known that the limiting distribution of the age A_i of $\{M_i(t)\}$ is given by

$$F_{A_i}(a) = \frac{\int_0^a (1 - F_{X_i}(u)) du}{\bar{x}_i} \tag{4}$$

(see, e.g., Kulkarni [19] Theorem 8.20). Moreover, the arrival rate of the customers who choose facility i is λq_i. Thus, the boarding probability conditioned on $C_i = j$ is

$$P_i(q_i, j) = \int_0^\infty p_j(\lambda q_i a) \frac{1 - F_{X_i}(a)}{\bar{x}_i} da, \tag{5}$$

with $p_j(s)$ given by (3). Unconditioning with respect to C, yields

$$P_i(q_i) = \sum_{j=1}^{m_i} g_i(j) \int_0^\infty p_j(\lambda q_i a) \frac{1 - F_{X_i}(a)}{\bar{x}_i} da.$$

The conditional expectation E_i is the mean residual lifetime of the process $\{M_i(t)\}$ at an arbitrary customer's arrival instant. Because of PASTA (see [32]), we have that it coincides with the corresponding mean residual lifetime in continuous time. Therefore,

$$E_i = \frac{\bar{x}_{i(2)}}{2\bar{x}_i}. \tag{6}$$

Inserting these expressions for $P_i(q_i)$ and E_i in (1) yields (2).   □

Note that formula (2) for the expected utility is valid for any other boarding discipline that does not affect the number of customers that are served. For example, the LCFS and the RO disciplines have the same expected utility formula. We can think of this result also algebraically: Using the renewal-reward theorem we have that the long-run proportion of customers who are served by the facility i equals the expected number of customers who are served in an inter-visit time over the expected number of customers that arrive in an inter-visit time. The former can be expressed as

$$\frac{E[\min(N_i(X_i), C_i)]}{E[N_i(X_i)]},$$

where $N_i(X_i)$ is the number of Poisson arrivals with rate λq_i in an inter-visit time X_i and this quantity is independent of the boarding discipline being the FCFS, the LCFS or the RO.

Corollary 1 *(Monotonicity of G_i).* *The functions G_i for $i \in \{1, 2, \ldots, n\}$ are strictly decreasing in q_i.*

Proof. This is immediate since the functions $p_j(s)$ for $j \geq 1$, are strictly decreasing in s.   □

3   Equilibrium Strategies

In the present section, we study the customer equilibrium strategies. First, in Subsect. 3.1, we show how Wardrop's equilibrium conditions are specified within the framework of this model. In Subsect. 3.2 we provide necessary and sufficient conditions for the existence of pure equilibrium strategies. Moreover, we report a sufficient condition so that no customers abandon the system. We also study the set of facilities that are used by some customers in equilibrium. Subsection 3.3 contains the core results of the paper. We prove the existence and uniqueness of an equilibrium strategy for each instant of the model. The equilibrium strategy assigns zero or positive probability to the balking option according to the value of a critical quantity $\tilde{l}$. After introducing this quantity, we show how the equilibrium strategy is obtained in these two cases.

3.1   Characterization of an Equilibrium Strategy

An equilibrium strategy, $\mathbf{q}^e$, is defined as a strategy where no player has an incentive to deviate from it unilaterally. In other words, a vector $\mathbf{q}^e = (q_0^e, q_1^e, \ldots, q_n^e)$ constitutes an equilibrium strategy if and only if there exists a set $\emptyset \neq K^e \subset \{0, 1, \ldots, n\}$ such that the following two conditions are met:

ES1: (i) $q_i^e > 0$, for $i \in K^e$, (ii) all $G_i(q_i^e)$, $i \in K^e$ are equal, and (iii) $\sum_{i \in K^e} q_i^e = 1$.

ES2: (i) $q_k^e = 0$, for $k \notin K^e$, and (ii) $G_i(q_i^e) \geq G_k(0)$ for $i \in K^e$ and $k \notin K^e$.

3.2   Unused Facilities

Depending on the parameter values, several facilities may remain unused. We will see how such facilities can be determined. We set $a_i = G_i(1)$, $b_i = G_i(0) = r_i - c\frac{\bar{x}_i(2)}{2\bar{x}_i}$, for $i \geq 1$. We assume that the facilities have been numbered so that

$$b_1 \geq b_2 \geq \cdots \geq b_n. \tag{7}$$

Then, facility 1 is considered the most attractive under ideal conditions (i.e., when the boarding probability is 1, for the tagged customer) which implies that it will be the sole candidate for being preferable by any pure equilibrium strategy. Indeed, if a pure equilibrium strategy dictates that the customers route to a station $i \neq 1$ only, the by ES2 we should have $a_i = G_i(1) \geq G_1(0) = b_1$ But $b_1 \geq b_i > a_i$, a contradiction.

The functions G_i are continuous and strictly decreasing and as a result, their image will be the sets $G_i([0,1]) = [a_i, b_i]$, $i \in \{1, 2, ..., n\}$. We also set

$$i^* = arg \max_{i=1,\ldots,n} a_i \quad \text{and} \quad i^{ef} = \max\{i \in \{1, ..., n\} : b_i \geq a_{i^*}\}. \tag{8}$$

Using the above quantities we can easily determine necessary and sufficient conditions for the existence of pure equilibrium strategies. Specifically we have the following proposition:

Proposition 2 (Pure equilibrium strategies). *The equilibrium strategy is*

(i) $\mathbf{q}^e = (1, 0, 0, 0, ..., 0)$ *if and only if* $b_1 \leq v = 0$.
(ii) $\mathbf{q}^e = (0, 1, 0, 0, ..., 0)$ *if and only if* $i^{ef} = 1$ *and* $a_1 > 0$.

Proof. (i) If $q_0^e = 1$, then $K^e = \{0\}$ and we immediately get from ES2(ii) that $G_i(0) \leq G_0(0) = v = 0$ for every $i \in \{1, 2, \ldots, n\}$ which implies that $b_1 \leq v = 0$. On the other hand, if $b_1 \leq v = 0$, then ES1 and ES2 are satisfied with $K^e = \{0\}$.

(ii) If $i^{ef} = 1$, then we necessarily have that $i^* = 1$. Indeed, if $i^* \geq 2$, then by using the definition of i^{ef} and the ordering of b_i, we will get that $b_1 > b_{i^*} > a_{i^*} > b_{i^{ef}} = b_1$, which is a contradiction. Therefore, $G_1(1) = a_1 > b_i > G_i(q)$ for every $q \in (0, 1]$ and $i \geq 2$. Also, $a_1 > 0$ implies that $G_1(1) > 0 = v =$

$G_0(0)$ and therefore the strategy $\mathbf{q}^e = (0,1,0,0,...,0)$ is equilibrium. For the converse, let $\mathbf{q}^e = (0,1,0,0,...,0)$. Then, $q_1^e = 1$ and $q_i^e = 0$ for $i \neq 1$ and also $G_1(1) > G_i(q)$ for every $i \neq 1$ and $q \in [0,1]$. We conclude that $a_1 = G_1(1) > G_0(0) = v = 0$ and $a_1 > G_i(0) = b_i \Rightarrow a_1 > b_i \Rightarrow i^{ef} = 1$.

$\square$

In view of Proposition 2, we will henceforth assume that

$$b_1 > v = 0.$$

This implies that $q_0^e < 1$, and thus, we exclude the situation where all passengers balk. In the following proposition we establish a simple sufficient condition so that no passenger balks.

Proposition 3 (No-balking condition). *Let q^e be the equilibrium strategy. If $a_{i^*} \geq v = 0$ then $q_0^e = 0$.*

Proof. Suppose that $q_0^e > 0$, i.e., $0 \in K^e$. In case that $i^* \in K^e$, from ES1 we have that $q_{i^*}^e > 0, G_{i^*}(q_{i^*}^e) = G_0(q_0^e)$ and also that $q_{i^*}^e + q_0^e \leq 1$ which implies that $q_{i^*}^e < 1$. However, $a_{i^*} \geq v = 0$ shows that $G_{i^*}(q_{i^*}^e) > G_{i^*}(1) = a_{i^*} \geq 0 = v = G_0(q_0^e)$. In case where $i^* \notin K^e$, ES2 implies that $v = G_0(q_0^e) \geq G_{i^*}(0) = b_{i^*} > a_{i^*}$ which contradicts the hypothesis. Therefore $q_0^e = 0$.

$\square$

Proposition 4 (Unused facilities). *Let $K^{ef} = \{0,1,...,i^{ef}\}$. Then, we have that $K^e \subset K^{ef}$. In particular, if the vector $\mathbf{q}^e$ is the equilibrium, then $\mathbf{q}^e = (q_0^e, q_1^e, ..., q_{i^{ef}}^e, 0, ..., 0)$, with $q_i^e \in [0,1)$ for every $i \in K^{ef}$. Moreover, if for some $\tilde{i} \in \{2,...,n\}$ we have that $q_{\tilde{i}}^e = 0$, then $q_i^e = 0$ for $i = \tilde{i}+1, ..., n$.*

Proof. Let $\mathbf{q}^e$ be the equilibrium strategy. We assume that $q_{\tilde{i}}^e = 0$ for some $\tilde{i} \in \{2,...,n\}$. The functions G_i are strictly decreasing and thus, by (7), we have that $b_{\tilde{i}} \geq b_i > a_i$ for every $i = \tilde{i}+1,...,n$. In this case

$$G_{\tilde{i}}(0) = b_{\tilde{i}} \geq b_i > G_i(q), \text{ for every } q \in (0,1] \text{ and for every } \quad i = \tilde{i}+1,...,n.$$

It is therefore clear from ES1 and ES2 that $q_i^e = 0$ for every $i = \tilde{i}+1,...,n$. It is only left to show that $q_{i^{ef}+1} = 0$. To this end, we start by showing that for the quantities i^* and i^{ef}, the following properties hold:

1. $i^{ef} \geq i^*$ and
2. $G_{i^*}(q) \neq G_{i^{ef}+1}(q)$ for every $q \in [0,1]$.

To show 1, let $i^{ef} < i^*$ or $i^* \geq i^{ef} + 1$. In this case $b_{i^*} \leq b_{i^{ef}+1}$ due to (7). On the other hand, from the definition of i^{ef} we have that $b_{i^{ef}+1} < a_{i^*}$ and thus, combining the above we conclude that $b_{i^*} < a_{i^*}$, a contradiction due the monotonicity of G.

To show 2, let $q \in [0,1]$. From the definition of i^{ef} we have that $b_{i^{ef}+1} < a_{i^*}$ and thus

$$G_{i^*}(q) \geq G_{i^*}(1) = a_{i^*} > b_{i^{ef}+1} = G_{i^{ef}+1}(0) \geq G_{i^{ef}+1}(q).$$

Therefore, if $i^* \in K^e$, then using 2, we note that condition ES1(ii) does not hold which implies that $i^{ef} + 1 \notin K^e$. Thus $q_{i^{ef}+1} = 0$. If $i^* \notin K^e$, then $q_{i^*} = 0$ and due to the fact that $i^* \leq i^{ef} < i^{ef} + 1$ we will have again that $q_{i^{ef}+1} = 0$.   □

The facilities that are included in the set $K^{ef} \setminus K^e$ will be those left unused, i.e. $q_i^e = 0$. Whenever $q_i^e > 0$, the corresponding facilities will be used.

3.3   Derivation of the Equilibrium Strategy

For the derivation of the equilibrium strategy, we need first to introduce some key quantities that are used for its computation.

By the strict monotonicity of $G_i(x)$, $i = 1, 2, \ldots, n$, it is easy to see that each equation $G_i(x) = b_j$, for $1 \leq i \leq j \leq i^{ef}$, has a unique solution which we denote by $q_{b_{ij}}$. In the sequel, we define the quantities

$$Q(k) = \sum_{i=1}^{k} q_{b_{ik}}, \quad k = 1, \ldots, i^{ef}$$

and we set

$$i^e = \max\{k \in \{1, \ldots, i^{ef}\} : Q(k) < 1\}. \tag{9}$$

We note that the set $\{k \in \{1, \ldots, i^{ef}\} : Q(k) < 1\}$ is non-empty due to $i^{ef} \geq 1$ and $Q(1) = q_{b_{11}} = 0 < 1$, hence i^e is well-defined.

An important function for the computation of the equilibrium strategy is the function

$$H(x) = \sum_{i=1}^{i^e} G_i^{-1}(x) - 1. \tag{10}$$

The following lemma is the basic step towards the derivation of the equilibrium strategy.

Lemma 1. *Let i^{ef} given by (8). Define $\tilde{l}$ as*

$$\tilde{l} = \begin{cases} H^{-1}(0) \in [b_{i^e+1}, b_{i^e}) & \text{if } i^e < i^{ef} \\ H^{-1}(0) \in [a_{i^*}, b_{i^e}) & \text{if } i^e = i^{ef} \end{cases} \tag{11}$$

Then, there exists a vector $(\tilde{q}_1, \tilde{q}_2, \ldots, \tilde{q}_{i^e})$ which satisfies the following conditions:

(i) $\tilde{q}_i \in (0,1]$, $i = 1, \ldots, i^e$
(ii) $\sum_{i=1}^{i^e} \tilde{q}_i = 1$
(iii) $G_i(\tilde{q}_i) = \tilde{l}$, for $i = 1, \ldots, i^e$.
(iv) $G_i(\tilde{q}_i) = \tilde{l} \geq G_j(0)$, for $i = 1, \ldots, i^e$ and for any $j = i^e + 1, \ldots, n$.

Proof. We distinguish two cases according to whether $i^e < i^{ef}$ or $i^e = i^{ef}$.

Case 1 ($i^e < i^{ef}$). We have that $Q(i^e) < 1$ and $Q(i^e + 1) \geq 1$. We consider the function $H(x)$ defined by (10) on $[b_{i^e+1}, b_{i^e}]$. We note that $b_{i^e+1} \neq b_{i^e}$. Indeed, if it were not the case we would have that $G_i(q_{b_{ii^e}}) = b_{i^e} = b_{i^e+1} = G_i(q_{b_{ii^e+1}}) \Rightarrow q_{b_{ii^e}} = q_{b_{ii^e+1}}$ for every $i \leq i^e$, implying that $1 > Q(i^e) = Q(i^e+1)$ which contradicts the definition of i^e. Thus, $H(x)$ is well defined continuous and decreasing as sum of continuous and decreasing functions. Moreover,

$$H(b_{i^e+1}) = \sum_{i=1}^{i^e} G_i^{-1}(b_{i^e+1}) - 1 = \sum_{i=1}^{i^e} q_{b_{ii^e+1}} - 1 = Q(i^e + 1) - 1 \geq 0 \quad \text{and}$$

$$H(b_{i^e}) = Q(i^e) - 1 < 0.$$

Therefore, $H(x)$ has a unique root in $[b_{i^e+1}, b_{i^e})$ which we denote by $\tilde{l}$. Setting $\tilde{q}_i = G_i^{-1}(\tilde{l})$ for $i = 1, ..., i^e$, which are uniquely defined as $G_i^{-1}(x)$ are also decreasing functions, we have that condition (i) of the Lemma is valid, i.e., $\tilde{q}_i \in (0, 1]$, $i = 1, ..., i^e$. Indeed,

$$\tilde{l} \in [b_{i^e+1}, b_{i^e}) \Rightarrow \tilde{q}_i = G_i^{-1}(\tilde{l}) \in G_i^{-1}([b_{i^e+1}, b_{i^e})) = (q_{b_{ii^e}}, q_{b_{ii^e+1}}],$$

and $q_{b_{ii^e}} \geq 0, q_{b_{ii^e+1}} \leq 1$ are the solutions in $[0, 1]$ of the equations $G_i(x) = b_{i^e}$ and $G_i(x) = b_{i^e+1}$ respectively.

Noting that $H(\tilde{l}) = 0$ implies immediately the validity of conditions (ii) and (iii) of the Lemma.

For proving (iv), we have only to show that $G_i(\tilde{q}_i) \geq G_{i^e+1}(0) = b_{i^e+1}$. But this is clear since $G_i(\tilde{q}_i) = \tilde{l} \geq b_{i^e+1}$ for every $i = 1, ..., i^e$.

Case 2 $i^e = i^{ef}$. We now have that $Q(i^e) = Q(i^{ef}) < 1$. We consider the function $H(x)$ given by (10) on $[a_{i^*}, b_{i^{ef}}]$. We have that $a_{i^*} \neq b_{i^{ef}}$. If this were not the case, then $Q(i^e) \geq q_{b_{i^* i^{ef}}} = 1$, because $q_{b_{i^* i^{ef}}}$ is the solution of the equation $G_{i^*}(x) = b_{i^{ef}} = a_{i^*}$ and thus $x = 1$. We set a_{ii^*} the unique solution of the equation $G_i(x) = a_{i^*}$. We get that

$$H(a_{i^*}) = \sum_{i=1}^{i^{ef}} G_i^{-1}(a_{i^*}) - 1 = \sum_{i=1}^{i^{ef}} q_{a_{ii^*}} - 1 = q_{a_{1i^*}} + ... + q_{a_{i^* i^*}} + ... + q_{a_{i^{ef} i^*}} - 1$$

$$= q_{a_{1i^*}} + ... + q_{a_{i^{ef} i^*}} \geq 0 \quad \text{due to} \quad q_{a_{i^* i^*}} = 1 \quad \text{and} \quad i^{ef} \geq i^*$$

$$H(b_{i^{ef}}) = \sum_{i=1}^{i^{ef}} G_i^{-1}(b_{i^{ef}}) - 1 = \sum_{i=1}^{i^{ef}} q_{b_{ii^{ef}}} - 1 = Q(i^{ef}) - 1 < 0.$$

Similarly to case 1, the function $H(x)$ has a unique root $\tilde{l} \in [a_{i^*}, b_{i^{ef}})$. We set $\tilde{q}_i = G_i^{-1}(\tilde{l})$ for $i = 1, ..., i^{ef}$. Then, we can check that conditions (i)–(iv) of the Lemma are valid, along the same lines with case 1. $\qquad\square$

Lemma 1 gives essentially the equilibrium strategy that satisfies conditions ES1 and ES2, if $\tilde{l} \geq 0$. Otherwise, a correction is needed, since some customers will balk in equilibrium. In this case, it turns out that the position of $v = 0$ in the ordering of the b_is is related to the number of facilities that remain unused. More specifically, consider the partition of $(\tilde{l}, b_1)$ as

$$(\tilde{l}, b_1) = (\tilde{l}, b_{i^e}) \cup [b_{i^e}, b_{i^e-1}) \cup \ldots \cup [b_2, b_1) = A_{i^e} \cup A_{i^e-1} \cup \ldots \cup A_1. \quad (12)$$

We define k^e to be the index i of the set A_i from the above partition that v belongs to. In other words,

$$k^e = \max\{i : b_i > v\}. \quad (13)$$

The following theorem presents the two cases ($\tilde{l} \geq 0$ and $\tilde{l} < 0$) for the equilibrium strategy of the customers. It constitutes the main result regarding the existence and uniqueness of the equilibrium strategy.

Theorem 1 *(Existence and uniqueness of the equilibrium strategy).*
A unique equilibrium strategy exists. We have the following cases:

1. *If $\tilde{l} \geq 0 = v$, then the unique equilibrium strategy is the vector $\mathbf{q}^e = (0, \tilde{q}_1, \ldots, \tilde{q}_{i^e}, 0, \ldots, 0)$, with $\tilde{q}_i$ given in Lemma 1.*
2. *If $\tilde{l} < 0 = v$ then the strategy $\mathbf{q}^e = (q_0^e, q_1^e, \ldots, q_{k^e}^e, 0, \ldots, 0)$, with $q_i^e = G_i^{-1}(0)$, $i = 1, 2, \ldots, k^e$ and $q_0^e = 1 - (q_1^e + \ldots + q_{k^e}^e)$, is the unique equilibrium strategy.*

Proof. Case 1 ($\tilde{l} \geq 0$): Let $i^{ef} \in \{1, \ldots, n\}$ and $i^e \in \{1, \ldots, i^{ef}\}$ be given by (8) and (9) respectively. Also, let $\tilde{q}_1, \tilde{q}_2, \ldots, \tilde{q}_{i^e}$ and $\tilde{l}$ be as in Lemma 1. If $\tilde{l} \geq 0 = v$, then we have that $G_0(0) = v \leq G_i(\tilde{q}_i)$ for every $i \in \{1, \ldots, i^e\}$. Combining this with the result of Lemma 1, we can clearly see that $K^e = \{1, 2, \ldots, i^e\}$ and that $q^e = (\tilde{q}_1, \ldots, \tilde{q}_{i^e}, 0, \ldots, 0)$ is the equilibrium strategy. The monotonicity of the functions $G_i^{-1}(x)$ implies the uniqueness of the equilibrium strategy.

Case 2 ($\tilde{l} < 0$): We now consider the case where $\tilde{l} < 0 = v$. Recall that $b_1 > 0 = v$. Thus, because of $\tilde{l} \in [b_{i^e+1}, b_{i^e})$, the value $v = 0$ will be in the partition of $(\tilde{l}, b_1)$ in the intervals $(\tilde{l}, b_{i^e}), [b_{i^e}, b_{i^e-1}), \ldots, [b_2, b_1)$ (noting that some of them may be empty due to equality of some b_i). By the definition of k^e, we have that $b_1 \geq b_2 \geq, \ldots, \geq b_{k^e} > v$. Moreover, $v > a_{i^*}$, so it is clear that $v \in \cap_{i=1}^{k^e}(a_i, b_i)$. By the definition of q_i^e for $i = 1, \ldots, k^e$, we have $0 = v = G_1(q_1^e) = \ldots = G_{k^e}(q_{k^e}^e)$. We will show that $K^e = \{0, 1, 2, \ldots, k^e\}$, that is, the vector $\mathbf{q}^e = (q_0^e, q_1^e, \ldots, q_{k^e}^e, 0, \ldots, 0)$ satisfies the conditions ES1 and ES2.

Indeed, for $i = 1, \ldots, k^e$, we have that $b_i > v > \tilde{l} \Rightarrow G_i(0) > G_i(q_i^e) > G_i(\tilde{q}_i)$. Therefore, $0 < q_i^e < \tilde{q}_i$ and $\sum_{i=1}^{k^e} q_i^e < \sum_{i=1}^{i^e} \tilde{q}_i = 1 \Rightarrow q_0^e \in (0, 1)$. The $G_i(q_i^e)$ are all equal by definition and $\sum_{i \in K^e} q_i^e = 1$. Finally, for any $j \in \{0, 1, 2, \ldots, n\} \setminus K^e$ we have that $q_j^e = 0$ and $G_i(q_i^e) \geq G_j(0)$ for every $i \in K^e$. $\qquad\square$

4   The Social Optimization Problem

We now study the problem of a central planner who imposes the routing probabilities with the objective of maximizing the social welfare, i.e., we seek for a probability vector $\mathbf{q}^{soc} = (q_0^{soc}, q_1^{soc}, ..., q_n^{soc})$, the (socially) optimal policy, that maximizes

$$S(\mathbf{q}) = \lambda \sum_{i=1}^{n} q_i G_i(q_i). \tag{14}$$

The function $S(q)$ is differentiable with partial derivatives

$$\frac{\partial S(q_i)}{\partial q_i} = \lambda \left(G_i(q_i) + q_i \frac{\partial G_i(q_i)}{\partial q_i} \right) = \lambda F_i(q_i).$$

Applying the Karush–Kuhn–Tucker (KKT) conditions (see for example [25]), we conclude that a vector $\mathbf{q}^{soc} = (q_0^{soc}, q_1^{soc}, \ldots, q_n^{soc})$ constitutes a KKT point (and therefore a candidate for optimal strategy) if and only if there exists a set $\emptyset \neq K^{soc} \subset \{0, 1, \ldots, n\}$ such that the following two conditions are met:

OS1: (i) $q_i^{soc} > 0$, for $i \in K^{soc}$, (ii) all $F_i(q_i^{soc})$, $i \in K^{soc}$ are equal, and (iii) $\sum_{i \in K^{soc}} q_i^{soc} = 1$.

OS2: (i) $q_k^{soc} = 0$, for $k \notin K^{soc}$, and (ii) $F_i(q_i^{soc}) \geq F_k(0)$ for $i \in K^{soc}$ and $k \notin K^{soc}$.

Therefore, the maximization problem is completely analogous to the equilibrium problem of the previous section. In particular, the same analysis can be carried out with the only difference being that, in this case we have the functions $F_i(x)$ in place of $G_i(x)$ and all related quantities (a_i, b_i, etc.) should be computed accordingly. However, there is a crucial difference: The functions $F_i(x)$ may not be monotone, in which case the overall analysis of the previous section breaks down. In particular, many KKT points may exist and in addition the structure that has been proved in the previous section will no longer be valid.

If all functions $F_i(x)$ are strictly decreasing, then the analysis can proceed exactly in the same lines with the equilibrium analysis. Therefore, we will have a counterpart of Theorem 1. A sufficient condition which ensures the monotonicity of the functions $F_i(x)$ in $[0, 1]$, is to assume concavity of the functions $P_i(x)$. In such cases, we can derive the unique social policy by following the same steps as we did for the equilibrium strategy. In the sequel, we will denote the corresponding quantities with the superscript 'soc' instead of 'e' that will refer to the equilibrium counterparts. We present a numerical example in Fig. 1 where both the equilibrium and the social policy are depicted. The idea of the algorithm for computing the equilibrium is clear in this figure: When there is no balking, plot all expected net benefit functions in the same grapgh. Then, draw a horizontal line at a given level. The intersection points give the corresponding candidate joining probabilities. When they sum to 1, these are the equilibrium joining probabilities. The appropriate level of the horizontal line is found by bisection procedure due to the monotonicity of the functions $G_j(q_j)$.

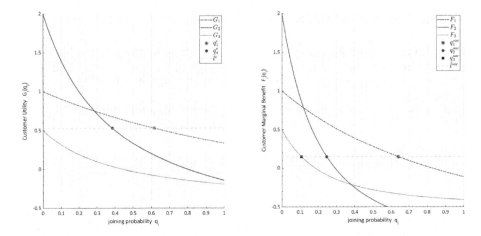

Fig. 1. Equilibrium strategy and social policy in the case of three substations with Exponential (μ_i) transportation visits and Geometric (ξ_i) available capacity for facility i. In this example, the economic parameters are set to $\mu = (2, 2, 4), \xi = (0.1, 0.5, 0.9), r = (2, 3, 1), c = 2, \lambda = 10, v = 0$.

We close this section by comparing the equilibrium strategy $\mathbf{q}^e$, and the optimal social policy $\mathbf{q}^{soc}$, when the functions $F_i(x)$ are decreasing. Of course, the key quantity for the comparison of the two strategies is the difference between $G_i(x)$ and $F_i(x)$. The following proposition states several results regarding the comparison of the two strategies.

Proposition 5 (Comparison of equilibrium and optimal strategies).
Suppose that the functions $F_i(x)$ are all strictly decreasing. Then,

(i) $\tilde{l}^e \geq \tilde{l}^{soc}$.
(ii) If $\tilde{l}^e \geq v$ then $q_0^e = 0$. If $\tilde{l}^{soc} \geq v$, then $q_0^e = q_0^{soc} = 0$.
(iii) If $v \geq \tilde{l}^e$, then $k^e = k^{soc}$ and $q_i^e > q_i^{soc}$ $i = 1, \ldots, k^e$.
(iv) $q_0^{soc} \geq q_0^e$.
(v) If $\tilde{l}^e > v$ then $i^{soc} \geq i^e$.

Proof. The functions F_i, F_i^{-1} are decreasing and $G_i(q) \geq F_i(q)$ for every $q \in [0,1]$, hence $G_i^{-1}(q) \geq F_i^{-1}(q)$ for every $q \in [0,1]$. Thus, $H^{soc}(x) \leq H^e(x)$. The $\tilde{l}^e$ and $\tilde{l}^{soc}$ being the unique solutions of the equations $H^e(x) = 0$ and $H^{soc}(x) = 0$ respectively, implies that $\tilde{l}^e \geq \tilde{l}^{soc}$. Then, using the relations $G_i(q) \geq F_i(q)$, $H^{soc}(x) \leq H^e(x)$ and $\tilde{l}^e \geq \tilde{l}^{soc}$, we can easily see the validity of (i)–(iv) by the construction of the probabilities q_i^e and q_i^{soc} in Theorem 1 - Case 1 and its social optimization counterpart.

To obtain (v), we first let $q_{b_{ij}}^G$ and $q_{b_{ij}}^F$ be the solutions of the equations $G_i(q) = b_j$ and $F_i(q) = b_j$ respectively. We also set $Q^G(k) = \sum_{i=1}^{k} q_{b_{ik}}^G$ and $Q^F(k) = \sum_{i=1}^{k} q_{b_{ik}}^G$. In light of $G_i(q) \geq F_i(q)$ for every $q \in [0,1]$, it holds that $q_{b_{ij}}^G \geq q_{b_{ij}}^F$ for every j. Therefore $Q^G(k) \geq Q^F(k)$ for every k and we have that $i^{soc} \geq i^e$.   □

In general, the above findings show that the equilibrium strategies do not coincide with the socially optimal policies. Moreover, according to (i), the individual welfare is always greater than the marginal profit under optimal social policy. Also, when the latter is greater than the value of the outside option, then (ii) shows that the station is fully utilized by the customers under any scenario. According to (iv), the customers are using the station instead of balking more than what is socially desirable.

Also, according to (iii), if the value of outside option is greater than or equal to the individual welfare, the number of active facilities under the two scenarios coincides. However, for each facility, the effective arrival rate under equilibrium strategy is strictly less than under social policy. In contrast, (v) shows that under the optimal policy, the number of active facilities would be at least equal to equilibrium in the case where the value of the outside option is strictly less than the individual welfare. Thus, regarding the number of active facilities under the two scenarios, in general we have that $K^e \subset K^{soc}$.

Finally, we note that if $k^{soc} > k^e$ then we have not necessarily that $q_i^e > q_i^s$ for $i \in K^e \cap K^{soc}$. For example, if we consider a station with 4 facilities, with $X_i \sim Exp(\mu_i)$ and $C_i \sim Geo(\xi_i)$ and with $r = (2.1, 2.1, 2.1, 2.1), \mu = (2, 3, 4, 10), \xi = (0.5, 0.5, 0.5, 0.5), v = 0$, we have that $K^e = \{3, 4\}$ and $K^s = \{2, 3, 4\}$ with $q_3^e = 0.0564 < 0.1047 = q_3^{soc}$ and with $q_4^e = 0.9498 > 0.8496 = q_4^{soc}$.

5   Discussion - Extensions

In this paper, we have studied the customer strategic behavior regarding routing in a transportation station, where different types of transportation facilities visit the station according to independent renewal processes. The primary focus was on determining the equilibrium strategy of the passengers and the optimal policy from the social planner's point of view. A main message of these results is that the passengers tend to overuse the station but use fewer facilities from what is social desirable. A common remedy for this situation is to impose admission fees for the available transportation facilities. Then, the customers will adopt the optimal policy as their joining strategy.

Of course, the study needs to be complemented by further considerations. For example, in real scenarios there are many passengers that are 'loyal' to certain facilities. These customers can be thought of as non-strategic users who do not change their preferences. A way to incorporate such cases into our model is to consider several additional independent Poisson arrival processes for the various facilities. The fixed vector $\lambda_R = (\lambda_1, ..., \lambda_n)$ may correspond to the arrival rates of the loyal users while strategic passengers may arrive according to another Poisson process with rate λ and can join any transportation facility. This situation is more involved and is currently under investigation. The solution of the associated profit-maximization problem for a monopoly is also of interest.

Some of the assumptions of the model seem substantially restrictive. However, as in all models, one makes simplifications in the seeking of the right balance between accuracy and tractability. As this work is our first endeavor towards a thorough study of equilibrium customer routing behavior in a transportation station from a queueing game perspective, we assumed the simplest possible framework. We now discuss our main simplifying assumptions and suggest possible generalizations and extensions:

- Clearing system.
  We have assumed that those customers who cannot be accommodated will not wait for the next facility, but they will abandon the station. This assumption is approximately valid in many transportation settings. Of course, there are pathological situations that do not occur in practice that violate this assumption, for example when the expected residual waiting time at an arbitrary epoch is larger than the expected waiting time at a renewal epoch (due to high coefficient of variation of the inter-visit times). In particular, the assumption is realistic in the case of quite large expected inter-visit times with small variances. For example, this is the case of remote stations that have only a few well-scheduled facility visits per day. If one removes this assumption, then the analysis becomes substantially more difficult, because the probability of a customer being served by the next facility of type i does not only depend on the number of customers that have been arrived during the ongoing inter-visit time, but also on the number of customers that arrived in previous inter-visit times and were not served. Moreover, to model the strategic behavior in this case, one should enhance the space of strategies to account for the possibility of reneging at the instants of the various facility visits.
- Unobservable system.
  The unobservability of the system is a strong assumption, but it is quite popular as a first level of analysis in the queueing game literature. This is the reason for adopting it in the present study. However, the consideration of more informative versions of the same problem seems a very challenging problem. Modern transportation stations can provide to the customers information about the arrival times of the facilities to come and on the congestion of the facilities. Such information influence customers' estimates for the probability of being served by the next facility and/or their expected waiting time in the station. If one drops the non-reneging feature of the model, it seems reasonable to assume that either users wait until they are served (possibly waiting for several facility arrivals), or they make a second decision (renege or stay), upon arrival to the platform of their desired facility and observing the number of waiting customers.
- Homogeneity of the customers.
  An aspect that has not been taken into account in the present work is passengers' heterogeneity. There are those passengers who arrive always the very last moment paying attention on not waiting too much and the others who want to be almost certain that it will be a seat for them. Moreover, there are customers that prefer to use a certain type of facility in comparison with others. Therefore, the consideration of a model with heterogeneous customers regarding r_i and c seems also interesting. Towards a tractable study, one can think that there are finite classes of customers, each one with its own vector

of r_is corresponding to customers that have different preferences for the various alternative facilities. And a customer's waiting cost per time unit can be sampled from a continuous distribution, as it is standard in the corresponding literature.

- Complete knowledge of operational and economic parameters of the model. In this study, we have assumed that the customers know exactly the operational and economic parameters of the model and can assess accurately the utility of the other customers. This is a standard assumption in the literature of queueing games, which is justified by the fact that a large population of customers reuse the service systems indefinitely and so acquire accurate estimates of its various parameters. However, some recent studies introduce learning processes in this area and their ideas can be used to relax this assumption in the present framework.

Acknowledgement. D. Logothetis was supported by the Hellenic Foundation for Research and Innovation (HFRI) under the HFRI PhD Fellowship grant (Fellowship Number: 1158.). We cordially thank the three anonymous reviewers for their constructive remarks on the initial revision of the paper.

References

1. Afimeimounga, H., Solomon, W., Ziedins, I.: The Downs-Thomson paradox: existence, uniqueness and stability of user equilibria. Queueing Syst. **49**(3–4), 321–334 (2005)
2. Afimeimounga, H., Solomon, W., Ziedins, I.: User equilibria for a parallel queueing system with state dependent routing. Queueing Syst. **66**(2), 169–193 (2010)
3. Altman, E., Ayesta, U., Prabhu, B.J.: Load balancing in processor sharing systems. Telecommun. Syst. **47**(1–2), 35–48 (2011)
4. Ayesta, U., Brun, O., Prabhu, B.J.: Price of anarchy in non-cooperative load balancing games. Perform. Eval. **68**(12), 1312–1332 (2011)
5. Bell, C.E., Stidham, S., Jr.: Individual versus social optimization in the allocation of customers to alternative servers. Manag. Sci. **29**, 831–839 (1983)
6. Bountali, O., Economou, A.: Equilibrium joining strategies in batch service queueing systems. Eur. J. Oper. Res. **260**, 1142–1151 (2017)
7. Bountali, O., Economou, A.: Equilibrium threshold joining strategies in partially observable batch service queueing systems. Ann. Oper. Res. **277**, 231–253 (2019)
8. Bountali, O., Economou, A.: Strategic customer behavior in a two-stage batch processing system. Queueing Syst. **93**, 3–29 (2019)
9. Calvert, B.: The Downs-Thomson effect in a Markov process. Probab. Eng. Inf. Sci. **11**(3), 327–340 (1997)
10. Canbolat, P.G.: Bounded rationality in clearing service systems. Eur. J. Oper. Res. **282**, 614–626 (2020)
11. Cohen, J.E., Kelly, F.P.: A paradox of congestion in a queuing network. J. Appl. Probab. **27**(3), 730–734 (1990)
12. Czerny, A.I., Guo, P., Hassin, R.: Hide or advertise: the carrier's choice of waiting time information strategies (2018). Preprint available at https://papers.ssrn.com/sol3/papers.cfm?abstract_id=3282276

13. Economou, A., Manou, A.: Equilibrium balking strategies for a clearing queueing system in alternating environment. Ann. Oper. Res. **208**, 489–514 (2013)
14. Edelson, N.M., Hildebrand, K.: Congestion tolls for Poisson queueing processes. Econometrica **43**, 81–92 (1975)
15. Grossman, T.A., Jr., Brandeau, M.L.: Optimal pricing for service facilities with self-optimizing customers. Eur. J. Oper. Res. **141**(1), 39–57 (2002)
16. Hassin, R.: Rational Queueing. CRC Press, Taylor and Francis Group, Boca Raton (2016)
17. Hassin, R., Haviv, M.: To Queue Or Not to Queue: Equilibrium Behavior in Queueing Systems. Kluwer Academic Publishers, Boston (2003)
18. Logothetis, D., Economou, A.: The Impact of Information on Transportation Systems with Strategic Customers (2020). Preprint
19. Kulkarni, V.: Modeling and Analysis of Stochastic Systems, 2nd edn. CRC Press, Boca Raton (2010)
20. Manou, A., Canbolat, P.G., Karaesmen, F.: Pricing in a transportation station with strategic customers. Prod. Oper. Manag. **26**, 1632–1645 (2017)
21. Manou, A., Economou, A., Karaesmen, F.: Strategic customers in a transportation station: when is it optimal to wait? Oper. Res. **62**, 910–925 (2014)
22. Naor, P.: The regulation of queue size by levying tolls. Econometrica **37**, 15–24 (1969)
23. Orda, A., Rom, R., Shimkin, N.: Competitive routing in multiuser communication networks. IEEE/ACM Trans. Netw. **1**(5), 510–521 (1993)
24. Patriksson, M.: The Traffic Assignment Problem: Models and Methods. Courier Dover Publications, New York (2015)
25. Peressini, A.L., Sullivan, F.E., Uhl, J.J.: The Mathematics of Nonlinear Programming, pp. 10–13. Springer-Verlag, New York (1988)
26. Richman, O., Shimkin, N.: Topological uniqueness of the Nash equilibrium for selfish routing with atomic users. Math. Oper. Res. **32**(1), 215–232 (2007)
27. Roughgarden, T., Tardos, É.: How bad is selfish routing? J. ACM (JACM) **49**(2), 236–259 (2002)
28. Stidham, S., Jr.: Stochastic clearing systems. Stoch. Process. Appl. **2**, 85–113 (1974)
29. Stidham, S., Jr.: Optimal Design of Queueing Systems. CRC Press, Taylor and Francis Group, Boca Raton (2009)
30. Wardrop, J.G.: Some theoretical aspects of road traffic research. Proc. Inst. Civ. Eng. **1**(3), 325–362 (1952)
31. Whitt, W.: Deciding which queue to join: some counterexamples. Oper. Res. **34**(1), 55–62 (1986)
32. Wolff, R.: Poisson arrivals see time averages. Oper. Res. **30**, 223–231 (1982)

Coupled Queueing and Charging Game Model with Energy Capacity Optimization

Alix Dupont[1], Yezekael Hayel[1], Tania Jiménez[1(✉)], Olivier Beaude[2], and Cheng Wan[2]

[1] LIA, Avignon, France
tania.jimenez@univ-avignon.fr
[2] EDF Lab' Paris-Saclay, MIRE & OSIRIS Department, Palaiseau, France

Abstract. With the rise of Electric Vehicles (EV), the demand in the stations - both for a parking spots equipped with plugging devices and for electricity to charge the batteries - is increasing. This motivates us to study how we can improve the quality of service at the stations. We model the arrival of EV in several stations and consider a criterium of choice of the station for the users. In our model, EV arrive at the stations according to a Poisson process with a random quantity of energy needed to have a fully charged battery and a random parking time, both following an exponential distribution. We quantify the quality of the service as the probability of leaving the station with a fully charged battery. The stations are characterized by their number of spots and the way they share the power between the EV users. In this model, we study the best way to share power in order to improve this quality of service.

Keywords: Stochastic model · Congestion games · Electric vehicles

1 Introduction

The huge and quick development of electric mobility raises many challenges among research communities. Governments and political authorities have imposed regulatory policies worldwide. Their target is that EVs should account for 7% of the global vehicle fleet by 2030 [1].

This increase on EV traffic poses several problems for the supply of energy. The providers offer efficient and smart charging mechanisms. Some of them can be studied through the angle of queueing models due to the inherent stochastic nature of events: arrival/departure of EV, power level available at a charging point, etc. Therefore, performance evaluation methodologies can help to design incentive and/or evaluate the impact of charging stations configuration (number of charging points, scheduling power management, pricing schemes, ...) on the quality of service of EV users [2]. Recent surveys have shown that the charging time and the 'risk' of being out of energy during travel, are two of the main

© Springer Nature Switzerland AG 2021
P. Ballarini et al. (Eds.): EPEW 2021/ASMTA 2021, LNCS 13104, pp. 325–344, 2021.
https://doi.org/10.1007/978-3-030-91825-5_20

reluctant factors for drivers to invest in an EV. Performance evaluation models help to improve smart charging in order to reduce this 'risk'.

A simple queueing model based on M/M/S queue is proposed in [3]. In this paper, the system is centralized and based on information about charging need of each EV, a central authority routes each EV to charging station in a given area in order to minimize the total average waiting time to charge completion. Tsang et al. [4] developed a mixed queueing network model with an open queue of EVs and a closed queue of batteries. Gusrialdi et al. in [5] addressed both the system-level scheduling problem and the individual EVs decisions about their choice of charging locations, while requiring only distributed information about EVs and their charging at service stations along a highway. Another recent queueing model is proposed in [6] where parking spots and a power allocation function are taken into account in the model. This later model is based on an Erlang loss system. Bounds and fluid limits of the number of EV that get fully charged are described. A deeper investigation of this model has been published in [7]. All these papers are not dealing with individual decisions of EV drivers such as which service station to choose, how much energy to recharge, how long staying at the parking spot, etc. Our framework is the first one that studies this kind of question based on a queueing model. Only a very recent paper [8] integrates the notion of EV user decision, but the charging time is considered to be deterministic, whereas in our framework, it depends on the available power and other EV requirements.

In this paper we address the decision-making problem of Electric Vehicles users to choose a charging station, in order to get their battery fully charged (i.e. their State of Charge, or SoC, at capacity) with a maximum probability at the end of their parking time.

As a simple beginning, and without losing generality for the following theoretical analysis, we consider only two stations, and we suppose that these stations are geographically close. For example, this could fit the situation of two parking in the same university campus, or in a business district. The two charging stations have different characteristics but are not independent in terms of electrical constraints because they are supposed to be connected to the same electricity network node (the "distribution network", to be specific). In the proposed model, this implies that they share a maximal power not to be exceeded when adding their consumption. In each station, EVs share the power available based on some scheduling rules which mainly depend on the number of EVs plugged in at this location. Based on these rules, the probability of leaving the stations with a fully charged battery can be explicitly computed. We analyze this setting as a strategic game model, where users choose the best station for them: the strategic aspect comes from the fact that the charging power delivered at a station depends on the decision of other EV users. The novelty of this work is to make the choice of a charging station an "active" EV users' decision. An interesting insight of our framework is the existence of a continuity of equilibrium, as it is the case in capacitated networks [9]. Moreover, it is possible to determine simple bounds on the Price of Anarchy (PoA) which helps to understand how bad selfish routing behavior is in our smart charging context.

This paper is organized as follows: in Sect. 2, we present the model and preliminary results needed for the game model. In Sect. 3, we introduce a nonatomic queueing game with some results on the associated equilibria. We also compute the price of anarchy of the queueing game. In Sect. 4, we study an optimizing problem on top of the equilibria. Finally, in Sect. 5, some numerical simulations corresponding to realistic cases are presented.

2 Model Description

2.1 Mathematical Description

The EV arrivals to the stations follow a Poisson process with rate λ as the two charging stations are supposed to be used by the same population. Each EV chooses to join charging station i, for $i \in \{1, 2\}$, with a probability denoted by θ_i. Then the arrival process of EVs at charging station i follows a Poisson process with rate $\lambda_i := \theta_i \lambda$. Each EV has a battery capacity denoted by $Capa$ and a State of Charge (SoC) when arriving at the stations: SoC_{arr}. The SoC_{arr} for each EV depends mainly on the distance traveled in order to reach the parking spots, but also on some exogenous parameters like ambient temperature, or traffic conditions [10]. Then, each arriving EV has a random quantity B of energy (in kWh) required in order to get a fully charged battery defined by:

$$B = Capa - SoC_{\mathrm{arr}}.$$

All EVs have different capacities and different states of charge when arriving at the stations, then the quantity B of energy is approximated by an exponential random variable of parameter μ. EV stays at their charging station a random amount of time which is independent of the charging process. The random parking time D of any EV follows an exponential distribution with parameter ν. Then, at each instant, among all EVs parked at a charging station, some are charging and some are not (see Fig. 1 for an illustration of the model).

Charging Stations Description. We follow the same nomenclature shown in [7]. Charging station $i \in \{1, 2\}$ has $K_i > 0$ parking spots; each one of them being equipped with a charging point. At each charging station i and time t, Z_t^i denotes the number of EVs charging and C_t^i the number of EVs that are parked but not charging at this time (given the definition of power scheduling rules, it will be seen that those EVs have a fully charged battery). Then, the total number of EVs at station i and time t is $Q_t^i := Z_t^i + C_t^i$.

Charging Power Scheduling. The available power is shared between the EVs present at a time in each station, following a Processor Sharing policy. A power level of $p_{\max}$ corresponds to the maximum power available for an EV at any charging station. In the following, this power level is considered to be the same

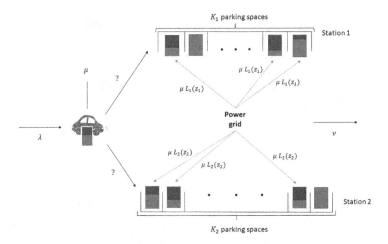

Fig. 1. Arriving users choose between the two stations connected to the same grid. The proportion of green corresponds to the percentage of charge of the batteries. Power is shared only between the EVs without a fully charged battery.

for both charging stations, which corresponds to the most common case for the first charging stations installed.

Another key aspect strongly impacting power scheduling is the parameter $\alpha_i \in [0,1]$, which determines the proportion of charging points that can be used simultaneously at power level $p_{\max}$. This coefficient corresponds to a sizing decision of the operator - out of the scope of this paper. When the number of EVs charging at station i is above this quantity $\alpha_i K_i$, the maximum power $p_{\max}$ is equally shared among all EVs. Denoting $L_i(.)$ the allocation function for charging station i, the power available for an EV at time t charging at this station i depends on Z_t^i through:

$$L_i(Z_t^i) = p_{\max} \frac{\min(Z_t^i, \alpha_i K_i)}{Z_t^i}. \tag{1}$$

Based on the energy requirement of any EV b (an exponential realization of the random variable B with rate μ) and arriving time t_0 at station i, it is possible to determine the finishing charging time t_1 which corresponds to the time at which the battery is fully charged. This time t_1 satisfies the following equation depending on b and t_0:

$$\int_{t_0}^{t_1} L_i(Z_t^i)\mathrm{dt} = p_{\max} \int_{t_0}^{t_1} \frac{\min(Z_t^i, \alpha_i K_i)}{Z_t^i}\mathrm{dt} = b.$$

It is clear from our context that, for each station i, the two-dimensional process (Q_t^i, Z_t^i) is Markov. Then, EV drivers choose their charging station in order to maximize the probability of having a fully charged battery when leaving ($SoC = Capa$). Below, we present an analytical form of the expected probability for an EV to leave a station with a full battery. Then, the strategic choice of the charging station is done in order to maximize this expected probability.

2.2 Expected Number of EV Charging

In stationary regime of the Markov process, it has been proved in [7] that the expected number $\bar{z}_i$ of EVs charging at station i can be approximated by the unique solution of Eq. (2):

$$\bar{z}_i = \min(\lambda_i, \nu K_i)\, \mathbb{E}\left[\min\left(D, \frac{B}{p_{\max}}\max(1, \frac{\bar{z}_i}{\alpha_i K_i})\right)\right]. \tag{2}$$

Considering the PASTA property, which holds that arriving customers find on average the same situation in the queueing system as an outside observer looking at the system at an arbitrary point in time [11], an arriving EV at station i will face an expected number $\bar{z}_i$ of EV already parked at charging station i. Based on that, the expected time $\bar{t}_i$ for such EV to gain 1kw of energy is approximated by:

$$\bar{t}_i = \frac{1}{L_i(\bar{z}_i)} = \frac{\bar{z}_i}{p_{\max}\min(\bar{z}_i, \alpha_i K_i)} = \frac{\max(1, \frac{\bar{z}_i}{\alpha_i K_i})}{p_{\max}}.$$

Then, the random duration $\bar{T}_i$ for an arriving EV at charging station i to completely recharge its battery, in stationary regime of the system, is equal to $B \times \bar{t}_i$ and follows an exponential distribution with parameter $\frac{p_{\max}\mu}{\max(1, \frac{\bar{z}_i}{\alpha_i K_i})}$.

Note that the solution of the fixed point Eq. (2) exists and is unique, so that $\bar{z}_i$ is well defined and its explicit expression is given in the following proposition.

Proposition 1. *In stationary regime, the expected number $\bar{z}_i$ of charging EVs at station i is given by:*

$$\bar{z}_i = \begin{cases} \frac{\min(\lambda_i, \nu K_i)}{\nu + p_{\max}\mu} & \text{if } \frac{\nu}{\nu + p_{\max}\mu} \le \alpha_i \text{ or } \frac{\lambda_i}{\nu + p_{\max}\mu} \le \alpha_i K_i, \\ \frac{\min(\lambda_i, \nu K_i) - p_{\max}\mu\alpha_i K_i}{\nu} & \text{else.} \end{cases} \tag{3}$$

Proof of Proposition 1. The random variable $\min(D, \bar{T}_i)$ follows an exponential distribution with parameter $\nu + \frac{\mu p_{\max}}{\max(1, \frac{\bar{z}_i}{\alpha_i K_i})}$ as it is the minimum of two independent exponential random variables. The fixed point Eq. (2) can then be re-written as

$$\bar{z}_i = \min(\lambda_i, \nu K_i)\frac{\max(1, \frac{\bar{z}_i}{\alpha_i K_i})}{\nu\max(1, \frac{\bar{z}_i}{\alpha_i K_i}) + p_{\max}\mu}. \tag{4}$$

Let f_i be the function from $[0, K_i]$ to itself such that $\forall z \in [0, K_i]$, $f_i(z) = \min(\lambda_i, \nu K_i)\frac{\max(1, \frac{z}{\alpha_i K_i})}{\nu\max(1, \frac{z}{\alpha_i K_i}) + p_{\max}\mu}$. In order to have an explicit expression of the solutions, $f_i(z)$ is expressed as

$$f_i(z) = \begin{cases} \frac{\min(\lambda_i, \nu K_1)}{\nu + p_{\max}\mu} & \forall z \in [0, \alpha_i K_i], \\ \frac{\min(\lambda_i, \nu K)}{\nu} \times \frac{z}{z + \frac{p_{\max}\mu}{\nu}\alpha K} & \forall z \in]\alpha_i K_i, K_i], \end{cases}$$

and we look for the solutions of the equation:

$$f_i(z) = z. \tag{5}$$

Then, two different cases appear:

- If $\frac{\min(\lambda_i, \nu K_i)}{\nu + p_{\max}\mu} \leq \alpha_i K_i$, then Eq. (5) admits a unique solution on $[0, \alpha_i K_i]$. A short analysis on the function f_i shows that there is no solution of (5) on $]\alpha_i K_i, K_i]$ because $\forall z \in]\alpha_i K_i, K_i]$, $f_i(z) < z$. Then the unique solution of (5) is:

$$z^* = \frac{\min(\lambda_i, \nu K_i)}{\nu + p_{\max}\mu} \quad \text{if} \quad \frac{\min(\lambda_i, \nu K_1)}{\nu + p_{\max}\mu} \leq \alpha_i K_i.$$

- If $\frac{\min(\lambda_i, \nu K_i)}{\nu + p_{\max}\mu} > \alpha_i K_i$, then there is no solution of (5) on $[0, \alpha_i K_i]$. As $f_i(\alpha_i K_i) > \alpha_i K_i$, $f_i(K_i) \leq K_i$, f_i is continuous on $[\alpha_i K_i, K_i]$, and $\forall z \in]\alpha_i K_i, K_i[$, $f_i'(z) < z$, there is a unique solution of (5) on $]\alpha_i K_i, K_i[$. After some algebras, the unique solution of (5) is given by:

$$z^* = \frac{\min(\lambda_i, \nu K_i) - p_{\max}\mu\alpha_i K_i}{\nu} \quad \text{if} \quad \frac{\min(\lambda_i, \nu K_1)}{\nu + p_{\max}\mu} > \alpha_i K_i.$$

Finally, the condition $\frac{\min(\lambda_i, \nu K_i)}{\nu + p_{\max}\mu} \leq \alpha_i K_i$ is equivalent to the following condition:

$$\frac{\nu}{\nu + p_{\max}\mu} \leq \alpha_i \quad \text{or} \quad \frac{\lambda_i}{\nu + p_{\max}\mu} \leq \alpha_i K_i,$$

which concludes the proof. ∎

Proposition 1 allows us to know explicitly when the stationary average number $\bar{z}_i$ of charging EVs at station i is greater than the number $\alpha_i K_i$ of parking spots that can charge at maximum power rate, that is :

$$\bar{z}_i > \alpha_i K_i \Leftrightarrow \begin{cases} \frac{\nu}{\nu + p_{\max}\mu} > \alpha_i, \\ \frac{\lambda_i}{\nu + p_{\max}\mu} > \alpha_i K_i. \end{cases}$$

2.3 Stationary Probability of Fully Charged Battery

It is therefore possible to determine the stationary probability $\bar{P}_i$ for an EV of leaving charging station i with a fully charged battery, which is the probability that the time to charge entirely the battery is less than the parking time.

Proposition 2. *In stationary regime, the probability $\bar{P}_i$ of leaving the station i with a fully charged battery can be approximated by:*

$$\bar{P}_i = \mathbb{P}\left(D > \frac{B}{p_{\max}}\max(1, \frac{\bar{z}_i}{\alpha K_i})\right) = \frac{p_{\max}\mu}{\nu \max(1, \frac{\bar{z}_i}{\alpha_i K_i}) + p_{\max}\mu}. \tag{6}$$

Note that if the expected number of EV charging $\bar{z}_i$ is less than the charging capacity $\alpha_i K_i$ at maximum power level at charging station i, then, an EV is charging with power level $p_{\max}$. In this case, the probability $\bar{P}_i$ that an EV leaves station i with a fully charged battery is simply given by the ratio $\frac{p_{\max}\mu}{\nu + p_{\max}\mu}$.

Proof of Proposition 2. $\bar{P}_i$ is the probability that the parking time D is greater than the time $\bar{T}_i$ to have a fully charged battery: $\bar{P}_i = P(D > \bar{T}_i)$. As D follows a exponential distribution of parameter ν (the distribution is the same regardless the station), and $\bar{T}_i$ follow as exponential distribution of parameter $\eta_i = \frac{p_{\max}\mu}{\max(1, \frac{\bar{z}_i}{\alpha_i K_i})}$,

$$\bar{P}_i = P(D > \bar{T}_i) = \int_0^{+\infty} \nu e^{-\nu t}\left(\int_0^t \eta_i e^{-\eta_i s}ds\right)dt = \int_0^{+\infty} \nu e^{-\nu t}(1 - e^{-\eta_i t})dt$$

$$= 1 - \frac{\nu}{\nu + \eta_i}\int_0^{+\infty}(\nu + \eta_i)e^{-(\nu+\eta_i)s}ds = 1 - \frac{\nu}{\nu + \eta_i}$$

$$= \frac{\eta_i}{\nu + \eta_i} = \frac{p_{\max}\mu}{\nu \times \max(1, \frac{\bar{z}_i}{\alpha_i K_i}) + p_{\max}\mu}.$$

∎

Given (6), we study in next section the non-cooperative game between EVs that strategically decide which charging station to join in order to maximize this probability.

3 Queueing Game with Two Stations

3.1 Definition of the Nonatomic Game

A possibly very large number of EVs interact in the system and then nonatomic games (games with infinite number of players or population of players) are typical game theoretic frameworks that are used to study equilibrium concepts in such situations [12]. In particular, interactions between EVs are performed in queueing systems, and therefore our approach is inspired from the rich literature of queueing games [13,16]. In nonatomic games, each player has an infinitesimal impact on the utility of the others. Precisely, as stated in the seminal paper [14] about nonatomic games: one player has no influence but only an aggregate behavior of large sets of players can change the utility. Let us first define mixed strategies in such games.

Definition 1 (Mixed-strategy nonatomic queueing game). *Let Γ denote a nonatomic mixed-strategy game defined by*

- *A continuum of nonatomic players, which are EVs, of total weight 1;*
- *Each player has two pure actions: station 1 and station 2. Players take mixed strategies that are probability distributions over the two pure actions.*
- *A profile of mixed strategies of all the players induce a probability law on the proportion of players choosing station 1 in the population.*

Based on previous definition, the utility of a player using mixed strategy θ (choosing station 1 with probability θ) depends on the average strategy used in the population, and this utility is defined by:

$$u(\theta, \bar{\theta}) = \theta \bar{P}_1(\bar{\theta}) + (1 - \theta)\bar{P}_2(1 - \bar{\theta}),$$

where $\bar{\theta}$ is the average strategy used in the population.

The number of EVs at the stations is assumed to be unobservable. An equilibrium is defined as a situation where any player cannot strictly increase her utility by changing her strategy. All EV users are assumed to be indistinguishable and then the equilibrium is symmetric in the sense that, at equilibrium, all EVs choose to join charging station 1 with the same probability denoted $\tilde{\theta}$.

Definition 2 (Symmetric Mixed Strategy equilibrium). *A symmetric mixed strategy equilibrium is a profile of identical mixed strategies, i.e. every player choosing going to station 1 with probability $\tilde{\theta}$, such that*

$$u(\tilde{\theta}, \tilde{\theta}) \geq u(\theta, \tilde{\theta}), \quad \forall \theta \in [0, 1].$$

The set of equilibria is denoted by

$$S_{eq} = \left\{ \tilde{\theta} \in [0, 1] : u(\theta, \tilde{\theta}) \leq u(\tilde{\theta}, \tilde{\theta}), \, \forall \theta \in [0, 1] \right\}. \tag{7}$$

There are two main congestion effects in our nonatomic game. First, EVs in the stations occupy parking spots and thus reduce the chance for other EVs to find a free parking spot. This can be called the 'parking congestion'. Second, the power allocation depends on the number of EVs already charging, so that there is also an 'energy congestion' effect. These two kinds of congestion have an impact on the players' utility.

3.2   Properties of Equilibrium

The following proposition says that, at any equilibrium, the probability to leave the station with a full battery is equal for both stations. This is mainly because the maximum charging power p_{max} is the same in both stations.

Proposition 3. $S_{eq} = \left\{ \theta \in [0, 1], \bar{P}_1(\theta) = \bar{P}_2(1 - \theta) \right\}.$

Proof of Proposition 3. A short analysis of Eq. (3) gives that $\bar{z}_i$ is increasing in λ_i, and Eq. (6) gives that $\bar{P}_i$ is decreasing in $\bar{z}_i$. Then $\bar{P}_1$ is decreasing in θ and $\bar{P}_2$ is increasing in θ. Let $g : [0, 1] \to [-1, 1]$ such that $\forall \theta \in [0, 1], g(\theta) = \bar{P}_1(\theta) - \bar{P}_2(1 - \theta)$. Then g is decreasing. Besides, Eqs. (3) and (6) gives that $\bar{P}_1(0) = \bar{P}_2(0)$. Thus, $g(0) \geq 0$ and $g(1) \leq 0$. Suppose there exist $\theta^* \in S_{eq}$ such that $\bar{P}_1(\theta^*) \neq \bar{P}_1(1 - \theta^*)$. Then, by the indifference principle, $\theta^* \in \{0, 1\}$. If $\theta^* = 0$, Eq. (7) gives that $\bar{P}_2(1 - \theta^*) > \bar{P}_1(\theta^*)$, which is impossible because $g(0) \geq 0$. The reasoning is the same for $\theta^* = 1$. Then we have shown that $S_{eq} \subset \left\{ \theta \in [0, 1], \bar{P}_1(\theta) = \bar{P}_2(1 - \theta) \right\}$. The other inclusion is immediate. ∎

The next proposition establishes that Wardrop equilibria exist, the set of such equilibria is not empty, and the probability for the players to leave their chosen station with a full battery is the same for all the Wardrop equilibria.

Proposition 4. *The set of Wardrop equilibria S_{eq} is non-empty and convex. Besides, there is a constant $\bar{P}_{eq}$ such that for any Wardrop equilibrium $\tilde{\theta} \in S_{eq}$, $\bar{P}_1(\tilde{\theta}) = \bar{P}_2(1 - \tilde{\theta}) = \bar{P}_{eq}$.*

Based on previous proposition, we observe that the utility of any EV at equilibrium is equal to the probability $\bar{P}_{eq}$, i.e. for any equilibrium strategy $\tilde{\theta}$, $u(\tilde{\theta}, \tilde{\theta}) = \tilde{\theta}\bar{P}_1(\tilde{\theta}) + (1 - \tilde{\theta})\bar{P}_2(1 - \tilde{\theta}) = \bar{P}_{eq}$.

Proof of Proposition 4. $S_{eq} = \{\theta \in [0, 1], g(\theta) = 0\}$ with g defined as in the proof of proposition 3. Equation (6) gives the continuity of $\bar{P}_i$ in θ, as a composition of continuous functions. Then, g is continuous, decreasing, and $g(0) \geq 0$ and $g(1) \leq 0$, so there exist $\theta \in [0, 1]$ such that $g(\theta) = 0$ and S_{eq} is convex.

If S_{eq} is a singleton, then the second part of the proposition holds. Otherwise, let $(\theta, \tilde{\theta}) \in S_{eq}^2$ such that $\tilde{\theta} > \theta$. Suppose $\bar{P}_1(\theta) \neq \bar{P}_1(\tilde{\theta})$ (which implies $\bar{P}_2(1-\theta) \neq \bar{P}_2(1-\tilde{\theta})$). Then, because $\bar{P}_1$ is decreasing and $\bar{P}_2$ is increasing in θ, $\bar{P}_1(\theta) > \bar{P}_1(\tilde{\theta})$ and $\bar{P}_2(1 - \theta) < \bar{P}_2(1 - \tilde{\theta})$. Then, $g(\theta) = \bar{P}_1(\theta) - \bar{P}_2(1 - \theta) > \bar{P}_1(\tilde{\theta}) - \bar{P}_2(1 - \tilde{\theta}) = g(\tilde{\theta})$, which contradicts $g(\theta) = g(\tilde{\theta}) = 0$. This shows the second part of the proposition. ∎

The following proposition gives an explicit expression of the probability $\bar{P}_{eq}$ of any EV to leave its station at equilibrium with a full battery. This metric will be used for determining which station to join.

Proposition 5. *At equilibrium, the probability $\bar{P}_{eq}$ for any EV to leave a station with a full battery is given by:*

$$
\bar{P}_{eq} = \begin{cases} \frac{p_{\max}\mu}{\nu + p_{\max}\mu} & \text{if } \max(\alpha_1, \alpha_2) \geq \frac{\nu}{\nu + p_{\max}\nu} \text{ or } \alpha_1 K_1 + \alpha_2 K_2 \geq \frac{\lambda}{\nu + p_{\max}\mu}, \\[2ex] \frac{p_{\max}\mu}{\lambda}(\alpha_1 K_1 + \alpha_2 K_2) & \text{if } \begin{cases} \max(\alpha_1, \alpha_2) < \frac{\nu}{\nu + p_{\max}\nu}, \\ \alpha_1 K_1 + \alpha_2 K_2 < \frac{\lambda}{\nu + p_{\max}\mu}, \\ K_1 + K_2 > \frac{\lambda}{\nu}, \\ \max(\alpha_1, \alpha_2) \leq \frac{\nu}{\lambda}(\alpha_1 K_1 + \alpha_2 K_2), \end{cases} \\[2ex] \frac{p_{\max}\mu}{\nu}\max(\alpha_1, \alpha_2) & \text{else.} \end{cases}
$$

(8)

This expression of $\bar{P}_{eq}$ can take three different values depending on system parameters.

– The first one $\bar{P}_{eq} = \frac{p_{\max}\mu}{\nu + p_{\max}\mu}$ is the largest $\bar{P}_{eq}$ possible. It happens when, for some $\theta \in [0, 1]$, the average number of charging EVs $\bar{z}_i$ at each station i is lower than the threshold $\alpha_i K_i$. This occurs when the profusion term α_i of one of the stations is large ($\max(\alpha_1, \alpha_2) \geq \frac{\nu}{\nu + p_{\max}\nu}$) or when the total number of simultaneous maximum rate charging points (sum of the two stations) is large $(\alpha_1 K_1 + \alpha_2 K_2 \geq \frac{\lambda}{\nu + p_{\max}\mu})$.

- The second expression $\bar{P}_{eq} = \frac{p_{\max}\mu}{\lambda}(\alpha_1 K_1 + \alpha_2 K_2)$ happens basically when the total resource of energy is low compared to the demand, but the total number of parking spots is large enough $(K_1 + K_2 > \frac{\lambda}{\nu})$. Then the maximum average number of charging EVs at the stations is not reached at equilibrium: $\bar{z}_i < K_i(1 - p_{\max}\frac{\mu}{\nu}\alpha_i), \forall i \in \{1,2\}$.

- The last expression $\bar{P}_{eq} = \frac{p_{\max}\mu}{\nu}\max(\alpha_1, \alpha_2)$ is when the total number of parking spots is low, so that at equilibrium the maximum average number of charging EVs is reached for at least one station : $\exists i \in \{1,2\}, \bar{z}_i = K_i(1 - p_{\max}\frac{\mu}{\nu}\alpha_i)$.

Proof of Proposition 5. Equation (6) with Eq. (3) gives an expression of $\bar{P}_1$ and $\bar{P}_2$ in function of θ. In the proof, the goal is to solve the solution of $\bar{P}_1(\theta) = \bar{P}_2(1-\theta)$. For this, the curves of $\bar{P}_1$ and $\bar{P}_2$ are analysed.

The expressions of $\bar{P}_1$ and $\bar{P}_2$ in function of θ_i are decomposed into three parts. When $\bar{z}_i \leq \alpha_i K_i$, then $\bar{P}_i$ is at its highest value $\frac{p_{\max}\mu}{\nu + p_{\max}\mu}$. When $\bar{z}_i > \alpha_i K_i$, if the maximum average number $K_i(1 - p_{\max}\frac{\mu}{\nu}\alpha_i)$ of charging EVs is not reached, $\bar{P}_i = \frac{p_{\max}\mu\alpha_i K_i}{\theta_i \lambda}\alpha_i$ strictly decreases in θ_i, and if the maximum average number $K_i(1 - p_{\max}\frac{\mu}{\nu}\alpha_i)$ of charging EVs is reached, then $\bar{P}_i = \frac{p_{\max}\mu}{\nu}\alpha_i$ is constant again.

When $\max(\alpha_1, \alpha_2) \geq \frac{\nu}{\nu + p_{\max}\nu}$, then in at least one of the two stations, say station i, the average number of charging EVs is always lower that the threshold $\alpha_i K_i$, so that the equilibrium happens when the average number of charging EVs at the other station, say station j, is also lower than the threshold $\alpha_j K_j$. Then in this case $\bar{P}_{eq} = \frac{p_{\max}\mu}{\nu + p_{\max}\mu}$.

In the following, $\max(\alpha_1, \alpha_2) < \frac{\nu}{\nu + p_{\max}\nu}$ is assumed. Then, $\bar{z}_1 \leq \alpha_1 K_1 \Leftrightarrow$ $\theta \leq \frac{\alpha_1 K_1(\nu + p_{\max}\mu)}{\lambda}$ and $\bar{z}_2 \leq \alpha_2 K_2 \Leftrightarrow \theta \geq 1 - \frac{\alpha_2 K_2(\nu + p_{\max}\mu)}{\lambda}$. Thus, $\alpha_1 K_1 + \alpha_2 K_2 \geq \frac{\lambda}{\nu + p_{\max}\mu} \Rightarrow \bar{P}_1 = \bar{P}_2 = \frac{p_{\max}\mu}{\nu + p_{\max}\mu}, \forall \theta \in [\max(0, 1 - \frac{\alpha_2 K_2(\nu + p_{\max}\mu)}{\lambda}), \min(1, \frac{\alpha_1 K_1(\nu + p_{\max}\mu)}{\lambda})]$.

In the following, $\alpha_1 K_1 + \alpha_2 K_2 < \frac{\lambda}{\nu + p_{\max}\mu}$ is also assumed. Then, denoting $\bar{z}_i^{max}$ the maximum value of $\bar{z}_i \forall i \in \{1,2\}$, $\bar{z}_1 = \bar{z}_1^{max} \Leftrightarrow \theta \geq \frac{\nu K_1}{\lambda}$ and $\bar{z}_2 = \bar{z}_2^{max} \Leftrightarrow \theta \leq 1 - \frac{\nu K_2}{\lambda}$. Thus, $K_1 + K_2 \leq \frac{\lambda}{\nu} \Rightarrow \bar{P}_i = \frac{p_{\max}\mu}{\nu}\alpha_i \forall \theta \in [\frac{\nu K_1}{\lambda}, 1 - \frac{\nu K_2}{\lambda}]$. Then, if $\alpha_1 = \alpha_2$, $\bar{P}_{eq} = \frac{p_{\max}\mu}{\nu}\alpha_1 = \frac{p_{\max}\mu}{\nu}\alpha_2 = \frac{p_{\max}\mu}{\nu}\max(\alpha_1, \alpha_2)$. Otherwise, suppose for instance $\alpha_1 < \alpha_2$. Then, $\forall \theta \in [\frac{\alpha_1 K_1(\nu + p_{\max}\mu)}{\lambda}, \frac{\nu K_1}{\lambda}], \bar{P}_2 = \frac{p_{\max}\mu}{\nu}\alpha_2$. Also, $\bar{P}_1(\frac{\alpha_1 K_1(\nu + p_{\max}\mu)}{\lambda}) = \frac{p_{\max}\mu}{\nu + p_{\max}\mu} > \frac{p_{\max}\mu}{\nu}\alpha_2$, $\bar{P}_1(\frac{\nu K_1}{\lambda}) = \frac{p_{\max}\mu}{\nu}\alpha_1 < \frac{p_{\max}\mu}{\nu}\alpha_2$ and $\bar{P}_1$ is continuous in θ. Thus, there exist $\theta^* \in [\frac{\alpha_1 K_1(\nu + p_{\max}\mu)}{\lambda}, \frac{\nu K_1}{\lambda}]$ such that $\bar{P}_1 = \bar{P}_2 = \frac{p_{\max}\mu}{\nu}\alpha_2 = \frac{p_{\max}\mu}{\nu}\max(\alpha_1, \alpha_2)$. The reasoning is the same if $\alpha_1 > \alpha_2$.

In the following, $K_1 + K_2 > \frac{\lambda}{\nu}$ is also assumed. Suppose also $\alpha_1 > \alpha_2$. Then $(\frac{\nu K_1}{\lambda} < 1$ and $\bar{P}_2(\frac{\nu K_1}{\lambda}) < \bar{P}_1(\frac{\nu K_1}{\lambda})) \Leftrightarrow \alpha_1 > \frac{\nu}{\lambda}(\alpha_1 K_1 + \alpha_2 K_2)$. In this case, $\bar{P}_1(\theta) = \frac{p_{\max}\mu}{\nu}\alpha_1, \forall \theta \in [\frac{\nu K_1}{\lambda}, 1]$. Besides, $\bar{P}_2(\frac{\nu K_1}{\lambda}) < \frac{p_{\max}\mu}{\nu}\alpha_1$, $\bar{P}_2(1) >$

$\frac{p_{\max}\mu}{\nu}\alpha_1$ and $\bar{P}_2$ is continuous in θ. Thus there exist $\theta^* \in [\frac{\nu K_1}{\lambda}, 1]$ such that $\bar{P}_2(\theta^*) = \bar{P}_1(\theta^*) = \frac{p_{\max}\mu}{\nu}\alpha_1 = \frac{p_{\max}\mu}{\nu}\max(\alpha_1, \alpha_2)$. Suppose now $\alpha_1 \le \frac{\nu}{\lambda}(\alpha_1 K_1 + \alpha_2 K_2)$. Then, $\bar{P}_2(\min(1 - \alpha_2 K_2\frac{\nu+p_{\max}\mu}{\lambda}, \frac{\nu K_1}{\lambda})) > \bar{P}_1(\min(1 - \alpha_2 K_2\frac{\nu+p_{\max}\mu}{\lambda}, \frac{\nu K_1}{\lambda}))$, $\bar{P}_2(\max(\alpha_1 K_1\frac{\nu+p_{\max}\mu}{\lambda}, 1 - \frac{\nu K_2}{\lambda})) < \bar{P}_1(\max(\alpha_1 K_1\frac{\nu+p_{\max}\mu}{\lambda}, 1 - \frac{\nu K_2}{\lambda}))$ and $\bar{P}_i(\theta) = \frac{p_{\max}\alpha_i K_i}{\theta_i \lambda}$, where $\theta_1 = \theta$ and $\theta_2 = 1 - \theta$. As $\bar{P}_1$ and $\bar{P}_2$ are continuous in θ and $\frac{p_{\max}\alpha_1 K_1}{\theta\lambda} = \frac{p_{\max}\alpha_2 K_2}{(1-\theta)\lambda} \Leftrightarrow \theta = \frac{\alpha_1 K_1}{\alpha_1 K_1+\alpha_2 K_2}$. Thus, $\bar{P}_{eq} = \bar{P}_1(\frac{\alpha_1 K_1}{\alpha_1 K_1+\alpha_2 K_2}) = \bar{P}_2(\frac{\alpha_1 K_1}{\alpha_1 K_1+\alpha_2 K_2}) = \frac{p_{\max}\mu}{\lambda}(\alpha_1 K_1 + \alpha_2 K_2)$. The reasoning is identical for $\alpha_1 < \alpha_2$. Note that $\alpha_1 = \alpha_2$ is impossible if $K_1 + K_2 > \frac{\lambda}{\nu}$ and $\max(\alpha_1, \alpha_2) > \frac{\nu}{\lambda}(\alpha_1 K_1 + \alpha_2 K_2)$.   ∎

3.3 Price of Anarchy

The Price of Anarchy (PoA) is a metric which evaluates how bad selfish behavior is compared to an optimal centralized decision making. Indeed, the PoA is defined as the ratio between the optimal social utility and the (worst) utility at equilibrium [15]. The former is determined as if a central authority is controlling the behavior, it is to say, the station choice of each EV. The social utility U is defined when θ proportion of players choosing 1 as the average utility by:

$$U(\theta) := u(\theta, \theta) = \theta \bar{P}_1(\theta) + (1 - \theta)\bar{P}_2(1 - \theta).$$

The optimal social utility U_{opt} is then defined by:

$$U_{opt} := \max_{\theta \in [0,1]} U(\theta).$$

Based on the two metrics which correspond to a centralized point of view with the social utility, and a decentralized point of view with the equilibrium utility, we can determine the PoA as the ratio between this two metrics.

Definition 3 (Price of anarchy [15]). *The price of anarchy of the game Γ is the ratio between the socially optimal utility and the social utility at the equilibria:*

$$PoA := \frac{U_{opt}}{u(\tilde{\theta}, \tilde{\theta})} = \frac{U_{opt}}{\bar{P}_{eq}}. \tag{9}$$

Since $\bar{P}_{eq}$ is always strictly positive, the PoA (9) is well defined. It is clear that $PoA \ge 1$ and, the closer it is to 1, the more efficient the equilibria are in the sense that decentralized selfish decision leads to the same performance for users as a centralized decision. Denote, for any couple $(i, j) \in \{(1, 2), (2, 1)\}$, the following values:

$$\gamma_i = \alpha_i K_i + \frac{\nu}{\nu + p_{\max}\mu}K_j, \quad A_i = \frac{p_{\max}\mu\alpha_j}{\nu}(1 - \alpha_i K_i\frac{\nu + p_{\max}\mu}{\lambda}) + \alpha_i K_i\frac{p_{\max}\mu}{\lambda}.$$

The following proposition gives sufficient conditions on system parameters such that the PoA is equal to 1, meaning that the decentralized system in which EV makes their decision selfishly induces the same performance in terms of probability of leaving a station with full battery compared to a centralized solution.

Proposition 6. *If* $\max(\alpha_1, \alpha_2) \geq \frac{\nu}{\nu + p_{\max}\mu}$ *or* $\min(\gamma_1, \gamma_2) \geq \frac{\lambda}{\nu + p_{\max}\mu}$, *then* $PoA = 1$.

The next proposition gives interesting bounds on the price of anarchy.

Proposition 7. *If* $\max(\alpha_1, \alpha_2) < \frac{\nu}{\nu + p_{\max}\mu}$ *and* $\min(\gamma_1, \gamma_2) < \frac{\lambda}{\nu + p_{\max}\mu}$, *then the PoA can take one of the values:* $\left\{ \frac{\lambda A_j}{p_{\max}\mu(\alpha_1 K_1 + \alpha_2 K_2)}, \frac{\nu A_j}{p_{\max}\,\mu\,\max(\alpha_1,\alpha_2)}, j = 1, 2 \right\}$.
In particular,

- *If the PoA is of the form* $PoA = \frac{\lambda A_j}{p_{\max}\mu(\alpha_1 K_1 + \alpha_2 K_2)}$ *then*

$$PoA \leq \frac{\frac{\nu}{\nu + p_{\max}\mu}}{\min(\alpha_1, \alpha_2)}.$$

- *If PoA is of the form* $PoA = \frac{\nu A_j}{p_{\max}\mu\,\max(\alpha_1,\alpha_2)}$ *then*

$$PoA \leq 2.$$

Proofs of Proposition 6 and 7. The proof mainly rely on Proposition 5. When $\max(\alpha_1, \alpha_2) \geq \frac{\nu}{\nu + p_{\max}\mu}$ or $\alpha_1 K_1 + \alpha_2 K_2 \geq \frac{\lambda}{\nu + p_{\max}\mu}$, $\bar{P}_{eq}$ takes the highest value $\bar{P}_1$ or $\bar{P}_2$ can have. u being an average between $\bar{P}_1$ and $\bar{P}_2$, it is clear that $U_{opt} \leq \bar{P}_{eq}$.

In the following, $\max(\alpha_1, \alpha_2) \geq \frac{\nu}{\nu + p_{\max}\mu}$ and $\alpha_1 K_1 + \alpha_2 K_2 < \frac{\lambda}{\nu + p_{\max}\mu}$ are assumed. Then, with a short analisys of the function u, it can be shown that u is strictly increasing for $\theta \in]0, \alpha_1 K_1 \frac{\nu + p_{\max}\mu}{\lambda}[$, then convex for $\theta \in [\alpha_1 K_1 \frac{\nu + p_{\max}\mu}{\lambda}, 1 - \alpha_2 K_2 \frac{\nu + p_{\max}\mu}{\lambda}]$, and then strictly decreasing for $\theta \in]1 - \alpha_2 K_2 \frac{\nu + p_{\max}\mu}{\lambda}, 1[$. It gives $\arg\max(u) \in [\alpha_1 K_1 \frac{\nu + p_{\max}\mu}{\lambda}, 1 - \alpha_2 K_2 \frac{\nu + p_{\max}\mu}{\lambda}]$.

- Suppose $K_1 + K_2 > \frac{\lambda}{\nu}$.
  - $\min(\gamma_1, \gamma_2) \geq \frac{\lambda}{\nu + p_{\max}\mu}$ implies that the equilibrium is reached for $\theta \in [\alpha_1 K_1 \frac{\nu + p_{\max}\mu}{\lambda}, 1 - \alpha_2 K_2 \frac{\nu + p_{\max}\mu}{\lambda}]$ and u is constant for $\theta \in [\alpha_1 K_1 \frac{\nu + p_{\max}\mu}{\lambda}, 1 - \alpha_2 K_2 \frac{\nu + p_{\max}\mu}{\lambda}]$. Then $PoA = 1$.
  - $\gamma_1 < \frac{\lambda}{\nu + p_{\max}\mu} \leq \gamma_2$ implies that u decreases and is then constant, so that $U_{opt} = A_1$. Two cases are possible for $\bar{P}_{eq}$: $\bar{P}_{eq} = \frac{p_{\max}\mu}{\lambda}(\alpha_1 K_1 + \alpha_2 K_2)$ if $\alpha_2 \leq \frac{\nu}{\lambda}(\alpha_1 K_1 + \alpha_2 K_2)$ and $\bar{P}_{eq} = \frac{p_{\max}\mu}{\nu}\alpha_2 = \frac{p_{\max}\mu}{\nu}\max(\alpha_1, \alpha_2)$ otherwise. The case $\gamma_2 < \frac{\lambda}{\nu + p_{\max}\mu} \leq \gamma_1$ is symmetrical.
  - $\max(\gamma_1, \gamma_2) < \frac{\lambda}{\nu + p_{\max}\mu}$ implies that u decreases, then is constant, and then increases, so that $U_{opt} = \max(A_1, A_2)$. The value of $\bar{P}_{eq}$ is the same as in the previous case.
- Suppose now $K_1 + K_2 \leq \frac{\lambda}{\nu}$, then the convex curve of u shows that $U_{opt} = \max(A_1, A_2)$, and $\bar{P}_{eq} = \frac{p_{\max}\mu}{\nu}\max(\alpha_1, \alpha_2)$.

The following gives the computations for the bounds of the PoA.

– When PoA is of the form $\frac{\lambda A_j}{p_{\max}\mu(\alpha_1 K_1 + \alpha_2 K_2)}$,

$$PoA \leq \frac{\lambda \max(A_1, A_2)}{p_{\max}\mu(\alpha_1 K_1 + \alpha_2 K_2)}$$

$$< \frac{\max(\alpha_1(1 - \alpha_2 K_2 \frac{\nu + p_{\max}\mu}{\lambda}) + \alpha_2 K_2 \frac{\nu}{\lambda}, \alpha_2(1 - \alpha_1 K_1 \frac{\nu + p_{\max}\mu}{\lambda}) + \alpha_1 K_1 \frac{\nu}{\lambda})}{\min(\alpha_1, \alpha_2)}$$

$$< \frac{\frac{\nu}{\nu + p_{\max}\mu}}{\min(\alpha_1, \alpha_2)}.$$

where for the second inequality $\alpha_1 K_1 + \alpha_2 K_2 \geq \min(\alpha_1, \alpha_2)(K_1 + K_2) > \min(\alpha_1, \alpha_2)\frac{\lambda}{\nu}$ has been used. Indeed this form of the PoA can only happen when $K_1 + K_2 > \frac{\lambda}{\nu}$. For the last inequality $\alpha_2 K_2 \leq \alpha_1 K_1 + \alpha_2 K_2 < \frac{\lambda}{\nu + p_{\max}\mu}, \forall i \in \{1, 2\}$ has been used.

– When PoA is of the form $\frac{\nu A_j}{p_{\max}\mu \max(\alpha_1, \alpha_2)}$,

$$PoA \leq \frac{\nu \max(A_1, A_2)}{p_{\max}\mu \max(\alpha_1, \alpha_2)}$$

$$\leq \max\left(1 - \alpha_1 K_1 \frac{\nu + p_{\max}\mu}{\lambda} + K_1 \frac{\nu}{\lambda}, 1 - \alpha_2 K_2 \frac{\nu + p_{\max}\mu}{\lambda} + K_2 \frac{\nu}{\lambda}\right)$$

$$= 1 + \frac{\nu}{\lambda} \max\left(K_1(1 - \alpha_1 \frac{\nu + p_{\max}\mu}{\nu}), K_2(1 - \alpha_2 \frac{\nu + p_{\max}\mu}{\nu})\right)$$

$$\leq 2 - \frac{\nu + p_{\max}\mu}{\nu} \min(\alpha_1, \alpha_2) < 2 - \min(\alpha_1, \alpha_2)^2 \leq 2.$$

where for the second inequality $\forall i \in \{1, 2\}, \alpha_i \leq \max(\alpha_1, \alpha_2)$ have been used, for the third one $\forall i \in \{1, 2\}, K_i \leq K_1 + K_2 \leq \frac{\lambda}{\nu}$ and $\alpha_i \geq \min(\alpha_1, \alpha_2)$, for the fourth one and last one only $0 \leq \min(\alpha_1, \alpha_2) < \frac{\nu}{\nu + p_{\max}\mu}$.

∎

4 Energy Capacity

As it has been mentioned in Sect. 2, the two stations are considered geographically close to each other and connected to the same node of the grid (see Fig. 1). In this context, the current section gives an illustration on how the considered model could serve for a charging station operator (CSO) responsible for both stations to take a fundamental sizing decision with a common power 'capacity' for both stations which is described by the value $\alpha_1 K_1 + \alpha_2 K_2$. Without giving all the practical details, note that this 'capacity' choice can correspond to: either an economical perspective if considering that the CSO has to choose for an electricity contract in which a maximal power has to be subscribed or; a physical perspective if this power limit corresponds to some upgrades to be made on the charging stations site (e.g. with a local transformer to be installed). The CSO decision-making problem is then expressed as the maximization (in α_1 and α_2)

of the probability $\bar{P}_{eq}$ of leaving the stations with a full battery at equilibrium (EV users' QoS metric) under the constraint of a given power 'capacity'. This constraint optimization problem can be formulated as follows:

$$\max_{\alpha_1,\alpha_2} \bar{P}_{eq} \tag{10}$$

$$\text{s.t } \alpha_1 K_1 + \alpha_2 K_2 \leq C, \tag{11}$$

where $C \in \mathbb{N}$ is the maximum number of charging points (parking spots) that can be used at maximum power level simultaneously. The following proposition gives the solution P_{eq}^{opt} of the optimization problem (10) under constraint (11).

Proposition 8. *Suppose, without loss of generalization, $K_1 \leq K_2$. Let P_{eq}^{opt} be the optimal value of $\bar{P}_{eq}$ under under the constraint (11), and S_{eq}^{opt} the set of optimal values of α_1 when the constraint (11) is active. Then, the solution $(P_{eq}^{opt}, S_{eq}^{opt})$ of (10) under (11) is:*

$$(P_{eq}^{opt}, S_{eq}^{opt}) = \begin{cases} (\frac{p_{max}\mu}{\nu+p_{max}\mu}, S_1), & \text{if } C \geq \frac{\min(\lambda,K_1\nu)}{\nu+p_{max}\mu}, \quad (a) \\[2mm] (\frac{p_{max}}{\lambda}C, I_1), & \text{if } \begin{cases} C < \frac{\lambda}{\nu+p_{max}\mu}, \\ K_1 > \frac{\lambda}{\nu}, \end{cases} \quad (b) \\[3mm] (\frac{p_{max}\mu}{\nu}\frac{C}{K_1}, S_2) & \text{if } \begin{cases} C < \frac{K_1\nu}{\nu+p_{max}\mu}, \quad (c) \\ K_1 \leq \frac{\lambda}{\nu}, \end{cases} \end{cases} \tag{12}$$

where

$$I_1 = \left[\max(0, \frac{C-K_2}{K_1}, \min(1, \frac{C}{K_1})\right],$$

$$S_1 = \begin{cases} S_1^1 & \text{if } K_1\frac{\nu}{\nu+\tilde{\mu}} \leq C < K_2\frac{\nu}{\nu+\tilde{\mu}}, \\[2mm] S_1^2 \cup S_1^1 & \text{if } \begin{cases} K_2 < \frac{\lambda}{\nu}, \\ K_2\frac{\nu}{\nu+\tilde{\mu}} \leq C < \frac{\min(\lambda,(K_1+K_2)\nu)}{\nu+\tilde{\mu}}, \end{cases} \\[3mm] I_1 & \text{if } \frac{\min(\lambda,(K_1+K_2)\nu)}{\nu+\tilde{\mu}} \leq C, \end{cases}$$

with

$$S_1^1 = \left[\frac{\nu}{\nu+\tilde{\mu}}, \min(1, \frac{C}{K_1})\right],$$
$$S_1^2 = \left[\max(0, \frac{C-K_2}{K_1}), \frac{C}{K_1} - \frac{K_2\nu}{K_1(\nu+\tilde{\mu})}\right],$$

and

$$S_2 = \begin{cases} \{\frac{C}{K_1}\} & \text{if } K_1 < K_2, \\[2mm] \{0, \frac{C}{K_1}\} & \text{if } K_1 = K_2. \end{cases}$$

Remark 1. The opposite case $K_2 \leq K_1$ is symmetric and straightforward.

Proof of Proposition 8. The proof mainly rely on Proposition 5. Suppose the constraint is $\alpha_1 K_1 + \alpha_2 K_2 = C$ instead of (11). Then, $\alpha_2 = \frac{C}{K_2} - \alpha_1 \frac{K_1}{K_2}$ and I_1 is the set of admissible α_1: $I_1 = \{\alpha_1 \in [0,1], \frac{C}{K_2} - \alpha_1 \frac{K_1}{K_2} \in [0,1]\}$.

Denoting X_1 the set of α_1 for which $\bar{P}_{eq} = \frac{p_{\max}\mu}{\nu + p_{\max}\mu}$, $X_1 = I_1$ if $C \geq \frac{\lambda}{\nu + p_{\max}\mu}$ and,

$$X_1 = \{\alpha_1 \in I_1, \max(\alpha_1, \frac{C}{K_2} - \alpha_1 \frac{K_1}{K_2}) \geq \frac{\nu}{\nu + p_{\max}\nu}\}$$

$$= \left[\max(0, \frac{C - K_2}{K_1}), \min(1, \frac{C}{K_1} - \frac{K_2\nu}{K_1(\nu + p_{\max}\mu)})\right] \cup \left[\frac{\nu}{\nu + p_{\max}\mu}, \min(1, \frac{C}{K_1})\right],$$

otherwise. Then X_1 is non-empty if and only if $C \geq \frac{\min(\lambda, K_1\nu)}{\nu + p_{\max}\mu}$. With a short analysis of the set X_1, it gives part (a) of Eq. (12).

In the following, suppose $C < \frac{\min(\lambda, K_1\nu)}{\nu + p_{\max}\mu}$. Then $\bar{P}_{eq} = \frac{p_{\max}\mu}{\lambda}C$ or $\bar{P}_{eq} = \frac{p_{\max}\mu}{\nu}\max(\alpha_1, \alpha_2)$.

– Suppose $K_1 + K_2 > \frac{\lambda}{\nu}$. Let

$$X_2 = \{\alpha_1 \in I_1, \max(\alpha_1, \frac{C}{K_2} - \alpha_1 \frac{K_1}{K_2}) \geq \frac{\nu}{\lambda}C\}$$

$$= \left[0, \frac{C}{K_1} - C\frac{\nu K_2}{\lambda K_1}\right] \cup \left]C\frac{\nu}{\lambda}, \frac{C}{K_1}\right].$$

A short analysis of the set X_2 gives that X_2 is empty if and only if $K_1 > \frac{\lambda}{\nu}$. In this case, $\bar{P}_{eq} = \frac{p_{\max}\mu}{\lambda}C$ independently of α_1, which gives part (b) of Eq. (12). Otherwise if $K_1 \leq \frac{\lambda}{\nu}$, a short analysis of X_2 gives different behaviors of $\bar{P}_{eq}$:

• constant equal to $\frac{p_{\max}\mu}{\lambda}C$ and then increasing equal to $\frac{p_{\max}\mu}{\nu}\alpha_1$ if $K_2 \geq \frac{\lambda}{\nu}$, which imply that

$$\begin{cases} S_{eq}^{opt} = \{\frac{C}{K_1}\}, \\ \max_{\alpha_1} \bar{P}_{eq}(\alpha_1) = \frac{p_{\max}\mu}{\nu}\frac{C}{K_1}. \end{cases}$$

This assymetric case can only happen if $K_1 < K_2$;

• decreasing equal $\frac{p_{\max}\mu}{\nu}(\frac{C}{K_2} - \alpha_1\frac{K_1}{K_2})$ then constant equal to $\frac{p_{\max}\mu}{\lambda}C$ and then increasing equal to $\frac{p_{\max}\mu}{\nu}\alpha_1$, which implies that

$$\begin{cases} S_{eq}^{opt} = \begin{cases} \{\frac{C}{K_1}\} & \text{if } K_1 < K_2, \\ \{0, \frac{C}{K_1}\} & \text{if } K_1 = K_2, \end{cases} \\ \max_{\alpha_1} \bar{P}_{eq}(\alpha_1) = \max(\frac{p_{\max}\mu}{\nu}\frac{C}{K_2}, \frac{p_{\max}\mu}{\nu}\frac{C}{K_1}) = \frac{p_{\max}\mu}{\nu}\frac{C}{K_1}; \end{cases}$$

• decreasing then increasing and the result is the same as in the previous case.

– If $K_1 + K_2 \leq \frac{\lambda}{\nu}$, then $\forall\alpha_1 \in I_1$, $\bar{P}_{eq} = \max(\alpha_1, \frac{C}{K_2} - \alpha_1\frac{K_1}{K_2})$, which gives the same result as in the two last subcases.

Part (c) of Eq. (12) is thus obtained.

Because $\bar{P}_{eq}^{opt}$ under constraint $\alpha_1 K_1 + \alpha_2 K_2 = C$ is increasing in C, its value cannot be strictly greater with a smaller value of C. Thus the optimal value $\bar{P}_{eq}^{opt}$ is the same with constraint (11). ∎

Note that, at the optimum solution, the constraint (11) is active and then all the capacity is used, i.e. $\alpha_1^* K_1 + \alpha_2^* K_2 = C$.

Next section illustrates all the previous results about equilibrium, PoA and optimization of the probability depending on energy capacity C.

5   Numerical Illustrations

In this section some numerical examples illustrate the results obtained in previous sections in order to deepen the understanding of the dependency on system parameters. These parameters are given in Table 1.

Table 1. Chosen parameters for the numerical examples.

$K_1 = 5$	5 parking spots in station 1
$K_2 = 30$	30 parking spots in station 2
$C = 19$	Maximum number of EVs that can charge at maximum power simultaneously at both stations
$p_{\max} = 7$	Maximum power available for an EV is 7 kW
$\lambda = 20$	EVs arrive in average every 3 min
$\nu = 1/2$	EVs stay in average 2 h at the chosen station
$\mu = 1/30$	EVs need in average 30 kWh to have a fully charged battery

5.1   Probability of Full Battery at Equilibrium

In this use case, the number of parking spots is considered small, meaning that: $K_1 + K_2 \leq \frac{\lambda}{\nu}$. As Proposition 5 states, there are two different equilibria depending on the maximum of the proportion of parking spots charging at maximum rate of stations 1 and 2 $\max(\alpha_1, \alpha_2)$. The first example considers the following parameters $\alpha_1 = \alpha_2 = 0.6$ which are inline with realistic scenario applied by a typical french CSO (60% of the charging points can deliver maximum level of power simultaneously).

- If $\max(\alpha_1, \alpha_2) < \frac{\nu}{\nu + p_{\max}\mu} \approx 0.68$ then the value at equilibrium of the probability of leaving the station with a full battery is $\bar{P}_{eq} = \frac{p_{\max}\mu}{\nu} \max(\alpha_1, \alpha_2)$. This first case is illustrated on Fig. 2a where the probability at equilibrium is $\bar{P}_{eq} = 0.28$. Note that in this particular case ($\alpha_1 = \alpha_2$) the equilibrium interval is $\left[\frac{\nu K_1}{\lambda}, 1 - \frac{\nu K_2}{\lambda}\right] = [0.125, 0.25]$. It means that around 20% of EVs join station 1 at equilibrium.

- Else $\max(\alpha_1, \alpha_2) > \frac{\nu}{\nu + p_{\max}\mu} \approx 0.68$ then $\bar{P}_{eq} = \frac{p_{\max}\mu}{\nu + p_{\max}\mu} = 0.31$. This case is illustrated on Fig. 2b where $\alpha_1 = 0.8$. In this case, the probability at equilibrium is not significantly higher but at least 35% of EVs join station 1 which has a higher capacity than station 2, even with lower parking spots.

Note that in the first case the probability $\bar{P}_{eq}$ of leaving the chosen station with a fully charged battery at equilibrium strictly increases with α_1 when $\alpha_1 > \alpha_2$, whereas it stays constant in α_1 in the second case.

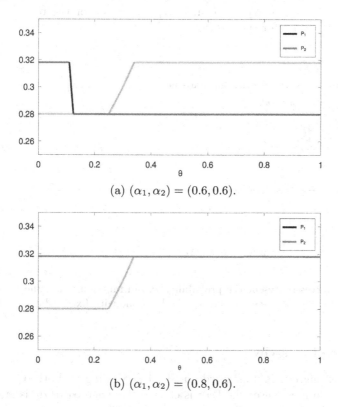

(a) $(\alpha_1, \alpha_2) = (0.6, 0.6)$.

(b) $(\alpha_1, \alpha_2) = (0.8, 0.6)$.

Fig. 2. The blue curve (resp. red curve) corresponds to the probability to leave station 1 (resp. station 2) with a fully charged battery. These probabilities are computed in function on the probability θ for a user to choose station 1. The intersection of the two curves correspond to the equilibrium.

The value of this probability at equilibrium is relatively low. But in fact EVs remain parked at a station in average 2 h and the maximum power $p_{\max}$ is pretty low 7kW. On Fig. 3 we illustrate the probability of leaving with a full battery at equilibrium depending on the capacity parameter α_1 of station 1. The capacity parameter α_2 for the second station is either fixed with the standard value of 60% (blue line) or depends on α_1 so that the constraint (11) is active (red line).

Interestingly, when α_1 is less than 20% the probability is higher when the energy capacity α_2 of the second station is correlated with α_1. Whereas, for the other cases when this energy parameter is higher, it is better to have fixed value of α_2. Note that the expression of $\bar{P}_{eq}$ is the same for both cases (blue and red lines). In particular, $\bar{P}_{eq} = \frac{p_{max}\mu}{\nu} \max(\alpha_1, \alpha_2)$ for $\alpha_1 \leq \frac{\nu}{\nu + p_{max}\mu} \approx 0.68$. The difference when $\alpha_1 < \alpha_2 = 0.6$ comes mathematically from two reasons:

- For $\alpha_1 \leq \frac{C}{K_1 + K_2} \approx 0.54$, $\max(\alpha_1, \alpha_2) = \alpha_2$ for both curves but α_2 stays constant for the blue line whereas it decreases in α_1 for the red line.
- For $\frac{C}{K_1 + K_2} \leq \alpha_1 \leq \alpha_2 = 0.6$, $\max(\alpha_1, \alpha_2) = \alpha_2$ for the blue line whereas $\max(\alpha_1, \alpha_2) = \alpha_1$ for the red line.

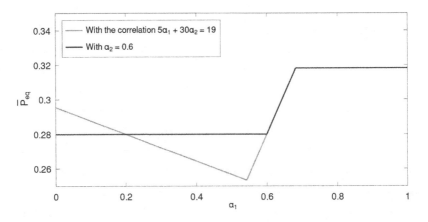

Fig. 3. The curves represent the probability $\bar{P}_{eq}$ in function of α_1: the red one in the case of dependency between α_1 and α_2, the blue one with fixed value $\alpha_2 = 0.6$.

5.2   Price of Anarchy

The price of anarchy is illustrated on Fig. 4 depending on both α_1 and α_2. We observe that in most cases the PoA is close to 1 : when α_1 or α_2 is greater than the threshold $\frac{\nu}{\nu + p_{max}\mu} \approx 0.68$. If not, for every fixed value of α_2, the price of anarchy in α_1 is at its largest when $\alpha_1 = \alpha_2$. The observation is the same when α_1 is fixed and α_2 varies. The price of anarchy takes only large values close to 1.7 when both energy capacities α_1 and α_2 are small and the two stations have the same sharing policy $\alpha_1 = \alpha_2$. In this case there are more interactions in terms of energy sharing between EVs and then a centralized point of view is better.

Figure 4 also shows that in case of dependency $5\alpha_1 + 30\alpha_2 = 19$, the price of anarchy will stay close to 1 because the maximum between α_1 and α_2 is always relatively large ($\max(\alpha_1, \alpha_2) \geq \frac{C}{K_1 + K_2} \approx 0.54$).

Note also that the asymmetry between α_1 and α_2 comes from the different number of parking spots in the stations: $K_1 \neq K_2$.

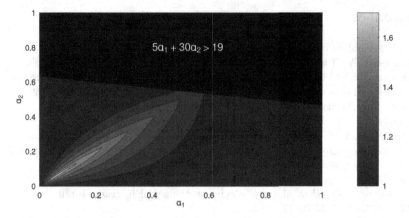

Fig. 4. Level lines and values of the price of anarchy in function of α_1 and α_2. The area in black is the set of (α_1, α_2) which do not respect the constraint $5\alpha_1 + 30\alpha_2 \leq 19$.

6 Conclusions and Future Works

In this paper a nonatomic game model for an electric vehicle (EV) charging system is proposed in order to study the EV best strategy about the choice of charging station in order to maximize the probability of fully charging its battery. EVs interact at the parking spots which have limited capacity in terms of the number of charging points, EVs may stay parked even if not charging, and also interact at the energy level because plugged EVs share a global amount of energy to be shared at a station. These different levels of interactions make the analysis of the equibrium of the nonatomic game complex but some properties of the equilibrium are obtained theoretically. Moreover, we have been able to compare the performance with a centralized point of view by giving explicit bounds on the Price of Anarchy. Finally, a deep study of energy capacity parameters for each station is proposed in order to maximize the probability of any EV to leave teh service station with a full battery. Numerical examples based on realistic scenario parameters illustrate our contributions.

In future works we will focus on the information aspect. In the framework of this paper, EVs do not have any information about the stations occupancy, and make their decision based on expected number of EV charging. Then, we plan to study the impact of online information about station occupancy and/or remaining charging time on the EV equilibrium decision. Another point of interest is the integration of the rejection into the strategic decision of EV. In fact, if all parking spots are occupied, a new EV is rejected and there is no waiting list for the moment.

References

1. IEA: Global ev outlook 2020, Technical report, IEA (2020)
2. Khaksari, A., Tsaousoglou, G., Makris, P., Steriotis, K., Efthymiopoulos, N., Varvarigos, E.: Sizing of electric vehicle charging stations with smart charging capabilities and quality of service requirements. Sustain. Cities Soc. **70** (2021)
3. Said, D., Cherkaoui, S., Khoukhi, L: Queuing model for EVs charging at public supply stations. In: Proceedings of the IWCNC (2013)
4. Tan, X., Sun, B., Tsang, D.H.: Queueing network models for electric vehicle charging station with battery swapping. In: IEEE International Conference on Smart Grid Communications, pp. 1–6 (2014)
5. Gusrialdi, A., Qu, Z., Simaan, M.: Scheduling and cooperative control for charging of electric vehicles' at highway service stations. IEEE Trans. Intell. Transp. Syst. **18**(10), 2713–2727 (2017)
6. Aveklouris, A., Nakahira, Y., Vlasiou, M., Zwart, B.: Electric vehicle charging: a queueing approach. In: MAMA workshop (2017)
7. Aveklouris, A., Vlasiou, M., Zwart, B.: Bounds and limit theorems for a layered queueing model in electric vehicle charging. Queueing Syst. **93**, 83–137 (2019)
8. Liu, B., Pantelidis, T., Tam, S.J.: Chow An electric vehicle charging station access equilibrium model with M/D/C queueing, ArXiv, 2102.05851 (2021)
9. Correa, J.R., Schulz, A.S., Stier-Moses, N.E.: Selfish routing in capacitated networks. Math. Oper. Res. **29**(4), 961–976 (2004)
10. Yi, Z., Bauer, P.: Effects of environmental factors on electric vehicle energy consumption: a sensitivity analysis. IET Electr. Syst. Transp. **7**(1), 3–13 (2017)
11. Kleinrock, L.: Queueing Systems. Wiley, Hoboken (1974)
12. Aumann, R., Shapley, L.: Value of Non-atomic Games. Princeton University Press, Princeton (1974)
13. Hassin, R.: Rational Queueing. CRC Press, Boca Raton (2016)
14. Schmeidler, D.: Equilibrium Points in nonatomic games. J. Stat. Phys. **7**(4), 295–300 (1973)
15. Roughgarden, T.: Selfish Routing and the Price of Anarchy. MIT Press, Cambridge (2005)
16. Hassin, R., Haviv, M.: To Queue or Not to Queue. Kluwer Press, Amsterdam (2003)

Hybrid Simulation of Energy Management in IoT Edge Computing Surveillance Systems

Lelio Campanile[1], Marco Gribaudo[2], Mauro Iacono[1],
and Michele Mastroianni[1(✉)]

[1] Dipartimento di Matematica e Fisica, Università degli Studi della Campania,
"L. Vanvitelli", viale Lincoln 5, 81100 Caserta, Caserta, Italy
{lelio.campanile,mauro.iacono,michele.mastroianni}@unicampania.it
[2] Dipartimento di Elettronica, Informatica e Bioingegneria, Politecnico di Milano,
via Ponzio 5, 20133 Milan, Italy
marco.gribaudo@polimi.it

Abstract. Internet of Things (IoT) is a well established approach used for the implementation of surveillance systems that are suitable for monitoring large portions of territory. Current developments allow the design of battery powered IoT nodes that can communicate over the network with low energy requirements and locally perform some computing and coordination task, besides running sensing and related processing: it is thus possible to implement edge computing oriented solutions on IoT, if the design encompasses both hardware and software elements in terms of sensing, processing, computing, communications and routing energy costs as one of the quality indices of the system. In this paper we propose a modeling approach for edge computing IoT-based monitoring systems energy related characteristics, suitable for the analysis of energy levels of large battery powered monitoring systems with dynamic and reactive computing workloads.

Keywords: Performance evaluation · IoT · Surveillance · Edge computing

1 Introduction

The use of IoT is a commodity in many application fields. Since the introduction of this technology, its appeal and the large flexibility attracted the interest of researchers, practitioners and industry, opening a wide number of different research directions. From the design and engineering point of view, IoT present many challenges, that encompasses all aspects related to the electronics of the node and of the sensors, to its software layer, to data (pre)processing, to communications, to network management, and to some aspects that are simultaneously related to both the node and the network levels, such as the ones that deal with energy management in battery powered nodes. Battery powered IoT architectures allow a high degree of flexibility in deployment, as they do not need any

P. Ballarini et al. (Eds.): EPEW 2021/ASMTA 2021, LNCS 13104, pp. 345–359, 2021.
https://doi.org/10.1007/978-3-030-91825-5_21

infrastructure, so they are ideal to be used for applications that target scenarios in historical buildings that cannot be wired, in natural environments where deployment happens in the wild, and in risky contexts that do not allow free access and in which nodes are, for example, thrown from drones or planes: conversely, they add a constraint in the design and development process, that diminishes the lifetime of each node and, at the same time, challenges the integrity of the whole network, that is based on the contribution of nodes that also act as intermediate interconnections to deliver data from all the network to the network sink managing the link towards the rest of the computing architecture.

The progress in battery technology and the lowering costs and power requirements of hardware resources such as system-on-chip devices and memories pave the way to a new generation of IoT, that can support edge computing technologies to provide higher level functionalities and on site distributed computing, or more advanced management to make IoT more resilient and robust even when the continuity of the connections of part of the nodes toward the sink is threatened or momentarily unavailable: this makes the role of software on the nodes more relevant, and also requires to consider the possibility of transmission of larger data chunks, to support migration on tasks and offloading when the energy level is critical on a node or to interact with the cloud; consequently, also scheduling of software activities on the node and event-driven activation of tasks play a significant role.

In this paper we present a modeling approach for battery powered, IoT based, edge computing enabled monitoring architectures, aiming at considering the energy aspects in scenarios in which sensing and local computing react to what happens in each part of the covered environment. The approach is based on discrete events simulation and is demonstrated with a realistic case study that deals with safety of natural sites.

This paper is organized as follows: Sect. 2 presents related work, Sect. 3 introduces the general architecture of the reference monitoring system, Sect. 4 describes the modeling approach, Sect. 5 describes a scenario to which the approach is applied, Sect. 6 presents the result of the simulation and conclusions close the paper.

2   Related Work

2.1   Analytical Methods

An approach quite used for minimize energy consumption in low-power devices is based on queuing networks modeling, resolving the energy minimization by using Lyapunov-based optimization algorithms. An early example may be found in [1], in which offloading is used as an effective method for extending the lifetime of handheld mobile devices by executing some components of applications remotely (e.g., on the server in a data center or in a cloud). In order to achieve energy saving while satisfying given application execution time requirement, a dynamic offloading algorithm is proposed, which is based on Lyapunov optimization. An example of this approach may be found in [2]. The authors deep into the problem of power consumption in a multiuser Mobile Edge Computing system with energy harvesting devices, and the system power consumption, which includes the local execution power and the offloading transmission power, is designated as the main system performance index.

Another approach is based on game theory. In [3] the multi-user computation offloading problem between mobile-edge and cloud is addressed. In this paper, the authors make use of a multi-user computation offloading game, analyzing the structural properties of the game and arguing that the game admits a Nash equilibrium. A similar approach may be found in [4] and [5].

2.2 Simulation of Energy Management

In [6] the expected performance of Mobile Edge Computing is evaluated to address application issues related to energy management on constrained IoT devices with limited power sources, while also providing low-latency processing of visual data being generated at high resolutions. In this paper an algorithm is proposed that analyzes the tradeoffs in computing policies to offload visual data processing to an edge cloud or a core cloud at low-to-high workloads, also analyzing the impact on energy consumption in the decision-making under different visual data consumption requirements (i.e., users with thick clients or thin clients). The algorithm is simulated using the ns-3 [7] network simulator.

In [8] a new simulator is proposed, named IoTSim-Edge, that extends the capability of CloudSim to incorporate the different features of edge and IoT devices. In particular, this simulator also includes support for modeling heterogeneous IoT protocols along with their energy consumption profile. In this paper there are also two useful tables that summarize the main characteristics of different communication protocols (e.g., WiFi, 4G-LTE, Bluetooth and so on) and different messaging application-layer protocols (e.g. HTTP, XMPP etc.).

2.3 Simulation and Analysis of Animal Movements

There is interesting literature which examines animal movements in small and large scale, for individuals, groups or crowds and in different contexts, and proposing analysis and simulation techniques. In [9] simulation of ecosystem data, including animal movements, is examined, in a "virtual ecologist" perspective. [10] specially deals with animal path observation, sampling and modeling, with particular focus on the effects of actual distance correct estimation. In [11] the authors introduce a 'stochastic movement simulator' for estimating habitat connectivity which is suitable for the analysis and modeling of animals and groups movement, including the effects of correlations and environmental factors. [12] presents a R package, namely *Animal movement tools*, for the analysis and synthesis of tracking data about animals, with features that connect data to stochastic models. In [13] individual movements are simulated as a correlated random walk in order to model animal dispersal in a fragmented landscape taking into account landscape structure and animal behavior. [14] introduces spatially explicit individual-based models and compares four distinct movement approaches on a fish related case study.

[15] proposes a point-to-point random trajectory generator for animal movement, to be used in simulation scenarios with different geographical scales and for different species, including physical limitations.

In this paper we opted for a simpler approach, which takes into account the essential elements pointed out by literature, due to the purposes of our work.

3   Characteristics of the System

The general architecture of the systems we aim to model is composed of IoT nodes that can run additional workloads besides what is needed to execute sensing, (pre)processing and communications. Nodes are scattered in the environment and have a basic monitoring mode, in which they alternate a sleeping phase and, periodically, a sensing phase, or a low-energy always-on sensing state in which they can react to events happening in the monitored environment. In both cases, when events happen, an alert condition is raised and computing tasks may be launched locally or on the near group of nodes, to analyze the situation on a single or cooperative way: in this case, the node goes to a higher energy need state, as it activates additional computing hardware and launches one or more software tasks, eventually generating additional communications on the network, with non negligible weight in the energy economy of the node. The node may also have more than one higher energy need states, that may be used to compute at different speeds if needed to face computing deadlines.

3.1   Activities

Activities may be classified into 4 categories: normal monitoring, alert, alarm and reaction.

The *normal monitoring* category includes minimal activities, in which nodes interact rarely to update the knowledge about the distributed state of the IoT and perform basic sensing and related data processing, logging information about the environment and waiting for anomalous elementary signals.

If such a signal is detected, the nodes that detected it execute *alert* activities, in which the signal is locally compared to known ordinary events to understand if it is not a threat, also coordinating with each other to obtain a better evaluation, and reports the condition to the cloud, that may also check the situation.

In case the condition is not considered normal, *alarm* activities are performed, that may require real time or computationally intensive tasks, to be executed in edge mode in collaboration with the cloud section of the system, such as image recognition, classification, tracking of moving object by analyzing audio signals or other privacy compliant information available by sensors or other node devices, generally operated by the nodes also in other conditions or specifically activated during alert or alarm activities.

If the alert activities identified a dangerous situation or a violation, *reaction* activities are executed on nodes, that may include video monitoring of the area surrounding the tracked offender or the dangerous phenomenon by following the target activating and deactivating cameras along its path or by activating all available cameras needed, depending on the area, the configuration and the nature of the surveilled environment.

While normal monitoring activities are executed in a low energy state or by periodically switching to it from a sleep state, all other activities are executed at higher energy need states.

3.2 Organization

If allowed by the deployment mechanism, nodes may be installed in groups that are meant to collaborate. In this case, nodes in the same group may be equipped with different sensors, in order to operate in organized groups while executing alert, alarm or reaction activities. Groups may be preconfigured to facilitate offloading of computing tasks or to execute distributed computing tasks to natively distribute energy utilization across a group. Groups may be reconfigurable during the lifetime of the IoT and may have redundant sensors to keep them as cold or hot spares. Groups may also collaborate to efficiently detect movements of targets that walk or run across the surveilled area. In the rest of this paper groups based features will generally not be applied.

3.3 Energy Management

In order to guarantee the survival of the system, the overall use of energy should be properly tuned and dynamically adjusted according to needs: the goal is to keep all the IoT connected to its sink, not to lose coverage of the surveilled area. In order to cope with this need, energy should be spent as a common resource. The nodes that are more solicited by acting as *intermediate carrier nodes* between other nodes and the sink will probably incur in a significantly faster energy consumption with respect to the ones that are positioned at the margin of the covered area, as, in a calm scenario, communications, that require transmission power, will anyway happen periodically, differently from alert and alarm computing activities. These nodes will serve as computing backup resources to which alert and alarm computing activities can be offloaded when intermediate carrier nodes will reach a given *attention threshold*: in this situation, the node the battery of which has a decrease of the energy level that reaches the attention threshold will have to transfer the most energy needing computing tasks on a near node, in its neighborhood, to the node that, according to a heuristic or to some optimization algorithm, is the best candidate to let the network use its energy in the most useful way.

4   Modeling Approach

We consider systems that can be described by a tuple (E, Σ, V) where $E = (N_x, N_y, \Delta_x, \Delta_y, \sigma_x, \sigma_y, r)$ describe the *environment*, Σ the sensor behavior, and V the patterns of agents visiting the environment. In particular, the considered environment is composed by $N_x \times N_y$ sensor uniformly spread over a rectangular region. Sensors are distributed with an average distance Δ_x and Δ_y from each other, with an error of σ_x and σ_y. Figure 1 shows an example of a region where $N_x = 40$, $N_y = 10$, $\Delta_x = \Delta_y = 80$ m, and $\sigma_x = \sigma_y = 20$ m, which will be used a running example in the following. Each sensor is described by a state machine $\Sigma = (S, T, A)$ composed by $S = \{s_1, \ldots, s_{|S|}\}$ states, and $T \subseteq \Sigma \times \Sigma$ transitions. Each transition is triggered by an action $a \in A$, according to events occurring in the environment. In the considered area, we have a set of possible *visitors* patterns

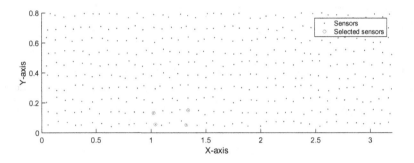

Fig. 1. Example sensors positioning map (unit for both dimensions is km): four sensors, that will be considered more in depth in the following, have been highlighted with a red circle

$V = \{v_1, \ldots, |V|\}$ that randomly pass through the environment. The proposed technique is independent on the way in which the visitor pattern is simulated, and such description, which is a research topic itself, is outside the scope of this paper. We then resort to a simplified motion pattern description, where:

$$v_i = \{P_i, \lambda_i, \pi_i, l_i\}, \quad \text{with} \tag{1}$$

$$P_i = \{p_{i1}, \ldots, p_{iK}\}, \quad \text{and} \tag{2}$$

$$p_{ij} = (x_{ij}, y_{ij}, t_{ij}, \mathbf{s}_{ij}) \tag{3}$$

Each motion pattern v_i is triggered by a Poisson process of rate λ_i, is characterized by an alert priority $l_i \in \mathcal{N}$ and by a detection probability $\pi_i \in [0, 1]$: l_i and π_i will be discussed more in depth in the following. Each path is described by an ordered set P_i (Eq. 2) of key-points p_{ij} (Eq. 3), and is composed by a set of linear trajectories travelled at uniform and constant speed. In particular, each segment $p_{ij} - p_{ij+1}$ connects the point of coordinates (x_{ij}, y_{ij}) to (x_{ij}, y_{ij}), which are respectively visited after t_{ij} and t_{ij+1} time units from the beginning of the pattern. Figure 5, 6 and 7 show the path that will be used in the considered example. To make simulation more realistic, each time a pattern is repeated, its key-points are perturbed both in position and timing with a Gaussian noise of zero mean, and variance $\mathbf{s}_{ij} = (\sigma_x, \sigma_y, \sigma_t)$. Since Gaussian noise can be negative, particular attention is given to the time component: in particular, time instants are sorted after the perturbation, so to ensure a monotone flow of time. Figure 2 shows how basic path *Hunting* of Fig. 5 is perturbed to produce 10 different instances, showing their variation in space and time.

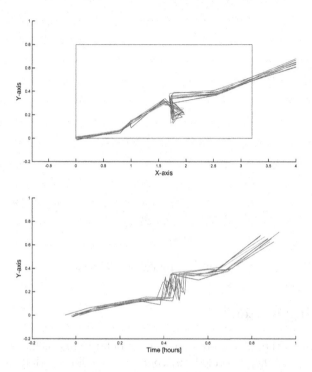

Fig. 2. Variation of trajectories in space and time

Simulation is performed by first generating a Poisson event stream for each possible pattern $v_i \in V$, and replicating a perturbed instance of v_i at each event, as shown in Fig. 3. Each trajectory and sensor are then considered separately, and intersection events are generated: in particular, times t_k^I and t_k^O in which an object enters in the sensor range r of a sensor is computed. Note that, as shown in Fig. 4, an agent might change direction while in the range of a sensor. In this case, the exit time t_k^O must be computed on the correct output segment.

Events generated by pattern v_i will be considered with probability π_i, and discarded otherwise: this can be used to simulate the fact that some events might not be detected by a sensor, or to model possible intermittent malfunctions. Events can cause the state change of the corresponding sensor: to simplify its logic, events are grouped in alarm levels l_i. The next state of a sensor will be determined by actions a considering both the current sensor state, and the alarm level l_i. Since actions can be very case specific, we will not describe them more formally, and we will focus on the one used in this paper in the following section.

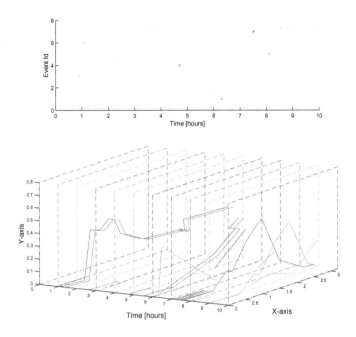

Fig. 3. Execution of trajectories

5   Modeling a Scenario

In order to show the effectiveness of the modeling approach, we analyzed an example scenario. The surveilled environment is a forest, that is usually free from human presence. The surveillance is enacted to ensure early intervention in case of spontaneous fires or in case of a pyromaniac or an arsonist. A sufficient number of nodes is installed in the forest that can detect events by sensing noises and temperature peaks, identify moving living beings, and make video footage or stream live video (an example positioning used for our simulations is depicted in Fig. 1). Each node executes activities from (some of) the categories presented in Sect. 3.

Events like a branch that falls solicit the system for a short time, specifically activating the group of nodes that is the nearest to the place in which the event happens; events like a spontaneous fire solicit the system for a short time but raise alert, alarm and reaction activities; events like wild animals that enter and run across the forest solicit the groups that monitor the path in sequence, thus allowing following the trajectory, identifying the signal as wildlife or humans to be distinguished from each other by a proper local analysis, in case with the assistance of the cloud; events like humans that enter and run across the forest also require that the kind of behavior is identified as normal hiking, hunting or other allowed activities versus suspicious behaviors, that may raise alarm activities and/or reaction activities on part or all the IoT. In consequence of

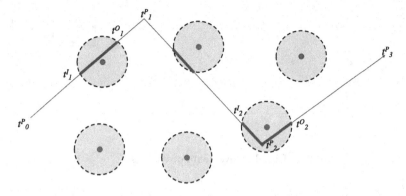

Fig. 4. Intersections between trajectories and sensing devices

these events, energy is used by the nodes in reaction to the events and differently from node to node.

For the sake of simplicity, we consider in the following as solicitations by living beings the activities of a fox and of a human. For the fox, three different typical behaviors are considered: a fox running across the forest, a fox that marks the forest as part of its own territory and a hunting fox that finds a prey; for the human, four different typical behaviors are considered: a human doing hiking in the forest, a human doing hiking with a stop to rest a bit, a hunter and an arsonist. For each of these cases, a single sequence of movements and stops in time has been considered as a base on which analogous cases have been generated with scaling, flipping and rotating the sequence. Example base movement sequences are in Fig. 5, 6 and 7. Speed profiles used for these simulated behavior are not reported for the sake of space.

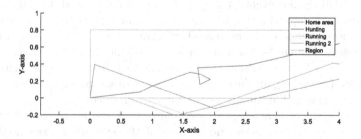

Fig. 5. Movements of foxes

Movement detected by the sensors are grouped in two different intensity levels: $l_i = 2$ for the arsonist (Fig. 7), to denote its critical effect on the system, and $l_i = 1$ for all the other movements (Fig. 5 and 6). Sensors are characterized by four states $S = \{\text{Idle, Low energy, High energy, Critical}\}$, as shown in Fig. 8 together with their ID number.

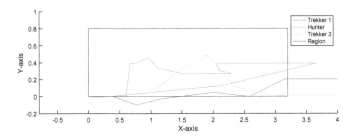

Fig. 6. Movements of visitors

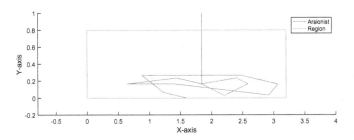

Fig. 7. Movements of an arsonist

Sensors starts in the "Idle" state. Each time t_j^I an agent enters it sensing range r, a counter is incremented. If the event is discarded, this leads to a "Low energy" state, and a counter of discarded events is incremented. If the entrance event is not discarded, it leads either to a "High energy" or to a "Critical" state depending on the level l_i of the considered event. Each time t_j^O an agent leaves the sensing range, the corresponding counter is decremented. If there are no more critical events, but still regular events, the system returns either to state "High energy" or "Low energy" depending on whether the reaming event was detected or not. When no more events are being sensed, it returns the the "Idle" state. Figure 9 shows the state change for the four sensors highlighted in Fig. 1. Each sensor is equipped with a battery, which in our model starts with an arbitrarily placed value of 150. Depending on the state, the battery discharges at a different speed, as shown in Table 1. The simulator then linearly reduces the charge of the battery, according to the time spent in each state, as shown for the four highlighted sensors in Fig. 10. To increase the lifetime of the system, sensors might relay their task on neighbour nodes if their battery level is below a threshold χ. In this example, this has been arbitrarily set to $\chi = 50$. In this case, the energy consumed by the considered sensor is greatly reduced (see right column of Table 1), and its workload is shared among the neighbours whose battery level is over the considered threshold. The latter sensors will then experience a discharge that is proportional to the type of detected event, and inversely proportional to the number of nodes sharing this task.

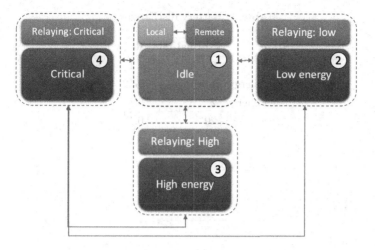

Fig. 8. State diagram of the sensors

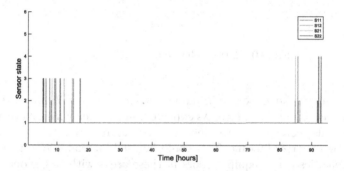

Fig. 9. Changes in the state for four sensors

6   Results

Results where computed on a 2016 MacBook Pro laptop, with Intel $i5$ CPU and 16 GB of RAM using a custom built script in MatLab. Each simulation run required about 5s of execution time. A total of 100 simulation runs were executed: where relevant, both the average and the variance of the results will be presented, showing the possibility of computing confidence intervals if desired.

Figure 11 shows the locations in which some events were missed by the system (i.e. when the missed event counter described in Sect. 5 was incremented). Dots represent standard events, while circles correspond to the critical ones. This type of plot can be used to better place sensors over the area: where more missed events, especially critical ones, happen, it would be desirable to have more IoT devices.

Table 1. Battery discharge rate

State	Normal rate	Relay rate
Idle	0.1	0.1
Low energy	1	0.12
High energy	2	0.12
Critical	20	0.12

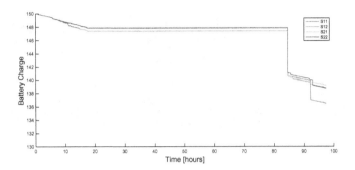

Fig. 10. Battery charge as function of time

The lifetime of the sensors is instead considered in Fig. 12, for what concerns both its average and its variance. As expected, in the area where more events are missed, that also corresponds to the zone where more action takes place, sensors tend to have a shorter lifetime. This could be used for example to plan different battery capacities, so to equip sensors in these areas with larger ones. What is also interesting is the high-variance area for $y > 0.4$ and $x \approx 1.5$. That particular section monitors the path used by the arsonist to escape the area: this higher variance might denote that in many cases events occurring there might not be detected, probably due to the speed at which they occur. This could guide in deploying IoT devices with better sensing equipment in that area, to reduce the probability of missing events.

Figure 13 shows instead the effect of the load distribution between neighbours on the first three rows of sensors at time $t = 500$, comparing it against a scenario in which all IoT only process data locally. As it can be seen, loading increases the battery of the most targeted nodes, at the expense of some of the neighbour devices. However, due to the large number of nodes among which load can be shared, the effect on the neighbour is almost negligible.

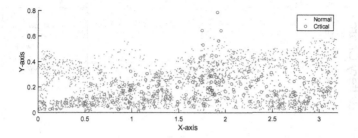

Fig. 11. Missed events

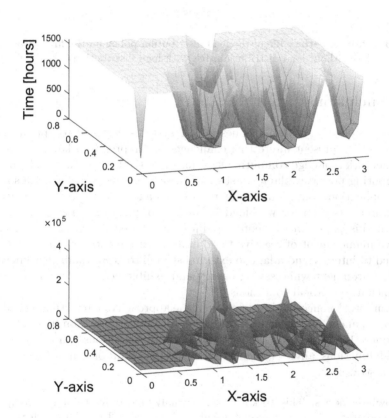

Fig. 12. Average and variance of failure time for each sensor

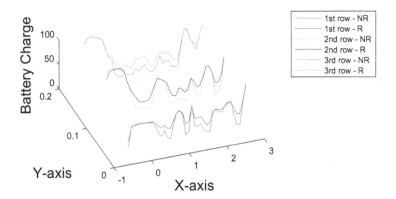

Fig. 13. Effect on battery life of the load distribution policy on the first three rows of sensors: NR - without load distribution, R - with load distribution

7   Conclusions

The energy efficiency related issues in IoT systems designed to operate in isolated locations present interesting challenges and represent a key factor for the implementation of edge computing systems on IoT infrastructures. The increase in computing power available on state-of-the-art devices calls for more sophisticated applications, but requires a more careful management of on-board available energy and attention for workload balancing to extend the maximum mission time. In this paper we presented a simulation based approach for an energy effective management of reactive tasks in an environmental monitoring system, designed to interact and adapt to external stimuli to keep a sufficient knowledge of the environment while saving energy and exhibit peak performances in case of suspicious or critical conditions.

Future work concerns the inclusion in the approach of harvesting mechanisms (e.g. solar panels) and of smarter migration techniques, based on fuzzy offloading techniques [16], as well as the study of network conditions that allow keeping selected functions alive even in critical energy conditions and selective node pruning strategies.

Acknowledgments. This work has been partially funded by the internal competitive funding program "VALERE: VAnviteLli pEr la RicErca" of Università degli Studi della Campania "Luigi Vanvitelli".

References

1. Huang, D., Wang, P., Niyato, D.: A dynamic offloading algorithm for mobile computing. IEEE Trans. Wireless Commun. **11**(6), 1991–1995 (2012)
2. Zhang, G., Chen, Y., Shen, Z., Wang, L.: Distributed energy management for multiuser mobile-edge computing systems with energy harvesting devices and QoS constraints. IEEE Internet Things J. **6**(3), 4035–4048 (2019)

3. Chen, X., Jiao, L., Li, W., Fu, X.: Efficient multi-user computation offloading for mobile-edge cloud computing. IEEE/ACM Trans. Networking **24**(5), 2795–2808 (2016)
4. Shah-Mansouri, H., Wong, V.W.S.: Hierarchical fog-cloud computing for IoT systems: a computation offloading game. IEEE Internet Things J. **5**(4), 3246–3257 (2018)
5. Liu, L., Chang, Z., Guo, X.: Socially aware dynamic computation offloading scheme for fog computing system with energy harvesting devices. IEEE Internet Things J. **5**(3), 1869–1879 (2018)
6. Trinh, H., et al.: Energy-aware mobile edge computing and routing for low-latency visual data processing. IEEE Trans. Multimedia **20**(10), 2562–2577 (2018)
7. Campanile, L., Gribaudo, M., Iacono, M., Marulli, F., Mastroianni, M.: Computer network simulation with ns-3: a systematic literature review. Electronics **9**(2), 272 (2020). https://doi.org/10.3390/electronics9020272
8. Jha, D.N., et al.: IoTSim-edge: a simulation framework for modeling the behavior of internet of things and edge computing environments. Softw.: Pract. Exp. **50**(6), 844–867 (2020)
9. Zurell, D., et al.: The virtual ecologist approach: simulating data and observers. Oikos **119**(4), 622–635 (2010)
10. Rowcliffe, J., Carbone, C., Kays, R., Kranstauber, B., Jansen, P.: Bias in estimating animal travel distance: the effect of sampling frequency. Methods Ecol. Evol. **3**(4), 653–662 (2012)
11. Palmer, S., Coulon, A., Travis, J.: Introducing a 'stochastic movement simulator' for estimating habitat connectivity. Methods Ecol. Evol. **2**(3), 258–268 (2011)
12. Signer, J., Fieberg, J., Avgar, T.: Animal movement tools (amt): R package for managing tracking data and conducting habitat selection analyses. Ecol. Evol. **9**(2), 880–890 (2019)
13. Vuilleumier, S., Metzger, R.: Animal dispersal modelling: handling landscape features and related animal choices. Ecol. Model. **190**(1–2), 159–170 (2006)
14. Watkins, K., Rose, K.: Evaluating the performance of individual-based animal movement models in novel environments. Ecol. Model. **250**, 214–234 (2013)
15. Technitis, G., Othman, W., Safi, K., Weibel, R.: From A to B, randomly: a point-to-point random trajectory generator for animal movement. Int. J. Geogr. Inf. Sci. **29**(6), 912–934 (2015)
16. Campanile, L., Iacono, M., Marulli, F., Mastroianni, M., Mazzocca, N.: Toward a fuzzy-based approach for computational load offloading of IoT devices. J. Univ. Comput. Sci. **26**(11), 1455–1474 (2020)

Highly Accurate, Explicit Approximation for Erlang B

Paul Kevin Reeser$^{(\boxtimes)}$ (iD)

AT&T Labs, Middletown, NJ 07748, USA
preeser@att.com

Abstract. Exact solutions for distributions in the M/M/N/Q system are known, but computational speed and stability can still be problematic. In this work, we develop new expressions for the waiting time and response time distributions in terms of the Erlang B formula, where its numerically stable recursion facilitates efficient solution for tail probabilities without any summations. We next develop a highly accurate, numerically stable, explicit closed-form approximation for Erlang B that requires no recursion. Used separately or in combination, these new foundational results can dramatically reduce the computational burden in simulation and optimization algorithms, when system performance measures must be computed many times, or when solutions are required in real time (e.g., SDN reconfiguration). In addition, the Erlang B approximation facilitates rapid back-of-the-envelope system analyses (e.g., when tail probabilities are inputs).

Keywords: Waiting time · Response time · Distribution · Erlang B · Approximation

1 Introduction

Analysis of continuous-time Markov chains, birth-and-death (B&D) processes, and Markovian queues dates to the early 20th century. Markov [1] first introduced the framework in 1907, and Erlang [2] first derived his classic B and C formulae in 1917. Exact solutions to all performance measures for M/M/N/Q finite-server, finite-buffer Markovian queues have been widely documented in queueing theory texts (c.f. [3–8] and references therein). Seemingly no stone has been left unturned.

Yet though the exact solutions are known, numerical stability and speed of computation remain persistent problems at times. Solving for the simplest of performance measures requires computing a normalization constant based on a sum of terms that can become numerically unstable even for systems of modest size. This problem becomes even more difficult when computing complicated performance measures such as distributions, where both sums and sums of sums are required.

In this paper, we revisit a known expression for the state space normalization constant and develop new expressions for the waiting time and response time distributions, in terms of the Erlang B formula, without any summation (even in the case of distributions). Although Erlang B is itself expressed as a sum of terms, there is a well-known, efficient, numerically stable recursion that avoids sums of terms entirely. Next, we develop a new approximation for Erlang B that allows us to avoid iteration or summation altogether.

© Springer Nature Switzerland AG 2021
P. Ballarini et al. (Eds.): EPEW 2021/ASMTA 2021, LNCS 13104, pp. 360–374, 2021.
https://doi.org/10.1007/978-3-030-91825-5_22

This new approximation is derived by first 'anchoring' the solution at known corners (offered utilization $\rho = 0$, $\rho = 1$, number of servers $N = 1$, and $N \to \infty$.) and then 'shaping' the approximation to mimic the known behavior in between. The result is explicit, closed form, numerically stable, and extremely accurate.

The ability to express the normalization constant explicitly in terms of Erlang B means that what could be a numerically challenging problem is now a simple, stable N-step recursion. Furthermore, the ability to express waiting time and response time distributions explicitly in terms of Erlang B means that what could be an extremely difficult computational problem is now replaced by simple N- and Q-step recursions. Finally, availability of an extremely accurate, explicit closed form, numerically stable approximation for Erlang B allows us to avoid recursion or summation altogether.

All M/M/N/Q system performance measures are given exactly as a function of the normalization constant. Thus, an accurate approximation for Erlang B yields accurate approximations for all metrics. Still, one may ask why develop approximations when we have the exact solutions via recursion. There are myriad reasons why fast, efficient, numerically stable computational forms for the normalization constant, the waiting time and response time distributions, and Erlang B itself are of value. For example, such techniques can dramatically reduce computational burden in simulation and optimization analyses, when key performance measures must be computed many times, or when solutions are required in real time. Optimization problems may be easily solved using approximation, whereas their solution may be difficult or impossible with the exact function. Also, a continuous Erlang B approximation aids solution of problems involving non-integer numbers of servers.

In addition, such approximations facilitate rapid back-of-the-envelope real-world system sizing, when 'normal' performance outputs (mean blocking, delay, tail probabilities, etc.) are specified as inputs, and the required system size (number of processors, number of buffers, etc.) must be determined recursively (reverse engineering). These capabilities are invaluable during the system requirements phase for vendor proposal evaluation, or during the architecture phase for solution design evaluation.

2 M/M/N/Q Queueing System Refresher

2.1 Formulation

Using Kendall notation, we consider the classic M/M/N/Q finite-server, finite-buffer, first-come-first-served Markovian queueing system with Poisson arrivals (rate λ), exponential service (rate μ), N servers ($0 < N \leq \infty$), and Q buffers ($0 \leq Q \leq \infty$), so that $N + Q$ customers are allowed in the system. Let $\tau = \mu^{-1}$ denote the mean service time, let $A = \lambda\tau$ denote the offered load, and let $\rho = A/N$ denote the utilization (assume $\rho \leq 1$).

Typical boundary cases include: single-server systems where $N = 1$ and (typically) $Q = \infty$, infinite-server systems where $N = \infty$ and $Q = 0$ by default, and multi-server systems where $1 < N < \infty$ and $Q = \infty$ (Erlang C *delay* system) or $Q = 0$ (Erlang B *loss* system). Although the Erlang B formula arose within this M/M/N/0 framework, applications of an accurate approximation extend well beyond Markovian systems. We consider only the infinite-source case but note that Erlang also derived the classic finite-source Engset formula, and much work exists addressing its numerical evaluation (c.f. [9]).

Let P_i $(0 \le i \le N + Q)$ denote the steady-state probability of i customers in the system. Then the stationary distribution can be expressed as

$$
P_i = \begin{cases} \frac{A^i}{i!}P_0 & (0 \le i \le N) \\ \frac{N^N \rho^i}{N!}P_0 & (N \le i \le N+Q) \end{cases}, \text{ where } P_0 = \left[\sum_{i=0}^{N} \frac{A^i}{i!} + \left(\frac{A^N}{N!} \right) \frac{\rho(1 - \rho^Q)}{(1 - \rho)} \right]^{-1}.
$$
(1)

Let $G \equiv P_N$ denote the normalization constant. Then

$$
G \equiv P_N = \left(\frac{A^N}{N!} \right) P_0 = \left[\left(\frac{N!}{A^N} \right) \sum_{i=0}^{N} \frac{A^i}{i!} + \frac{\rho(1 - \rho^Q)}{(1 - \rho)} \right]^{-1}.
$$
(2)

We define the normalization constant in terms of P_N in order to facilitate its expression as a function of the Erlang B formula. It is immediately obvious from (2) that

$$
G = \left[\frac{1}{B(N,A)} + \frac{\rho(1 - \rho^Q)}{(1 - \rho)} \right]^{-1}, \text{ where } B(N, A) \equiv \left[\left(\frac{N!}{A^N} \right) \sum_{i=0}^{N} \frac{A^i}{i!} \right]^{-1}
$$
(3)

is the well-known Erlang B formula. $B(N, A)$ represents the probability of blocking (loss) in the M/M/N/0 system.

This result in (3) is not new (c.f. [8]). However, as mentioned previously, there are several reasons to express the normalization constant G in terms of Erlang B. Most notably, $B(N, A)$ can be computed using the well-known recursion

$$
B(N, A) = \left[\frac{AB(N-1, A)}{N + AB(N-1, A)} \right], \text{ starting with } B(0, A) = 1.
$$
(4)

In all cases, N iterations or terms are still required, but computing $B(N, A)$ recursively is more stable than computing G directly by summation. As a prelude to the results to follow, we develop an extremely accurate, explicit approximation for $B(N, A)$ that avoids iteration or summation altogether and allows for the direct and accurate evaluation of G using the relationship in (3).

2.2 Key Performance Indicators

All M/M/N/Q system performance measures are given exactly as a function of G, or equivalently as a function of $B(N, A)$ using (3). For example:

- Probability of blocking $\beta = G\rho^Q$
- Probability of queueing $\pi = G(1 - \rho^Q)/(1 - \rho)$
- Probability of waiting (given customer is admitted) $P(W > 0) = \pi/(1 - \beta)$, where the random variable (r.v.) W denotes the waiting time
- Mean number in service (carried load) $n = A(1 - \beta)$
- Mean number in queue $q = G\rho[1 - (Q - Q\rho + 1)\rho^Q]/(1 - \rho)^2$
- Mean waiting time (given admitted) $w = E(W) = \tau q/n$
- Mean response time (given admitted) $r = E(R) = w + \tau$, where the r.v. R denotes the response (sojourn) time.

Next, consider the waiting time and response time distributions in the M/M/N/Q system. Let $W(t) \equiv P(W \le t | admitted)$ denote the waiting time distribution and let $R(t) \equiv P(R \le t | admitted)$ denote the response time distribution. We prove the following relationships in the Appendix. For $0 < Q < \infty$ and $\rho < 1$,

$$P(W > t \mid admitted) = \frac{G(At)^{Q-1}e^{-Nt}}{(1-\beta)(1-\rho)(Q-1)!}\left[\frac{1}{B(Q-1, At)} - \frac{\rho}{B(Q-1, Nt)}\right].$$

$$(5)$$

For $N > 1$ and $N - A \ne 0, 1$,

$$P(R > t \mid admitted) = \left\{1 - \pi - \beta + \frac{GN}{N-A-1}\left[1 - \left(\frac{A}{N-1}\right)^Q\right]\right\}\frac{e^{-t}}{1-\beta}$$

$$+ \frac{G(At)^{Q-1}e^{-Nt}}{(1-\beta)(N-1)(Q-1)!}\left\{\left(\frac{N-1}{N-A}\right)\left[\frac{N}{B(Q-1,At)} - \frac{A}{B(Q-1,Nt)}\right] - \right.$$
$$\left. \left(\frac{N}{N-A-1}\right)\left[\frac{N-1}{B(Q-1,At)} - \frac{A}{B(Q-1,(N-1)t)}\right]\right\}. \quad (6)$$

Special cases for $N = 1, N = 1 - A, N = A, Q = 0$, and $Q = \infty$ are also derived in the Appendix.

To the best of our knowledge, the results (5–6) are new. These representations of $W(t)$ and $R(t)$ in terms of Erlang B eliminate the multiple layers of summation and recursion altogether. The significance of these representations is that $W(t)$ and $R(t)$ can now be expressed explicitly in terms of Erlang B, resulting in the ability to compute waiting time and response time tail probabilities quickly via simple traditional recursion, or directly using the approximation to follow that avoids recursion altogether.

These new representations (5–6) for $W(t)$ and $R(t)$ contain a coefficient of the form $\{(Nt)^Q e^{-Nt}/Q!\}$. This expression is the probability mass function at Q of the Poisson r.v. with rate Nt, hence this coefficient $\in [0, 1]$. Nevertheless, for large N or Q or t, direct computation of its individual parts may still become numerically unstable. In this case, simple Q-step recursion can be used concurrent with the Q-step Erlang B recursion. Thus, these results are numerically stable even for large systems.

3   Prior Work

As noted earlier, Markovian queueing systems have been widely studied and documented (but it is rare to find the derivations of the waiting time and response time distributions explicitly included in standard texts). Jagerman, Whitt, and many others at Bell Laboratories performed seminal analyses of the characteristics of Erlang B beginning in the 1970s (c.f. [10–15]). Over the years, numerous bounds and approximations for Erlang B have been proposed. Although many of these approximations are useful to provide ballpark initial sizing estimates, none achieve the level of accuracy required for embedding in complex optimization problems such as real-time software defined network (SDN) route reconfiguration. Harel [16] developed bounds and approximations in 1988 that are still widely cited. Figure 1 shows the exact value for $B(N, N\rho)$ vs. Harel's bounds and approximation as a function of $\rho \in [0, 1]$ for $N = 10$ and $N = 100$. As can be seen, although the fit improves as N increases, the bounds are not tight, and the approximation does not become accurate until $\rho \to 1$.

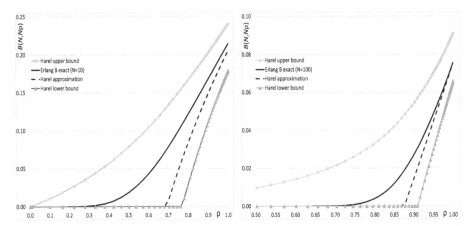

Fig. 1. Exact $B(N, N\rho)$ vs. Harel bounds and approximation for $N = 10$ and 100 as a function of ρ.

More recently, others (c.f. [17–20] and references therein) have proposed various approximations for Erlang B that achieve good accuracy under certain conditions (e.g., low B or very large N or heavy traffic $\rho > 1$). For example, Pinsky [17] proposes a simple approximation for the low blocking range $B \in [10^{-5}, 10^{-1}]$. Salles and Barria [18] and Shakhov [19] propose simple heavy-traffic approximations that are asymptotically exact as $N \to \infty$ or $\rho \to \infty$. Janssen, et al. [20] develop bounds for large N that are asymptotically exact as $N \to \infty$. Unfortunately, these approximations fail to achieve the required level of accuracy across the full practical range of N and $\rho \le 1$.

4   Erlang B Approximation

4.1   Properties

For simplicity, let $B \equiv B(N, N\rho)$ denote the exact Erlang B value for N servers and ρ offered utilization, and let $B_a \equiv B_a(N, N\rho)$ denote the corresponding approximation. The obvious properties of an accurate Erlang B approximation B_a are that it matches the exact result for known boundary conditions (corners), and that it has the right shape in between. The known boundary conditions are $\rho = 0$, $\rho = 1$, $N = 1$, and $N \to \infty$.

Obviously we need $B_a = 0$ at $\rho = 0$ $\forall N \ge 1$. Next, we want $B_a =$ some function $F(N)$ at $\rho = 1$ $\forall N \ge 1$. The solution to $F(N) = B(N, N)$ will be provided shortly. Then we want $B_a = \rho/(1 + \rho)$ at $N = 1$ for $0 \le \rho \le 1$. Finally, we want $B_a \to 0$ via some function $G(\rho, N)$ as $N \to \infty$ for $0 \le \rho < 1$. The solution to $G(\rho, N)$ will be provided shortly. In terms of the shape (as a function of ρ), we want B_a to be concave for $N = 1$, given by $\rho/(1 + \rho)$, and we want B_a to be convex, with an inflection point, then concave for $N > 1$.

The foundation for our Erlang B approximation consists of three pillars:

1. the solution to $F(N) = B(N, N)$ at $\rho = 1$,
2. the solution to $G(\rho, N) = \lim_{N \to \infty} B(N, N\rho)$ for fixed ρ, and
3. the solution to the shaping and fitting function $H(\rho, N)$ at and in between corners.

Thus, $B_a = F(N)G(\rho, N)H(\rho, N)$, where the derivations of F, G, and H follow.

4.2 $B(N, N)$

The first crucial pillar is the solution to $F(N) = B(N, N)$. Through asymptotic expansion of the incomplete Gamma function, Jagerman [10] and later van Leeuwaarden and Temme [21] show that

$$\frac{6}{4 + 3\sqrt{2\pi N} + \sqrt{2\pi}/4\sqrt{N}} \leq B(N, N) \leq \frac{6}{4 + 3\sqrt{2\pi N}}, \tag{7}$$

$$\text{and that } \lim_{N \to \infty} B(N, N) = \frac{6}{4 + 3\sqrt{2\pi N}}. \tag{8}$$

Figure 2 shows the exact value and *limit* (upper bound) as $N \to \infty$ for $B(N, N)$ as a function of N. As can be seen, the *limit* is surprisingly accurate even for small values of N.

Next, we can easily correct the inaccuracy for small N. Figure 3 shows the ratio of the exact value of $B(N, N)$ over the *limit* for N between 1 and 1000 (log scale). As can be seen, the *correction* is visually of the form $1 - \frac{k}{f(N)}$ where $(1) = 1$, $\lim_{N \to \infty} f(N) \to \infty$, and

$$k = 1 - \frac{B(1, 1)}{limit} = 1 - \frac{0.5}{6/\left(4 + 3\sqrt{2\pi}\right)} = \frac{8 - 3\sqrt{2\pi}}{12} \approx 0.04. \tag{9}$$

The *limit* formula already includes the power of N term $N^{0.5}$, so choosing $f(N) = N^x$ for some x simplifies the final form for $F(N)$. The best (least squares) *correction* is achieved with $(N) = N^{0.8}$. Thus, the first approximation pillar $F(N)$ is given by

$$F(N) \approx correction \times limit = \left(1 - \frac{8 - 3\sqrt{2\pi}}{12N^{0.8}}\right)\left(\frac{6}{4 + 3\sqrt{2\pi N}}\right) = \left(\frac{12N^{0.8} - 8 + 3\sqrt{2\pi}}{8N^{0.8} + 6\sqrt{2\pi}N^{1.3}}\right). \tag{10}$$

For future use, note that $F(1) = \frac{1}{2}$ and $\lim_{N \to \infty} F(N) = \lim_{N \to \infty} \frac{2}{\sqrt{2\pi N}}$.

4.3 $\lim_{N \to \infty} B(N, N\rho)$

The next crucial pillar is the solution to $G(\rho, N) = \lim_{N \to \infty} B(N, N\rho)$ for fixed ρ. We have

$$\lim_{N \to \infty} B(N, N\rho) = \lim_{N \to \infty} \left[\left(\frac{N!}{N^N \rho^N}\right) \sum_{i=0}^{N} \frac{(\rho N)^i}{i!}\right]^{-1}. \tag{11}$$

Using the relationships $\lim_{N \to \infty}\left[\sum_{i=0}^{N} \frac{(\rho N)^i}{i!}\right] = e^{\rho N}$ (Taylor) then $\lim_{N \to \infty} N! = \sqrt{2\pi N}N^N e^{-N}$ (Sterling), we have

$$\lim_{N \to \infty} B(N, N\rho) = \lim_{N \to \infty} \frac{N^N \rho^N e^{-\rho N}}{N!} = \lim_{N \to \infty} \frac{\left[\rho e^{(1-\rho)}\right]^N}{\sqrt{2\pi N}}. \tag{12}$$

The $\sqrt{2\pi N}$ factor in the denominator of (12) is already contributed by $F(N)$ as $N \to \infty$, so this factor is redundant. Thus, the second approximation pillar is

$$G(\rho, N) = \left[\rho e^{(1-\rho)}\right]^N. \tag{13}$$

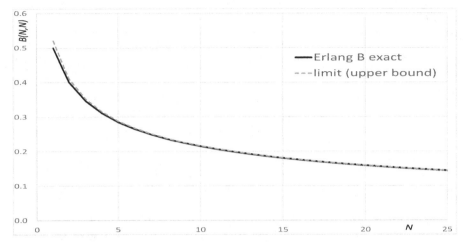

Fig. 2. Exact $B(N, N)$ vs. *limit* $\dfrac{6}{4+3\sqrt{2\pi N}}$ as a function of N.

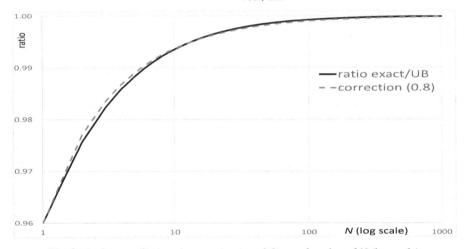

Fig. 3. Ratio exact/*limit* and *correction* ($x = 0.8$) as a function of N (log scale).

4.4 Shape and Fit

Thus far, our approximation is given by

$$B_a \equiv F(N)G(\rho, N)H(\rho, N) = \left(\frac{12N^{0.8} - 8 + 3\sqrt{2\pi}}{8N^{0.8} + 6\sqrt{2\pi}N^{1.3}}\right)\left[\rho e^{(1-\rho)}\right]^N H(\rho, N). \quad (14)$$

The shaping/fitting function $H(\rho, N)$ must ensure that the approximation matches at the corners and achieves the proper fit in between. To illuminate the desired shape, Fig. 4 shows the exact value for $B(N, N\rho)$ and $F(N)G(\rho, N)$ as functions of $\rho \in [0, 1]$ for $N = 1,\ldots,10000$, where all functions are normalized to 1 when $\rho = 1$. As can be seen, by anchoring the approximation at $\rho = 1$ via $F(N)$ and using the theoretical asymptotic decay rate $G(\rho, N)$ as $N \to \infty$, our results (thus far) significantly underestimate the exact value, and there is opportunity for improvement via the fitting function $H(\rho, N)$.

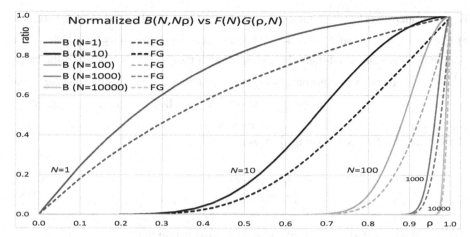

Fig. 4. Normalized exact $B(N, N\rho)$ vs. $F(N)G(\rho, N)$ for $N = 1,\ldots,10000$ as a function of ρ.

The first step is to compare the exact solution $B(N, N\rho)$ to the approximation at the corners. When $\rho = 1$, we need

$$B(N, N) = F(N) \approx F(N)G(1, N)H(1, N) = F(N)H(1, N).$$

$$\text{Thus, } H(1, N) = 1. \tag{15}$$

Next, when $N = 1$, we need

$$B(1, \rho) = \rho/(1 + \rho) \approx F(1)G(\rho, 1)H(\rho, 1) = \frac{1}{2}\rho e^{(1-\rho)}H(\rho, 1).$$

$$\text{Thus, } H(\rho, 1) = \left(\frac{2}{1+\rho}\right)e^{\rho-1}. \tag{16}$$

Finally, as $N \to \infty$, we need

$$\lim_{N\to\infty} B(N, N\rho) = \lim_{N\to\infty} \frac{\left[\rho e^{(1-\rho)}\right]^N}{\sqrt{2\pi N}} \approx \lim_{N\to\infty} F(N)G(\rho, N)H(\rho, N) = \lim_{N\to\infty} \frac{2\left[\rho e^{(1-\rho)}\right]^N}{\sqrt{2\pi N}}H(\rho, N).$$

$$\text{Thus, } \lim_{N\to\infty} H(\rho, N) = \frac{1}{2}. \tag{17}$$

There are many approaches to designing a function $H(\rho, N)$ that matches these three corners and achieves the proper fit in between. The approach adopted here is to first gradually 'neutralize' the most complex corner result (16). In particular, we desire the two terms of $H(\rho, 1)$ in (16) to evolve as follows as N goes from 1 to ∞: $\left(\frac{2}{1+\rho}\right) \to 1$ and $\left(e^{\rho-1}\right) \to 1$. With this desired behavior in mind, a first term of the form $\left(\frac{1+\rho^{x-1}}{1+\rho^y}\right)$ and a second term of the form $e^{(\rho-1)/z}$ meet the needs, where (for simplicity) x, y, and z are limited to either N or $\sqrt{N}$. Based on trial and error, $x = y = N$ and $z = \sqrt{N}$ provide the best fit in between. Thus, $H(\rho, 1) = \left(\frac{2}{1+\rho}\right)e^{\rho-1}$ generalizes well to

$$\left(\frac{1+\rho^{N-1}}{1+\rho^N}\right)e^{(\rho-1)/\sqrt{N}} \text{ for } N \geq 1. \tag{18}$$

Next considering (15, 17), a number of functional forms can be used to fit the corners (1 at $\rho = 1$ or $N = 1$ and $\frac{1}{2}$ as $N \to \infty$). For instance, the forms $\frac{1}{2-\rho^{x-1}}$ or $\left(\frac{1+\rho^y}{2}\right)^{1-1/z}$ both fit the boundary conditions. Though somewhat more complex, we prefer the latter $\left(\frac{1+\rho^y}{2}\right)^{1-1/z}$ form since it is continuous and more computationally stable, whereas the former $\frac{1}{2-\rho^{x-1}}$ form has a discontinuity at $\rho^{x-1} = 2$ (in heavy traffic). Again based on trial and error, $y = \sqrt{\pi N}$ and $z = \sqrt{N}$ provide the best fit. Thus, an excellent fit and shape is achieved by generalizing $\lim_{N\to\infty} H(\rho, N) = \frac{1}{2}$ to

$$\left(\frac{1 + \rho^{\sqrt{\pi N}}}{2}\right)^{1-1/\sqrt{N}}. \tag{19}$$

Thus combining (18–19), the third and final approximation pillar is

$$H(\rho, N) = \left(\frac{1 + \rho^{N-1}}{1 + \rho^N}\right)e^{(\rho-1)/\sqrt{N}}\left(\frac{1 + \rho^{\sqrt{\pi N}}}{2}\right)^{1-1/\sqrt{N}}. \tag{20}$$

Combining (10, 13, 20), our Erlang B approximation $B_a = F(N)G(\rho, N)H(\rho, N)$ is given by

$$B_a = \left(\frac{12N^{0.8} - 8 + 3\sqrt{2\pi}}{8N^{0.8} + 6\sqrt{2\pi}N^{1.3}}\right)\left[\rho e^{(1-\rho)}\right]^N\left(\frac{1 + \rho^{N-1}}{1 + \rho^N}\right)e^{(\rho-1)/\sqrt{N}}\left(\frac{1 + \rho^{\sqrt{\pi N}}}{2}\right)^{1-1/\sqrt{N}}. \tag{21}$$

While this approximation in (21) appears to be quite complicated, its five parts are all computationally stable over the range $\rho \in [0, 1] \,\forall\, N > 0$, and straightforward to compute directly without any recursion. Specifically, $F(N) = \left(\frac{12N^{0.8}-8+3\sqrt{2\pi}}{8N^{0.8}+6\sqrt{2\pi}N^{1.3}}\right) \in [0, 1] \,\forall\, N$ and $\lim_{N\to\infty} F(N) = \lim_{N\to\infty} \sqrt{\frac{2}{\pi N}} = 0$. Next, $G(\rho, N) = \left[\rho e^{(1-\rho)}\right]^N \in [0, 1] \,\forall\, N$ and ρ (including heavy traffic $\rho > 1$). Finally, $\left(\frac{1+\rho^{N-1}}{1+\rho^N}\right) \in [0,2] \,\forall\, N$ and ρ (including $\rho > 1$), $e^{(\rho-1)/\sqrt{N}} \in [0, 1] \,\forall\, N$ and $\forall\, \rho \in [0, 1]$, and $\left(\frac{1+\rho^{\sqrt{\pi N}}}{2}\right)^{1-1/\sqrt{N}} \in [\frac{1}{2},1] \,\forall\, N$ and $\forall\, \rho \in [0, 1]$.

5 Results and Application

Figures 5, 6 and 7 show our approximation $B_a(N, N\rho)$ compared to the exact value of Erlang B $B(N, N\rho)$ as functions of $\rho \in [0, 1]$ for $N = 1, 2, 5, 10$ (Fig. 5), $N = 50$, 100, 500 (Fig. 6), and $N = 1000, 10000$ (Fig. 7). As can be seen, the approximation is remarkably accurate, even for the small values of $N > 1$ in Fig. 5, where the approximate *correction* to the *limit* as $N \to \infty$ for $B(N, N)$ introduces some error. Note that the x-axis in Fig. 6 starts at $\rho = 0.5$, and the x-axis in Fig. 7 starts at $\rho = 0.9$, in order to better highlight the accuracy of the approximation in the range where $B(N, N) > 0$.

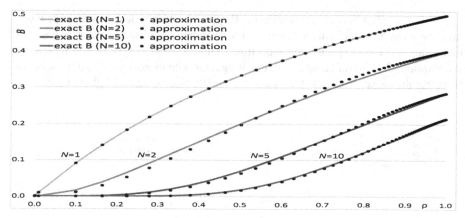

Fig. 5. Exact $B(N, N\rho)$ vs. approximation $B_a(N, N\rho)$ for $N = 1, 2, 5, 10$ as a function of ρ.

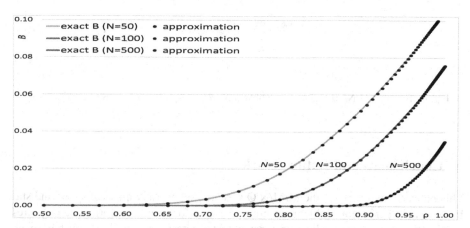

Fig. 6. Exact $B(N, N\rho)$ vs. approximation $B_a(N, N\rho)$ for $N = 50, 100, 500$ as a function of ρ.

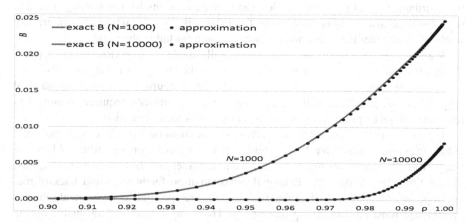

Fig. 7. Exact $B(N, N\rho)$ vs. approximation $B_a(N, N\rho)$ for $N = 1000, 10000$ as a function of ρ.

As discussed earlier, an accurate Erlang B approximation yields accurate approximations for all system performance metrics. For example, Fig. 8 shows exact and approximate values for the expected response time in the M/M/N/∞ system as a function of ρ for various N. As can be seen, the resulting approximations are extremely accurate. Note that the x-axes in Fig. 8 start at ρ = 0.5 and 0.95 to better highlight the accuracy.

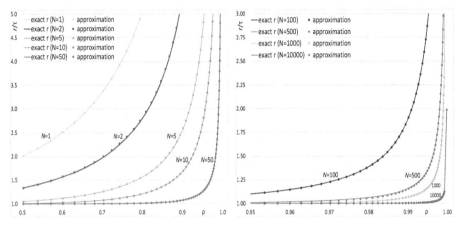

Fig. 8. Exact M/M/N/∞ expected response time r/τ vs. approximation as a function of ρ.

6 Conclusions and Future Work

Though exact solutions for the M/M/N/Q system are known, numerical speed and stability remain persistent problems at times. Solving for simple performance measures requires computing a normalization constant that can become numerically unstable even for systems of modest size. This problem becomes even more difficult when computing distributions. In this work, we develop new expressions for the waiting time and response time distributions in terms of the Erlang B formula, where its numerically stable recursion facilitates fast and efficient solution for tail probabilities.

We next develop a highly accurate, numerically stable, explicit closed-form approximation for Erlang B that requires no recursion. Unlike most prior Erlang B approximations, our new result achieves both the required level of accuracy across the full practical range of N and ρ ≤ 1 as well as the computational simplicity required to embed in real-time network optimization algorithms (e.g., SDN reconfiguration).

Used separately or in combination, these new foundational results can dramatically reduce the computational burden in simulation and optimization algorithms, when system performance measures must be computed many times, or when solutions are required in real time. In addition, the Erlang B approximation facilitates rapid back-of-the-envelope inverse analyses, when the usual outputs are specified as inputs, and required system size must be determined recursively. These capabilities are invaluable during the requirements phase for vendor proposal evaluation, or during the architecture phase for solution design evaluation. Additional applications of an accurate approximation for Erlang B extend well beyond Markovian systems.

Although the goal of this work has always been to develop an approximation for the practical range $\rho \in [0, 1]$, possible future work includes assessing the accuracy of the approximation in heavy traffic ($\rho > 1$), together with exploring techniques to ensure computational stability when $\rho > 1$. The heavy traffic behavior of Erlang B has been widely studied (c.f., Whitt [12] and references therein, as well as [18–20]). Various heavy traffic 'anchor' points have been proposed for $\lim_{N \to \infty} B(N, N\rho)$ for fixed $\rho > 1$, and for $\lim_{\rho \to \infty} B(N, N\rho)$ for fixed N. Leveraging these additional asymptotic behaviors may lead to an entirely different (and perhaps simpler) shaping/fitting function $H(\rho, N)$.

Furthermore, as noted previously, there are many approaches to designing a shaping/fitting function that matches the corners and achieves the proper fit in between, and the approach adopted here was to first 'neutralize' the most complex corner result. While beneficial in achieving excellent accuracy for the practical range $\rho \in [0, 1]$, this approach resulted in terms $e^{(\rho-1)/\sqrt{N}}$ and $\left(\frac{1+\rho^{\sqrt{\pi N}}}{2}\right)^{1-1/\sqrt{N}}$ that both eventually become unstable when $\rho > 1$, thus constraining this particular result to the range $\rho \leq 1$. Taking an entirely different approach to designing $H(\rho, N)$ may lead to a different result that is both accurate and computationally stable for both light and heavy traffic.

Acknowledgements. We wish to thank D. Hoeflin for his invaluable assistance with parts of the Appendix, and W. Whitt for his insightful comments on the manuscript.

Appendix

Assume that $\mu = 1$. Let $W(t) \equiv P(W \leq t \mid admitted)$ denote the waiting time distribution in the M/M/N/Q system, and let $1 - W(t) = P(W > t \mid admitted)$ denote its tail probability. Let $R(t) \equiv P(R \leq t \mid admitted)$ denote the response time distribution, and let $1 - R(t) = P(R > t \mid admitted)$ denote its tail probability.

$W(t)$ and $R(t)$ are typically expressed in terms of the Poisson distribution. Let

$$P_{\lambda x}(k) \equiv \sum_{i=0}^{k} \left[\frac{(\lambda x)^i}{i!} e^{-\lambda x} \right] = \frac{(\lambda x)^k e^{-\lambda x}}{k! B(k, \lambda x)} \tag{22}$$

denote the CDF of the Poisson r.v. with rate λx, and note that $P_{\lambda x}(k)$ can be expressed as a function of the Erlang B formula (the foundation of the proofs to follow).

For $0 < Q < \infty$ and $\rho < 1$, it has been shown (c.f. [7] pp. 97–99) that

$$P(W > t \mid admitted) = \frac{G}{1 - \beta} \sum_{n=0}^{Q-1} \rho^n [P_{Nt}(n)]. \tag{23}$$

Substituting (22) into (23) and rearranging (omitting straightforward steps) yields

$$
\begin{aligned}
P(W > t \mid admitted) &= \frac{G}{1-\beta} \sum_{n=0}^{Q-1} \rho^n [P_{Nt}(n)] \\
&= \frac{G e^{-Nt}}{1-\beta} \sum_{n=0}^{Q-1} \rho^n \sum_{i=0}^{n} \frac{(Nt)^i}{i!} = \frac{G e^{-Nt}}{1-\beta} \sum_{i=0}^{Q-1} \frac{(Nt)^i}{i!} \sum_{n=i}^{Q-1} \rho^n \\
&= \frac{G e^{-Nt}}{1-\beta} \sum_{i=0}^{Q-1} \frac{(At)^i}{i!} \left[\frac{1-\rho^{Q-i}}{1-\rho} \right] = \frac{G e^{-Nt}}{(1-\beta)(1-\rho)} \left[\sum_{i=0}^{Q-1} \frac{(At)^i}{i!} - \rho^Q \sum_{i=0}^{Q-1} \frac{(Nt)^i}{i!} \right] \\
&= \frac{G (At)^{Q-1} e^{-Nt}}{(1-\beta)(1-\rho)(Q-1)!} \left[\frac{1}{B(Q-1, At)} - \frac{\rho}{B(Q-1, Nt)} \right].
\end{aligned}
\tag{24}
$$

For $\rho = 1$,

$$
\begin{aligned}
P(W > t) &= \frac{G}{1-\beta} \sum_{n=0}^{Q-1} [P_{Nt}(n)] = \frac{1}{[B(N,N)^{-1}+Q-1]} \sum_{n=0}^{Q-1} \sum_{i=0}^{n} \frac{(Nt)^i}{i!} e^{-Nt} \\
&= \frac{e^{-Nt}}{[B(N,N)^{-1}+Q-1]} \left[Q \sum_{i=0}^{Q-1} \frac{(Nt)^i}{i!} - Nt \sum_{i=0}^{Q-2} \frac{(Nt)^i}{i!} \right] \\
&= \frac{e^{-Nt}}{[B(N,N)^{-1}+Q-1]} \left[\frac{Q(Nt)^{Q-1}}{(Q-1)!} B(Q-1, Nt)^{-1} - \frac{(Nt)^{Q-1}}{(Q-2)!} B(Q-2, Nt)^{-1} \right] \\
&= \frac{(Nt)^{Q-1} e^{-Nt}}{[B(N,N)^{-1}+Q-1](Q-1)!} \left[Nt + \frac{Q-Nt}{B(Q-1,Nt)} \right].
\end{aligned}
\tag{25}
$$

For $Q = 0$, $P(W > t) = 0$, and for $Q = \infty$, $P(W > t) = e^{-\pi t/w}$.

For $N > 1$ and $N - A \neq 0$ or 1, it has been shown (c.f. [7] pp. 97–99) that

$$
\begin{aligned}
P(R > t \mid admitted) &= \left[\frac{1-\pi-\beta}{1-\beta} \right] e^{-t} \\
&+ \frac{G}{1-\beta} \sum_{n=0}^{Q-1} \rho^n \left[P_{Nt}(n) + e^{-t} \left(\frac{N}{N-1} \right)^{n+1} [1 - P_{(N-1)t}(n)] \right].
\end{aligned}
\tag{26}
$$

When $N > 1$ and $N - A \neq 0$ or 1, substituting (22) into (26) and rearranging (again, omitting straightforward steps) yields

$$
\begin{aligned}
P(R > t \mid admitted) &= \left[\frac{1-\pi-\beta}{1-\beta} \right] e^{-t} \\
&+ \frac{G}{1-\beta} \sum_{n=0}^{Q-1} \rho^n \left[P_{Nt}(n) + e^{-t} \left(\frac{N}{N-1} \right)^{n+1} [1 - P_{(N-1)t}(n)] \right] \\
&= \left[\frac{1-\pi-\beta}{1-\beta} \right] e^{-t} + \frac{G}{1-\beta} \left\{ \begin{array}{l} \left[\frac{N}{N-A-1} \right] \left[1 - \left(\frac{A}{N-1} \right)^Q \right] e^{-t} \\ + \sum_{n=0}^{Q-1} \rho^n P_{Nt}(n) - \frac{Ne^{-t}}{N-1} \sum_{n=0}^{Q-1} \left(\frac{A}{N-1} \right)^n P_{(N-1)t}(n) \end{array} \right\} \\
&= \left\{ 1 - \pi - \beta + \frac{GN}{N-A-1} \left[1 - \left(\frac{A}{N-1} \right)^Q \right] \right\} \frac{e^{-t}}{1-\beta} \\
&+ \frac{G}{1-\beta} \left\{ \sum_{n=0}^{Q-1} \rho^n P_{Nt}(n) - \frac{Ne^{-Nt}}{N-1} \sum_{n=0}^{Q-1} \left(\frac{A}{N-1} \right)^n \sum_{i=0}^{n} \frac{[(N-1)t]^i}{i!} \right\} \\
&= \left\{ 1 - \pi - \beta + \frac{GN}{N-A-1} \left[1 - \left(\frac{A}{N-1} \right)^Q \right] \right\} \frac{e^{-t}}{1-\beta} \\
&+ \frac{G}{1-\beta} \left\{ \sum_{n=0}^{Q-1} \rho^n P_{Nt}(n) - \frac{Ne^{-Nt}}{N-A-1} \sum_{i=0}^{Q-1} \frac{(At)^i}{i!} \left[1 - \left(\frac{A}{N-1} \right)^{Q-i} \right] \right\} \\
&= \left\{ 1 - \pi - \beta + \frac{GN}{N-A-1} \left[1 - \left(\frac{A}{N-1} \right)^Q \right] \right\} \frac{e^{-t}}{1-\beta} \\
&+ \frac{G(At)^{Q-1} e^{-Nt}}{(1-\beta)(N-1)(Q-1)!} \left\{ \begin{array}{l} \left(\frac{N-1}{N-A} \right) \left[\frac{N}{B(Q-1,At)} - \frac{A}{B(Q-1,Nt)} \right] - \\ \left(\frac{N}{N-A-1} \right) \left[\frac{N-1}{B(Q-1,At)} - \frac{A}{B(Q-1,(N-1)t)} \right] \end{array} \right\}.
\end{aligned}
\tag{27}
$$

For $N = 1$ and $\rho < 1$, $P(R > t) = \frac{(1-\pi-\beta)(At)^Q e^{-t}}{(1-\beta)(1-A)Q!} \left[\frac{1}{B(Q,At)} - \frac{A}{B(Q,t)} \right]$.

For $N = 1$ and $\rho = 1$, $P(R > t) = \frac{t^Q e^{-t}}{(Q+1)!} \left[\frac{Q+1}{B(Q,t)} - \frac{Q}{B(Q-1,t)} \right]$.

For $N > 1$ and $N - A = 1$,

$$P(R > t) = \left\{ 1 - \pi - \beta + \frac{GQ}{\rho} \right\} \frac{e^{-t}}{1 - \beta} + \frac{GN(At)^{Q-1}e^{-Nt}}{(1-\beta)(Q-1)!} \left\{ \frac{1+t-Q/A}{B(Q-1,At)} - \frac{\rho}{B(Q-1,Nt)} - t \right\}.$$

For $N > 1$ and $\rho = 1$,

$$P(R > t) = \left\{ 1 - \pi - \beta - GN\left[1 - \left(\frac{N}{N-1} \right)^Q \right] \right\} \frac{e^{-t}}{1-\beta} + \frac{G(Nt)^{Q-1}e^{-Nt}}{(1-\beta)(Q-1)!} \left\{ \begin{array}{l} Nt + \frac{Q+N-Nt}{B(Q-1,Nt)} \\ - \frac{N^2/(N-1)}{B(Q-1,(N-1)t)} \end{array} \right\}.$$

For $Q = 0$, $P(R > t) = e^{-t}$.

For $N > 1$, $N - A \neq 1$, and $Q = \infty$, $P(R > t) = \left[1 - \frac{\pi}{1-N+A} \right] e^{-t} + \left[\frac{\pi}{1-N+A} \right] e^{-(N-A)t}$.

For $N > 1$, $N - A = 1$, and $Q = \infty$, $P(R > t) = (\pi t + 1)e^{-t}$.

Finally, for $N = 1$ and $Q = \infty$, $P(R > t) = e^{-t/r}$.

References

1. Markov, A.: Extension of the limit theorems of probability theory to a sum of variables connected in a chain. Notes of the Imperial Academy of Sciences of St. Petersburg, VIII series, XXII(9) (1907)
2. Erlang, A.: Solution of some problems in the theory of probabilities of significance in automatic telephone exchanges. P.O. Electr. Eng. J. **10**, 189–197 (1917, 1918)
3. Cox, D., Smith, W.: Queues. Methuen, London (1961)
4. Saaty, T.: Elements of Queueing Theory. McGraw-Hill, New York (1961)
5. Takács, L.: Introduction to the Theory of Queues. Oxford Press, New York (1962)
6. Cooper, R.: Introduction to Queueing Theory. Macmillan, New York (1972)
7. Gross, D., Harris, C.: Fundamentals of Queueing Theory, 2nd edn. Wiley, New York (1985)
8. Zukerman, M.: Introduction to queueing theory and stochastic teletraffic models. © Moshe Zuckerman, Hong Kong (2000–2020). https://arxiv.org/pdf/1307.2968.pdf
9. Stevens, R., Sinclair, M.: Finite-source analysis of traffic on private mobile radio systems. Electron. Lett. **33**(15), 1291–1293 (1997)
10. Jagerman, D.: Some properties of the Erlang loss function. Bell Syst. Tech. J. **53**(3), 525–551 (1974). https://doi.org/10.1002/j.1538-7305.1974.tb02756.x
11. Fredericks, A.: Congestion in blocking systems – a simple approximation technique. Bell Syst. Tech. J. **59**(6), 805–827 (1980). https://doi.org/10.1002/j.1538-7305.1980.tb03034.x
12. Whitt, W.: Heavy traffic approximations for service systems with blocking. AT&T Bell Lab. Tech. J. **63**(4), 689–708 (1984). https://doi.org/10.1002/j.1538-7305.1984.tb00102.x
13. Jagerman, D.: Methods in traffic calculations. AT&T Bell Lab. Tech. J. **63**(7), 1283–1310 (1984). https://doi.org/10.1002/j.1538-7305.1984.tb00037.x
14. Whitt, W.: Approximations for GI/G/m queue. Prod. Oper. Manag. **2**(2), 114–161 (1993). https://doi.org/10.1111/j.1937-5956.1993.tb00094.x
15. Whitt, W.: The Erlang B and C formulas: problems and solutions. Columbia University course notes (2002). http://www.columbia.edu/~ww2040/ErlangBandCFormulas.pdf
16. Harel, A.: Sharp bounds and simple approximations for the Erlang delay and loss formulas. Manag. Sci. **34**(8), 959–972 (1988). https://doi.org/10.1287/mnsc.34.8.959
17. Pinsky, E.: A simple approximation for the Erlang loss function. Perform. Eval. **15**(3), 155–161 (1992)

18. Salles, R., Barria, J.: Proportional differentiated admission control. IEEE Commun. Lett. **8**(5), 320–322 (2004). https://doi.org/10.1109/LCOMM.2004.827384
19. Shakhov, V.: Simple approximation for Erlang B formula. In: Proceedings IEEE Region 8 SIBIRCON, Listvyanka, Russia, pp. 220–222 (2010). https://doi.org/10.1109/SIBIRCON.2010.5555345
20. Janssen, A., van Leeuwaarden, J., Zwart, B.: Gaussian expansions and bounds for the Poisson distribution applied to the Erlang B formula. Adv. Appl. Probab. **40**, 122–143 (2008). https://doi.org/10.1239/aap/1208358889
21. van Leeuwaarden, J., Temme, N.: Asymptotic inversion of the Erlang B formula. SIAM J. Appl. Math. **70**(1), 1–23 (2009). https://doi.org/10.1137/080722795

The Join the Shortest Orbit Queue System with a Finite Priority Line

Ioannis Dimitriou$^{(\boxtimes)}$

Department of Mathematics, University of Patras, 26504 Patras, Greece
idimit@math.upatras.gr

Abstract. We consider a Markovian single server retrial system with two infinite capacity orbits and a finite capacity priority line. Arriving job are primarily routed to the priority line. If an arriving job finds the priority line fully occupied, it is forwarded to the least loaded orbit queue with ties broken randomly. Orbiting jobs of either type retry to access the server independently. We investigate the stability condition, and the tail decay problem. Moreover, we obtain the equilibrium distribution by using the compensation method.

Keywords: Join the shortest orbit queue · Retrials · Priority

1 Introduction

We consider a single server system with two infinite capacity orbit queues and a finite capacity priority line, operating under the *join the shortest queue* (JSQ) policy in the presence of retrials. Arriving jobs are initially routed to the priority line (in case there is available space). If they find the service station fully occupied, they join the least loaded orbit queue, and in case of a tie, the job joins either orbit queue with probability $1/2$. Orbiting jobs retry independently to connect with the single server after a random time period. We assume that orbiting jobs are not allowed to enter the priority line, and their only chance to get served is to find the server idle upon a retrial time. Our aim is a) to show that the tail probabilities for the shortest orbit queue are asymptotically geometric when the difference of orbit queue sizes, and the size of the priority line are fixed, b) to investigate its stationary behaviour by using the compensation method (CM).

The standard (i.e., without retrials) two-dimensional JSQ problem was initially studied in [9,10], and further developed in [5,8]. In [2–4], the authors introduced the CM, an elegant and direct method to obtain explicitly the equilibrium join queue-length distribution as infinite series of product form terms, see also [14] (not exhaustive list). In [1], the CM was applied in the symmetric two-queue system fed by Erlang arrivals under the JSQ policy.

In this work, we provide an exact analysis that incorporates priorities in a retrial system with two orbits operating under the JSQ policy. Our primary aim is to extend the applicability of the CM to random walks in the quarter plane

© Springer Nature Switzerland AG 2021
P. Ballarini et al. (Eds.): EPEW 2021/ASMTA 2021, LNCS 13104, pp. 375–395, 2021.
https://doi.org/10.1007/978-3-030-91825-5_23

modulated by a $(K + 1)$-state Markov process. For this modulated $(K + 1)$-dimensional random walk we investigate its stationary behaviour by using the CM. We also study the *stability condition* and investigate its *stationary tail decay rate*. Our work, joint with [1,7], imply that CM can be extended to multi-layered two-dimensional random walks that satisfy similar criteria as in [4].

Applications of this model can be found in relay-assisted cooperative communication systems that operate as follows: There is a source user that transmits packets to a common destination node (i.e., the single service station), and a finite number of relay nodes (i.e., the orbit queues) that assist the source user by retransmitting its blocked packets, e.g., [6,12]. The JSQ policy serves as a typical cooperation protocol among the source and the relays.

The paper is organized as follows. In Sect. 2 we describe the model in detail, and investigate the necessary stability condition. Some useful preliminary results along with the decay rate problem is presented in are also presented. The main result that refers to the three-dimensional CM applied in a retrial network is given in Sect. 3. A numerical example is presented in Sect. 4.

2   Model Description

Consider a single server retrial system with two infinite capacity orbit queues and a single priority line of finite capacity K. Jobs arrive at the system according to a Poisson process with rate $\lambda > 0$.

If an arriving customer finds the server free, it immediately occupies the server and leaves the system after service. Otherwise, it joins the priority queue provided there is available space. If an arriving job finds the priority queue fully occupied, it joins the least loaded orbit queue. In case of a tie, the job joins either orbit queue with probability $1/2$. Orbiting jobs of either type retry independently to occupy the server after an exponentially distributed time period with rate α, i.e., we consider the *constant retrial policy*. Note that orbiting jobs are *not* allowed to enter the priority line, and their only chance to get served is to find the server idle upon a retrial time. Service times are independent and exponentially distributed with rate μ.

Let $Y(t) = \{(N_1(t), N_2(t), C(t)), t \geq 0\}$, where $N_l(t)$ the number of jobs stored in orbit l, $l = 1, 2$, at time t, and by $C(t)$ the number of jobs in the service station, i.e., the number of jobs in the priority line and in service at time t, respectively. $Y(t)$ is an irreducible Markov process on $\{0, 1, \ldots\} \times \{0, 1, \ldots\} \times \{0, 1, \ldots, K\}$. Denote by $Y = \{(N_1, N_2, C)\}$ its stationary version. Define the set of stationary probabilities for $i, j = 0, 1, \ldots$, $k = 0, 1, \ldots, K$,

$$p_{i,j}(k) = \mathbb{P}(N_1 = i, N_2 = j, C = k),$$

and let the marginal probabilities $p_{i,.}(k) = \sum_{j=0}^{\infty} p_{i,j}(k)$, $p_{.,j}(k) = \sum_{i=0}^{\infty} p_{i,j}(k)$. Note that due to the symmetry of the model $p_{i,.}(k) = p_{.,i}(k)$, $i = 0, 1, \ldots$.

Let J be a random variable indicating the orbit queue which an arriving blocked job joins. Clearly, J is dependent on the vector $\bar{N} = (N_1, N_2, C)$. The conditional distribution of J given the vector $\bar{N}$ and satisfies:

$$\mathbb{P}(J = 1|\bar{N} = (N_1, N_2, K), N_1 > N_2) = \mathbb{P}(J = 2|\bar{N} = (N_1, N_2, K), N_1 < N_2) = 0,$$
$$\mathbb{P}(J = l|\bar{N} = (N_1, N_2, K)) = \frac{1_{\{N_l \le N_k\}}}{[\sum_{k=1}^{2} 1_{\{N_k = N_l\}}]}, \mathbb{P}(J = l|\bar{N} = (N_1, N_2, 0)) = 0, \, l = 1, 2,$$

where, $1_{\{E\}}$ the indicator function of the event E.

2.1 Equilibrium Equations and Stability Condition

The equilibrium equations read as follows:

$$p_{i,j}(0)(\lambda + \alpha(1_{\{i>0\}} + 1_{\{j>0\}})) = \mu p_{i,j}(1), \tag{1}$$

$$p_{i,j}(1)(\lambda + \mu) = \lambda p_{i,j}(0) + \alpha[p_{i+1,j}(0) + p_{i,j+1}(0)] + \mu p_{i,j}(2), \tag{2}$$

$$\begin{aligned} p_{i,j}(k)(\lambda + \mu) &= \lambda p_{i,j}(k-1) + \mu p_{i,j}(k+1)1_{\{k \ne K\}} \\ &+ \lambda[p_{i-1,j}(K)\mathbb{P}(J = 1|Q = (i-1,j,K))1_{\{i>0\}} \\ &+ p_{i,j-1}(K)\mathbb{P}(J = 2|Q = (i, j-1, K))1_{\{j>0\}}]1_{\{k=K\}}. \end{aligned} \tag{3}$$

Proposition 1.

$$\begin{aligned} \mathbb{P}(C = 0) &= 1 - \tfrac{\lambda}{\mu}, \mathbb{P}(C = 1) = \tfrac{\lambda}{\mu}\left(\frac{1 - \frac{\lambda}{\mu}}{1 - (\frac{\lambda}{\mu})^K}\right), \\ \mathbb{P}(C = k) &= (\tfrac{\lambda}{\mu})^{k-1}\mathbb{P}(C = 1), \, k = 2, \dots, K, \end{aligned} \tag{4}$$

$$p_{0,.}(0) = p_{.,0}(0) = [\frac{1 - \frac{\lambda}{\mu}}{1 - (\frac{\lambda}{\mu})^K}](1 - \left(\frac{\lambda}{\mu}\right)^K \left(\frac{\lambda + 2\alpha}{2\alpha}\right)). \tag{5}$$

Proof. For each $i = 0, 1, 2, \dots$ we consider the cut between the states $\{N_1 = i, C = K\}$ and $\{N_1 = i + 1, C = 0\}$. According to the local balance approach, we have

$$\lambda \mathbb{P}(J = 1, N_1 = i, C = K) = \alpha p_{i+1,.}(0). \tag{6}$$

Summing (6) for all $i \ge 0$ yields

$$\lambda \mathbb{P}(J = 1, C = K) = \alpha[\mathbb{P}(C = 0) - p_{0,.}(0)]. \tag{7}$$

Note that (7) is a conservation of flow relation since it equates the flow of jobs into orbit queue 1, with the flow of jobs out orbit queue 1. By repeating the procedure for $\{N_2 = j, C = K\}$ and $\{N_2 = j + 1, C = 0\}$ yields

$$\lambda \mathbb{P}(J = 2, C = K) = \alpha[\mathbb{P}(C = 0) - p_{.,0}(0)]. \tag{8}$$

Summing (7), (8), and having in mind the symmetry of the model yields

$$\lambda \mathbb{P}(C = K) = 2\alpha \mathbb{P}(C = 0) - 2\alpha p_{0,.}(0)]. \tag{9}$$

Summing (1) for all $i, j \ge 0$ we obtain

$$(\lambda + 2\alpha)\mathbb{P}(C = 0) - \mu \mathbb{P}(C = 1) = 2\alpha p_{0,.}(0). \tag{10}$$

Then, using (9), (10) we obtain

$$\lambda\mathbb{P}(C = K) + \lambda\mathbb{P}(C = 0) = \mu\mathbb{P}(C = 1). \tag{11}$$

Now, summing (2), (3) for all $i, j \geq 0$ it is readily seen that

$$\mathbb{P}(C = k) = \tfrac{\lambda}{\mu}\mathbb{P}(C = k - 1) = \ldots = \left(\tfrac{\lambda}{\mu}\right)^{k-1}\mathbb{P}(C = 1), \ k = 2, 3, \ldots, K. \tag{12}$$

Then, using the normalization condition, along with (11)–(12), we obtain (4), and substituting back in (9), we obtain (5). Equation (5) indicates that $\rho := \left(\tfrac{\lambda}{\mu}\right)^{K}\left(\tfrac{\lambda+2\alpha}{2\alpha}\right) < 1$, is necessary for the system to be stable.

Simulation experiments justify the theoretical findings. In Fig. 1 we observe that as long as $\rho < 1$, the system is stable, while when $\rho > 1$, both orbits become unstable. In all cases we observe that the external arrivals balance the orbit queue lengths as expected.

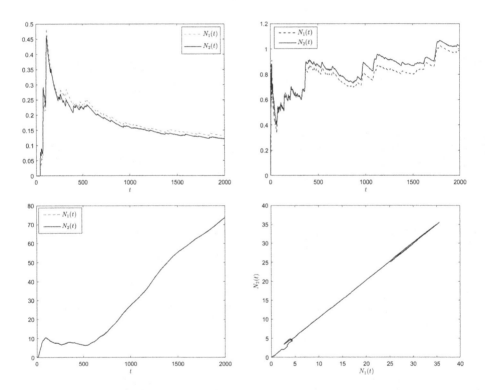

Fig. 1. Orbit dynamics when $K = 3$ for the case $\rho = 0.1$ (up-left), $\rho = 0.5$ (up-right), and for the case $\rho = 1.1$ (below-left and right)

To apply the CM, we consider the transformed process $X(t) = \{(X_1(t), X_2(t), C(t)), t \geq 0\}$, where $X_1(t) = min\{N_1(t), N_2(t)\}$, $X_2(t) =$

$|N_2(t) - N_1(t)|$, and state space $S = \{(m, n, k) : m, n = 0, 1, \ldots, k = 0, 1, \ldots K\}$; see Fig. 2. Let

$$q_{m,n}(k) = \lim_{t \to \infty} \mathbb{P}((X_1(t), X_2(t), C(t)) = (m, n, k)), \ (m, n, k) \in S.$$

Let the column vector $\mathbf{q}(m, n) := (q_{m,n}(0), q_{m,n}(1), \ldots, q_{m,n}(K))^T$, where $\mathbf{x}^T$ denotes the transpose of vector (or matrix) $\mathbf{x}$. The equilibrium equations read:

$$\mathbf{A}_{0,0}\mathbf{q}(0,0) + \mathbf{A}_{0,-1}\mathbf{q}(0,1) = \mathbf{0}, \tag{13}$$

$$\mathbf{B}_{0,0}\mathbf{q}(0,1) + \mathbf{A}_{0,-1}\mathbf{q}(0,2) + 2\mathbf{A}_{-1,1}\mathbf{q}(1,0) + \mathbf{A}_{0,1}\mathbf{q}(0,0) = \mathbf{0}, \tag{14}$$

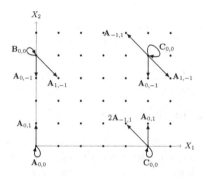

Fig. 2. Transition diagram of the transformed process.

corresponding to the states $(0, 0)$, $(0, 1)$. The vertical boundary equations are

$$\mathbf{B}_{0,0}\mathbf{q}(0,n) + \mathbf{A}_{0,-1}\mathbf{q}(0,n+1) + \mathbf{A}_{-1,1}\mathbf{q}(1,n-1) = \mathbf{0}, \ n \geq 2 \tag{15}$$

The equations corresponding to the horizontal boundary are

$$\mathbf{C}_{0,0}\mathbf{q}(m,0) + \mathbf{A}_{0,-1}\mathbf{q}(m,1) + \mathbf{A}_{1,-1}\mathbf{q}(m-1,1) = \mathbf{0}, \ m \geq 1 \tag{16}$$

$$\mathbf{C}_{0,0}\mathbf{q}(m,1) + \mathbf{A}_{0,-1}\mathbf{q}(m,2) + 2\mathbf{A}_{-1,1}\mathbf{q}(m+1,0)$$
$$+ \mathbf{A}_{1,-1}\mathbf{q}(m-1,2) + \mathbf{A}_{0,1}\mathbf{q}(m,0) = \mathbf{0}, \ m \geq 1. \tag{17}$$

The equations corresponding to the states in the interior of the positive quadrant read

$$\mathbf{C}_{0,0}\mathbf{q}(m,n) + \mathbf{A}_{0,-1}\mathbf{q}(m,n+1) + \mathbf{A}_{1,-1}\mathbf{q}(m-1,n+1)$$
$$+ \mathbf{A}_{-1,1}\mathbf{q}(m+1,n-1) = \mathbf{0}, \ m \geq 1, n \geq 2, \tag{18}$$

where $\mathbf{B}_{0,0} = \mathbf{A}_{0,0} - \mathbf{H}$, $\mathbf{C}_{0,0} = \mathbf{A}_{0,0} - 2\mathbf{H}$, are all $(K+1) \times (K+1)$ matrices. Moreover, $\mathbf{H}$ has the element $(1, 1)$ equal to α and all the rest equal to zero and

$$\mathbf{A}_{0,0} = \begin{pmatrix} -\lambda & \mu & & & \\ \lambda & -(\lambda+\mu) & \mu & & \\ & \ddots & \ddots & \ddots & \\ & & \lambda & -(\lambda+\mu) & \mu \\ & & & \lambda & -(\lambda+\mu) \end{pmatrix},$$

$\mathbf{A}_{0,-1} = \mathbf{A}_{-1,1}$ are $(K+1) \times (K+1)$ matrices with element $(2,1)$ equal to α and zeros elsewhere, and $\mathbf{A}_{0,1} = \mathbf{A}_{1,-1}$ are also $(K+1) \times (K+1)$ matrices with element $(K+1, K+1)$ equal to λ and zeros elsewhere. The transformed process $\{X(t)\}$ is a quasi birth death (QBD) process with repeated blocks $\mathbf{T}_{-1}, \mathbf{T}_0, \mathbf{T}_1$

$$\mathbf{T}_0 = \begin{pmatrix} \mathbf{C}_{0,0}^T & \mathbf{A}_{0,1}^T & \\ \mathbf{A}_{0,-1}^T & \mathbf{C}_{0,0}^T & \\ & \mathbf{A}_{0,-1}^T & \mathbf{C}_{0,0}^T \\ & & \ddots & \ddots \end{pmatrix}, \mathbf{T}_{-1} = \begin{pmatrix} \mathbf{O}_{K+1} & 2\mathbf{A}_{-1,1}^T & \\ & \mathbf{O}_{K+1} & \mathbf{A}_{-1,1}^T \\ & & \mathbf{O}_{K+1} & \mathbf{A}_{-1,1}^T \\ & & & \ddots & \ddots \end{pmatrix},$$

$$\mathbf{T}_1 = \begin{pmatrix} \mathbf{O}_{K+1} & \mathbf{O}_{K+1} & \\ \mathbf{A}_{1,-1}^T & \mathbf{O}_{K+1} & \\ & \mathbf{A}_{1,-1}^T & \mathbf{O}_{K+1} \\ & & \ddots & \ddots \end{pmatrix},$$

where $\mathbf{O}_{K+1}$ is the $(K+1) \times (K+1)$ zero matrix.

2.2   Preliminary Results and Decay Rate

Simulation experiments indicate that in the heavy traffic case (i.e., as $\rho \to 1$) orbit queue lengths become indistinguishable, thus it seems that our model employ a state space collapse property, i.e., the original model in heavy-traffic would collapse to a one-dimensional line where the orbit queue lengths are equal; see Fig. 1 (below-right).

State-space collapse occurs because the join the shortest orbit queue (JSOQ) policy "forces" the two orbit queues to be equal. Thus, it seams that in the heavy traffic regime the sum of the orbit queue lengths can be approximated by a single orbit queue system with a double retrial rate, say the *reference model*; see below. The reference model behaves as if there is only a single orbit queue with all the "servers" (i.e., the retrial servers) pooled together as an aggregated "server" (in standard JSQ systems this is called *complete resource pooling*). This result implies that JSOQ is asymptotically optimal, i.e., heavy-traffic delay optimal, since the response time in the pooled single-orbit system is stochastically less than that of a typical load balancing system. Note that the reference model seams to serve as a lower bound (in the stochastic sense) on the total orbit queue length of the original model. This means that the JSOQ policy is heavy-traffic optimal as expected. However, a more formal justification is needed.

We conjecture that the equilibrium probabilities $q_{m,n}(k)$ can be written as a series of product forms. To provide intuition for this conjecture we first study its tail asymptotics. Following intuition from standard (i.e., without retrials) JSQ systems, we show that when the difference of the orbit queue sizes, and the state of the priority line are fixed, the decay rate for the shortest orbit queue equals η^2, where $\eta < 1$ is the decay rate for the orbit length of the *reference model*. We prove this conjecture following the lines in [13]; see also [7].

The reference model operates as follows: jobs arrive according to a Poisson process with rate λ, and service times are exponentially distributed with rate μ.

Arriving jobs are primarily directed to the priority line. If there is available space, the arriving job joins the priority line. If the priority line is fully occupied, the arriving job is routed to the orbit. Orbiting jobs retry to access the server after an exponentially distributed time with rate 2α. Denote by Q the equilibrium orbit queue length of this model.

Lemma 1. *For the reference model, and for $\rho < 1$, $k = 0, 1, \ldots, K$, $\lim_{m \to \infty} \eta^{-m} \mathbb{P}(Q = m, C = k) = dx_k$, where*

$$\eta := \eta(K) = \frac{\lambda^K(\lambda+2\alpha)+2\lambda\alpha\mu\sum_{i=0}^{K-2}\lambda^{K-2-i}\mu^i}{2\alpha\mu\sum_{i=0}^{K-1}\lambda^{K-1-i}\mu^i}, \quad K \geq 1, \tag{19}$$

and $\mathbf{x} = (x_k)_{k=0,1,\ldots,K}$ is the left eigenvector of the rate matrix $\mathbf{R}$ (see Appendix A) associated with η, and d is a multiplicative constant.

Proof. See Appendix A.

As already indicated in Fig. 1 (below-right), it is expected that $\mathbb{P}(N_1 + N_2 = m)$, and $\mathbb{P}(Q = m)$ have the same decay rate, since both models will work at full capacity when the number of jobs grow. On the other hand, since the JSOQ policy always aims to balance the orbit queue lengths, we expect that as $m \to \infty$,

$$\mathbb{P}(min\{N_1, N_2\} = m) \approx \mathbb{P}(N_1 + N_2 = 2m) \approx \mathbb{P}(Q = 2m).$$

Lemma 1 states that $\mathbb{P}(Q = m) \approx u\eta^m$. Therefore, we conjecture that $\mathbb{P}(min\{N_1, N_2\} = m) \approx u_1\eta^{2m}$. We now proceed with the analysis of the original model by following [13].

Lemma 2. *1) Define the $S_* \times S_*$ matrix $(S_* = \mathbb{N}_0 \times \{0, 1, \ldots, K\})$ $\mathbf{K} = \eta^2\mathbf{T}_{-1} + \mathbf{T}_0 + \eta^{-2}\mathbf{T}_1$. Let $\mathbf{v} = (1, v_1, \ldots, v_K)^T$, where*

$$v_1 = \frac{\lambda+2\alpha}{\lambda+2\alpha\eta}, \quad v_K = \frac{\mu\eta}{(\lambda+\mu)\eta-\lambda}v_{K-1},$$
$$v_k = \frac{\lambda^{k-1}(\lambda+2\alpha)+2\alpha\mu(1-\eta)\sum_{j=0}^{k-2}\lambda^{k-2-j}\mu^j}{\lambda^{k-1}(\lambda+2\alpha\eta)}, \quad k = 2, \ldots, K-1,$$

and $\mathbf{z} = (\eta^{-n}\mathbf{v}, n \geq 0)$. Then $\mathbf{z}$ is positive such that $\mathbf{Kz} = \mathbf{0}$.

2) Let $\mathbf{p} = \{p(n, k); (n, k) \in S_*\} = \{\eta\boldsymbol{\xi}_n\Delta_v^{-1}; l \geq 0\}$. Then, $\mathbf{p}$ is such that $\mathbf{pK} = \mathbf{0}$, $\mathbf{pz} < \infty$, where $\bar{\boldsymbol{\xi}} = \{\boldsymbol{\xi}_n; n \geq 0\}$, $\bar{\boldsymbol{\xi}}\mathbf{D}^{-1}\mathbf{KD} = \mathbf{0}$, and $\mathbf{D} = diag(\Delta_v, \eta^{-1}\Delta_v, \eta^{-2}\Delta_v, \ldots)$ and $\Delta_v = diag(1, v_1, \ldots, v_{K-1}, v_K)$.

Proof. See Appendix B.

Theorem 1. *For $\rho < 1$, and fixed (n, k), $\lim_{m\to\infty} \eta^{-2m}q_{m,n}(k) = Wp(n,k)$, where $W > 0$, and $\mathbf{p} = \{p(n, k); (n, k) \in S_*\}$ as given in Lemma 2.*

Proof. The proof of Theorem 1 is lengthy and composed of several steps. Lemma 2 is crucial for its proof. The rest of the proof is done following the lines in [13, pp. 195–199]. Due to space constraints we omit the rest of the details.

Among others, Theorem 1 provides intuition about the (product) form that would have the equilibrium probabilities $q_{m,n}(k)$.

3   The Compensation Method

We now focus on obtaining the equilibrium probabilities of the transformed
process $\{X(t); t \geq 0\}$ by applying the CM. The CM aims to obtain explicitly the
equilibrium probabilities as infinite series of product form terms, and consists of
four basic steps:

1. Construct a basis of product form terms that satisfy the inner and the horizontal boundary equilibrium equations.
2. We then construct a formal solution of the equilibrium equations as follows:
   i) Construct an initial solution that satisfies the equilibrium equations (16)–
   (18). ii) The initial solution does not satisfy the vertical boundary Eqs. (15),
   thus we need to compensate for this error by adding additional terms from the
   basis. iii) The new solution derived by the vertical compensation step does
   not satisfy the horizontal boundary Eqs. (16), (17). Thus, another horizontal
   compensation step is needed, etc. By repeating the above steps, a formal
   solution is obtained.
3. The formal solution must converge. Thus, we must ensure that the sequences
   of the terms derived in step 1 should converge to zero exponentially fast.
4. Finally the normalization constant should be obtained.

Before we formally applying the CM, we exploit the intuition developed in Subsect. 2.2 about the tail asymptotics for the shortest orbit queue, namely, information about the initial term that we need to construct the formal solution
through CM.

   To this end, we create a *modified* model, which has equilibrium equations
(16)–(18), i.e., the inner and the horizontal boundary equilibrium equations of
the original one.

Lemma 3. *For $\rho < 1$, the equilibrium distribution of the modified model has the
form*

$$\widehat{\mathbf{q}}(m,n) = \eta^{2m}\mathbf{w}(n), \; m \geq -n, n \geq 0, \tag{20}$$

*where η is as in (19), and $\mathbf{w}(n) = (w_0(n), \ldots, w_K(n))^T$ with $\{w_k(n)\}_{n\in\mathbb{N}_0}$
the unique solution (up to a multiplicative constant) of (16)–(18) such that
$\sum_{n\geq 0}\eta^{-2n}w_k(n) < \infty$.*

Proof. See Appendix C.

Lemma 3 states that the solution of the equilibrium equations (16)–(18), i.e.,
corresponding to the horizontal boundary and the inner states has a product-
form solution as given in (20). Moreover, this product-form is unique, since the
equilibrium distribution of the modified model is unique. A direct consequence
of Lemma 3 is the following lemma, which implies that the CM starts with
an initial solution that satisfies the inner and horizontal boundary equilibrium
equations.

Lemma 4. *For $\rho < 1$, the equilibrium equations (16)–(18) have a unique up to a multiplicative constant solution of the form*

$$\mathbf{q}(m,n) = \eta^{2m}\mathbf{v}(n), \ m,n \geq 0, \tag{21}$$

where η is as in (19), and $\mathbf{v}(n) = (v_0(n),\ldots,v_K(n))^T$ with $\{v_k(n)\}_{n\in\mathbb{N}_0}$ such that $\sum_{n\geq 0}\eta^{-2n}v_k(n) < \infty$, $k = 0,1,\ldots,K$.

As stated above, CM starts with a product form satisfying the inner and the horizontal equilibrium equations (as in all the related models that employ the JSQ feature). Following the lines in [1, Proposition 4.2] we can show that CM cannot start with a product form satisfying the inner and the vertical equilibrium equations.

3.1 Construction of the Basis of Product Forms

We characterize the set of product forms $\gamma^m\delta^n\boldsymbol{\theta}$ satisfying (18).

Lemma 5. *The product form $\mathbf{q}(m,n) = \gamma^m\delta^n\boldsymbol{\theta}$, $m \geq 0$, $n \geq 1$, satisfies (18) if*

$$\mathbf{D}(\gamma,\delta)\boldsymbol{\theta} = \mathbf{0}, \tag{22}$$

where $\mathbf{D}(\gamma,\delta) = \gamma\delta\mathbf{C}_{0,0} + \gamma\delta^2\mathbf{A}_{0,-1} + \gamma^2\mathbf{A}_{-1,1} + \delta^2\mathbf{A}_{1,-1}$, and the eigenvector $\boldsymbol{\theta} := \boldsymbol{\theta}_{K+1}(\gamma,\delta) = (\theta_0(\gamma,\delta), \theta_1(\gamma,\delta),\ldots,\theta_K(\gamma,\delta))^T$, satisfies

$$\frac{\theta_1(\gamma,\delta)}{\theta_0(\gamma,\delta)} = \frac{\lambda+2\alpha}{\mu}, \ \frac{\theta_k(\gamma,\delta)}{\theta_{k-1}(\gamma,\delta)} = \frac{\lambda[\gamma s^{(k)}-\lambda\delta s^{(k+1)}]}{\gamma s^{(k-1)}-\lambda\delta s^{(k)}}, \ k = 2,3,\ldots,K, \tag{23}$$

where $s^{(k)} = \sum_{j=0}^{K-k}\lambda^{K-k-j}\mu^j$, $k = 1,\ldots,K$, $s^{(K+1)} = 0$.

Proof. Substituting the product form $\mathbf{q}(m,n) = \gamma^m\delta^n\boldsymbol{\theta}$ in (18) and dividing with $\gamma^{m-1}\delta^{n-1}$ yields (22). Since we seek for non-zero solution $\boldsymbol{\theta}$, $det(\mathbf{D}(\gamma,\delta)) = 0$. The rank of $\mathbf{D}(\gamma,\delta)$ is K and allows to express $\theta_k(\gamma,\delta)$ in terms of $\theta_0(\gamma,\delta)$, $k = 1,\ldots,K$. The eigenvector $\boldsymbol{\theta}$ is obtained by using (22). Simple calculations yields in (23).

Asking for a non-zero solution of $\boldsymbol{\theta}$, $det(\mathbf{D}(\gamma,\delta)) = 0$ assumes the form

$$\gamma\delta[2(\eta+1) - \lambda\tfrac{s^{(2)}}{s^{(1)}}] + \lambda\tfrac{s^{(2)}}{s^{(1)}}\delta^3 = 2\eta\delta^2 + \gamma\delta^2 + \gamma^2. \tag{24}$$

Lemma 6. 1) *For fixed γ such that $|\gamma| \in (0,1)$ Eq. (24) has exactly a root δ, such that $|\delta| \in (0,|\gamma|)$; see Fig. 3.*

2) *For fixed δ such that $|\delta| \in (0,1)$ Eq. (24) has exactly a root γ, such that $|\gamma| \in (0,|\delta|)$; see Fig. 3.*

Proof. Due to space constraints we prove only assertion 1). Assertion 2) is proved analogously. Divide both sides of (24) with γ^2 and set $z = \delta/\gamma$. Then,

$$f(z) := \lambda\frac{s^{(2)}}{s^{(1)}}\gamma z^3 + z(2(\eta+1) - \lambda\frac{s^{(2)}}{s^{(1)}}) = (2\eta+\gamma)z^2 + 1 =: g(z). \tag{25}$$

Note that after some algebra $f(z) = z\lambda\frac{s^{(2)}}{s^{(1)}}\gamma[z^2 + \frac{\lambda^K(\lambda+2\alpha)+\alpha\mu(\lambda s^{(2)}+2s^{(1)})}{\lambda\alpha\mu s^{(2)}\gamma}]$, and since $\left|\frac{\lambda^K(\lambda+2\alpha)+\alpha\mu(\lambda s^{(2)}+2s^{(1)})}{\lambda\alpha\mu s^{(2)}\gamma}\right| > 1$, $f(z) = 0$ has a single root in $|z| < 1$, and two roots in $|z| > 1$. Then, for $|z| = 1$,

$$|f(z)| = |z||\lambda\frac{s^{(2)}}{s^{(1)}}\gamma z^2 + 2(\eta+1) - \lambda\frac{s^{(2)}}{s^{(1)}}| \geq |\lambda\frac{s^{(2)}}{s^{(1)}}|\gamma| + 2(\eta+1) - \lambda\frac{s^{(2)}}{s^{(1)}}|$$
$$= |\lambda\frac{s^{(2)}}{s^{(1)}}(|\gamma| - 1) + 2(\eta+1)| = \lambda\frac{s^{(2)}}{s^{(1)}}(|\gamma| - 1) + 2(\eta+1),$$

since $|\gamma|, \lambda\frac{s^{(2)}}{s^{(1)}} \in (0,1)$. Moreover, $|g(z)| = |(2\eta+\gamma)z^2 + 1| \leq 2\eta + |\gamma| + 1$. Note that

$$2\eta + |\gamma| + 1 < \lambda\frac{s^{(2)}}{s^{(1)}}(|\gamma| - 1) + 2(\eta+1) \Leftrightarrow |\gamma|(1 - \lambda\frac{s^{(2)}}{s^{(1)}}) < 1 - \lambda\frac{s^{(2)}}{s^{(1)}} \Leftrightarrow |\gamma| < 1,$$

which is true. Thus, for $|z| = 1$, $|f(z)| > |g(z)|$, and by Rouché's theorem $f(z)+g(z)$ (i.e., (25)) has a single root in $|z| < 1$. This means that for $|\gamma| \in (0,1)$, (24) has a unique root $|\delta| \in (0,|\gamma|)$.

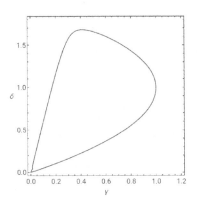

Fig. 3. $det(\mathbf{D}(\gamma,\delta)) = 0$ in $\mathbb{R}_+^2$ for $K = 3$, $\lambda = 1$, $\mu = 6$, $\alpha = 8$.

3.2   The Formal Solution

Lemmas 5, 6 characterize the basic solutions that satisfy the inner equilibrium equations (18). In the following, based on the basis of terms that satisfy (18) we construct the formal solution of (13)–(18). We start with a suitable initial term satisfying (16)–(18). In Sect. 2.2, we shown that $\gamma_0 = \eta^2$, and from Lemma 5 we obtain a unique δ_0, such that $|\delta_0| < |\gamma_0|$. In Lemma 7 we specify the form of the vector $\mathbf{u}(n)$.

Lemma 7. *For* $\gamma = \gamma_0 \in (0,1)$ *let* δ *be the unique root of* (24) *such that* $|\delta| < |\gamma_0|$ *and let* $\boldsymbol{\theta}$ *the corresponding zon-zero vector satisfying* (22). *Then, the solution*

$$\mathbf{q}(m,n) = \begin{cases} h_0 \gamma_0^m \delta^n \boldsymbol{\theta}, & m \geq 0, n \geq 1, \\ \gamma_0^m \boldsymbol{\xi}_0, & m \geq 1, n = 0, \end{cases} \tag{26}$$

$$\boldsymbol{\xi} = -\tfrac{h_0}{\gamma_0} \mathbf{C}_{0,0}^{-1}[\mathbf{A}_{1,-1} + \gamma_0 \mathbf{A}_{0,-1}]\delta_0 \boldsymbol{\theta}, \tag{27}$$

satisfies the equilibrium equations (16)–(18).

Proof. Substituting (26) in (16) to immediately obtain (27). By substituting (16) to (17), then, using (26), and finally taking into account (22) the lemma has been proved.

The initial solution (26) does not satisfy the vertical boundary equilibrium equations (15). Thus, to compensate for this error, we add a new term $c_1 \gamma^m \delta^n \boldsymbol{\theta}(\gamma, \delta)$ such that the sum $h_0 \gamma_0^m \delta_0^n \boldsymbol{\theta}(\gamma_0, \delta_0) + c_1 \gamma^m \delta^n \boldsymbol{\theta}(\gamma, \delta)$ satisfies (15), (18). Substituting the last expression in (15) yields

$$h_0 \delta_0^{n-1} V(\gamma_0, \delta_0)\boldsymbol{\theta}(\gamma_0, \delta_0) + c_1 \delta^{n-1} V(\gamma, \delta)\boldsymbol{\theta}(\gamma, \delta) = \mathbf{0}_{K+1}, \ n \geq 2, \tag{28}$$

where $V(\gamma, \delta) = \delta \mathbf{B}_{0,0} + \delta^2 \mathbf{A}_{0,-1} + \gamma \mathbf{A}_{-1,1}$. Using (22), we realize that $V(\gamma, \delta)\boldsymbol{\theta}(\gamma, \delta) = (\mathbf{H} - \tfrac{\delta}{\gamma}\mathbf{A}_{1,-1})\boldsymbol{\theta}(\gamma, \delta)$. Thus, $\delta = \delta_0$, and $\gamma = \gamma_1$ obtained in Lemma 5 such that $|\gamma_1| < |\delta_0|$, and leaving c_1 for fulfilling $K+1$ requirements. Thus, this choice does not provide sufficient freedom to adapt the compensating term to the requirements.

Lemma 8. *(Vertical compensation) For* $|\delta| \in (0,1)$*, let* γ_0, γ_1 *be roots of* (24) *such that* $|\gamma_1| < |\delta| < |\gamma_0|$*, and let* $\boldsymbol{\theta} := \boldsymbol{\theta}(\gamma_0, \delta)$*,* $\widehat{\boldsymbol{\theta}} := \boldsymbol{\theta}(\gamma_1, \delta)$*, the non-zero vectors satisfying* (22). *Then, there exists a coefficient* c_1 *such that*

$$\mathbf{q}(m,n) = \begin{cases} h_0 \gamma_0^m \delta^n \boldsymbol{\theta} + c_1 \gamma_1^m \delta^n \widehat{\boldsymbol{\theta}}, & m > 0, n \geq 1, \\ h_0 \delta^n \boldsymbol{p} \cdot \boldsymbol{\theta} + c_1 \delta^n \widehat{\boldsymbol{p}} \cdot \widehat{\boldsymbol{\theta}}, & m = 0, n \geq 1, \end{cases}$$

where "·" denotes internal vector product, $\boldsymbol{p} = (p_0, \dots, p_K)^T$*,* $\widehat{\boldsymbol{p}} = (\widehat{p}_0, \dots, \widehat{p}_K)^T$*, where* $p_k = \frac{\gamma_0 s^{(k)}}{\gamma_0 s^{(k)} - \lambda \delta s^{(k+1)}}$*,* $k = 1, \dots, K$*,* $p_0 = \frac{\lambda + 2\alpha}{\lambda + \alpha} \frac{\gamma_0 s^{(1)}}{\gamma_0 s^{(1)} - \lambda \delta s^{(2)}}$*,* $\widehat{p}_k = \frac{\gamma_1 s^{(k)}}{\gamma_1 s^{(k)} - \lambda \delta s^{(k+1)}}$*,* $k = 1, \dots, K$*,* $\widehat{p}_0 = \frac{\lambda + 2\alpha}{\lambda + \alpha} \frac{\gamma_1 s^{(1)}}{\gamma_1 s^{(1)} - \lambda \delta s^{(2)}}$*, and satisfies* (18), (15). *The coefficient* c_1 *equals*

$$c_1 = -h_0 \frac{\delta L(\delta) - \gamma_0}{\delta L(\delta) - \gamma_1} \frac{\theta_K}{\widehat{\theta}_K}, \tag{29}$$

where $L(\delta) = \frac{(\lambda + 2\alpha)[\lambda(\lambda + \alpha)(s^{(1)} - \mu s^{(2)}) + \alpha \mu (1 - \delta)s^{(1)}] + \lambda \alpha \mu (\lambda + \alpha)s^{(2)}}{\alpha \mu (\lambda + \alpha)s^{(1)}}$.

Proof. The proof is similar to the one given in [1, Lemma 4.5] and further details are omitted due to space constraints.

The vertical compensation step described in Lemma 8 creates an error, and the new solution does not satisfy (16), (17). Consider the term $c_1\gamma_1^m\delta_0^m\boldsymbol{\theta}(\gamma_1,\delta_0)$, $|\gamma_1| < |\delta_0|$ (that was added previously in the vertical boundary compensation step). To compensate for the error of this term on the horizontal boundary, we add the term $h_1\gamma^m\delta^n\boldsymbol{\theta}(\gamma,\delta)$ such that the sum $c_1\gamma_1^m\delta_0^m\boldsymbol{\theta}(\gamma_1,\delta_0)+h_1\gamma^m\delta^n\boldsymbol{\theta}(\gamma,\delta)$ satisfies (16)–(18). Substituting (16) to (17) yields for $m \geq 1$,

$$[\mathbf{C}_{0,0} - \mathbf{A}_{0,1}\mathbf{C}_{0,0}^{-1}\mathbf{A}_{0,-1} - 2\mathbf{A}_{-1,1}\mathbf{C}_{0,0}^{-1}\mathbf{A}_{1,-1}]\mathbf{q}(m,1) + \mathbf{A}_{0,-1}\mathbf{q}(m,2)$$
$$+\mathbf{A}_{1,-1}\mathbf{q}(m-1,2) - \mathbf{A}_{0,1}\mathbf{C}_{0,0}^{-1}\mathbf{A}_{0,-1}\mathbf{q}(m-1,1) - 2\mathbf{A}_{-1,1}\mathbf{C}_{0,0}^{-1}\mathbf{A}_{0,-1}\mathbf{q}(m+1,1) = \mathbf{0}.$$
$$(30)$$

Substituting this linear combination into (30) gives for $m \geq 1$,

$$c_1\gamma_1^{m-1}\delta_0\mathbf{L}(\gamma_1,\delta_0)\boldsymbol{\theta}(\gamma_1,\delta_0) + h_1\gamma^{m-1}\delta\mathbf{L}(\gamma,\delta)\boldsymbol{\theta}(\gamma,\delta) = \mathbf{0}_{K+1}, \ m \geq 1,$$

where $\mathbf{L}(\gamma,\delta) = \gamma(\mathbf{C}_{0,0} - \mathbf{A}_{0,1}\mathbf{C}_{0,0}^{-1}\mathbf{A}_{0,-1} - 2\mathbf{A}_{-1,1}\mathbf{C}_{0,0}^{-1}\mathbf{A}_{1,-1}) + \mathbf{A}_{1,-1}\delta - \mathbf{A}_{0,1}\mathbf{C}_{0,0}^{-1}\mathbf{A}_{1,-1} + \mathbf{A}_{0,-1}\gamma\delta - 2\mathbf{A}_{-1,1}\mathbf{C}_{0,0}^{-1}\mathbf{A}_{0,1}\gamma^2$. Since the above equation holds for $m \geq 1$, we choose $\gamma = \gamma_1$. Moreover, since the pair (γ_1,δ) must satisfy also the equilibrium equations (18), we must have $\delta = \delta_1$ as the only root of (24) such that $|\delta_1| < |\gamma_1|$. Therefore, the linear combination $c_1\gamma_1^m\delta_0^m\boldsymbol{\theta}(\gamma_1,\delta_0)+h_1\gamma_1^m\delta_1^n\boldsymbol{\theta}(\gamma_1,\delta_1)$ satisfies (18). The following lemma provides the coefficient h_1 so that this linear combination to satisfy (30).

Lemma 9. *(Horizontal compensation) Consider the product form* $c_1\gamma_1^m\delta_0^n\boldsymbol{\theta}$ (γ_1,δ_0) *with* $0 < |\gamma_1| < |\delta_0| < 1$ *and some coefficient* c_1*, that satisfies* (18) *and stems from a solution that satisfies* (18)*,* (15)*. For this fixed* γ_1*, let* δ_1 *be the root that satisfies* (24) *with* $|\delta_1| < |\gamma_1|$*. Then, there exists a non-zero vector* $\boldsymbol{\xi}_1$ *and a coefficient* h_1 *such that*

$$\mathbf{q}(m,n) = \begin{cases} c_1\gamma_1^m\delta_0^n\boldsymbol{\theta}(\gamma_1,\delta_0) + h_1\gamma_1^m\delta_1^n\boldsymbol{\theta}(\gamma_1,\delta_1), & m \geq 0, n \geq 1, \\ \gamma_0^m\boldsymbol{\xi}_0 + \gamma_1^m\boldsymbol{\xi}_1, & m \geq 1, n = 0, \end{cases} \quad (31)$$

where the vector $\boldsymbol{\xi}$ *as in* (27) *and* $\boldsymbol{\xi}_1$ *is given by*

$$\boldsymbol{\xi}_1 = -\frac{\gamma_0}{\gamma_1}\boldsymbol{\xi} - \frac{1}{\gamma_1}\mathbf{C}_{0,0}^{-1}[\mathbf{A}_{0,-1}\gamma_1 + \mathbf{A}_{1,-1}](c_1\delta_0\boldsymbol{\theta}(\gamma_1,\delta_0) + h_1\delta_1\boldsymbol{\theta}(\gamma_1,\delta_1)), \quad (32)$$

and the constant h_1 *satisfies*

$$c_1\mathbf{K}(\gamma_1,\delta_0) + h_1\mathbf{K}(\gamma_1,\delta_1) = \mathbf{0}_{K+1}, \quad (33)$$

where for $j = 0,1$*,* $\mathbf{K}(\gamma,\delta_j) = [\gamma^2\mathbf{A}_{-1,1} + (2\mathbf{A}_{-1,1}\gamma + \mathbf{A}_{0,1})\mathbf{C}_{0,0}^{-1}(\mathbf{A}_{0,-1}\gamma + \mathbf{A}_{1,-1})\delta_j]\boldsymbol{\theta}(\gamma,\delta_j)$.

Proof. Substituting (31) in (16) gives (32). Then, substituting (31) in (30), and using (22) we obtain after some algebra (33).

Adding the compensation term $h_1\gamma_1^m\delta_1^n\boldsymbol{\theta}(\gamma_1,\delta_1)$ we introduce an error on the vertical boundary equations (15), for which another vertical compensation step needs to be performed. Therefore the approach is as follows: after an initial product form solution is constructed, CM alternates between horizontal (resp.

vertical) compensation step to compensate for the error introduced on the vertical (resp. horizontal) boundary in the previous compensation step. The CM leads to an expression for the $\mathbf{q}(m,n)$ in terms of an infinite series of product forms. These terms are obtained in an iterative fashion using Lemmas 6, 8, 9. Continuing this procedure leads to all pairs of γ, δ, and the corresponding nonzero vectors $\boldsymbol{\theta}(\gamma,\delta)$, the vectors $\boldsymbol{\xi}$'s and the coefficients c_i's, h_i's, for which we construct a formal expression for the equilibrium probability vectors $\mathbf{q}(m,n)$ as follows (where $\propto$ means directly proportional).

Theorem 2. *For all states* (m,n,k), $k = 0, 1, \ldots, K$,

- *For* $m > 0$, $n > 0$, $\mathbf{q}(m,n) \propto \sum_{i=0}^{\infty} h_i \gamma_i^m \delta_i^n \boldsymbol{\theta}(\gamma_i, \delta_i) + \sum_{i=0}^{\infty} c_{i+1} \gamma_{i+1}^m \delta_i^n$
  $\boldsymbol{\theta}(\gamma_{i+1}, \delta_i)$.
- *For* $m > 0$, $n = 0$, $\mathbf{q}(m,0) \propto \sum_{i=0}^{\infty} \gamma_i^m \boldsymbol{\xi}_i$.
- *For* $m = 0$, $n > 0$, $\mathbf{q}(m,n) \propto \sum_{i=0}^{\infty} h_i \delta_i^n \boldsymbol{p}_i \cdot \boldsymbol{\theta}(\gamma_i, \delta_i) + \sum_{i=0}^{\infty} c_{i+1} \delta_i^n \widehat{\boldsymbol{p}}_i \cdot$
  $\boldsymbol{\theta}(\gamma_{i+1}, \delta_i)$.
- *For* $m = n = 0$, $\mathbf{q}(0,0) \propto -\mathbf{A}_{0,0}^{-1} \mathbf{A}_{0,-1} \mathbf{q}(0,1)$.

3.3 Absolute Convergence and the Normalization Constant

It remains to show that these formal expressions are absolutely convergent. This means that we have to show that the error terms converge sufficiently fast to zero. This is accomplished in two steps: *i*) to show that the sequences $\{\gamma_i\}_{i\in\mathbb{N}}$, $\{\delta_i\}_{i\in\mathbb{N}}$ converge to zero exponentially fast, and *ii*) the formal solution converges absolutely.

Proposition 2. *The sequences* $\{\gamma_i\}_{i\in\mathbb{N}}$, $\{\delta_i\}_{i\in\mathbb{N}}$ *satisfy:* 1) $1 > \eta^2 = |\gamma_0| > |\delta_0| > |\gamma_1| > |\delta_1| > \ldots$ 2) $0 \leq |\gamma_i| \leq (\frac{3}{8})^i \eta^2$, and $0 \leq |\delta_i| \leq \frac{1}{2}(\frac{3}{8})^i \eta^2$.

Proof. 1) Note that each γ_i generates a δ_i through (24) that satisfies $|\delta_i| < |\gamma_i|$, and each δ_i generates an γ_{i+1} through (24) that satisfies $|\gamma_{i+1}| < |\delta_i|$. Thus, we have the ordering $|\gamma_0| > |\delta_0| > |\gamma_1| > |\delta_1| > \ldots$.
2) We prove this assertion by firstly showing that *a*) for a fixed γ, with $|\gamma| < \gamma_0$, $|\delta| < \frac{|\gamma|}{2}$, *b*) for a fixed δ, with $|\delta| \leq \gamma_0/2$, we have $|\gamma| < \frac{2}{3}|\delta|$. For a fixed γ, set $z = \delta/\gamma$ on $|z| = 1/2$. Under this transform, (24) reads $f(z) + g(z) = 0$ with $f(z), g(z)$ as given in Lemma 6. Then, for $|z| = 1/2$,

$$|f(z)| \geq |z| \left| \lambda \frac{s^{(2)}}{s^{(1)}} |\gamma| |z|^2 + 2(\eta + 1) - \lambda \frac{s^{(2)}}{s^{(1)}} \right| = \frac{1}{2} \left| \lambda \frac{s^{(2)}}{s^{(1)}} (\frac{|\gamma|}{4} - 1) + 2(\eta + 1) \right|$$
$$= \frac{1}{2} [\lambda \frac{s^{(2)}}{s^{(1)}} (\frac{|\gamma|}{4} - 1) + 2(\eta + 1)].$$

Moreover, $|g(z)| \leq (2\eta + |\gamma|)\frac{1}{4} + 1 < \frac{1}{2}[\lambda \frac{s^{(2)}}{s^{(1)}}(\frac{|\gamma|}{4} - 1) + 2(\eta + 1)] \Leftrightarrow |\gamma| < \frac{2(\eta - \lambda \frac{s^{(2)}}{s^{(1)}})}{1 - \lambda \frac{s^{(2)}}{2s^{(1)}}}$.

Note that $|\gamma| \leq \gamma_0 = \eta^2 < \frac{2(\eta - \lambda \frac{s^{(2)}}{s^{(1)}})}{1 - \lambda \frac{s^{(2)}}{2s^{(1)}}}$. This completes the proof that $|f(z)| > |g(z)|$ on $|z| = 1/2$, and Rouché's theorem proves the corresponding assertion.

Set $z = \gamma/\delta$, $|z| = 3/4$. Equation (24) reads $f(z) + g(z) = 0$, with $f(z) = z(\delta - (B+2)) + B + \lambda\frac{s^{(2)}}{s^{(1)}}(1-\delta)$, $g(z) = z^2$, where $B = 2\eta - \lambda\frac{s^{(2)}}{s^{(1)}}$. Then,

$$|f(z)| = |z(\delta - (B+2)) + B + \lambda\tfrac{s^{(2)}}{s^{(1)}}(1-\delta)| > |z(\delta - (B+2))|$$
$$\geq |z|||\delta| - (B+2)| = \tfrac{3}{4}(B+2-|\delta|),$$
$$|g(z)| = |z|^2 = \tfrac{9}{16} < \tfrac{3}{4}(B+2-|\delta|) < |f(z)| \Leftrightarrow |\delta| < B + \tfrac{5}{4},$$

which is true since $|\delta| \leq \frac{|\gamma_0|}{2} = \frac{\eta^2}{2} < 1 < B + \frac{5}{4}$.

Having these results,

$$|\delta_i| \leq \tfrac{|\gamma_i|}{2} \leq \tfrac{1}{2}\tfrac{3}{4}|\delta_{i-1}| \leq \dots \leq \left(\tfrac{1}{2}\tfrac{3}{4}\right)^i |\delta_0| = \tfrac{1}{2}\left(\tfrac{1}{2}\tfrac{3}{4}\right)^i |\gamma_0| = \tfrac{1}{2}\left(\tfrac{3}{8}\right)^i \eta^2,$$
$$|\gamma_i| \leq \tfrac{3}{4}|\delta_{i-1}| \leq \tfrac{3}{4}\tfrac{1}{2}|\gamma_{i-1}| \leq \dots \leq \left(\tfrac{3}{4}\tfrac{1}{2}\right)^i |\gamma_0| = \left(\tfrac{3}{8}\right)^i \eta^2$$

In Proposition 2 we established that the sequences $\{\gamma_i\}_{i\in\mathbb{N}}$, $\{\delta_i\}_{i\in\mathbb{N}}$ tend to zero as $i \to \infty$. This means that letting $\gamma \to 0$, $\delta \to 0$ is equivalent to letting $i \to \infty$. We now focus on the limiting behaviour of the compensation parameters and their associated eigenvectors.

Lemma 10. 1) *For a fixed γ_i, let δ_i be the root of (24) with $|\delta_i| < |\gamma_i|$. Then, as $\gamma_i \to 0$, the ratio $\delta_i/\gamma_i \to v_- < 1$, where v_- the smallest root of*

$$2\eta z^2 - z(B+2) + 1 = 0. \tag{34}$$

The roots of (34) are given by $v_\pm = \dfrac{B+2\pm\sqrt{B^2+4(1-\lambda\frac{s^{(2)}}{s^{(1)}})}}{4\eta}$.

2) *For a fixed δ_i, let γ_{i+1} be the root of (24) such that $|\gamma_{i+1}| < |\delta_i|$. Then, as $\delta_i \to 0$, the ratio $\gamma_{i+1}/\delta_i \to 1/v_+$, where $v_+ > 1$, the larger root of (34).*

Proof. 1) Set $z = \frac{\delta_i}{\gamma_i}$ in (24), divide with γ_i^2 and let $\gamma_i \to 0$, to obtain (34). Note that since $\eta < 1$, $v_+ = \dfrac{B+2+\sqrt{B^2+4(1-\lambda\frac{s^{(2)}}{s^{(1)}})}}{4\eta} > \frac{1}{2\eta} > 1$, and $v_+v_- = \frac{1}{2\eta}$. Thus, $v_- < 1$.

2) Setting $w = \gamma_{i+1}/\delta_i$ in (24), dividing with δ_i^2 and letting $\delta_i \to 0$ we obtain

$$w^2 - (B+2)w + 2\eta = 0. \tag{35}$$

Note that (35) is derived from (34) for $z = 1/w$. We are interested for the root of (35) inside the unit disk, say $w_- = 1/v_+$, where $v_+ > 1$, the largest root of (34).

Lemma 11. 1) *Let γ_i, δ_i the roots of (24) for fixed γ_i, with $1 > |\gamma_i| > |\delta_i|$. Then, the eigenvector $\boldsymbol{\theta}(\gamma_i, \delta_i)$ converges to $\boldsymbol{\theta} = (\theta_0, \theta_1, \dots, \theta_K)^T$ such that*

$$\frac{\theta_1}{\theta_0} \to \frac{\lambda+2\alpha}{\mu}, \quad \frac{\theta_k}{\theta_{k-1}} \to \frac{\lambda[s^{(k)}-\lambda v_- s^{(k+1)}]}{s^{(k-1)}-\lambda v_- s^{(k)}}, \; k = 2,3,\dots,K, \tag{36}$$

2) *Let* γ_{i+1}, δ_i *the roots of* (24) *for fixed* δ_i, *with* $1 > |\delta_i| > |\gamma_{i+1}|$. *Then, the eigenvector* $\boldsymbol{\theta}(\gamma_{i+1}, \delta_i)$ *converges to* $\tilde{\boldsymbol{\theta}} = (\tilde{\theta}_0, \tilde{\theta}_1, \ldots, \tilde{\theta}_K)^T$ *such that*

$$\frac{\tilde{\theta}_1}{\tilde{\theta}_0} \to \frac{\lambda + 2\alpha}{\mu}, \quad \frac{\tilde{\theta}_k}{\tilde{\theta}_{k-1}} \to \frac{\lambda[w_- s^{(k)} - \lambda s^{(k+1)}]}{w_- s^{(k-1)} - \lambda s^{(k)}}, \quad k = 2, 3, \ldots, K, \qquad (37)$$

3) *As* $i \to \infty$, $\boldsymbol{\xi} \to -h_0 \mathbf{C}_{0,0}^{-1} v_- \mathbf{A}_{1,-1} \boldsymbol{\theta}$.

4) *For* $i \geq 1$, *and* $\boldsymbol{\xi}_0 := \boldsymbol{\xi}$, *the vector* $\boldsymbol{\xi}_i$, *is such that* $\boldsymbol{\xi}_i \to -[\frac{v^+}{v^-}\boldsymbol{\xi}_{i-1} + \mathbf{C}_{0,0}^{-1}\mathbf{A}_{1,-1}[h_i v_- \boldsymbol{\theta} + v_+ c_i \tilde{\boldsymbol{\theta}}]]$, *as* $i \to \infty$.

5) *As* $i \to \infty$, $\frac{c_{i+1}}{h_i} \to \frac{v_+}{v_-}\left(\frac{qv_--1}{1-qv_+}\right)\frac{\theta_K}{\tilde{\theta}_K} := \zeta$, *where* $q = \frac{(\lambda+2\alpha)(\lambda(\lambda+\alpha)+\alpha\mu)s^{(1)} - \lambda(\lambda+\alpha)^2\mu s^{(2)}}{\alpha\mu(\lambda+\alpha)s^{(2)}}$.

6) *As* $i \to \infty$, $\frac{h_i}{c_i} \to -\frac{v_+}{v_-}\left(\frac{\lambda+\mu-\lambda v_-}{\lambda+\mu-\lambda v_+}\right) := \tau$.

Proof. 1) Using (23) it is easy to see that for a fixed γ_i, with $1 > |\gamma_i| > |\delta_i|$, as $\gamma_i \to 0$,

$$\frac{\theta_k}{\theta_{k-1}} \to \frac{\lambda(s^{(k)} - \lambda v_- s^{(k+1)})}{s^{(k-1)} - \lambda v_- s^{(k)}}, \quad k = 2, \ldots, K-1.$$

2) The proof is similar to the one in assertion 3.

3) Note that (27) and Lemma 10 implies that $\boldsymbol{\xi} \to -h_0 \mathbf{C}_{0,0}^{-1} \mathbf{A}_{1,-1} v_- \boldsymbol{\theta}$.

4) The indexing in (32) implies for $i \geq 1$ that

$$\begin{aligned}
\boldsymbol{\xi}_i &= -\frac{1}{\gamma_i}[\gamma_{i-1}\boldsymbol{\xi}_{i-1} + \mathbf{C}_{0,0}^{-1}(\mathbf{A}_{1,-1} + \gamma_i \mathbf{A}_{0,-1})(c_i \delta_{i-1}\tilde{\boldsymbol{\theta}} + h_i \delta_i \boldsymbol{\theta})] \\
&= -[\frac{\gamma_{i-1}}{\delta_i}\frac{\delta_i}{\gamma_i}\boldsymbol{\xi}_{i-1} + \mathbf{C}_{0,0}^{-1}(\mathbf{A}_{1,-1} + \gamma_i \mathbf{A}_{0,-1})(c_i \frac{\delta_{i-1}}{\gamma_i}\tilde{\boldsymbol{\theta}} + h_i \frac{\delta_i}{\gamma_i}\boldsymbol{\theta})] \\
&\to -[\frac{v_+}{v_-}\boldsymbol{\xi}_{i-1} + \mathbf{C}_{0,0}^{-1}\mathbf{A}_{1,-1}(h_i v_- \boldsymbol{\theta} + v_+ c_i \tilde{\boldsymbol{\theta}})].
\end{aligned}$$

The rest of the assertions are proved by using the indexing of the compensation parameters and (33), (29).

Theorem 3. *There exists a positive integer* N *such that:*

1) *The series* $\sum_{i=0}^{\infty} h_i \gamma_i^m \delta_i^n \theta_{i,k}$, $\sum_{i=0}^{\infty} c_{i+1}\gamma_{i+1}^m \delta_i^n \tilde{\theta}_{i+1,k}$ *(where* $\boldsymbol{\theta}_i := \boldsymbol{\theta}(\gamma_i, \delta_i)$, $\tilde{\boldsymbol{\theta}} := \boldsymbol{\theta}(\gamma_{i+1}, \delta_i)$*), the sum of which defines* $q_{m,n}(k)$ *for* $m, n > 0$, *converge absolutely for* $m, n > 0$, *such that* $m + n > N$.
2) *The series* $\sum_{i=0}^{\infty} \gamma_i^m \xi_{i,k}$ *where for* $\boldsymbol{\xi}_0 := \boldsymbol{\xi} = (\xi_{0,0}, \xi_{0,1}, \ldots, \xi_{0,K})^T$, $\boldsymbol{\xi}_i := (\xi_{i,0}, \xi_{i,1}, \ldots, \xi_{i,K})^T$, $i \geq 1$, *defines* $q_{m,0}(k)$ *for* $m > 0$, *converge absolutely for* $m \geq N$.
3) *The series* $\sum_{i=0}^{\infty} h_i \delta_i^n p_{i,k}\theta_{i,k}$, $\sum_{i=0}^{\infty} c_{i+1}\delta_i^n \widehat{p}_{i+1,k}\tilde{\theta}_{i+1,k}$, *the sum of which defines* $q_{0,n}(k)$ *for* $n > 0$, *converge absolutely for* $n > N$.
4) *The series* $\sum_{m+n>N} q_{m,n}(k)$, *converges absolutely.*

Proof. Note that we can assume $\theta_0(\gamma_i, \delta_i) = \mu$, for all γ_i, δ_i. Then, using results from Lemma 5 we can show that $\frac{\theta_k(\gamma_{i+1},\delta_{i+1})}{\theta_k(\gamma_i,\delta_i)} = \frac{\theta_k(\gamma_{i+1},\delta_{i+1})}{\theta_0(\gamma_{i+1},\delta_{i+1})} \frac{\theta_0(\gamma_{i+1},\delta_{i+1})}{\theta_k(\gamma_i,\delta_i)} \to 1$. So, to establish absolute convergence, we do not need to take into account the eigenvectors $\boldsymbol{\theta}$, $\tilde{\boldsymbol{\theta}}$. Due to space constraint we will consider only the first case. Set for $m \geq 0$, $n \geq 1$,

$$R_1(m,n) := \lim_{i\to\infty} \left| \frac{h_{i+1}\gamma_{i+1}^m \delta_{i+1}^n}{h_i \gamma_i^m \delta_i^n} \right| = \lim_{i\to\infty} \frac{\begin{vmatrix} \frac{h_{i+1}}{c_{i+1}} & \frac{\gamma_{i+1}^m}{\delta_{i+1}^m} & \frac{\delta_{i+1}^{m+n}}{\gamma_i^{m+n}} \\ \frac{h_i}{c_{i+1}} & \frac{\gamma_i^m}{\delta_i^m} & \frac{\delta_i^{m+n}}{\gamma_i^{m+n}} \end{vmatrix}}{} = |\tau\zeta| \left(\frac{|v_-|}{|v_+|} \right)^{m+n-1}.$$

$R_1(m,n)$ depends on m, n through their sum $m + n$. Similarly we can handle the other term that refers to $\sum_{i=0}^{\infty} c_{i+1}\gamma_{i+1}^m \delta_i^n \tilde{\theta}_{i+1,k}$, as well as the rest cases. So by defining N to be the minimum integer for which all the ratios are less than one, we know that the series that refers to the corresponding terms converge absolutely. The rest of the proof is similar to the one in [1, Theorem 5.2], and due to space constraints further details are omitted (the index N is small. Numerical experiments indicate that $N = 0$, as in the standard JSQ).

4   A Simple Numerical Example

As a simple numerical example, we aim to compare the JSOQ policy with Bernoulli routing policy. Under the Bernoulli routing, a blocked arriving job joins either orbit with probability $1/2$. Our aim is to compare $E(N_1 + N_2)$ when we apply the JSOQ policy (by using the CM implemented through the Algorithm 1), and the Bernoulli routing. We consider the case where $K = 2$. To obtain $E(N_1 + N_2)$ using the Bernoulli routing, we apply the generating function method, and by exploiting the symmetry of the model we can explicitly obtain it. Let $g_{i,j}(k)$ the corresponding stationary probabilities for the Bernoulli routing scheme, denote $H^{(k)}(x,y) = \sum_{i=0}^{\infty}\sum_{j=0}^{\infty} g_{i,j}(k)x^i y^j$, $|x|, |y| \leq 1$. Writing down the balance equations and forming the generating functions, we obtain for $K = 2$:

$$H^{(0)}(x,y)(\lambda + 2\alpha) - \alpha(H^{(0)}(x,0) + H^{(0)}(0,y)) = \mu H^{(1)}(x,y),$$
$$(\lambda + \mu)H^{(1)}(x,y) = (\lambda + \tfrac{\alpha}{y} + \tfrac{\alpha}{x})H^{(0)}(x,y) - \tfrac{\alpha}{y}H^{(0)}(x,0) - \tfrac{\alpha}{x}H^{(0)}(0,y) + \mu H^{(2)}(x,y)$$
$$H^{(2)}(x,y) = \frac{\lambda}{\mu+\frac{\lambda}{2}(2-x-y)}H^{(1)}(x,y).$$

The structure of the model allows to express all the pgfs in terms of $H^{(0)}(x,y)$, which is a solution to the following functional equation:

$$R(x,y)H^{(0)}(x,y) = A(x,y)H^{(0)}(x,0) + B(x,y)H^{(0)}(0,y),$$
$$R(x,y) = (1-x)y(\frac{\lambda^2(\lambda+2\alpha)}{2}x - \alpha\mu^2) + (1-y)x(\frac{\lambda^2(\lambda+2\alpha)}{2}y - \alpha\mu^2)$$
$$- \frac{\lambda\alpha\mu(2-x-y)^2}{2},$$
$$A(x,y) = \alpha x[\frac{\lambda^2 y}{2}(2-x-y) + \mu(y-1)(\mu+\tfrac{\lambda}{2}(2-x-y))],$$
$$B(x,y) = \alpha y[\frac{\lambda^2 x}{2}(2-x-y) + \mu(x-1)(\mu+\tfrac{\lambda}{2}(2-x-y))].$$

Symmetry implies that $H^{(0)}(1,0) = H^{(0)}(0,1)$, $H_1^{(0)}(1,1) = H_2^{(0)}(1,1)$, $H_1^{(0)}(1,0) = H_2^{(0)}(0,1)$, where $E(N_j^{(k)}) = H_j^{(k)}(1,1)$, $j = 1,2$, $k = 0,1,2$. It is easy to show that

$$E(N_1^{(1)}) = \frac{\lambda+2\alpha}{\mu}E(N_1^{(0)}) - \frac{\alpha}{\mu}H_1^{(0)}(1,0), \ E(N_1^{(2)}) = \frac{\lambda^2}{2\mu^2}\mathbb{P}(C=1) + \frac{\lambda}{\mu}E(N_1^{(0)}),$$

and $E(N_1) = \sum_{k=0}^{2} E(N_1^{(k)})$, $E(N_1^{(k)}) = E(N_2^{(k)})$. By applying a similar approach as in [6, Section 6], we obtain

$$E(N_1^{(0)}) = \frac{\lambda^2(1-\lambda/\mu)}{2\mu^2(1-\rho)}[\lambda^2(\lambda+2\alpha) + 2\alpha\mu(\lambda+\mu) - \frac{\alpha(1-(\lambda/\mu)^2)(4(\lambda^2+\mu^2)-\lambda\mu)}{1-\rho}],$$

$$H_1^{(0)}(1,0) = \frac{2(1-\lambda/\mu)[\frac{\lambda^2(\lambda+2\alpha)}{2}+\frac{\alpha\lambda\mu(1-\frac{1-(\lambda/\mu)^2}{1-\rho})}{2}+\frac{\alpha\mu^2(1-(\lambda/\mu)^2)}{1-\rho}]}{\alpha\lambda^2}.$$

Thus, after tedious calculations we obtain

$$E(N_1+N_2) = \frac{\lambda^3}{\mu^3}\frac{1-\lambda/\mu}{1-(\lambda/\mu)^2} - \frac{2\alpha}{\mu}H_1^{(0)}(1,0) + E(N_1^{(0)})[1 + \frac{\lambda+2\alpha}{\mu}(1+\lambda/\mu)].$$

Figure 4 compares JSOQ to random routing for increasing values of λ, in terms of the expected total number of orbiting jobs. It is easy to observe that JSOQ routing is superior to Bernoulli routing, especially for large values of λ. For small λ, an arriving job usually finds an available position in the priority line, so in such a case, both routing policies operate equally well.

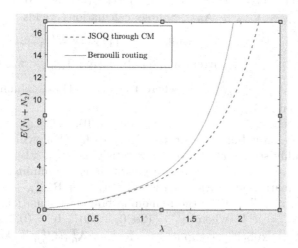

Fig. 4. JSOQ vs Bernoulli Routing, $\mu = 3$, $\alpha = 5$, $K = 2$.

Recall that as we observed in Sect. 2, as $\rho \to 1$ the JSOQ system behaves as a pooled system, which means that the sum of the orbit queue lengths can be approximated by the reference model.

Algorithm 1. CA algorithmic implementation

1: Inputs λ, α, μ, K, and the precision ϵ.
2: Set $\gamma_0 = \eta^2$, $h_0 = 1$ and $N_{ca} = 1$.
3: Compute δ_0 from Equation (24), and set $T_{ca} = \max\{\lceil \log(\epsilon)/\log(\delta_0)\rceil, 3\}$.
4: Compute recursively γ_i, δ_i, $\boldsymbol{\theta}(\gamma_i, \delta_i)$ for $i = 1, \ldots, N_{ca}$, from (24), and (23), respectively.
5: Compute the coefficients h_i, c_i, $i = 0, 1, \ldots, N_{ca}$, starting from $h_0 = 1$ and using the compensation steps. Then, calculate the vectors $\boldsymbol{\xi}_i$, $i = 0, 1, \ldots, N_{ca}$.
6: For all $\lfloor T_{ca}/2 \rfloor < m, n \le T_{ca}$, compute $q_{m,n}^{(N_{ca})}(k)$, $k = 0, 1, \ldots, K$, from Theorem 2.
7: For all $0 \le m, n \le \lfloor T_{ca}/2 \rfloor$, solve the linear system of the balance equations (13)-(18) and compute $q_{m,n}^{(N_{ca})}(k)$.
8: Normalize $q_{m,n}^{(N_{ca})}(k)$.
9: Stop if $\left| \dfrac{\sum_{m,n=0}^{T_{ca}} q_{m,n}^{(N_{ca})}(k) - \sum_{m,n=0}^{T_{ca}} q_{m,n}^{(N_{ca}-1)}(k)}{\sum_{m,n=0}^{T_{ca}} q_{m,n}^{(N_{ca})}(k)} \right| < \epsilon$, else $N_{ca} = N_{ca} + 1$ and go to Step 4.

A   Proof of Lemma 1

Let $X_0(t) = \{(Q(t), C(t)); t \ge 0\}$ the Markovian process that describes the *reference model*. Note that $\{X_0(t)\}$ is a QBD with state space $\mathbb{Z}^+ \times \{0, 1, \ldots, K\}$ and a block tri-diagonal infinitesimal generator $\mathbf{Q}_0$ with repeated blocks $2\mathbf{A}_{0,-1}^T$, $\mathbf{C}_{0,0}^T$, $\mathbf{A}_{0,1}^T$.

Following [11], let $\mathbf{z} = (z_0, z_1, \ldots, z_K)$ the stationary probability vector of $\mathbf{\Lambda} = 2\mathbf{A}_{0,-1}^T + \mathbf{C}_{0,0}^T + \mathbf{A}_{0,1}^T$. Simple calculations yields $z_1 = \frac{\lambda+2\alpha}{\mu}z_0$, $z_i = \left(\frac{\lambda}{\mu}\right)^{i-1} z_1$, $i = 2, \ldots, K$, while $z_0 = (1 + \frac{\lambda+2\alpha}{\mu}\left(\frac{1-(\lambda/\mu)^K}{1-(\lambda/\mu)}\right))^{-1}$. Then, $\{X_0(t); t \ge 0\}$ is positive recurrent if and only if $\mathbf{z}\mathbf{A}_{0,1}^T \mathbf{1}_{K+1} < \mathbf{z}2\mathbf{A}_{0,-1}^T \mathbf{1}_{K+1}$, or equivalently $\rho = \left(\frac{\lambda}{\mu}\right)^K \frac{(\lambda+2\alpha)}{2\alpha} < 1$, where $\mathbf{1}$ is a $(K+1) \times 1$ column vector with all elements equal to 1.

Let $\rho < 1$, and $\pi = (\pi_0, \pi_1, \ldots)$, $\pi_m = (\pi_m(0), \pi_m(1), \ldots, \pi_m(K))$, $m \ge 0$, the stationary probability vector associated to $\mathbf{Q}_0$. Following [11, Theorem 3.1.1], it is readily seen that the vector π satisfying $\pi\mathbf{Q}_0$, $\pi\mathbf{1} = 1$ is given by $\pi_m = \pi_1 \mathbf{R}^{m-1}$, $m \ge 1$, where the rate matrix $\mathbf{R}$ is the minimal non-negative solution of the matrix-quadratic equation $\mathbf{R}^2 2\mathbf{A}_{0,-1}^T + \mathbf{R}\mathbf{C}_{0,0}^T + \mathbf{A}_{0,1}^T = \mathbf{O}_{K+1}$. The vectors π_0, π_1 follow from the remaining boundary equations. Due to the fact that $2\mathbf{A}_{0,-1}^T = \mathbf{a}^T\mathbf{b}$, where $\mathbf{a}^T = (2\alpha, 0, \ldots, 0)^T$, $\mathbf{b} = (0, 1, 0, \ldots, 0)$, the rate matrix $\mathbf{R}$ is explicitly computed as $\mathbf{R} = -\mathbf{A}_{0,1}^T(\mathbf{C}_{0,0}^T + \mathbf{A}_{0,1}^T \mathbf{1}_{K+1}\mathbf{b})^{-1}$. After simple but tedious algebraic computations, the maximal eigenvalue (and the only positive) of $\mathbf{R} = (r_{i,j})_{i,j=0,\ldots,K}$, is given in (19). Following known results [11], $\mathbf{x}\mathbf{R} = \eta\mathbf{x}$.

Alternatively, one can realize that $\mathbf{A}_{0,1}^T = \mathbf{t}^T\mathbf{l}$, where $\mathbf{t}^T = (0, 0, \ldots, 0, \lambda)^T$, $\mathbf{l} = (0, 0, \ldots, 0, 1)$, and thus, $\mathbf{R} = -\mathbf{A}_{0,1}^T(\mathbf{C}_{0,0}^T + \eta 2\mathbf{A}_{0,-1}^T)^{-1}$, where the decay rate η is the unique root in $(0, 1)$ of the determinant equation $det(\mathbf{A}_{0,1}^T + \eta\mathbf{C}_{0,0}^T + \eta^2 2\mathbf{A}_{0,-1}^T) = 0$. Tedious algebraic calculations yields (19).

B   Proof of Lemma 2

Note that $\mathbf{Ky} = \mathbf{0}$ implies

$$\eta^{-1}[\mathbf{C}_{0,0}^T\eta + 2\eta^2\mathbf{A}_{-1,1}^T + \mathbf{A}_{0,1}^T]\mathbf{v} = \mathbf{0}, \; n = 0,$$
$$\eta^{-n-1}[\mathbf{A}_{0,1}^T + 2\eta^2\mathbf{A}_{-1,1}^T + \eta\mathbf{C}_{0,0}]\mathbf{v} = \mathbf{0}, \; n \geq 1.$$

Simple computations imply that, $\mathbf{Ky} = \mathbf{0}$, and $\mathbf{y}$ is positive since $\mathbf{v}$ is positive. We now construct a positive vector $\mathbf{p} = \{\mathbf{p}_n, n \geq 0\}$ such that $\mathbf{pK} = \mathbf{0}$. Let Δ_v be the diagonal matrix whose diagonal elements are the corresponding elements of $\mathbf{v}$. Let the diagonal matrix $\mathbf{D} = diag(\Delta_v, \eta^{-1}\Delta_v, \eta^{-2}\Delta_v, \ldots)$, and denote $\mathbf{K}_D = \mathbf{D}^{-1}\mathbf{KD}$.

Note that $\mathbf{K}_D$ is a transition rate matrix of a QBD with finite phases at each level. Moreover, note that $\mathbf{K}$ is a block triangular matrix with repeated blocks $\bar{K}_1 = 2\eta^2\mathbf{A}_{-1,1}^T + \mathbf{A}_{0,1}^T$, $K_1 = \eta^2\mathbf{A}_{-1,1}^T$, $K_{-1} = \eta^{-2}\mathbf{A}_{1,-1}^T + \mathbf{A}_{0,-1}^T$. To check the ergodicity o $\mathbb{K}_D$ let $\mathbf{u}$ the stationary probability vector of $\Delta_v^{-1}[\eta K_{-1} + K_0 + \eta^{-1}K_1]$. After simple computations, the mean drift at internal states reads

$$\mathbf{u}[\eta^{-1}\Delta_v^{-1}K_1\Delta_v - \eta\Delta_v^{-1}K_{-1}\Delta_v] = \eta\mathbf{u}\Delta_v^{-1}[\eta^{-2}K_1 - K_{-1}]\mathbf{v} < 0.$$

Since $\mathbf{K}_D$ is ergodic [11], there exists a stationary distribution $\bar{\boldsymbol{\xi}} = \{\bar{\boldsymbol{\xi}}_n, n \geq 0\}$ such that $\bar{\boldsymbol{\xi}}\mathbf{K}_D = \mathbf{0}$, or equivalently, since $\mathbf{D}$ is invertible, $\bar{\boldsymbol{\xi}}\mathbf{D}^{-1}\mathbf{K} = \mathbf{0}$. Thus, $\mathbf{p} = \{\eta^n\bar{\boldsymbol{\xi}}_n\Delta_v^{-1}; n \geq 0\}$ satisfies $\mathbf{pK} = \mathbf{0}$, and $\mathbf{pz} < \infty$, since $\mathbf{pz} = \bar{\boldsymbol{\xi}}\mathbf{1} = 1 < \infty$.

C   Proof of Lemma 3

We consider a modified model, closely related to the original one, which has the same asymptotic behaviour, and is considered on a slightly different grid, namely $\{(m,n,k) : m \geq 0, n \geq 0, k = 0, 1, \ldots, K\} \cup \{(m,n,k) : m < 0, 2m + n \geq 0, k = 0, 1, \ldots, K\}$.

In the interior and on the horizontal boundary, the modified model has the same transition rates as the original model. Moreover, its balance equations for $2m + n = 0$ are exactly the same as the ones in the interior (i.e., the modified model has no "vertical" boundary equations) and both models have the same stability region. Therefore, the balance equations for the modified model are given by (16)–(18) for all $2m + n \geq 0$, $m \in \mathbb{Z}$ with only the equation for state $(0,0,k)$, $k = 0, 1, \ldots, K$ being different due to the incoming rates from the states with $2m + n = 0$, $m \in \mathbb{Z}$.

The modified model, restricted the area $\{(m,n,k) : 2m \geq m_0 - n, n \geq 0, m_0 = 1, 2, \ldots, k = 0, 1, \ldots, K\}$ embarked by a line parallel to the $2m + n = 0$ axis, yields the exact same process. Hence, we can conclude that the equilibrium distribution of the modified model, say $\hat{\mathbf{q}}(m,n) := (\hat{q}_{m,n}(0), \hat{q}_{m,n}(1), \ldots, \hat{q}_{m,n}(K))^T$, satisfies $\hat{\mathbf{q}}(m+1, n) = \gamma\hat{\mathbf{q}}(m,n)$, $2m \geq -n$, $n \geq 0$, and therefore

$$\hat{\mathbf{q}}(m,n) = \gamma^m\hat{\mathbf{q}}(0,n), \; 2m \geq -n, \; n \geq 0. \tag{38}$$

We further observe that $\sum_{n=0}^{\infty} \hat{q}_{-n,n}(k) = \sum_{n=0}^{\infty} \gamma^{-n} \hat{q}_{0,n}(k) < 1$. To determine the term γ we consider levels of the form $(L,k) = \{(m,n,k), : 2m+n = L\}$, $k = 0, 1, \ldots, K$ and let $\hat{\mathbf{q}}_L = \sum_{2m+n=L} \hat{\mathbf{q}}(m,n)$. The balance equations among the levels are:

$$\mathbf{C}_{0,0}\hat{\mathbf{q}}_L + \mathbf{A}_{1,-1}\hat{\mathbf{q}}_{L-1} + 2\mathbf{A}_{0,-1}\hat{\mathbf{q}}_{L+1} = 0, \quad L \geq 1, \tag{39}$$

Moreover, Eq. (38) yields

$$\hat{\mathbf{q}}_{L+1} = \sum_{2k+l=L+1} \gamma^k \hat{\mathbf{q}}(0,n) = \gamma \sum_{2k+l=L-1} \gamma^k \hat{\mathbf{q}}(0,n) = \gamma \hat{\mathbf{q}}_{L-1}. \tag{40}$$

Substituting (40) into (39) yields $\hat{\mathbf{q}}_{L+1} = -[\gamma(\mathbf{A}_{1,-1}+2\gamma\mathbf{A}_{0,-1})^{-1}\mathbf{C}_{0,0}]\hat{\mathbf{q}}_L$. Combining (40) with (39) with $\gamma = \eta^2$ yields $det(\eta\mathbf{C}_{0,0} + \mathbf{A}_{1,-1} + \eta^2 2\mathbf{A}_{0,-1}) = 0$, or equivalently,

$$2\alpha\mu s^{(1)}(1-\eta)\left(\frac{\lambda^K(\lambda+2\alpha)+2\alpha\lambda\mu\sum_{i=0}^{K-1}\lambda^{K-2-i}\mu^i}{2\alpha\mu\sum_{i=0}^{K-1}\lambda^{K-1-i}\mu^i} - \eta\right) = 0,$$

which implies that indeed $\gamma = \eta^2$.

Thus, it is shown that the equilibrium distribution of the modified model has a product-form solution which is unique up to a positive multiplicative constant. Returning to the original process $X(t)$, we immediately assume that the solution of (16)–(18) is identical to the expression for the modified model as given in (21). Moreover, the above analysis implies that this product-form is unique, since the equilibrium distribution of the modified model is unique.

References

1. Adan, I.J.B.F., Kapodistria, S., van Leeuwaarden, J.S.H.: Erlang arrivals joining the shorter queue. Queueing Syst. **74**(2–3), 273–302 (2013). https://doi.org/10.1007/s11134-012-9324-8
2. Adan, I.J.B.F., Wessels, J., Zijm, W.H.M.: Analysis of the symmetric shortest queue problem. Stoch. Model. **6**(1), 691–713 (1990)
3. Adan, I.J.B.F., Wessels, J., Zijm, W.H.M.: Analysis of the asymmetric shortest queue problem. Queueing Syst. **8**(1), 1–58 (1991). https://doi.org/10.1007/BF02412240
4. Adan, I.J.B.F., Wessels, J., Zijm, W.H.M.: A compensation approach for two-dimensional Markov processes. Adv. Appl. Probab. **25**(4), 783–817 (1993)
5. Cohen, J., Boxma, O.: Boundary Value Problems in Queueing Systems Analysis. North Holland Publishing Company, Amsterdam (1983)
6. Dimitriou, I.: A queueing system for modeling cooperative wireless networks with coupled relay nodes and synchronized packet arrivals. Perform. Eval. **114**, 16–31 (2017)
7. Dimitriou, I.: Analysis of the symmetric join the shortest orbit queue. Oper. Res. Lett. **49**(1), 23–29 (2021)
8. Fayolle, G., Iasnogorodski, R., Malyshev, V.: Random Walks in the Quarter Plane: Algebraic Methods, Boundary Value Problems, Applications to Queueing Systems and Analytic Combinatorics. PTSM, vol. 40. Springer, Cham (2017). https://doi.org/10.1007/978-3-319-50930-3

9. Haight, F.A.: Two queues in parallel. Biometrika **45**(3/4), 401–410 (1958)
10. Kingman, J.F.C.: Two similar queues in parallel. Ann. Math. Stat. **32**(4), 1314–1323 (1961)
11. Neuts, M.F.: Matrix-Geometric Solutions in Stochastic Models: An Algorithmic Approach. Johns Hopkins University Press, Baltimore (1981)
12. Pappas, N., Kountouris, M., Ephremides, A., Traganitis, A.: Relay-assisted multiple access with full-duplex multi-packet reception. IEEE Trans. Wireless Commun. **14**(7), 3544–3558 (2015)
13. Sakuma, Y., Miyazawa, M., Zhao, Y.: Decay rate for a PH/M/2 queue with shortest queue discipline. Queueing Syst. **53**, 189–201 (2006). https://doi.org/10.1007/s11134-006-7634-4
14. Saxena, M., Dimitriou, I., Kapodistria, S.: Analysis of the shortest relay queue policy in a cooperative random access network with collisions. Queueing Syst. **94**(1–2), 39–75 (2020). https://doi.org/10.1007/s11134-019-09636-9

A Short Note on the System-Length Distribution in a Finite-Buffer $GI^X/C\text{-}MSP/1/N$ Queue Using Roots

Abhijit Datta Banik[1]([✉]) [iD], Mohan L. Chaudhry[2], Sabine Wittevrongel[3] [iD], and Herwig Bruneel[3] [iD]

[1] School of Basic Sciences, Indian Institute of Technology Bhubaneswar, Permanent Campus Argul, Jatni, Khurda 752050, Odisha, India
adattabanik@iitbbs.ac.in
[2] Department of Mathematics and Computer Science, Royal Military College of Canada, P.O. Box 17000, STN Forces, Kingston, ON K7K 7B4, Canada
chaudhry-ml@rmc.ca
[3] Department of Telecommunications and Information Processing, Ghent University, UGent Sint-Pietersnieuwstraat 41, 9000 Gent, Belgium
{Sabine.Wittevrongel,Herwig.Bruneel}@ugent.be

Abstract. This paper deals with a renewal input finite-buffer single-server queue, where the arrivals occur in batches and the server serves the customers singly. It is assumed that the inter-batch arrival times are generally distributed and the successive service times are correlated. The correlated single-service process is exhibited by a continuous-time Markovian service process ($C\text{-}MSP$). As the buffer capacity N (including the one in service) is finite, the partial-batch rejection policy is considered here. Steady-state distributions at different epochs, namely pre-arrival and arbitrary epochs are obtained. These distributions are used to obtain some important performance measures, e.g. the blocking probability of the first, an arbitrary, and the last customer of a batch, the average number of customers in the system and the mean waiting time in the system. The proposed analysis is based on the roots of a characteristic equation which is derived from the balance equations of an embedded Markov chain at pre-arrival epochs of a batch. For this non-renewal service finite-buffer queueing model, we implement a novel as well as simple procedure for deriving the characteristic equation and then finding the stationary probability vectors in terms of the roots of the characteristic equation. Finally, some numerical results are presented in the form of tables for the case of a phase-type inter-batch arrival distribution.

Keywords: Renewal input · Batch arrival · Finite-buffer queue · Continuous-time Markovian service process ($C\text{-}MSP$) · Roots

1 Introduction

A correlated nature of arrival and service processes has been frequently observed in application areas such as manufacturing systems, production systems, and

© Springer Nature Switzerland AG 2021
P. Ballarini et al. (Eds.): EPEW 2021/ASMTA 2021, LNCS 13104, pp. 396–410, 2021.
https://doi.org/10.1007/978-3-030-91825-5_24

communication networks. Such correlation is what makes queueing theory interesting to study. The correlation among the inter-arrival and inter-batch arrival times can be modelled by the continuous-time Markovian arrival process ($C\text{-}MAP$) and the continuous-time batch Markovian arrival process ($C\text{-}BMAP$), respectively. Queueing models with $C\text{-}MAP$ arrivals were studied by many researches, see Lucantoni et al. [15], Kasahara et al. [12] and Choi et al. [9]. It is to be remarked here that the $C\text{-}BMAP$ can suitably represent the versatile Markovian point process ($VMPP$), see Neuts [18] and Ramaswami [20].

Analogous to the correlated inter-arrival times, the correlation among the service times can be represented by the continuous-time Markovian service process ($C\text{-}MSP$), see Bocharov [4] as well as Albores and Tajonar [1]. The $GI/C\text{-}MSP/1$ queue with infinite-buffer capacity was examined by Bocharov et al. [5], while Albores and Tajonar [1] studied the multi-server $GI/C\text{-}MSP/c/r$ queue. The $GI/C\text{-}MSP/1$ queue was also analyzed by Gupta and Banik [11] using a combination of an embedded Markov chain process, the supplementary variable technique, and the matrix-geometric method. Applying perturbation theory, the asymptotic analysis of $GI/C\text{-}MSP/1$ queue was also explained by Alfa et al. [2]. Allowing the server to take vacations, Machihara [17] investigated the $G/SM/1/\infty$ queueing model. Chaudhry et al. [6] used roots and carried out the analytic analysis of the $GI^X/C\text{-}MSP/1/\infty$ model. Simultaneously, Chaudhry et al. [8] discussed the $GI/C\text{-}MSP^{(a,b)}/1/\infty$ queueing model using roots. Queueing systems with finite buffers are more realistic than the corresponding infinite-buffer models. For a detailed study of various finite-buffer queues with or without vacation(s) of the server, the readers are referred to the book by Takagi [21]. Later, Banik and Gupta [3] investigated a finite-buffer batch arrival queue under both the partial and total batch acceptance policies.

In this paper, it is assumed that the arrivals occur in batches of random sizes and the service is governed by a $C\text{-}MSP$. The model can be denoted by $GI^X/C\text{-}MSP/1/N$. As the system capacity (N) is considered to be finite, it may happen that an arriving batch cannot be accommodated fully in the system. If an arriving batch contains more customers than the available space in the system, then only the number of customers required for filling the remaining space are allowed to join the system, i.e., a portion of the arriving batch is accepted which is known as the partial batch rejection (PBR) policy. The system-length distributions at pre-arrival and arbitrary epochs of the $GI^X/C\text{-}MSP/1/N$ model with the PBR policy are considered here. It may be noted that the $GI^X/C\text{-}MSP/1/N$ model was analyzed by Banik and Gupta [3] by solving simultaneous equations. The current paper analyzes the same queueing model under slightly different assumptions and with a quite different methodology, i.e., using the roots of a characteristic equation. In view of this, the method presented here unifies the approach used here and in the paper by Chaudhry et al. [6] who deal with the infinite-buffer queue, i.e., both the finite-capacity and infinite-capacity models can be solved simultaneously by using the same approach. Kim and Chaudhry [13] first used the roots of a characteristic equation to find the stationary probabilities in a finite-buffer $GI/M/c/N$ queue. In this paper, we implement a similar idea of obtaining each phase of the stationary probability vectors in terms of roots of the characteristic equation that is derived from the vector-difference equations of the embedded Markov chain at pre-arrival epochs in a GI^X/C-$MSP/1/N$ queue with a non-renewal type of service process (it consists of several phases of service). Further, the present paper makes several developments over some of the existing results (see Banik and Gupta [3]) in the literature, e.g., the stationary probability vectors are obtained using the balance equations associated with the embedded Markov chain at pre-arrival epochs without evaluating the structure of the corresponding transition probability matrix. Further, it may be mentioned here that in [3], the C-MSP process was dependent on the idle restart service phase distribution, whereas in the present paper we consider that the service phase of the C-MSP process remains the same phase in which the last busy period ended.

2   Description of the Model GI^X/C-$MSP/1/N$

A single-server finite-buffer queue with batch arrivals is discussed here. The batch size of the arriving batches is assumed to be of random size and the corresponding random variable is symbolized by X with $P(X = i) = g_i, i \geq 1$. The mean batch size is determined as $E(X) = \sum_{i=1}^{\infty} i g_i = \bar{g}$ and further g_i' is defined as $g_i' = \sum_{r=i}^{\infty} g_r, i \geq 1$. The probability generating function of the mass function g_i ($i \geq 1$) is given by $G(z) = \sum_{i=1}^{\infty} g_i z^i$. The time intervals between the occurrence of two successive batches, i.e. the inter-batch arrival times, are considered to be independent and identically distributed (i.i.d.) random variables (r.v.'s) A with mean $1/\lambda$. The arrival process is also considered to be independent of the service process. Let the cumulative distribution function (D.F.), probability density function (p.d.f.) and the Laplace-Stieltjes transform (LST) of the p.d.f. of a random variable Y be symbolized by $F_Y(y)$, $f_Y(y)$ and $f_Y^*(s)$, respectively. Further, it is assumed that the capacity (N) of the system includes the customer that is being served.

In this model, the m-state C-MSP is represented by the matrices L_0 and L_1, where the (i, j)th ($1 \leq i, j \leq m$) entry of the matrix L_0 denotes the state transition rates among the underlying m states of the C-MSP without a service completion and the (i, j)th ($1 \leq i, j \leq m$) entry of the matrix L_1 denotes the state transition rates among the underlying m states of the C-MSP with a service completion. All entries of the matrices L_0 and L_1 are positive except that $[L_0]_{i,i}$ ($1 \leq i \leq m$) is negative and $|[L_0]_{i,i}| = \sum_{j=1,j\neq i}^{m}[L_0]_{i,j} + \sum_{j=1}^{m}[L_1]_{i,j}$, where $|\alpha|$ indicates the modulus of α. This means the server can serve one customer at a time. Further details on the C-MSP can be found in Chaudhry et al. [6].

3   Analysis of the System-Length Distributions

If $N(t)$ and $J(t)$ represent the number of customers served and the state of the underlying Markov chain at time t, respectively, then $\{N(t), J(t)\}$ can be defined as a two-dimensional Markov process with state space $\{(\ell, i) : 0 \leq \ell \leq N, 1 \leq i \leq m\}$. If $\tilde{\pi}_j$ denotes the probability that a customer is getting service in steady state with the server in phase j ($1 \leq j \leq m$), then the stationary probability row vector is given by $\tilde{\boldsymbol{\pi}} = [\tilde{\pi}_1, \tilde{\pi}_2, \ldots, \tilde{\pi}_m]$ and can be calculated from the following relations:

$$\tilde{\boldsymbol{\pi}} \boldsymbol{L} = \boldsymbol{0} \quad \text{and} \quad \tilde{\boldsymbol{\pi}} \boldsymbol{e} = 1, \tag{1}$$

where $\boldsymbol{e}$ is a column vector of suitable size with all the entries 1 and $\boldsymbol{L} = \boldsymbol{L}_0 + \boldsymbol{L}_1$. We write the dimension of $\boldsymbol{e}$ in its suffix when it is other than m. Hence, the fundamental service rate or the mean service rate of customers is calculated by $\mu^* = \tilde{\boldsymbol{\pi}} \boldsymbol{L}_1 \boldsymbol{e}$.

For $n \geq 0$, $t \geq 0$ and $1 \leq i, j \leq m$, if we define the conditional probability as

$$P_{i,j}(n,t) = Pr\{N(t) = n, J(t) = j | N(0) = 0, J(0) = i\},$$

then $\boldsymbol{P}(n,t)$ represents an $m \times m$ matrix whose (i,j)th component is $P_{i,j}(n,t)$. Hence, using matrix notation the system may be expressed as

$$\frac{d}{dt}\boldsymbol{P}(0,t) = \boldsymbol{P}(0,t)\boldsymbol{L}_0, \tag{2}$$

$$\frac{d}{dt}\boldsymbol{P}(n,t) = \boldsymbol{P}(n,t)\boldsymbol{L}_0 + \boldsymbol{P}(n-1,t)\boldsymbol{L}_1, \ 1 \leq n \leq N-1, \tag{3}$$

$$\frac{d}{dt}\boldsymbol{P}(N,t) = \boldsymbol{P}(N,t)\boldsymbol{L} + \boldsymbol{P}(N-1,t)\boldsymbol{L}_1, \tag{4}$$

with $\boldsymbol{P}(0,0) = \boldsymbol{I}_m$, where $\boldsymbol{I}_m$ is the identity matrix of dimension $m \times m$ and $\boldsymbol{P}(n,0) = \boldsymbol{0}$ for $n \geq 1$. In the sequel the suffix m is dropped from the identity matrix $\boldsymbol{I}_m$ and in all other cases the dimension of an identity matrix $\boldsymbol{I}$ is written in its suffix. The traffic intensity or offered load is given by $\rho = \lambda \bar{g}/\mu^*$ and let the carried load, i.e., the probability that the server is busy be denoted by ρ'. For $n \geq 0$, let the (i,j)th ($1 \leq i, j \leq m$) element of the matrix $\boldsymbol{S}_n$ with dimension $m \times m$ be defined as the conditional probability of service completion of n customers in an inter-batch arrival span, during which the phase of the service process starting from phase i passes to phase j after completion of the n-th service, provided that at the previous batch arrival instant there were at least n customers in the system. Then

$$\boldsymbol{S}_n = \int_0^\infty \boldsymbol{P}(n,t)dF_A(t), \quad 0 \leq n \leq N. \tag{5}$$

Let us denote $\boldsymbol{S}_n'$ as

$$\boldsymbol{S}_n' = \sum_{k=n}^N \boldsymbol{S}_k, \quad 0 \leq n \leq N.$$

The $\boldsymbol{S}_n$ matrices can be evaluated using the arguments given by Lucantoni [14] and Neuts [19]. Following the renewal theory of semi-Markov processes, $\hat{A}$ and

$\widetilde{A}$ are defined as the stationary remaining and elapsed times of an inter-batch time, respectively, which satisfy the following equation.

$$F_{\widehat{A}}(x) = F_{\widetilde{A}}(x) = \int_0^x \lambda(1 - F_A(y)) \, dy. \tag{6}$$

An $m \times m$ order matrix $\boldsymbol{\Omega}_n$ ($n \geq 0$) is defined such that the (i,j)th entry of the matrix represents the limiting probability that n services are completed during an elapsed inter-batch arrival period while starting from phase i, the phase of the service process passes to phase j, provided the inter-batch arrival period started with at least $(n+1)$ customers in the system. Hence, applying Markov renewal theory and following Chaudhry and Templeton [7, p. 74–77], the following equation can be deduced:

$$\boldsymbol{\Omega}_n = \lambda \int_0^\infty \boldsymbol{P}(n,x)(1 - F_A(x)) \, dx, \quad 0 \leq n \leq N. \tag{7}$$

The matrices $\boldsymbol{\Omega}_n$ can be expressed in terms of the matrices $\boldsymbol{S}_n$ and their relationship discussed by Chaudhry et al. [6] is as follows:

$$\boldsymbol{\Omega}_0 = \lambda \Big(\boldsymbol{I}_m - \boldsymbol{S}_0 \Big)(-\boldsymbol{L}_0)^{-1}, \tag{8}$$

and

$$\boldsymbol{\Omega}_n = \Big(\boldsymbol{\Omega}_{n-1}\boldsymbol{L}_1 - \lambda\boldsymbol{S}_n \Big)(-\boldsymbol{L}_0)^{-1}, \quad 1 \leq n \leq N-1, \tag{9}$$

$$\boldsymbol{\Omega}_N = \Big(\boldsymbol{\Omega}_{N-1}\boldsymbol{L}_1 - \lambda\boldsymbol{S}_N \Big)(-\boldsymbol{L})^{-1}. \tag{10}$$

It may be noted here that $\boldsymbol{L}$ being a singular matrix we are not able to evaluate $\boldsymbol{\Omega}_N$ from Eq. (10). In view of this, we use $\boldsymbol{\Omega}_N = \Big(\boldsymbol{\Omega}_{N-1}\boldsymbol{L}_1 - \lambda\boldsymbol{S}_N \Big)(-\boldsymbol{L}_0)^{-1}$ as a close approximation. Further, one may note that the expression for $\boldsymbol{\Omega}_N$ is given just for the sake completeness and it is not required to compute the stationary probabilities in subsequent sections of the paper.

Further, $\widetilde{P}_{ij}(n,t)$ is defined as the conditional probability that starting with n customers in the system, at least n customers are served (which includes only the possibility of all potential service completions of customers) in the time interval $(0,t]$ while the phase of the service process is i and j at the start and at the end of the time interval, respectively. Hence, using the definition, we can obtain the following relation:

$$\widetilde{P}_{ij}(n,t+\Delta t) = \widetilde{P}_{ij}(n,t) + \sum_{k=1}^m P_{ik}(n-1,t)[L_1]_{kj}\Delta t + o(\Delta t), \quad 1 \leq n \leq N,$$

where the initial condition is $\widetilde{P}_{ij}(n,0) = 0$, $n \geq 1$. Now, for $t \geq 0$, $1 \leq i,j \leq m$, rearranging the terms and taking the limit as $\Delta t \to 0$, we can reduce this equation to

$$\frac{d}{dt}\widetilde{P}_{ij}(n,t) = \sum_{k=1}^m P_{ik}(n-1,t)[L_1]_{kj}, \quad 1 \leq n \leq N,$$

where the initial condition is given by $\widetilde{P}_{ij}(n,0) = 0$. The matrix form of the system may be written as

$$\frac{d}{dt}\widetilde{P}(n,t) = P(n-1,t)L_1, \quad 1 \le n \le N, \tag{11}$$

with $\widetilde{P}(n,0) = \mathbf{0}$ ($1 \le n \le N$). Let the (i,j)th element of the matrix S_n^* with dimension $m \times m$ denote the probability that having exactly n customers in the system at a batch arrival instant with the service phase being i, at least n customers are served during the inter-batch arrival period and the service phase at the next batch-arrival instant changes to phase j at the end of the n-th service completion. Then using the above definition of S_n^* and Eq. (11), we obtain (see Chaudhry et al. [6]).

$$S_n^* = \int_0^\infty \widetilde{P}(n,t)dF_A(t)$$
$$= \frac{1}{\lambda}\boldsymbol{\Omega}_{n-1}L_1, \ 1 \le n \le N. \tag{12}$$

For $n \ge 1$, the matrix $\boldsymbol{\Omega}_n^*$ of order $m \times m$ is defined such that the (i,j)th element of the matrix represents the limiting probability that n or more services are completed during an elapsed inter-batch arrival cycle while the phase of the service process is i and j at the beginning of the cycle and at the n-th service completion epoch, respectively, provided that at the batch arrival instant the arrivals in the batch join to make a total of n customers already present in the system. Then, from Markov renewal theory, it can be formulated that

$$\boldsymbol{\Omega}_n^* = \lambda \int_0^\infty \widetilde{P}(n,x)(1 - F_A(x)) \, dx, \quad 1 \le n \le N. \tag{13}$$

As derived by Chaudhry et al. [6], it may be shown that

$$\boldsymbol{\Omega}_1^* = (I_m - \boldsymbol{\Omega}_0)(-L_0)^{-1}L_1, \tag{14}$$
$$\text{and } \boldsymbol{\Omega}_{n+1}^* = (\boldsymbol{\Omega}_n^* - \boldsymbol{\Omega}_n)(-L_0)^{-1}L_1, \quad 1 \le n \le N-1. \tag{15}$$

3.1   Steady-State Distribution at a Pre-arrival Epoch

Let us consider the embedded points as the time epochs just before the batches arrive in the system and denote the embedded points as t_i^- ($i \ge 0$), where t_i's are the time epochs at which batch arrivals are about to occur. If we let $N_{t_i^-}$ and $\xi_{t_i^-}$ denote the number of customers in the system (including the one in service) and the phase of the service process at time t_i^- ($i \ge 0$), respectively, then at that time the state of the system is given by $\{N_{t_i^-}, \xi_{t_i^-}\}$. Let $\pi_j^-(n)$ denote the limiting probabiloity that at a pre-arrival epoch of a batch there are n number of customers in the system and the phase of the service is j, that is $\pi_j^-(n) = \lim_{i\to\infty} P(N_{t_i^-} = n, \ \xi_{t_i^-} = j), \ 0 \le n \le N, \ 1 \le j \le m$. For $0 \le n \le N$,

$\boldsymbol{\pi}^-(n)$ denotes a row vector of order $1 \times m$ with i-th component $\pi_i^-(n)$. We then form the following set of vector-difference equations by relating the state of the system at two consecutive embedded Markov points (i.e., at two successive pre-arrival epochs):

$$\boldsymbol{\pi}^-(0) = \sum_{i=0}^{N-1} \boldsymbol{\pi}^-(i) \sum_{r=1}^{N-i} g_r \boldsymbol{S}_{i+r}^* + \sum_{i=0}^{N-1} \boldsymbol{\pi}^-(i) g'_{N-i} \boldsymbol{S}_N^* + \boldsymbol{\pi}^-(N) \boldsymbol{S}_N^*, \qquad (16)$$

$$\boldsymbol{\pi}^-(n) = \sum_{i=0}^{n-2} \boldsymbol{\pi}^-(i) \sum_{r=n-i}^{N-i} g_r \boldsymbol{S}_{i+r-n} + \sum_{i=n-1}^{N-1} \boldsymbol{\pi}^-(i) \sum_{r=1}^{N-i} g_r \boldsymbol{S}_{i+r-n}$$

$$+ \sum_{i=0}^{N-1} \boldsymbol{\pi}^-(i) g'_{N-i} \boldsymbol{S}_{N-n} + \boldsymbol{\pi}^-(N) \boldsymbol{S}_{N-n}, \quad 1 \le n \le N-1, \qquad (17)$$

$$\boldsymbol{\pi}^-(N) = \sum_{i=0}^{N-1} \boldsymbol{\pi}^-(i) g'_{N-i} \boldsymbol{S}_0 + \boldsymbol{\pi}^-(N) \boldsymbol{S}_0. \qquad (18)$$

The system of vector-difference Eqs. (16)–(18) may be solved using roots of a characteristic equation, see Kim and Chaudhry [13]. For this, we assume the solution to be of a general form (see Remark 1 below)

$$\boldsymbol{\pi}^-(n) = \left[c_1 z^n, c_2 z^n, c_3 z^n, \ldots, c_m z^n \right], \quad 0 \le n \le N, \ c_1, \ c_2, \ c_3, \ \ldots \ c_m \ne 0. \qquad (19)$$

Substituting this general solution form (19) into (16), (17), and (18), we obtain

$$\left[c_1, c_2, c_3, \ldots, c_m \right] = \sum_{i=0}^{N-1} \left[c_1 z^i, c_2 z^i, c_3 z^i, \ldots, c_m z^i \right] \sum_{r=1}^{N-i} g_r \boldsymbol{S}_{i+r}^*$$

$$+ \sum_{i=0}^{N-1} \left[c_1 z^i, c_2 z^i, c_3 z^i, \ldots, c_m z^i \right] g'_{N-i} \boldsymbol{S}_N^* + \boldsymbol{\pi}^-(N) \boldsymbol{S}_N^*, \qquad (20)$$

$$\left[c_1 z^n, c_2 z^n, c_3 z^n, \ldots, c_m z^n \right] = \sum_{i=0}^{n-2} \left[c_1 z^i, c_2 z^i, c_3 z^i, \ldots, c_m z^i \right] \sum_{r=n-i}^{N-i} g_r \boldsymbol{S}_{i+r-n}$$

$$+ \sum_{i=n-1}^{N-1} \left[c_1 z^i, c_2 z^i, c_3 z^i, \ldots, c_m z^i \right] \sum_{r=1}^{N-i} g_r \boldsymbol{S}_{i+r-n}$$

$$+ \sum_{i=0}^{N-1} \left[c_1 z^i, c_2 z^i, c_3 z^i, \ldots, c_m z^i \right] g'_{N-i} \boldsymbol{S}_{N-n}$$

$$+ \left[c_1 z^N, c_2 z^N, c_3 z^N, \ldots, c_m z^N \right] \boldsymbol{S}_{N-n}, \quad 1 \le n \le N-1, \qquad (21)$$

and
$$\left[c_1 z^N, c_2 z^N, c_3 z^N, \ldots, c_m z^N\right] = \sum_{i=0}^{N-1} \left[c_1 z^i, c_2 z^i, c_3 z^i, \ldots, c_m z^i\right] g'_{N-i} \mathbf{S}_0$$

$$+ \left[c_1 z^N, c_2 z^N, c_3 z^N, \ldots, c_m z^N\right] \mathbf{S}_0. \quad (22)$$

Now after calculation of the right-hand sides of Eqs. (20)–(22), we obtain N vector equations of the following form:

$$\left[c_1 z^n, c_2 z^n, c_3 z^n, \ldots, c_m z^n\right] = \left[c_1 F_{n1}^1(z) + c_2 F_{n1}^2(z) + c_3 F_{n1}^3(z) + \cdots + c_m F_{n1}^m(z),\right.$$

$$c_1 F_{n2}^1(z) + c_2 F_{n2}^2(z) + c_3 F_{n2}^3(z) + \cdots + c_m F_{n2}^m(z), \ldots,$$

$$\left. c_1 F_{nm}^1(z) + c_2 F_{nm}^2(z) + c_3 F_{nm}^3(z) + \cdots + c_m F_{nm}^m(z)\right],$$

$$0 \leq n \leq N-1, \quad (23)$$

and

$$\left[c_1 z^N, c_2 z^N, c_3 z^N, \ldots, c_m z^N\right] = \left[c_1 F_{N1}^1(z) + c_2 F_{N1}^2(z) + c_3 F_{N1}^3(z) + \cdots + c_m F_{N1}^m(z),\right.$$

$$c_1 F_{N2}^1(z) + c_2 F_{N2}^2(z) + c_3 F_{N2}^3(z) + \cdots + c_m F_{N2}^m(z), \ldots,$$

$$\left. c_1 F_{Nm}^1(z) + c_2 F_{Nm}^2(z) + c_3 F_{Nm}^3(z) + \cdots + c_m F_{Nm}^m(z)\right],$$

$$(24)$$

where the functions $F_{nj}^k(z)$, $0 \leq n \leq N$, $1 \leq j \leq m$, $1 \leq k \leq m$, are certain polynomials of degree up to N in z. From (20)–(22) these polynomials follow as

$$F_{0j}^k(z) = \sum_{i=0}^{N-1} \sum_{r=1}^{N-i} g_r z^i \left[S_{i+r}^*\right]_{k,j}$$

$$+ \sum_{i=0}^{N-1} g'_{N-i} z^i \left[S_N^*\right]_{k,j} + z^N \left[S_N^*\right]_{k,j}, \quad 1 \leq j \leq m, 1 \leq k \leq m, \quad (25)$$

$$F_{nj}^k(z) = \sum_{l=0}^{n-2} \sum_{r=n-l}^{N-l} g_r z^l \left[S_{l+r-n}\right]_{k,j} + \sum_{l=n-1}^{N-1} \sum_{r=1}^{N-l} g_r z^l \left[S_{l+r-n}\right]_{k,j}$$

$$+ \sum_{l=0}^{N-1} g'_{N-l} z^l \left[S_{N-n}\right]_{k,j}, \quad 1 \leq n < N, 1 \leq j \leq m, 1 \leq k \leq m, \quad (26)$$

$$F_{Nj}^k(z) = \sum_{l=0}^{N-1} g'_{N-i} z^l \left[S_0\right]_{k,j} + z^N \left[S_0\right]_{k,j}, \quad 1 \leq j \leq m, 1 \leq k \leq m, \quad (27)$$

where $\left[\boldsymbol{S}_n\right]_{k,j}$ are the (k,j)th elements of $\boldsymbol{S}_n$ $(0 \leq n \leq N)$ and $\left[\boldsymbol{S}_n^*\right]_{k,j}$ are the (k,j)th elements of $\boldsymbol{S}_n^*$ $(1 \leq n \leq N)$. Now, after adding the vector Eqs. (23) and (24), we consider the component-wise equality in the vector equations, and considering a unique set of values of $c_1, c_2, c_3, \ldots, c_m$, we obtain the following determinant which must be equal to zero.

$$
\boldsymbol{D}_N(z) = \begin{vmatrix}
\sum_{n=0}^{N}(F_{n1}^1(z)-z^n) & \sum_{n=0}^{N}F_{n1}^2(z) & \sum_{n=0}^{N}F_{n1}^3(z) & \cdots & \sum_{n=0}^{N}F_{n1}^{m-1}(z) & \sum_{n=0}^{N}F_{n1}^m(z) \\
\sum_{n=0}^{N}F_{n2}^1(z) & \sum_{n=0}^{N}(F_{n2}^2(z)-z^n) & \sum_{n=0}^{N}F_{n2}^3(z) & \cdots & \sum_{n=0}^{N}F_{n2}^{m-1}(z) & \sum_{n=0}^{N}F_{n2}^m(z) \\
\sum_{n=0}^{N}F_{n3}^1(z) & \sum_{n=0}^{N}F_{n3}^2(z) & \sum_{n=0}^{N}(F_{n3}^3(z)-z^n) & \cdots & \sum_{n=0}^{N}F_{n3}^{m-1}(z) & \sum_{n=0}^{N}F_{n3}^m(z) \\
\vdots & \vdots & \vdots & \ddots & \vdots & \vdots \\
\sum_{n=0}^{N}F_{nm-1}^1(z) & \sum_{n=0}^{N}F_{nm-1}^2(z) & \sum_{n=0}^{N}F_{nm-1}^3(z) & \cdots & \sum_{n=0}^{N}(F_{nm-1}^{m-1}(z)-z^n) & \sum_{n=0}^{N}F_{nm-1}^m(z) \\
\sum_{n=0}^{N}F_{nm}^1(z) & \sum_{n=0}^{N}F_{nm}^2(z) & \sum_{n=0}^{N}F_{nm}^3(z) & \cdots & \sum_{n=0}^{N}F_{nm}^{m-1}(z) & \sum_{n=0}^{N}(F_{nm}^m(z)-z^n)
\end{vmatrix}.
$$
(28)

Thus, we obtain the characteristic equation of the $GI^X/C\text{-}MSP/1/N$ queueing model as follows:

$$\boldsymbol{D}_N(z) = 0, \tag{29}$$

which is a polynomial in z of degree Nm giving Nm roots. Out of these Nm roots, we use $(N+1)$ roots (which may be chosen arbitrarily from those Nm roots) and the solution of the vector-difference Eqs. (16)–(18) can be written as

$$
\boldsymbol{\pi}^-(n) = \left[\sum_{i=1}^{N+1} d_{1i}z_i^n, \sum_{i=1}^{N+1} d_{2i}z_i^n, \sum_{i=1}^{N+1} d_{3i}z_i^n, \ldots, \sum_{i=1}^{N+1} d_{mi}z_i^n\right], \quad 0 \leq n \leq N,
$$
(30)

where $z_1, z_2, \ldots, z_{N+1}$ are $N+1$ arbitrarily chosen roots of the characteristic Eq. (29) and d_{ki} $(1 \leq k \leq m,\ 1 \leq i \leq N+1)$ are non-zero constants which can be calculated using the roots and the required number of vector-difference Eqs. (16)–(18). It is interesting to note that out of the Nm roots of Eq. (29), we use only $(N+1)$ roots. This is because each component of the probability vector has $N+1$ unknowns. Of course, if $(N+1)$ is odd, we need to use one real root and the other complex roots though the complex conjugates have to be used together. Another interesting point to note is that each component of the probability vector is a mixture of constant weighted geometric terms, where the constant weights are denoted by $d_{ki}(1 \leq k \leq m$ and $1 \leq i \leq N+1)$.

To determine these constants d_{ki} $(1 \leq k \leq m,\ 1 \leq i \leq N+1)$, we use the vector-difference Eqs. (16)–(18). First, we substitute the solution of $\boldsymbol{\pi}^-(n)$ as presented above (see Eq. (30)) in the right-hand side of each and every Eq. (16)–(18) and obtain component-wise values of each $\boldsymbol{\pi}^-(n)$ $(0 \leq n \leq N)$ in terms of those constants d_{ki} $(1 \leq k \leq m,\ 1 \leq i \leq N+1)$. Now equating these $\boldsymbol{\pi}^-(n)$ $(0 \leq n \leq N)$ with the corresponding components of $\boldsymbol{\pi}^-(n)$ $(0 \leq n \leq N)$ in Eq. (30) gives a total of $(N+1)m$ linear equations with $(N+1)m$ unknowns d_{ki} $(1 \leq k \leq m,\ 1 \leq i \leq N+1)$. Finally, one may remember that we must use the following normalizing condition by skipping one of those $(N+1)m$ linear equations, which is a redundant one. The normalizing condition is given by

$$\sum_{n=0}^{N} \left[\sum_{i=1}^{N+1} d_{1i}z_i^n + \sum_{i=1}^{N+1} d_{2i}z_i^n + \sum_{i=1}^{N+1} d_{3i}z_i^n + \cdots + \sum_{i=1}^{N+1} d_{mi}z_i^n \right] = 1. \quad (31)$$

Remark 1. It may be remarked here that the general solution of the form (19) is along the lines proposed by Kim and Chaudhry [13] for the case of exponential service time distribution where the number of service phases is '1', i.e., $m = 1$. There the assumption of general solution is cz^n for the case of the finite-buffer multi-server $GI^X/M/k/N$ queueing model, see Kim and Chaudhry [13] for details.

3.2 Stationary Distribution at Arbitrary Epoch

For $0 \leq n \leq N$, the steady-state system-length distribution at an arbitrary epoch is denoted by $\boldsymbol{\pi}(n) = [\pi_1(n), \pi_2(n), \ldots \pi_m(n)]$ and is derived in this section. The derivation requires the classical arguments of renewal theory and the relation between the steady-state system-length distribution at an arbitrary epoch and the corresponding pre-arrival epoch distribution. Hence, using Markov renewal theory and semi-Markov processes, see Cinlar [10] or Lucantoni and Neuts [16], it can be obtained as follows:

$$\boldsymbol{\pi}(0) = \sum_{i=0}^{N-1} \boldsymbol{\pi}^-(i) \sum_{r=1}^{N-i} g_r \boldsymbol{\Omega}_{i+r}^* + \sum_{i=0}^{N-1} \boldsymbol{\pi}^-(i) g_{N-i}' \boldsymbol{\Omega}_N^* + \boldsymbol{\pi}^-(N) \boldsymbol{\Omega}_N^*, \quad (32)$$

$$\boldsymbol{\pi}(n) = \sum_{i=0}^{n-2} \boldsymbol{\pi}^-(i) \sum_{r=n-i}^{N-i-1} g_r \boldsymbol{\Omega}_{i+r-n} + \sum_{i=n-1}^{N-1} \boldsymbol{\pi}^-(i) \sum_{r=1}^{N-i} g_r \boldsymbol{\Omega}_{i+r-n}$$

$$+ \sum_{i=0}^{N-1} \boldsymbol{\pi}^-(i) g_{N-i}' \boldsymbol{\Omega}_{N-n} + \boldsymbol{\pi}^-(N) \boldsymbol{\Omega}_{N-n}, \quad 1 \leq n \leq N-1, \quad (33)$$

$$\boldsymbol{\pi}(N) = \sum_{i=0}^{N-1} \boldsymbol{\pi}^-(i) g_{N-i}' \boldsymbol{\Omega}_0 + \boldsymbol{\pi}^-(N) \boldsymbol{\Omega}_0. \quad (34)$$

Note that since an empty system causes an interruption in the service process, it follows that $\sum_{n=1}^{N} \boldsymbol{\pi}(n)(\boldsymbol{L}_0 + \boldsymbol{L}_1) = \boldsymbol{0}$, which implies that

$$\sum_{n=1}^{N} \boldsymbol{\pi}(n) = C\tilde{\boldsymbol{\pi}}, \quad [\text{since from relation (1)}, \ \tilde{\boldsymbol{\pi}}\boldsymbol{L} = \tilde{\boldsymbol{\pi}}(\boldsymbol{L}_0 + \boldsymbol{L}_1) = \boldsymbol{0}] \quad (35)$$

where C is a positive constant. Now post-multiplying both sides of the Eq. (35) by $\boldsymbol{L}_1 \boldsymbol{e}$, it can be concluded that

$$\sum_{n=1}^{N} \boldsymbol{\pi}(n) \boldsymbol{L}_1 \boldsymbol{e} = C\mu^*. \quad (36)$$

The left-hand side of Eq. (36) is the departure rate in steady state, which is quantitatively the same as the steady-state effective arrival rate. If the blocking probability of an arbitrary customer of a batch is denoted by P_{BA}, then the effective arrival rate (λ') in steady state is $\lambda\bar{g}(1-P_{BA})$ and the positive constant can be evaluated as

$$C = \frac{\lambda\bar{g}(1-P_{BA})}{\mu^*} = \rho'. \tag{37}$$

Now, rearranging the equation $\sum\limits_{n=0}^{N} \boldsymbol{\pi}(n)\boldsymbol{e} = 1$ as $\boldsymbol{\pi}(0)\boldsymbol{e} + \sum\limits_{n=1}^{N} \boldsymbol{\pi}(n)\boldsymbol{e} = 1$ and noting that $\sum\limits_{n=1}^{N} \boldsymbol{\pi}(n)\boldsymbol{e} = C = \rho'$ (as follows from Eq. (35)), which also denotes the probability that the server is busy, we can also formulate the probability that the server is idle:

$$\boldsymbol{\pi}(0)\boldsymbol{e} = 1 - \rho'. \tag{38}$$

3.3   Performance Measures

Following the determination of the state probabilities at various epochs, the computation of some important performance measures may be obtained and is discussed in this section. The mean number of customers in the system and in the queue at an arbitrary epoch are given by

$$L = \sum_{i=0}^{N} i\boldsymbol{\pi}(i)\boldsymbol{e}, \quad L_q = \sum_{i=1}^{N}(i-1)\boldsymbol{\pi}(i)\boldsymbol{e}.$$

Let P_{Bf}, P_{Bl} and P_{Ba} denote the blocking probabilities of the first, last and an arbitrary customer in a batch. Then one can compute P_{Bf} as $\boldsymbol{\pi}^-(N)\boldsymbol{e}$. The probabilities P_{Bl} and P_{Ba} can be obtained by defining a r.v. G^- which denotes the number of customers in front of an arbitrary customer within the batch. Chaudhry and Templeton [7, p. 93] provide the distribution of G^- as

$$g_r^- = P[G^- = r] = \frac{1}{\bar{g}} \sum_{i=r+1}^{\infty} g_i, \quad r \geq 0.$$

Hence, P_{Ba} and P_{Bl} are given by

$$P_{Ba} = \sum_{i=0}^{N}\sum_{j=N-i}^{\infty} \boldsymbol{\pi}^-(i)g_j^- \boldsymbol{e}, \quad \text{and} \quad P_{Bl} = \sum_{i=0}^{N}\sum_{j=N-i+1}^{\infty} \boldsymbol{\pi}^-(i)g_j\boldsymbol{e}, \tag{39}$$

respectively. Further, if the mean waiting time of an arbitrary customer in the system (queue) is denoted by $\overline{w}_a$ ($\overline{w}_{qa}$), then Little's formula can be applied to get $\overline{w}_a = \frac{L}{\lambda'}$, and $\overline{w}_{qa} = \frac{L_q}{\lambda'}$, respectively, where λ' is the effective arrival rate and is evaluated as $\lambda' = \lambda\bar{g}(1-P_{Ba})$.

4 Numerical Results

Using the results formulated in previous sections, we present some numerical results in this section to illustrate the developed methodology. Although numerical calculations were carried out with high precision, due to lack of space the results are reported to 6 decimal places. The system length distribution at pre-arrival and arbitrary time epochs (using roots of a characteristic equation) of a $GI^X/C\text{-}MSP/1/10$ queue under the PBR policy is represented in Table 1. For this system, g_n $(n \geq 1)$ is considered to be the coefficient of z^n in the expansion of the probability generating function of the batch-size distribution $G(z) = 4z/(3 - z)^2$ with $\bar{g} = 2.0$. The density function of the inter-batch arrival time is considered as a 3-phase continuous phase-type $(C\text{-}PH)$ distribution with representation

$$\boldsymbol{\alpha}_1 = [0.2, 0.3, 0.5], \quad \boldsymbol{T}_1 = \begin{pmatrix} -2.5 & 0.5 & 0.2 \\ 0.1 & -2.0 & 0.1 \\ 0.2 & 0.4 & -3.7 \end{pmatrix},$$

and the 4-phase $C\text{-}MSP$ representation is taken as

$$\boldsymbol{L}_0 = \begin{pmatrix} -6.5 & 0.1 & 0.6 & 0.3 \\ 0.3 & -5.4 & 0.8 & 0.2 \\ 0.2 & 0.7 & -4.3 & 0.1 \\ 0.6 & 0.4 & 0.1 & -3.9 \end{pmatrix}, \quad \boldsymbol{L}_1 = \begin{pmatrix} 1.1 & 1.7 & 1.5 & 1.2 \\ 1.3 & 0.9 & 0.7 & 1.2 \\ 1.4 & 0.9 & 0.3 & 0.7 \\ 0.9 & 0.5 & 1.1 & 0.3 \end{pmatrix}.$$

Interested readers are referred to Neuts [19] for a probability vector and a non-singular matrix representation of a $C\text{-}PH$-distribution. For these representations of $C\text{-}PH$ and $C\text{-}MSP$, it is found that $\mu^* = 3.862496$, $\lambda = 2.221854$, $\rho = 1.150476$ and $\widetilde{\boldsymbol{\pi}} = [0.224993, 0.240930, 0.283610, 0.250467]$.

Table 1. Distribution of the system length for the $C\text{-}PH^X/C\text{-}MSP/1/10\text{-}PBR$ queue at pre-arrival and arbitrary epochs using roots

n	$\pi_1^-(n)$	$\pi_2^-(n)$	$\pi_3^-(n)$	$\pi_4^-(n)$	$\pi^-(n)e$	$\pi_1(n)$	$\pi_2(n)$	$\pi_3(n)$	$\pi_4(n)$	$\pi(n)e$
0	0.031660	0.026752	0.023690	0.022724	0.104826	0.034236	0.028905	0.025595	0.024553	0.113289
1	0.014403	0.014454	0.015758	0.014173	0.058788	0.014513	0.014623	0.016016	0.014387	0.059538
2	0.015804	0.016158	0.018006	0.016105	0.066073	0.015901	0.016300	0.018221	0.016284	0.066707
3	0.016746	0.017376	0.019695	0.017541	0.071358	0.016854	0.017517	0.019893	0.017710	0.071974
4	0.017787	0.018625	0.021335	0.018954	0.076700	0.017896	0.018759	0.021515	0.019109	0.077279
5	0.019003	0.020000	0.023048	0.020449	0.082500	0.019092	0.020109	0.023196	0.020575	0.082973
6	0.020416	0.021543	0.024906	0.022081	0.088946	0.020447	0.021593	0.024987	0.022147	0.089175
7	0.022075	0.023308	0.026978	0.023919	0.096280	0.021982	0.023233	0.026925	0.023865	0.096005
8	0.024009	0.025443	0.029443	0.026064	0.104958	0.023668	0.025122	0.029129	0.025773	0.103692
9	0.027158	0.027687	0.032041	0.028458	0.115344	0.026322	0.026903	0.031233	0.027720	0.112179
10	0.024216	0.031132	0.042509	0.036371	0.134227	0.022828	0.029476	0.040365	0.034520	0.127189
sum	0.233278	0.242477	0.277408	0.246838	1.000000	0.233740	0.242540	0.277075	0.246645	1.000000

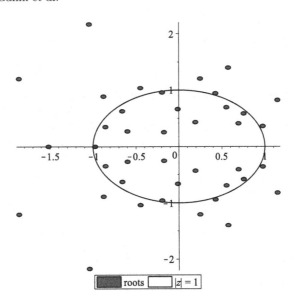

Fig. 1. The roots of the characteristic equation for the PBR model are marked as red dots in the complex plane (Color figure online)

The roots used to calculate the stationary probability vectors for the above Table 1 are presented next. The $Nm = 40$ roots of the characteristic Eq. (29) for this PBR model are given in Fig. 1 above.

We use $N + 1 = 11$ roots ($z_1 = -0.975907$, $z_2 = -0.855094 - 0.348176i$, $z_3 = -0.855094 + 0.348176i$, $z_4 = -0.660588 - 0.626857i$, $z_5 = -0.660588 + 0.626857i$, $z_6 = -0.600024 - 0.270433i$, $z_7 = -0.600024 + 0.270433i$, $z_8 = -0.193425 - 0.958256i$, $z_9 = -0.193425 + 0.958256i$, $z_{10} = -0.172185 - 0.254405i$, $z_{11} = -0.172185 + 0.254405i$) to get the 11 unknown constants involved in the 1st component of the probability vector $\pi^-(n)$ and subsequently the stationary probability vectors at a pre-arrival epoch. The constants corresponding to the first component of the stationary probability vectors at a pre-arrival epoch (for the PBR model) are found as $d_{1,1} = 132.111093$, $d_{1,2} = 149.655211 - 77.022170i$, $d_{1,3} = 149.655211 + 77.022170i$, $d_{1,4} = 32.811374 - 23.656608i$, $d_{1,5} = 32.811374 + 23.656608i$, $d_{1,6} = -174.340305 + 544.507012i$, $d_{1,7} = -174.340305 - 544.507012i$, $d_{1,8} = 0.041539 - 1.016160i$, $d_{1,9} = 0.041539 + 1.016160i$, $d_{1,10} = -74.207536 - 31.033273i$, $d_{1,11} = -74.207536 + 31.033273i$, and the corresponding probabilities (for the PBR model) are given in Table 1. The constants in the other components of the probability vector as well as the related probabilities are evaluated using the same 11 roots.

From Table 1, it is calculated that $\rho' = 0.886710$ and it can also be checked that $\sum_{n=1}^{N} \pi(n)/\rho' \simeq \tilde{\pi}$ for the PBR model. The mean waiting time of an arbitrary customer is also obtained from Little's law for the PBR model and it is found that $\overline{w}_a = L/\lambda' = 1.591564$ with $P_{Ba} = 0.229266$, $L = 5.450976$, and $\lambda' = 3.424917$.

5 Conclusion

In this paper, we have analyzed the $GI^X/C\text{-}MSP/1/N$ queue and obtained steady-state probability distributions at pre-arrival and arbitrary epochs. The proposed method of analysis is based on the roots of the characteristic equation which is derived from the balance equations obtained from an embedded Markov chain at pre-arrival epochs. Specifically, the pre-arrival epoch probability vectors are written in terms of the roots of the characteristic equation. Also, we have obtained stationary performance measures, such as the mean waiting time and the blocking probabilities for the first customer, an arbitrary customer and the last customer in an accepted batch. It should be pointed out that the roots method works very well for any type of inter-batch arrival distribution including heavy-tailed inter-batch arrival times which have been discussed by Chaudhry et al. [6] through numerical examples of the $C\text{-}PH^X/C\text{-}MSP/1/\infty$ and $Pareto^X/C\text{-}MSP/1/\infty$ queueing models. The Laplace-Stieltjes transforms (LST) of actual waiting time of the first, an arbitrary, and the last customer in an accepted batch can be derived as a further extension of this queueing model. The system can also adopt the total batch rejection (TBR) policy, i.e., where an entire batch is rejected if the batch is loaded with more customers than the available system space. The system-length distributions at pre-arrival and arbitrary epochs of the $GI^X/C\text{-}MSP/1/N$ model with a TBR policy can be considered in future work. Further, the corresponding finite-buffer discrete-time queues may be attempted using a similar methodology of roots and are left for future investigations.

Acknowledgment. The second author was supported partially by NSERC under research grant number RGPIN-2014-06604.

References

1. Albores-Velasco, F.J., Tajonar-Sanabria, F.S.: Analysis of the $GI/MSP/c/r$ queueing system. Inf. Theory Inf. Process. **4**(1), 46–57 (2004)
2. Alfa, A.S., Xue, J., Ye, Q.: Perturbation theory for the asymptotic decay rates in the queues with Markovian arrival process and/or Markovian service process. Queueing Syst. **36**(4), 287–301 (2000). https://doi.org/10.1023/A:1011032718715
3. Banik, A.D., Gupta, U.C.: Analyzing the finite buffer batch arrival queue under Markovian service process: $GI^X/MSP/1/N$. TOP **15**(1), 146–160 (2007). https://doi.org/10.1007/s11750-007-0007-2
4. Bocharov, P.P.: Stationary distribution of a finite queue with recurrent input and Markov service. Autom. Remote Control **57**(9), 1274–1283 (1996)
5. Bocharov, P.P., D'Apice, C., Pechinkin, A.V., Salerno, S.: The stationary characteristics of the $G/MSP/1/r$ queueing system. Autom. Remote Control **64**(2), 288–301 (2003). https://doi.org/10.1023/A:1022219232282

6. Chaudhry, M.L., Banik, A.D., Pacheco, A.: A simple analysis of the batch arrival queue with infinite-buffer and Markovian service process using roots method: $GI^{[X]}/C\text{-}MSP/1/\infty$. Ann. Oper. Res. **252**(1), 135–173 (2017). https://doi.org/10.1007/s10479-015-2026-y

7. Chaudhry, M.L., Templeton, J.G.C.: First Course in Bulk Queues. Wiley, Hoboken (1983)

8. Chaudhry, M., Banik, A.D., Pacheco, A., Ghosh, S.: A simple analysis of system characteristics in the batch service queue with infinite-buffer and Markovian service process using the roots method: $GI/C\text{-}MSP^{(a,b)}/1/\infty$. RAIRO-Oper. Res. **50**(3), 519–551 (2016)

9. Choi, B.D., Hwang, G.U., Han, D.H.: Supplementary variable method applied to the $MAP/G/1$ queueing system. J. Aust. Math. Soc. Ser. B. Appl. Math. **40**(01), 86–96 (1998)

10. Cinlar, E.: Introduction to Stochastic Processes. Courier Corporation, Chelmsford (2013)

11. Gupta, U.C., Banik, A.D.: Complete analysis of finite and infinite buffer $GI/MSP/1$ queue - a computational approach. Oper. Res. Lett. **35**(2), 273–280 (2007)

12. Kasahara, S., Takine, T., Takahashi, Y., Hasegawa, T.: $MAP/G/1$ queues under N-policy with and without vacations. J. Oper. Res. Soc. Jpn. **39**(2), 188–212 (1996)

13. Kim, J.J., Chaudhry, M.L.: A novel way of treating the finite-buffer queue $GI/M/c/N$ using roots. Int. J. Math. Models Methods Appl. Sci. **11**, 286–289 (2017)

14. Lucantoni, D.M.: New results on the single server queue with a batch Markovian arrival process. Commun. Stat. Stoch. Models **7**(1), 1–46 (1991)

15. Lucantoni, D.M., Meier-Hellstern, K.S., Neuts, M.F.: A single-server queue with server vacations and a class of non-renewal arrival processes. Adv. Appl. Probab. **22**, 676–705 (1990)

16. Lucantoni, D.M., Neuts, M.F.: Some steady-state distributions for the $MAP/SM/1$ queue. Stoch. Model. **10**(3), 575–598 (1994)

17. Machihara, F.: A $G/SM/1$ queue with vacations depending on service times. Stoch. Model. **11**(4), 671–690 (1995)

18. Neuts, M.F.: A versatile Markovian point process. J. Appl. Probab. **16**, 764–779 (1979)

19. Neuts, M.F.: Matrix-Geometric Solutions in Stochastic Models: An Algorithmic Approach. Courier Corporation, Chelmsford (1981)

20. Ramaswami, V.: The $N/G/1$ queue and its detailed analysis. Adv. Appl. Probab. **12**, 222–261 (1980)

21. Takagi, H.: Queueing Analysis: A Foundation of Performance Evaluation, vol. 2. North-Holland, New York (1993)

Stationary Analysis of Infinite Server Queue with Batch Service

Ayane Nakamura[1] and Tuan Phung-Duc[2(✉)]

[1] Graduate School of Science and Technology, University of Tsukuba,
Tsukuba, Ibaraki 305-8577, Japan
s2020431@s.tsukuba.ac.jp
[2] Faculty of Engineering, Information and Systems, University of Tsukuba,
Tsukuba, Ibaraki 305-8577, Japan
tuan@sk.tsukuba.ac.jp

Abstract. In this paper, we consider an infinite server queue with batch service. Although many studies on queues with finite number of servers and batch service have been conducted, there are only a few research related to the infinite server queue with batch service and they only consider the moments of the queue length. We present the derivation of the stationary distribution for the infinite server queue with batch service by the method based on factorial moment generating function. Furthermore, we derive some performance measures and show some numerical examples of the stationary distribution.

Keywords: Infinite server queue · Batch service · Factorial moments

1 Introduction

In our modern society, there exist various systems with batch service. For example, transportation services such as ride-sharing and demand-bus, logistics such as home-delivery can be cited. In the telecommunications field, research on batch services of data centers and optical burst switched networks is conducted from the perspectives of the effective use of the resources.

Batch service queueing models were originated by Bailey [1]. He derived the stationary distribution of queue with the fixed-batch service by using the embedded-Markov chain technique. After that, Neuts [2] introduced the general batch service rule. The general batch service rule states that the server will start to provide service only when at least 'a' units in the queue, and the maximum service capacity is 'b', which also includes the fixed-batch service. Downton [3] derived the waiting time distribution of batch service queues. Holman et al. [4] analyzed batch service queue with general service time using supplementary method. As studies on multiple servers queue with batch service, Arora [5] analyzed two server batch service queue and Ghare [6] generalized the results of Arora to 'c' servers. In addition, Cosmetatos [7] and Sim et al. [8,9] considered $M/M(a, \infty)/N$ queue where the number of customers exceeds some control limit

© Springer Nature Switzerland AG 2021
P. Ballarini et al. (Eds.): EPEW 2021/ASMTA 2021, LNCS 13104, pp. 411–424, 2021.
https://doi.org/10.1007/978-3-030-91825-5_25

'a', an idle server will take for service all the waiting customers in the queue. Besides, studies on batch service queues with vacation [10] and tandem queues with batch service [11] and so on have been conducted (see the survey [12] in details).

However, only few studies on an infinite server queue with batch service have been conducted. Ushakumari et al. [13] studied an infinite server queue in continuous time in which arrivals are batches of variable size and service is provided in groups of fixed size R. This paper derived the recursive relations of the binomial moment of the number of busy servers at arbitrary time. As a special case of the binomial moment, they derived the expected number of busy servers at arbitrary epoch. Liu et al. [14] also considered same approach to the same model as [13]. They discussed several cases and gave some insights. For example, about $GI^K/M^R/\infty$ system (the batch sizes of the arrival and the service are both constant), they found that the change of K has only a small effect on the amount of computational work while the computational complexity increases with the service batch size R. Besides, Liu et al. [15] considered an infinite server queue where customers have a choice of individual service or batch service is studied. They analyzed the model as two variable Markov chain of the busy servers and the waiting customers. They derived the first two moments of the time dependent number of busy servers by transforming the Kolmogorov equations and LST of the waiting time for the customers. Ushakumari et al. [16] considered an infinite server queue where customers can opt for single service of varying size with the restriction that single service is provided only when none is waiting in the queue. This paper derived the first two moments of busy servers using the similar solution with [15]. They also derived the waiting time distribution. We should note that all these studies do not consider the steady state probabilities of the system. In this paper, we analyze $M/M(b)/\infty$ queue, in other words, an infinite server queue with the fixed-batch service. The main contribution of this paper is the derivation of the steady state probabilities of the model. We also derive other performance measures such as the sojourn time distribution.

The rest of the paper is constructed as follows. In Sect. 2, we show the derivation of the stationary distribution for $M/M(b)/\infty$ queue. Section 3 presents some other performance measures and Sect. 4 shows some numerical examples for the stationary distribution. Finally, we present the conclusion of this paper and future work in Sect. 5.

2   Stationary Distribution of $M/M(b)/\infty$ Queue

This section presents the derivation of the stationary distribution for $M/M(b)/\infty$ queue. In $M/M(b)/\infty$ queue, customers arrive according to a Poisson process with rate λ. A service is executed in batch with a fixed size b. In that reason, there is a waiting time of customers for the batch of b customers to be completed although $M/M(b)/\infty$ queue has infinite servers. Besides, we assume that the service time follows the exponential distribution of parameter μ.

An example of the possible application of this model is shared transportation systems such as ridesharing [17]. We assume the situation that there are enough number of cars at a car station and a capacity of a car is b. Customers who want to visit same destination arrive the car station according to a Poisson process with rate λ. If there are b customers at the car station, the customers immediately ride on the car and start to drive the car, which takes the time following an exponential distribution of rate μ. We can also adapt other examples, such as the batch service of a huge data center, logistic system, inventory management system, etc., to this model.

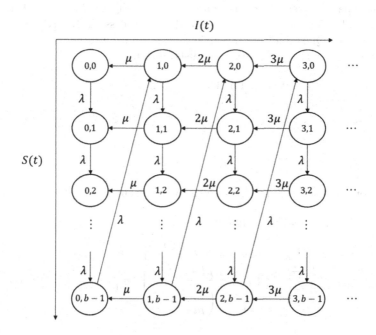

Fig. 1. The transition diagram of $M/M(b)/\infty$ queue.

Let $N(t)$ denote the number of customers in the system at time t, the stochastic process $\{N(t) \mid t \geq 0\}$ is an irreducible continuous-time Markov chain with state space $\mathbb{Z} = \{0, 1, 2, \dots\}$. Here, we can define the stationary distribution of the model p_j as,

$$p_j = \lim_{t \to \infty} \mathbf{P}(N(t) = j). \tag{1}$$

It should be noted that it is difficult to analyze the model as one variable Markov chain of $N(t)$. In the previous research of queues with batch service (e.g., [1]), the models were analyzed as a two variable Markov chain of the number of customers in the system and the number of busy servers. However, in $M/M(b)/\infty$ queue, it is not easy to adapt this technique itself because both variables have an infinite state space. Therefore, we present to analyze $M/M(b)/\infty$ queue as two variable Markov chain of the number of busy servers (i.e., the number of groups of customers) and the number of waiting customers in the system. Let $I(t)$ and $S(t)$ denote the number of busy servers and the number of waiting customers in the system at time t respectively, it is easy to find the following relational expression for $N(t)$, $I(t)$ and $S(t)$:

$$N(t) = bI(t) + S(t), \qquad (2)$$

Besides, defining as $\mathcal{S} = \{0, 1, 2, \ldots, b-1\}$, $\mathcal{L} = \mathbb{Z} \times \mathcal{S}$, it is clear that $\{(I(t), S(t)) \mid t \geq 0\}$ becomes an irreducible continuous-time Markov chain with state space $\mathcal{L}$ (see the transition diagram in Fig. 1). Note that this model becomes level-dependent quasi-birth-and-death process (QBD) [18]. However, a level-dependent QBD does not have close from solution in general.

We can also denote the stationary distribution as:

$$\pi_{i,k} = \lim_{t \to \infty} \mathbf{P}(I(t) = i, S(t) = k). \qquad (3)$$

In addition, from (2), it is obvious that the following relationship between the stationary distributions holds.

$$p_{bi+k} = \pi_{i,k}. \qquad (4)$$

The stationary distribution $\pi_{i,k}$ is given by Theorem 1.

Theorem 1. $\pi_{i,k}$ *is given as follows:*

$$\pi_{i,b-1} = \frac{1}{i!} \sum_{n=0}^{\infty} \frac{1}{b} \prod_{l=1}^{n+i} \left(\frac{l \left(\frac{\lambda}{\mu l + \lambda} \right)^b}{1 - \left(\frac{\lambda}{\mu l + \lambda} \right)^b} \right) \frac{(-1)^n}{n!}, \qquad k = b-1, \qquad (5)$$

$$\pi_{i,k} = \frac{1}{i!} \sum_{n=0}^{\infty} \frac{1}{b} \left(\frac{\lambda}{\mu(n+i)+\lambda} \right)^{k+1}$$

$$\times \left\{ \prod_{l=1}^{n+i} \left(\frac{l \left(\frac{\lambda}{\mu l + \lambda} \right)^b}{1 - \left(\frac{\lambda}{\mu l + \lambda} \right)^b} \right) + (n+i) \prod_{l=1}^{n+i-1} \left(\frac{l \left(\frac{\lambda}{\mu l + \lambda} \right)^b}{1 - \left(\frac{\lambda}{\mu l + \lambda} \right)^b} \right) \right\} \frac{(-1)^n}{n!},$$

$$0 \leq k \leq b-2, \qquad (6)$$

where $\prod_{l=1}^{0} = 1$.

Proof. The balance equations of the Markov chain $(I(t), S(t))$ are given by (7)–(9).

$$\lambda \pi_{0,0} = \mu \pi_{1,0}, \qquad i = 0, k = 0, \tag{7}$$

$$(\lambda + i\mu)\pi_{i,0} = \lambda \pi_{i-1,b-1} + (i+1)\mu \pi_{i+1,0}, \qquad i \geq 1, k = 0, \tag{8}$$

$$(\lambda + i\mu)\pi_{i,k} = \lambda \pi_{i,k-1} + (i+1)\mu \pi_{i+1,k}, \qquad 1 \leq k \leq b-1. \tag{9}$$

Here, we define the probability generating functions (PGF) $\Pi_k(z)$ $(0 \leq k \leq b-1)$ as follows:

$$\Pi_k(z) = \sum_{i=0}^{\infty} \pi_{i,k} z^i, \qquad 0 \leq k \leq b-1. \tag{10}$$

Multiplying (7)–(9) by z^i, taking the sum over $i \in \mathbb{Z}$, and rearranging the result, we obtain the following simultaneous differential equations:

$$(\mu z - \mu)\Pi_0'(z) = -\lambda \Pi_0(z) + \lambda z \Pi_{b-1}(z), \qquad k = 0, \tag{11}$$

$$(\mu z - \mu)\Pi_k'(z) = -\lambda \Pi_k(z) + \lambda \Pi_{k-1}(z), \qquad 1 \leq k \leq b-1. \tag{12}$$

However, it is not easy to obtain the solution of the simultaneous differential equations because the coefficients are not constant. Hence, we use the moment generating function method [19]. Let I and S denote the number of busy servers in the steady state, and the number of the waiting customers, respectively, we define the factorial moments as follows:

$$m_n^{(k)} = \mathbf{E}\big[I(I-1)(I-2)\ldots(I-n+1)\mathbb{1}_{\{S=k\}}\big], \qquad n \geq 1,$$

$$m_0^{(k)} = \sum_{i=0}^{\infty} \pi_{i,k},$$

where $\mathbb{1}_{\{A\}}$ is the indicator function. Moreover, we define the factorial moment generating function (FMGF) as:

$$M_k(z) = \sum_{n=0}^{\infty} m_n^{(k)} \frac{z^n}{n!} = \Pi_k(z+1). \tag{13}$$

Using FMGF, we can rearrange (11) and (12) as the following simultaneous differential equations.

$$\mu z M_0'(z) = -\lambda M_0(z) + \lambda(1+z)M_{b-1}(z), \qquad k = 0, \tag{14}$$

$$\mu z M_k'(z) = -\lambda M_k(z) + \lambda M_{k-1}(z), \qquad 1 \leq k \leq b-1. \tag{15}$$

Equating the coefficients of z^n on both sides of (14) and (15) yield

$$(\mu n + \lambda)m_n^{(0)} = \lambda m_n^{(b-1)} + \lambda n m_{n-1}^{(b-1)}, \qquad k = 0, \tag{16}$$

$$(\mu n + \lambda)m_n^{(k)} = \lambda m_n^{(k-1)}, \qquad 1 \leq k \leq b-1. \tag{17}$$

Furthermore, the following equation obviously holds by substituting 0 into (11) and (12):

$$m_0^{(k)} = \sum_{i=0}^{\infty} \pi_{i,k} = \frac{1}{b}, \qquad 0 \leq k \leq b-1. \tag{18}$$

Therefore, we obtain

$$m_n^{(b-1)} = \frac{1}{b} \prod_{l=1}^{n} \frac{l\left(\frac{\lambda}{\mu l + \lambda}\right)^b}{1 - \left(\frac{\lambda}{\mu l + \lambda}\right)^b}, \qquad k = b-1, \tag{19}$$

$$m_n^{(k)} = \left(\frac{\lambda}{\mu n + \lambda}\right)^{k+1} \left(m_n^{(b-1)} + n m_{n-1}^{(b-1)}\right)$$

$$= \frac{1}{b} \left(\frac{\lambda}{\mu n + \lambda}\right)^{k+1} \left\{ \prod_{l=1}^{n} \frac{l\left(\frac{\lambda}{\mu l + \lambda}\right)^b}{1 - \left(\frac{\lambda}{\mu l + \lambda}\right)^b} + n \prod_{l=1}^{n-1} \frac{l\left(\frac{\lambda}{\mu l + \lambda}\right)^b}{1 - \left(\frac{\lambda}{\mu l + \lambda}\right)^b} \right\},$$
$$0 \leq k \leq b-2. \tag{20}$$

Moreover, substituting $z \to z - 1$ into (13), we obtain

$$\pi_{i,k} = \sum_{n=i}^{\infty} \binom{n}{i} m_n^{(k)} \frac{(-1)^{n-i}}{n!} = \frac{1}{i!} \sum_{n=0}^{\infty} m_{n+i}^{(k)} \frac{(-1)^n}{n!}, \tag{21}$$

The stationary distribution is given by (5) and (6).

It should be noted that (5) and (6) always hold by the ratio test for FMGF as follows:

For $k = b - 1$,

$$\lim_{n \to \infty} \left| \frac{m_{n+1}^{(b-1)}/(n+1)!}{m_n^{(b-1)}/n!} \right|$$

$$= \lim_{n \to \infty} \left| \frac{1}{(n+1)} \frac{(n+1)\left(\frac{\lambda}{\mu(n+1)+\lambda}\right)^b}{1 - \left(\frac{\lambda}{\mu(n+1)+\lambda}\right)^b} \right| \tag{22}$$

$$= 0,$$

and for $k \neq b - 1$,

$$\lim_{n \to \infty} \left| \frac{m_{n+1}^{(k)}/(n+1)!}{m_n^{(k)}/n!} \right|$$

$$= \lim_{n \to \infty} \left| \frac{1}{(n+1)} \frac{\left(\dfrac{\lambda}{\mu(n+1)+\lambda} \right)^{k+1}}{\left(\dfrac{\lambda}{\mu n + \lambda} \right)^{k+1}} \right.$$

$$\left. \times \left\{ \frac{(n+1)\left(\dfrac{\lambda}{\mu(n+1)+\lambda} \right)^b}{1 - \left(\dfrac{\lambda}{\mu(n+1)+\lambda} \right)^b} + \frac{n(n+1)\left(\dfrac{\lambda}{\mu n + \lambda} \right)^b}{1 - \left(\dfrac{\lambda}{\mu n + \lambda} \right)^b} \right\} \right|$$

$$= 0.$$

(23)

These results readily guarantee the convergent radius of the FMGF is ∞.   □

Remark 1. *The stationary distribution of $M/M(1)/\infty$ queue is identical to a Poisson distribution of parameter ρ $(= \lambda/\mu)$ (i.e., the stationary distribution of $M/M/\infty$ queue).*

Proof. We obtain the following formula for $b = 1$ by considering (4) and substituting $b = 1$ into (5).

$$p_j = \pi_j$$

$$= \frac{1}{j!} \sum_{n=0}^{\infty} \prod_{l=1}^{n+j} \left(\frac{l\left(\dfrac{\lambda}{\mu l + \lambda} \right)}{1 - \left(\dfrac{\lambda}{\mu l + \lambda} \right)} \right) \frac{(-1)^n}{n!}$$

$$= \frac{\rho^j}{j!} \sum_{n=0}^{\infty} \frac{(-\rho)^n}{n!}$$

$$= \frac{\rho^j}{j!} e^{-\rho}.$$

(24)

□

3   Other Performance Measures for $M/M(b)/\infty$ Queue

Additionally, we present some performance measures in the following theorems. Let L, I, W, L_q and W_q denote the number of customers in the system, the number of busy servers in the system, the sojourn time in the system, the number of waiting customers, and the waiting time, at the steady state, respectively.

Theorem 2. *The mean number of customers in the system in steady state $\mathbf{E}[L]$ is given as follows:*

$$\mathbf{E}[L] = \frac{b-1}{2} + \frac{\lambda}{\mu}.$$

(25)

Proof. Using (4) and the factorial moments, we obtain the following formula manipulation.

$$
\begin{aligned}
\mathbf{E}[L] &= \sum_{j=0}^{\infty} p_j j \\
&= \sum_{k=0}^{b-1} \sum_{i=0}^{\infty} \pi_{i,k}(ib+k) \\
&= \sum_{k=0}^{b-1} \left(bm_1^{(k)} + \frac{k}{b} \right) \\
&= \frac{b-1}{2} + \frac{\left(\frac{\lambda}{\mu+\lambda} \right) \left\{ 1 - \left(\frac{\lambda}{\mu+\lambda} \right)^{b-1} \right\}}{1 - \left(\frac{\lambda}{\mu+\lambda} \right)} b \left(m_1^{(b-1)} + m_0^{(b-1)} \right) + bm_1^{(b-1)} \\
&= \frac{b-1}{2} + \frac{\lambda}{\mu}.
\end{aligned}
$$

$$(26)$$

□

Theorem 3. *The variance for number of busy servers in the system in steady state* $\mathbf{V}(I)$ *is given as follows:*

$$
\mathbf{V}(I) = \frac{\lambda}{b\mu} \frac{1}{1 - \left(\frac{\lambda}{\mu+\lambda} \right)^b} - \left(\frac{\lambda}{b\mu} \right)^2.
$$

$$(27)$$

Proof. The first and second order moments for number of busy servers in the system are given as follows:

$$
\mathbf{E}[I] = \frac{\lambda}{b\mu}.
$$

$$(28)$$

$$
\begin{aligned}
\mathbf{E}[I^2] &= \sum_{k=0}^{b-1} m_2^{(k)} + \sum_{k=0}^{b-1} m_1^{(k)} \\
&= \frac{\lambda}{b\mu} \frac{1}{1 - \left(\frac{\lambda}{\mu+\lambda} \right)^b}.
\end{aligned}
$$

$$(29)$$

Therefore, Theorem 3 holds.

□

Theorem 4. *The mean sojourn time (the sum of the waiting time and the service time) in steady state* $\mathbf{E}[W]$ *is given as follows:*

$$
\mathbf{E}[W] = \frac{b-1}{2\lambda} + \frac{1}{\mu}.
$$

$$(30)$$

Proof. Theorem 2 and Little's law yield Theorem 4. $\qquad \square$

Theorem 5. *The mean number of waiting customers in steady state* $\mathbf{E}[L_q]$ *is given as follows:*

$$\mathbf{E}[L_q] = \frac{b-1}{2}. \tag{31}$$

Proof. Using (4) and the factorial moments, we obtain the following formula manipulation.

$$\begin{aligned}
\mathbf{E}[L_q] &= \sum_{k=0}^{b-1}\sum_{i=0}^{\infty} \pi_{i,k}k \\
&= \sum_{k=0}^{b-1}\frac{k}{b} \\
&= \frac{b-1}{2}.
\end{aligned} \tag{32}$$

$\qquad \square$

Theorem 6. *The mean waiting time in steady state* $\mathbf{E}[W_q]$ *is given as follows:*

$$\mathbf{E}[W_q] = \frac{b-1}{2\lambda}. \tag{33}$$

Proof. Theorem 5 and Little's law yield Theorem 6. $\qquad \square$

Theorem 7. *The distribution function of the waiting time in steady state* $F_q(t)$ *is given as follows:*

$$F_q(t) = \frac{1}{b} + \frac{1}{b}\sum_{k=0}^{b-2}\left(1 - \sum_{n=0}^{b-k-2} e^{-\lambda t}\frac{(\lambda t)^n}{n!}\right). \tag{34}$$

Proof. The waiting time for a tagged customer who sees $b-1$ waiting customers upon its arrival is 0 due to the infinite number of servers. In other words, a group of customers is completed upon its arrival. On the other hand, the waiting time for a tagged customer who sees less than $b-1$ waiting customers upon its arrival follows Erlang distribution with parameter λ and phase which is derived by b minus the position of the tagged customer. Additionally, due to PASTA, (35) holds.

$$\begin{aligned}
F_q(t) &= \mathbf{P}(W_q \leq t) \\
&= \frac{1}{b}\times 1 + \sum_{k=0}^{b-2}\frac{1}{b}\left(1 - \sum_{n=0}^{b-k-2} e^{-\lambda t}\frac{(\lambda t)^n}{n!}\right).
\end{aligned} \tag{35}$$

$\qquad \square$

Theorem 8. *The distribution function of the sojourn time* $F(t)$ *is given as follows:*

For $\lambda \neq \mu$,

$$F(t) = 1 - e^{-\mu t} - \sum_{k=0}^{b-2} \frac{1}{b} \left(\sum_{n=0}^{b-k-2} \mu e^{-\lambda t} \lambda^n T_n \right), \tag{36}$$

and T_n *is explicitly given by*

$$T_n = \frac{1}{(\lambda - \mu)^{n+1}} \left\{ e^{(\lambda - \mu)t} - \sum_{k=0}^{n} \frac{\{t(\lambda - \mu)\}^k}{k!} \right\}, \qquad n = 0, 1, 2, \ldots, \tag{37}$$

and for $\lambda = \mu$,

$$F(t) = \frac{1}{b} \sum_{k=0}^{b-1} \left(1 - \sum_{n=0}^{b-k-1} e^{-\lambda t} \frac{(\lambda t)^n}{n!} \right). \tag{38}$$

Proof. The proof for $\lambda \neq \mu$ is as follows. The sojourn time for a tagged customer who sees $b-1$ waiting customers upon its arrival follows an exponential distribution of parameter μ. On the other hand, the sojourn time for a tagged customer who sees less than $b-1$ waiting customers upon its arrival is calculated by the sum of the Erlang distribution derived in Theorem 6 and an exponential distribution of parameter μ. Therefore, due to PASTA, we can obtain the following formula manipulation.

$$F(t) = \mathbf{P}(W \leqq t)$$
$$= \frac{1}{b}(1 - e^{-\mu t}) + \sum_{k=0}^{b-2} \frac{1}{b} \int_0^t \left(1 - \sum_{n=0}^{b-k-2} e^{-\lambda(t-y)} \frac{(\lambda(t-y))^n}{n!} \right) \mu e^{-\mu y} dy$$
$$= \frac{1}{b}(1 - e^{-\mu t})$$
$$+ \sum_{k=0}^{b-2} \frac{1}{b} \left\{ \int_0^t \mu e^{-\mu y} dy - \int_0^t \left(\sum_{n=0}^{b-k-2} \mu e^{-\lambda(t-y)-\mu y} \frac{(\lambda(t-y))^n}{n!} \right) dy \right\}$$
$$= 1 - e^{-\mu t} - \sum_{k=0}^{b-2} \frac{1}{b} \left(\sum_{n=0}^{b-k-2} \frac{\mu e^{-\lambda t} \lambda^n}{n!} \int_0^t e^{-(\mu-\lambda)y} (t-y)^n dy \right). \tag{39}$$

Defining $I_n = \int_0^t e^{-(\mu-\lambda)y} (t-y)^n dy$, the following holds.

$$I_n = \int_0^t e^{-(\mu-\lambda)y} (t-y)^n dy$$
$$= \int_0^t \frac{1}{-(\mu-\lambda)} (e^{-(\mu-\lambda)y})' (t-y)^n dy \tag{40}$$
$$= \frac{t^n}{\mu - \lambda} - \frac{n}{\mu - \lambda} I_{n-1}, \qquad n = 1, 2, \ldots,$$

where

$$I_0 = \frac{1}{\lambda - \mu} (e^{(\lambda-\mu)t} - 1). \tag{41}$$

Dividing (40) by $n!$, we obtain

$$\frac{I_n}{n!} = \frac{t^n}{n!(\mu - \lambda)} - \frac{1}{\mu - \lambda}\frac{I_{n-1}}{(n-1)!}, \tag{42}$$

and by defining $T_n = \dfrac{I_n}{n!}$, we gain

$$T_n = \frac{T_{n-1}}{\lambda - \mu} - \frac{t^n}{n!(\lambda - \mu)}. \tag{43}$$

From this equation and $T_0 = I_0$, we obtain (37).

The sojourn time distribution for $\lambda = \mu$ is obtained by adapting (35) as

$$F(t) = \frac{1}{b} \times (1 - e^{-\lambda t}) + \sum_{k=0}^{b-2}\frac{1}{b}\left(1 - \sum_{n=0}^{b-k-1}e^{-\lambda t}\frac{(\lambda t)^n}{n!}\right). \tag{44}$$

$\square$

4   Numerical Examples

In this section, we present some numerical examples for the stationary distribution of $M/M(b)/\infty$ queue. It should be noted that the stationary distribution (Theorem 1) contains infinite sum. That means that we have to use the approximation for the infinite sum to conduct the numerical experiments. Therefore, it is important to confirm the validity of the results compared to the simulation experiments.

In our numerical experiments, we calculate (45) and (46). Besides, we define a parameter ϵ (we set $\epsilon = 10^{-16}$ in this paper), and obtain the approximation value of the stationary distribution which satisfies (47).

$$\pi_{i,b-1}^{(s)} = \frac{1}{i!}\sum_{n=0}^{s}\frac{1}{b}\prod_{l=1}^{n+i}\left(\frac{l\left(\frac{\lambda}{\mu l + \lambda}\right)^b}{1 - \left(\frac{\lambda}{\mu l + \lambda}\right)^b}\right)\frac{(-1)^n}{n!}, \qquad k = b-1, \tag{45}$$

$$\pi_{i,k}^{(s)} = \frac{1}{i!}\sum_{n=0}^{s}\frac{1}{b}\left(\frac{\lambda}{\mu(n+i) + \lambda}\right)^{k+1}$$

$$\times \left\{\prod_{l=1}^{n+i}\left(\frac{l\left(\frac{\lambda}{\mu l + \lambda}\right)^b}{1 - \left(\frac{\lambda}{\mu l + \lambda}\right)^b}\right) + (n+i)\prod_{l=1}^{n+i-1}\left(\frac{l\left(\frac{\lambda}{\mu l + \lambda}\right)^b}{1 - \left(\frac{\lambda}{\mu l + \lambda}\right)^b}\right)\right\}\frac{(-1)^n}{n!},$$

$$0 \le k \le b-2, \tag{46}$$

$$\pi_{i,k} \approx \pi_{i,k}^{(s^*)}, \qquad s^* := \inf\{s \in \mathbb{Z} \mid |\pi_{i,k}^{(s)} - \pi_{i,k}^{(s-1)}| < \epsilon\}. \tag{47}$$

Figure 2 and Fig. 3 show the results of the calculation and the simulation for the stationary distribution of $M/M(b)/\infty$ queue. By these results, we can consider the approximation of the stationary distribution (i.e., (47)) is almost valid.

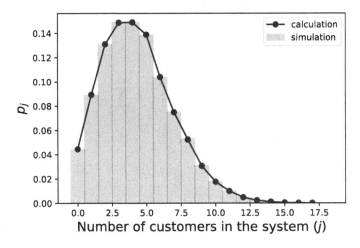

Fig. 2. The stationary distribution of $M/M(b)/\infty$ queue ($\lambda = 100, \mu = 30, b = 3$).

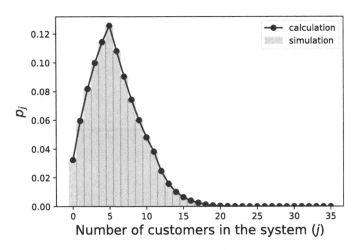

Fig. 3. The stationary distribution of $M/M(b)/\infty$ queue ($\lambda = 100, \mu = 30, b = 6$).

5   Conclusion

In this paper, we have considered $M/M(b)/\infty$ queue and have presented the analysis of the stationary distribution. Moreover, we have derived some performance measures such as the expected sojourn time. Through the numerical experiments, we have confirmed the results of the analysis generally match the results of simulation. As potential extensions, we are planning to conduct time-dependent analysis and also discuss $M/G(b)/\infty$ queue.

Acknowledgements. This work is partly supported by JSPS KAKENHI Nos. 21K11765, 18K18006. This study is supported by F-MIRAI: R&D Center for Frontiers of MIRAI in Policy and Technology, the University of Tsukuba and Toyota Motor Corporation collaborative R&D center.

References

1. Bailey, N.T.: On queueing processes with bulk service. J. Roy. Stat. Soc.: Ser. B (Methodol.) **16**(1), 80–87 (1954)
2. Neuts, M.F.: A general class of bulk queues with Poisson input. Ann. Math. Stat. **38**(3), 759–770 (1967)
3. Downton, F.: Waiting time in bulk service queues. J. Roy. Stat. Soc.: Ser. B (Methodol.) **17**(2), 256–261 (1955)
4. Holman, D.F., Chaudhry, M.L., Ghosal, A.: Some results for the general bulk service queueing system. Bull. Aust. Math. Soc. **23**(2), 161–179 (1981)
5. Arora, K.L.: Two-server bulk-service queuing process. Oper. Res. **12**(2), 286–294 (1964)
6. Ghare, P.M.: Letter to the editor - multichannel queuing system with bulk service. Oper. Res. **16**(1), 189–192 (1968)
7. Cosmetatos, G.P.: Closed-form equilibrium results for the $M/M(a,\infty)/N$ queue. Eur. J. Oper. Res. **12**(2), 203–204 (1983)
8. Sim, S.H., Templeton, J.G.C.: Computational procedures for steady-state characteristics of unscheduled multi-carrier shuttle systems. Eur. J. Oper. Res. **12**(2), 190–202 (1983)
9. Sim, S.H., Templeton, J.G.C.: Further results for the $M/M(a,\infty)/N$ batch-service system. Queueing Syst. **6**, 277–286 (1990)
10. Arumuganathan, R., Jeyakumar, S.: Analysis of a bulk queue with multiple vacations and closedown times. Int. J. Inf. Manage. Sci. **15**(1), 45–60 (2004)
11. Reddy, G.K., Nadarajan, R., Kandasamy, P.R.: Tandem queue with three multiserver units and bulk service in unit III. Microelectron. Reliab. **33**(4), 513–519 (1993)
12. Sasikala, S., Indhira, K.: Bulk service queueing models - a survey. Int. J. Pure Appl. Math. **106**(6), 43–56 (2016)
13. Ushakumari, P.V., Krishnamoorthy, A.: On a bulk arrival bulk service infinite service queue. Stoch. Anal. Appl. **16**(3), 585–595 (1998)
14. Liu, L., Kashyap, B.R.K., Templeton, J.G.C.: Queue lengths in the $GI^X/M^R/\infty$ service system. Queueing Syst. **22**(1), 129–144 (1996)
15. Li, L., Kashyap, B.R.K., Templeton, J.G.C.: The service system $M/M^R/\infty$ with impatient customers. Queueing Syst. **2**(4), 363–372 (1987)

16. Ushakumari, P.V., Krishnamoorthy, A.: The queueing system $M/M^{X(R)}/\infty$. Asia-Pac. J. Oper. Res. **15**(1), 17 (1998)
17. Nakamura, A., Phung-Duc, T., Ando, H.: Queueing analysis for a mixed model of carsharing and ridesharing. In: Gribaudo, M., Sopin, E., Kochetkova, I. (eds.) ASMTA 2019. LNCS, vol. 12023, pp. 42–56. Springer, Cham (2020). https://doi.org/10.1007/978-3-030-62885-7_4
18. Phung-Duc, T., Masuyama, H., Kasahara, S., Takahashi, Y.: A simple algorithm for the rate matrices of level-dependent QBD processes. In: Proceedings of The 5th International Conference on Queueing Theory and Network Applications, pp. 46–52 (2010)
19. Kapodistria, S., Phung-Duc, T., Resing, J.: Linear birth/immigration-death process with binomial catastrophes. Probab. Eng. Inf. Sci. **30**(1), 79–111 (2016)

Performance Evaluation of Stochastic Bipartite Matching Models

Céline Comte[1]([✉]) and Jan-Pieter Dorsman[2]

[1] Eindhoven University of Technology, Eindhoven, The Netherlands
c.m.comte@tue.nl
[2] University of Amsterdam, Amsterdam, The Netherlands
j.l.dorsman@uva.nl

Abstract. We consider a stochastic bipartite matching model consisting of multi-class customers and multi-class servers. Compatibility constraints between the customer and server classes are described by a bipartite graph. Each time slot, exactly one customer and one server arrive. The incoming customer (resp. server) is matched with the earliest arrived server (resp. customer) with a class that is compatible with its own class, if there is any, in which case the matched customer-server couple immediately leaves the system; otherwise, the incoming customer (resp. server) waits in the system until it is matched. Contrary to classical queueing models, both customers and servers may have to wait, so that their roles are interchangeable. While (the process underlying) this model was already known to have a product-form stationary distribution, this paper derives a new compact and manageable expression for the normalization constant of this distribution, as well as for the waiting probability and mean waiting time of customers and servers. We also provide a numerical example and make some important observations.

Keywords: Bipartite matching models · Order-independent queues · Performance analysis · Product-form stationary distribution

1 Introduction

Stochastic matching models typically consist of items of multiple classes that arrive at random instants to be matched with items of other classes. In the same spirit as classical (static) matching models, stochastic models encode compatibility constraints between items using a graph on the classes. This allows for the modeling of many matching applications that are stochastic in nature, such as organ transplants where not every patient is compatible with every donor organ.

In the literature on stochastic matching, a rough distinction is made between *bipartite* and *non-bipartite* models. In a bipartite matching model, the graph that describes compatibility relations between item classes is bipartite. In this way, item classes can be divided into two groups called *customers* and *servers*, so that customers (resp. servers) cannot be matched with one another. This is the

© Springer Nature Switzerland AG 2021
P. Ballarini et al. (Eds.): EPEW 2021/ASMTA 2021, LNCS 13104, pp. 425–440, 2021.
https://doi.org/10.1007/978-3-030-91825-5_26

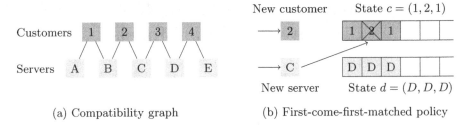

(a) Compatibility graph   (b) First-come-first-matched policy

Fig. 1. A stochastic bipartite matching model with a set $\mathcal{I} = \{1, 2, 3, 4\}$ of customer classes and a set $\mathcal{K} = \{A, B, C, D, E\}$ of server classes.

variant that we consider in this paper. It is discrete-time in nature and assumes that, every time unit, exactly one customer and one server arrive. The classes of incoming customers and servers are drawn independently from each other, and they are also independent and identically distributed across time units. Following [7,8], we adopt the common first-come-first-matched policy, whereby an arriving customer (resp. server) is matched with the earliest arriving compatible server (resp. customer). A toy example is shown in Fig. 1. This model is equivalent to the *first-come-first-served infinite bipartite matching model*, studied in [1,2,8], which can be used to describe the evolution of waiting lists in public-housing programs and adoption agencies for instance [8].

In contrast, in a stochastic *non*-bipartite matching model, item classes cannot be divided into two groups because the compatibility graph is non-bipartite. Another notable difference is that only one item arrives at each time slot, the classes of successive items being drawn independently from the same distribution. While [15] derived stability conditions for this non-bipartite model, [16] showed that the stationary distribution of the process underlying this model has a product form. The recent work [10] built on the latter result to derive closed-form expressions for several performance metrics, by also exploiting a connection with order-independent (loss) queues [5,14]. The present work seeks to provide a similar analysis for the above-mentioned bipartite model.

More specifically, we derive closed-form expressions for several performance metrics in the stochastic bipartite model studied in [1]. While earlier studies [3,8] were skeptical about the tractability of the stationary distribution corresponding to this variant, [1] showed that this stationary distribution in fact possesses the product-form property, thus paving the way for an analysis similar to that of [10]. That is, we use techniques from order-independent (loss) queues [5,14] and other product-form models with compatibility constraints (cf. [12] for a recent overview) to analyze the stochastic bipartite matching model.

The rest of this paper is structured as follows. In Sect. 2, we introduce the model and cast it in terms of a framework commonly used for analysis of product-form models. We then provide a performance evaluation of this model. More specifically, we first derive an alternative closed-form expression for the normalization constant of the stationary distribution. While the computational complexity of this expression is prohibitive for instances with many classes (as was the case for the expression derived in [1]), it draws the relation with product-form queues and paves the way for heavy-traffic analysis. Furthermore, it allows us to directly derive recursive expressions for several other performance metrics, such as the probability that incoming customers or servers have to wait and the mean number of customers and servers that are waiting. To the best of the authors' knowledge, this paper is the first to provide expressions for these performance metrics. This analysis is presented in Sect. 3. Finally, Sect. 4 numerically studies a model instance and makes some important observations.

2   Model and Preliminary Results

In Sect. 2.1, we describe a stochastic bipartite matching model in which items of two groups, called customers and servers, arrive randomly and are matched with one another. As mentioned earlier, this model is analogous to that introduced in [8] and further studied in [1–3,7]. Section 2.2 focuses on a discrete-time Markov chain that describes the evolution of this model. Finally, Sect. 2.3 recalls several results that are useful for the analysis of Sect. 3.

2.1   Model and Notation

Bipartite Compatibility Graph. Consider a finite set $\mathcal{I}$ of I customer classes and a finite set $\mathcal{K}$ of K server classes. Also consider a connected bipartite graph on the sets $\mathcal{I}$ and $\mathcal{K}$. For each $i \in \mathcal{I}$ and $k \in \mathcal{K}$, we write $i \sim k$ if there is an edge between nodes i and k in this graph, and $i \nsim k$ otherwise. This bipartite graph is called the *compatibility* graph of the model. It describes the compatibility relations between customers and servers in the sense that, for each $i \in \mathcal{I}$ and $k \in \mathcal{K}$, a class-i customer and a class-k server can be matched with one another if and only if there is an edge between the corresponding nodes in the compatibility graph. An example is shown in Fig. 1a. To simplify reading, we consistently use the letters $i, j \in \mathcal{I}$ for customer classes and $k, \ell \in \mathcal{K}$ for server classes.

Discrete-Time Stochastic Matching. Unmatched customers and servers are stored in two separate queues in their arrival order. In Fig. 1b, items are ordered from the oldest on the left to the newest on the right, and each item is labeled by its class. The two queues are initially empty. Time is slotted and, during each time slot, exactly one customer *and* exactly one server arrive. The incoming customer belongs to class i with probability $\lambda_i > 0$, for each $i \in \mathcal{I}$, and the incoming server belongs to class k with probability $\mu_k > 0$ for each $k \in \mathcal{K}$, with $\sum_{i \in \mathcal{I}} \lambda_i = \sum_{k \in \mathcal{K}} \mu_k = 1$. The classes of incoming customers and servers are independent within and across time slots. The matching policy, called *first-come-first-matched*, consists of applying the following four steps upon each arrival:

1. Match the incoming customer with the compatible unmatched server that has been in the queue the longest, if any.

2. Match the incoming server with the compatible unmatched customer that has been in the queue the longest, if any.
3. If neither the incoming customer nor the incoming server can be matched with unmatched items, match them together if they are compatible.
4. If an incoming customer and/or incoming server remains unmatched after the previous steps, it is appended to the back of its respective queue.

When two items are matched with one another, they immediately disappear. In the example of Fig. 1b, the couple $(2, C)$ arrives while the sequence of unmatched customer and server classes are $(1, 2, 1)$ and (D, D, D). According to the compatibility graph of Fig. 1a, class C is compatible with class 2 but not with class 1. Therefore, the incoming class-C server is matched with the second oldest unmatched customer, of class 2. The incoming class-2 customer is not matched with any present item (even if it is compatible with the incoming class-C server), therefore it is appended to the queue of unmatched customers. After this transition, the sequence of unmatched customer classes becomes $(1, 1, 2)$, while the sequence of unmatched server classes is unchanged.

Remark 1. If we would consider the random sequences of classes of incoming customers and servers, we would retrieve the state descriptor of the infinite bipartite matching model introduced in [8] and studied in [1–3,7]. For analysis purposes, we however adopted the above-introduced state descriptor consisting of the sequences of (waiting) unmatched customers and servers, corresponding to the *natural pair-by-pair FCFS Markov chain* introduced in [1, Sect. 2].

Set Notation. The following notation will be useful. Given two sets $\mathcal{A}$ and $\mathcal{B}$, we write $\mathcal{A} \subseteq \mathcal{B}$ if $\mathcal{A}$ is a subset of $\mathcal{B}$ and $\mathcal{A} \subsetneq \mathcal{B}$ if $\mathcal{A}$ is a proper subset of $\mathcal{B}$. For each $i \in \mathcal{I}$, we let $\mathcal{K}_i \subseteq \mathcal{K}$ denote the set of server classes that can be matched with class-i customers. Similarly, for each $k \in \mathcal{K}$, we let $\mathcal{I}_k \subseteq \mathcal{I}$ denote the set of customer classes that can be matched with class-k servers. For each $i \in \mathcal{I}$ and $k \in \mathcal{K}$, the statements $i \sim k$, $i \in \mathcal{I}_k$, and $k \in \mathcal{K}_i$ are equivalent. In Fig. 1a for instance, we have $\mathcal{K}_1 = \{A, B\}$, $\mathcal{K}_2 = \{B, C\}$, $\mathcal{K}_3 = \{C, D\}$, $\mathcal{K}_4 = \{D, E\}$ $\mathcal{I}_A = \{1\}$, $\mathcal{I}_B = \{1, 2\}$, $\mathcal{I}_C = \{2, 3\}$, $\mathcal{I}_D = \{3, 4\}$, and $\mathcal{I}_E = \{4\}$. With a slight abuse of notation, for each $\mathcal{A} \subseteq \mathcal{I}$, we let $\lambda(\mathcal{A}) = \sum_{i \in \mathcal{A}} \lambda_i$ denote the probability that the class of an incoming customer belongs to $\mathcal{A}$ and $\mathcal{K}(\mathcal{A}) = \bigcup_{i \in \mathcal{A}} \mathcal{K}_i$ the set of server classes that are compatible with customer classes in $\mathcal{A}$. Similarly, for each $\mathcal{A} \subseteq \mathcal{K}$, we write $\mu(\mathcal{A}) = \sum_{k \in \mathcal{A}} \mu_k$ and $\mathcal{I}(\mathcal{A}) = \bigcup_{k \in \mathcal{A}} \mathcal{I}_k$. In particular, we have $\lambda(\mathcal{I}) = \mu(\mathcal{K}) = 1$, $\mathcal{K}(\mathcal{I}) = \mathcal{K}$, and $\mathcal{I}(\mathcal{K}) = \mathcal{I}$.

2.2   Discrete-Time Markov Chain

We now consider a Markov chain that describes the evolution of the system.

System State. We consider the couple (c, d), where $c = (c_1, \dots, c_n) \in \mathcal{I}^*$ is the sequence of unmatched customer classes, ordered by arrival, and $d = (d_1, \dots, d_n) \in \mathcal{K}^*$ is the sequence of unmatched server classes, ordered by arrival.

In particular, c_1 is the class of the oldest unmatched customer, if any, and d_1 is the class of the oldest unmatched server, if any. The notation $\mathcal{I}^*$ (resp. $\mathcal{K}^*$) refers to the Kleene star on $\mathcal{I}$ (resp. $\mathcal{K}$), that is, the set of sequences of elements in $\mathcal{I}$ (resp. $\mathcal{K}$) with a length that is finite but arbitrarily large [11, Chapter 1, Sect. 2]. As we will see later, the matching policy guarantees that the numbers of unmatched customers and servers are always equal to each other, and consequently the integer n will be called the *length* of the state. The empty state, with $n = 0$, is denoted by $\varnothing$.

The evolution of this state over time defines a (discrete-time) Markov chain that is further detailed below. For each sequence $c = (c_1, \ldots, c_n) \in \mathcal{I}^*$, we let $|c| = n$ denote the length of sequence c, $|c|_i$ the number of occurrences of class i in sequence c, for each $i \in \mathcal{I}$, and, with a slight abuse of notation, $\{c_1, \ldots, c_n\}$ the set of classes that appear in sequence c (irrespective of their multiplicity). Analogous notation is introduced for each sequence $d = (d_1, \ldots, d_n) \in \mathcal{K}^*$.

Transitions. Each transition of the Markov chain is triggered by the arrival of a customer-server couple. We distinguish five types of transitions depending on their impact on the queues of unmatched customers and servers:

$-/-$ The incoming customer is matched with an unmatched server and the incoming server is matched with an unmatched customer.

$\pm/=$ The incoming customer cannot be matched with any present server but the incoming server is matched with an unmatched customer.

$=/\pm$ The incoming customer is matched with a present server but the incoming server cannot be matched with any present customer.

$=/=$ Neither the incoming customer nor the incoming server can be matched with an unmatched item, but they are matched with one another.

$+/+$ Neither the incoming customer nor the incoming server can be matched with an unmatched item, and they cannot be matched with one another.

Labels indicate the impact of the corresponding transition. For instance, a transition $-/-$ leads to a deletion ($-$) in the customer queue and a deletion ($-$) in the server queue, while a transition $\pm/=$ leads to a replacement ($\pm$) in the customer queue and no modification in the server queue ($=$). Transitions $-/-$ reduce the lengths of both queues by one, transitions $\pm/=$, $=/\pm$, and $=/=$ leave the queue lengths unchanged, and transitions $+/+$ increase the lengths of both queues by one. Note that the numbers of unmatched customers and servers are always equal to each other. We omit the transition probabilities, as we will rely on an existing result giving the stationary distribution of the Markov chain.

State Space. The greediness of the matching policy prevents the queues from containing an unmatched customer and an unmatched server that are compatible. Therefore, the state space of the Markov chain is the subset of $\mathcal{I}^* \times \mathcal{K}^*$ given by

$$\Pi = \bigcup_{n=0}^{\infty} \{(c, d) \in \mathcal{I}^n \times \mathcal{K}^n : c_p \nsim d_q \text{ for each } p, q \in \{1, \ldots, n\}\}.$$

The Markov chain is irreducible. Indeed, using the facts that the compatibility graph is connected, that $\lambda_i > 0$ for each $i \in \mathcal{I}$, and that $\mu_k > 0$ for each $k \in \mathcal{K}$, we can show that the Markov chain can go from any state $(c, d) \in \Pi$ to any state $(c', d') \in \Pi$ via state $\varnothing$ in $|c| + |c'| = |d| + |d'|$ jumps.

Remark 2. We can also consider the following continuous-time variant of the model introduced in Sect. 2.1. Instead of assuming that time is slotted, we can assume that customer-server couples arrive according to a Poisson process with unit rate. If the class of the incoming customers and servers are drawn independently at random, according to the probabilities λ_i for $i \in \mathcal{I}$ and μ_k for $k \in \mathcal{K}$, then the rate diagram of the continuous-time Markov chain describing the evolution of the sequences of unmatched items is identical to the transition diagram of the Markov chain introduced above. Consequently, the results recalled in Sect. 2.3 and those derived in Sect. 3 can be applied without any modification to this continuous-time Markov chain.

2.3 Stability Conditions and Stationary Distribution

For purposes of later analysis, we now state the following theorem, which was proved in [3, Theorem 3] and [1, Lemma 2 and Theorems 2 and 8].

Theorem 1. *The stationary measures of the Markov chain associated with the system state are of the form*

$$\pi(c, d) = \pi(\varnothing) \prod_{p=1}^{n} \frac{\lambda_{c_p}}{\mu(\mathcal{K}(\{c_1, \ldots, c_p\}))} \frac{\mu_{d_p}}{\lambda(\mathcal{I}(\{d_1, \ldots, d_p\}))}, \quad (c, d) \in \Pi. \quad (1)$$

The system is stable, in the sense that this Markov chain is ergodic, if and only if one of the following two equivalent conditions is satisfied:

$$\lambda(\mathcal{A}) < \mu(\mathcal{K}(\mathcal{A})) \text{ for each non-empty set } \mathcal{A} \subsetneq \mathcal{I}, \quad (2)$$

$$\mu(\mathcal{A}) < \lambda(\mathcal{I}(\mathcal{A})) \text{ for each non-empty set } \mathcal{A} \subsetneq \mathcal{K}. \quad (3)$$

In this case, the stationary distribution of the Markov chain associated with the system state is given by (1), with the normalization constant

$$\pi(\varnothing) = \left(\sum_{(c,d) \in \Pi} \prod_{p=1}^{n} \frac{\lambda_{c_p}}{\mu(\mathcal{K}(\{c_1, \ldots, c_p\}))} \frac{\mu_{d_p}}{\lambda(\mathcal{I}(\{d_1, \ldots, d_p\}))} \right)^{-1}. \quad (4)$$

The states of the two queues are not independent in general because their lengths are equal. However, (1) shows that these two queue states are *conditionally* independent *given* the number n of unmatched items. This property will contribute to simplify the analysis in Sect. 3.

Remark 3. The stationary measures (1) seem identical to the stationary measures associated with another queueing model, called an FCFS-ALIS parallel queueing model [2,4]. The only (crucial) difference lies in the definition of the state space of the corresponding Markov chain. In particular, our model imposes that the lengths of the two queues are equal to each other. In contrast, in the FCFS-ALIS parallel queueing model, there is an upper bound on the number of unmatched servers, while the number of customers can be arbitrarily large. This difference significantly changes the analysis. The analysis that we propose in Sect. 3 is based on the resemblance with another queueing model, called a multi-server queue for simplicity, that was introduced in [9,13].

3   Performance Evaluation by State Aggregation

We now assume that the stability conditions (2)–(3) are satisfied, and we let π denote the stationary distribution, recalled in Theorem 1, of the Markov chain of Sect. 2.2. Sections 3.2, 3.3 and 3.4 provide closed-form expressions for several performance metrics, based on a method explained in Sect. 3.1. The time complexity to implement these formulas and the relation with related works [1,3] are discussed in Sect. 3.5. The reader who is not interested in understanding the proofs can move directly to Sect. 3.2.

3.1   Partition of the State Space

A naive application of (4) does not allow calculation of the normalization constant, nor any other long-run performance metric as a result, because the state-space Π is infinite. To circumvent this, we define a partition of the state space.

Partition of the State Space Π. Let $\mathbb{I}$ denote the family of sets $\mathcal{A} \subseteq \mathcal{I} \cup \mathcal{K}$ such that $\mathcal{A}$ is an independent set of the compatibility graph and the sets $\mathcal{A} \cap \mathcal{I}$ and $\mathcal{A} \cap \mathcal{K}$ are non-empty. Also let $\mathbb{I}_0 = \mathbb{I} \cup \{\emptyset\}$. For each $\mathcal{A} \in \mathbb{I}_0$, we let $\Pi_{\mathcal{A}}$ denote the set of couples $(c, d) \in \Pi$ such that $\{c_1, \dots, c_n\} = \mathcal{A} \cap \mathcal{I}$ and $\{d_1, \dots, d_n\} = \mathcal{A} \cap \mathcal{K}$; in other words, $\Pi_{\mathcal{A}}$ is the set of states such that the set of unmatched classes is $\mathcal{A}$. We can show that $\{\Pi_{\mathcal{A}}, \mathcal{A} \in \mathbb{I}_0\}$ forms a partition of Π, and in particular

$$\Pi = \bigcup_{\mathcal{A} \in \mathbb{I}_0} \Pi_{\mathcal{A}}.$$

The first cornerstone of our analysis is the observation that, for each $(c, d) \in \Pi_{\mathcal{A}}$, we have $\mu(\mathcal{K}(\{c_1, \dots, c_n\})) = \mu(\mathcal{K}(\mathcal{A} \cap \mathcal{I}))$ and $\lambda(\mathcal{I}(\{d_1, \dots, d_n\})) = \lambda(\mathcal{I}(\mathcal{A} \cap \mathcal{K}))$. In anticipation of Sect. 3.2, for each $\mathcal{A} \in \mathbb{I}$, we let

$$\Delta(\mathcal{A}) = \mu(\mathcal{K}(\mathcal{A} \cap \mathcal{I}))\lambda(\mathcal{I}(\mathcal{A} \cap \mathcal{K})) - \lambda(\mathcal{A} \cap \mathcal{I})\mu(\mathcal{A} \cap \mathcal{K}). \tag{5}$$

One can verify that $\Delta(\mathcal{A}) > 0$ for each $\mathcal{A} \in \mathbb{I}$ if and only if the stability conditions (2)–(3) are satisfied. The product $\lambda(\mathcal{A} \cap \mathcal{I})\mu(\mathcal{A} \cap \mathcal{K})$ is the probability that an incoming client-server couple has its classes in $\mathcal{A}$, while the product $\mu(\mathcal{K}(\mathcal{A} \cap \mathcal{I}))\lambda(\mathcal{I}(\mathcal{A} \cap \mathcal{K}))$ is the probability that an incoming client-server couple can be matched with clients and servers whose classes belong to $\mathcal{A}$. By analogy with the queueing models in [10,13], the former product can be seen as the "arrival rate" of the classes in $\mathcal{A}$, while the latter product can be seen as the maximal "departure rate" of these classes.

Partition of the Subsets $\Pi_{\mathcal{A}}$. The second cornerstone of the analysis is a partition of the set $\Pi_{\mathcal{A}}$ for each $\mathcal{A} \in \mathbb{I}$. More specifically, for each $\mathcal{A} \in \mathbb{I}$, we have

$$\Pi_{\mathcal{A}} = \bigcup_{i \in \mathcal{A} \cap \mathcal{I}} \bigcup_{k \in \mathcal{A} \cap \mathcal{K}} \left(\Pi_{\mathcal{A}} \cup \Pi_{\mathcal{A} \setminus \{i\}} \cup \Pi_{\mathcal{A} \setminus \{k\}} \cup \Pi_{\mathcal{A} \setminus \{i,k\}} \right) \cdot (i,k), \qquad (6)$$

where $\mathcal{S} \cdot (i,k) = \{((c_1, \ldots, c_n, i), (d_1, \ldots, d_n, k)) : ((c_1, \ldots, c_n), (d_1, \ldots, d_n)) \in \mathcal{S}\}$ for each $\mathcal{S} \subseteq \Pi$, $i \in \mathcal{I}$, and $k \in \mathcal{K}$, and the unions are disjoint. Indeed, for each $(c,d) \in \Pi_{\mathcal{A}}$, the sequence $c = (c_1, \ldots, c_n)$ can be divided into a prefix $(c_1, \ldots, c_{n-1})$ and a suffix $i = c_n$; the suffix can take any value in $\mathcal{A} \cap \mathcal{I}$, while the prefix satisfies $\{c_1, \ldots, c_{n-1}\} = \mathcal{A} \cap \mathcal{I}$ or $\{c_1, \ldots, c_{n-1}\} = (\mathcal{A} \setminus \{i\}) \cap \mathcal{I}$. Similarly, for each $(c,d) \in \Pi_{\mathcal{A}}$, the sequence $d = (d_1, \ldots, d_n)$ can be divided into a prefix $(d_1, \ldots, d_{n-1})$ and a prefix $k = d_n$; the prefix can take any value in $\mathcal{A} \cap \mathcal{K}$, while the prefix satisfies $\{d_1, \ldots, d_{n-1}\} = \mathcal{A} \cap \mathcal{K}$ or $\{d_1, \ldots, d_{n-1}\} = (\mathcal{A} \setminus \{k\}) \cap \mathcal{K}$.

3.2   Normalization Constant

The first performance metric that we consider is the probability that the system is empty. According to (4), this is also the normalization constant. With a slight abuse of notation, we first let

$$\pi(\mathcal{A}) = \sum_{(c,d) \in \Pi_{\mathcal{A}}} \pi(c,d), \quad \mathcal{A} \in \mathbb{I}_0.$$

To simplify notation, we adopt the convention that $\pi(\mathcal{A}) = 0$ if $\mathcal{A} \notin \mathbb{I}_0$. The following proposition, combined with the normalization equation $\sum_{\mathcal{A} \in \mathbb{I}_0} \pi(\mathcal{A}) = 1$, allows us to calculate the probability $\pi(\emptyset) = \pi(\varnothing)$ that the system is empty.

Proposition 1. *The stationary distribution of the set of unmatched item classes satisfies the recursion*

$$\Delta(\mathcal{A})\pi(\mathcal{A}) = \mu(\mathcal{A} \cap \mathcal{K}) \sum_{i \in \mathcal{A} \cap \mathcal{I}} \lambda_i \pi(\mathcal{A} \setminus \{i\}) + \lambda(\mathcal{A} \cap \mathcal{I}) \sum_{k \in \mathcal{A} \cap \mathcal{K}} \mu_k \pi(\mathcal{A} \setminus \{k\})$$

$$+ \sum_{i \in \mathcal{A} \cap \mathcal{I}} \sum_{k \in \mathcal{A} \cap \mathcal{K}} \lambda_i \mu_k \pi(\mathcal{A} \setminus \{i,k\}), \quad \mathcal{A} \in \mathbb{I}. \qquad (7)$$

Proof. Let $\mathcal{A} \in \mathbb{I}$. Substituting (1) into the definition of $\pi(\mathcal{A})$ yields

$$\pi(\mathcal{A}) = \sum_{(c,d) \in \Pi_{\mathcal{A}}} \prod_{p=1}^{n} \frac{\lambda_{c_p}}{\mu(\mathcal{K}(\{c_1, \ldots, c_p\}))} \frac{\mu_{d_p}}{\lambda(\mathcal{I}(\{d_1, \ldots, d_p\}))},$$

$$= \sum_{(c,d) \in \Pi_{\mathcal{A}}} \frac{\lambda_{c_n}}{\mu(\mathcal{K}(\mathcal{A} \cap \mathcal{I}))} \frac{\mu_{d_n}}{\lambda(\mathcal{I}(\mathcal{A} \cap \mathcal{K}))} \pi((c_1, \ldots, c_{n-1}), (d_1, \ldots, d_{n-1})).$$

Then, by applying (6) and making a change of variable, we obtain

$$\mu(\mathcal{K}(\mathcal{A} \cap \mathcal{I}))\lambda(\mathcal{I}(\mathcal{A} \cap \mathcal{K}))\pi(\mathcal{A})$$

$$= \sum_{i \in \mathcal{A} \cap \mathcal{I}} \sum_{k \in \mathcal{A} \cap \mathcal{K}} \lambda_i \mu_k \left(\pi(\mathcal{A}) + \pi(\mathcal{A} \setminus \{i\}) + \pi(\mathcal{A} \setminus \{k\}) + \pi(\mathcal{A} \setminus \{i,k\}) \right). \qquad (8)$$

The result follows by rearranging the terms.

3.3   Waiting Probability

The second performance metric that we consider is the waiting probability, that is, the probability that an item cannot be matched with another item upon arrival. The waiting probabilities of the customers and servers of each class can again be calculated using Proposition 1, as they are given by

$$
\omega_i = \sum_{\mathcal{A} \in \mathbb{I}_0 : \mathcal{A} \cap \mathcal{K}_i = \emptyset} \left(1 - \sum_{k \in \mathcal{K}_i \setminus \mathcal{K}(\mathcal{A} \cap \mathcal{I})} \mu_k \right) \pi(\mathcal{A}), \quad i \in \mathcal{I},
$$

$$
\omega_k = \sum_{\mathcal{A} \in \mathbb{I}_0 : \mathcal{A} \cap \mathcal{I}_k = \emptyset} \left(1 - \sum_{i \in \mathcal{I}_k \setminus \mathcal{I}(\mathcal{A} \cap \mathcal{K})} \lambda_i \right) \pi(\mathcal{A}), \quad k \in \mathcal{K}.
$$

If we consider the continuous-time variant described in Remark 2, these equations follow directly from the PASTA property. That this result also holds for the discrete-time variant of the model follows from the fact that the transition diagrams and stationary distributions of both models are identical.

Corollary 1 below follows from Proposition 1. It shows that the probability that both the incoming customer and the incoming server can be matched with present items (corresponding to transitions $-/-$) is equal to the probability that both the incoming customer and the incoming server have to wait (corresponding to transitions $+/+$). The proof is given in the appendix.

Corollary 1. *The following equality is satisfied:*

$$
\sum_{(i,k) \in \mathcal{I} \times \mathcal{K}} \lambda_i \mu_k \sum_{\substack{\mathcal{A} \in \mathbb{I} : i \in \mathcal{I}(\mathcal{A} \cap \mathcal{K}), \\ k \in \mathcal{K}(\mathcal{A} \cap \mathcal{I})}} \pi(\mathcal{A}) = \sum_{\substack{(i,k) \in \mathcal{I} \times \mathcal{K} : \\ i \nsim k}} \lambda_i \mu_k \sum_{\substack{\mathcal{A} \in \mathbb{I}_0 : i \notin \mathcal{I}(\mathcal{A} \cap \mathcal{K}), \\ k \notin \mathcal{K}(\mathcal{A} \cap \mathcal{I})}} \pi(\mathcal{A}). \quad (9)
$$

This corollary means that, in the long run, the rate at which the queue lengths increase is equal to the rate at which the queue lengths decrease. Equation (9) is therefore satisfied by every matching policy that makes the system stable. This equation also has the following graphical interpretation. Consider a *busy sequence* of the system, consisting of a sequence of customer classes and a sequence of server classes that arrive between two consecutive instants when both queues are empty. We construct a bipartite graph, whose nodes are the elements of these two sequences, by adding an edge between customers and servers that arrive at the same time or are matched with one another. An example is shown in Fig. 2 for the compatibility graph of Fig. 1a. If we ignore the customer-server couples that arrive at the same time and are also matched with one another, we obtain a 2-regular graph, that is, a graph where all nodes have degree two. Such a graph consists of one or more disconnected cycles. We define a left (resp. right) extremity as a vertical edge adjacent only to edges moving to the right (resp. left); such an edge represents a $+/+$ (resp. $-/-$) transition. One can verify that each cycle contains as many left extremities as right extremities. In the example of Fig. 2, after eliminating the couple 1–A, we obtain two disconnected cycles.

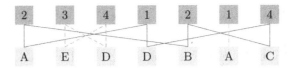

Fig. 2. A busy sequence associated with the compatibility graph of Fig. 1a. The arrival sequences are 2, 3, 4, 1, 2, 1, 4 and A, E, D, D, B, A, C. Each component of the corresponding bipartite graph is depicted with a different line style.

The cycle depicted with a solid line has one left extremity (2–A) and one right extremity (4–C). The cycle depicted with a dashed line also has one left extremity (3–E) and one right extremity (4–C). Since stability means that the mean length of a busy sequence is finite, combining this observation with classical results from renewal theory gives an alternative proof that (9) is satisfied by every matching policy that makes the system stable.

3.4 Mean Number of Unmatched Items and Mean Waiting Time

We now turn to the mean number of unmatched items. Proposition 2 gives a closed-form expression for the mean number of unmatched customers of each class. Proposition 3 gives a simpler expression for the mean number of unmatched customers (all classes included). The proofs are similar to that of Proposition 1, with a few technical complications, and are deferred to the appendix. Analogous results can be obtained for the servers by using the model symmetry.

Proposition 2. *For each* $i \in \mathcal{I}$, *the mean number of unmatched class-i customers is* $L_i = \sum_{\mathcal{A} \in \mathbb{I}_0} \ell_i(\mathcal{A})$, *where* $\ell_i(\mathcal{A})/\pi(\mathcal{A})$ *is the mean number of unmatched class-i customers given that the set of unmatched classes is $\mathcal{A}$, and satisfies the recursion*

$$\Delta(\mathcal{A})\ell_i(\mathcal{A}) = \lambda_i \mu(\mathcal{A} \cap \mathcal{K})(\pi(\mathcal{A}) + \pi(\mathcal{A}\setminus\{i\}))$$

$$+ \lambda_i \sum_{k \in \mathcal{A} \cap \mathcal{K}} \mu_k (\pi(\mathcal{A}\setminus\{k\}) + \pi(\mathcal{A}\setminus\{i,k\}))$$

$$+ \mu(\mathcal{A} \cap \mathcal{K}) \sum_{j \in \mathcal{A}} \lambda_j \ell_i(\mathcal{A}\setminus\{j\}) + \lambda(\mathcal{A} \cap \mathcal{I}) \sum_{k \in \mathcal{A} \cap \mathcal{K}} \mu_k \ell_i(\mathcal{A}\setminus\{k\})$$

$$+ \sum_{j \in \mathcal{A}} \sum_{k \in \mathcal{A} \cap \mathcal{K}} \lambda_j \mu_k \ell_i(\mathcal{A}\setminus\{j,k\}), \tag{10}$$

for each $\mathcal{A} \in \mathbb{I}$ *such that* $i \in \mathcal{A}$, *with the base case* $\ell_i(\mathcal{A}) = 0$ *if* $i \notin \mathcal{A}$ *and the convention that* $\ell_i(\mathcal{A}) = 0$ *if* $\mathcal{A} \notin \mathbb{I}_0$.

Proposition 3. *The mean number of unmatched customers is* $L_{\mathcal{I}} = \sum_{\mathcal{A} \in \mathbb{I}_0} \ell_{\mathcal{I}}(\mathcal{A})$, *where* $\ell_{\mathcal{I}}(\mathcal{A})/\pi(\mathcal{A})$ *is the mean number of unmatched customers given that the set of unmatched classes is $\mathcal{A}$, and satisfies the recursion*

$$\Delta(\mathcal{A})\ell_{\mathcal{I}}(\mathcal{A}) = \mu(\mathcal{K}(\mathcal{A} \cap \mathcal{I}))\lambda(\mathcal{I}(\mathcal{A} \cap \mathcal{K}))\pi(\mathcal{A})$$

$$+ \mu(\mathcal{A} \cap \mathcal{K}) \sum_{i \in \mathcal{A} \cap \mathcal{I}} \lambda_i \ell_{\mathcal{I}}(\mathcal{A} \backslash \{i\}) + \lambda(\mathcal{A} \cap \mathcal{I}) \sum_{k \in \mathcal{A} \cap \mathcal{K}} \mu_k \ell_{\mathcal{I}}(\mathcal{A} \backslash \{k\})$$

$$+ \sum_{i \in \mathcal{A} \cap \mathcal{I}} \sum_{k \in \mathcal{A} \cap \mathcal{K}} \lambda_i \mu_k \ell_{\mathcal{I}}(\mathcal{A} \backslash \{i, k\}). \tag{11}$$

for each $\mathcal{A} \in \mathbb{I}$, with the base case $\ell_{\mathcal{I}}(\emptyset) = 0$ and the convention that $\ell_{\mathcal{I}}(\mathcal{A}) = 0$ for each $\mathcal{A} \notin \mathbb{I}_0$.

By Little's law, the mean waiting time of class-i customers is L_i/λ_i, for each $i \in \mathcal{I}$, and the mean waiting time of customers (all classes included) is L. By following the same approach as [10, Propositions 9 and 10], we can derive, for each class, closed-form expressions for the distribution transforms of the number of unmatched items and waiting time. In the interest of space, and to avoid complicated notation, these results are omitted.

3.5   Time Complexity and Related Work

To conclude Sect. 3, we briefly discuss the merit of our approach compared to the expression derived in [3, Theorem 3] and rederived in [1, Theorem 7] for the normalization constant (equal to the inverse of the probability that the system is empty). This approach relies on a Markov chain called the *server-by-server FCFS augmented matching process* in [1, Sect. 5.4].

Flexibility. The first merit of our approach is that it can be almost straightforwardly applied to derive other relevant performance metrics. Sections 3.3 and 3.4 provide two examples: the expression of the waiting probability is a side-result of Proposition 1, while the mean waiting time follows by a derivation along the same lines. Performance metrics that can be calculated in a similar fashion include the variance of the stationary number of unmatched items of each class, the mean length of a busy sequence, and the fractions of transitions of types $-/-$, $\pm/=$, $=/\pm$, $=/=$, and $+/+$. Our approach may also be adapted to derive an alternative expression for the matching rates calculated in [3, Sect. 3]. Indeed, upon applying the PASTA property, it suffices to calculate the stationary distribution of the *order* of first occurrence of unmatched classes in the queues (rather than just the *set* of unmatched classes); this distribution can be evaluated by considering a refinement of the partition introduced in Sect. 3.1.

Time Complexity. Compared to the formula of [3, Theorem 3], our method leads to a lower time complexity if the number of independent sets in the compatibility graph is smaller than the cardinalities of the power sets of the sets $\mathcal{I}$ and $\mathcal{K}$. This is the case, for instance, in d-regular graphs, where the number of independent sets is at most $(2^{d+1} - 1)^{(I+K)/2d}$ [18]. To illustrate this, let us first recall how to compute the probability that the system is empty using Proposition 1. The idea is to first apply (7) recursively with the base case $\pi(\emptyset) = 1$, and then derive the value of $\pi(\emptyset)$ by applying the normalization equation. For each $\mathcal{A} \in \mathbb{I}$, assuming that the values of $\pi(\mathcal{A} \backslash \{i\})$, $\pi(\mathcal{A} \backslash \{k\})$, and $\pi(\mathcal{A} \backslash \{i, k\})$ are known for each $i \in \mathcal{A} \cap \mathcal{I}$ and $k \in \mathcal{A} \cap \mathcal{K}$, evaluating $\pi(\mathcal{A})$ using (7) requires $O(I \cdot K)$ operations, where I is the number of customer classes and K is the number of server classes. The time complexity to evaluate the probability that the system is empty is therefore given by $O(T + N \cdot I \cdot K)$, where N is the number of independent sets in the compatibility graph and T is the time complexity to enumerate all maximal independent sets. The result of [17] implies that the time complexity to enumerate all maximal independent sets in the (bipartite) compatibility graph $O((I + K) \cdot I \cdot K \cdot M)$, where M is the number of maximal independent sets. Overall, the time complexity to evaluate the normalization constant using Proposition 1 is $O(I \cdot K \cdot ((I + K) \cdot M + N))$.

In comparison, the time complexity to evaluate the normalization constant using [3, Theorem 3] is $O((I + K) \cdot 2^{\min(I,K)})$ if we implement these formulas recursively, in a similar way as in [6]. Our method thus leads to a lower time complexity if the number of independent sets of the compatibility graph is small.

4   Numerical Evaluation

To illustrate our results, we apply the formulas of Sect. 3 to the toy example of Fig. 1a. The arrival probabilities are chosen as follows: for any $\rho \in (0, 1)$,

$$\lambda_1 = \lambda_2 = \lambda_3 = \lambda_4 = \frac{1}{4}, \quad \mu_A = \frac{\rho}{4}, \quad \mu_B = \mu_C = \mu_D = \frac{1}{4}, \quad \mu_E = \frac{1 - \rho}{4}. \quad (12)$$

Figure 3 shows several performance metrics. The lines are plotted using the results of Sect. 3. To verify these results, we plotted marks representing simulated values based on averaging the results of 20 discrete-event simulation runs, each consisting of 10^6 transitions after a warm-up period of 10^6 transitions. The standard deviation of the simulated waiting times (resp. probabilities) never exceeded 1.9 (resp. 0.008) per experiment, validating the reliability of the results.

Due to the parameter settings, performance is symmetrical around $\rho = \frac{1}{2}$. Figure 3a and 3b show that classes 1, 2, 3, C, D, and E become unstable, in the sense that their mean waiting time tends to infinity, as $\rho \downarrow 0$. This is confirmed by observing that $\Delta(\mathcal{A}) \downarrow 0$ for $\mathcal{A} \in \{\{4, A\}, \{4, A, B\}, \{4, A, B, C\}, \{3, 4, A\}, \{3, 4, A, B\}, \{2, 3, 4, A\}\}$ when $\rho \downarrow 0$. We conjecture that this limiting regime can be studied by adapting the heavy-traffic analysis of [10, Section 6.2], although the behavior is different due to the concurrent arrivals of customers and servers.

Even if classes 1, 2, 3, C, D, and E all become unstable as $\rho \downarrow 0$, we can distinguish two qualitatively-different behaviors: the waiting probabilities of classes 1 and E tend to one, while for classes 2, 3, C, and D the limit is strictly less than one. This difference lies in the fact that the former classes have degree one in the compatibility graph, while the latter have degree two. Especially class C is intriguing, as the monotonicity of its waiting probability and mean waiting time are reversed, and would be worth further investigation.

Figure 3c shows that the probabilities of transitions $-/-$ and $+/+$ are equal to each other (as announced by Corollary 1) and are approximately constant. The probabilities of transitions $\pm/=$, $=/\pm$, and $=/=$, which impact the imbalance

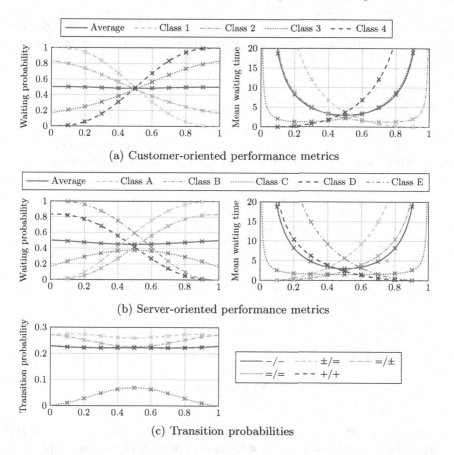

(a) Customer-oriented performance metrics

(b) Server-oriented performance metrics

(c) Transition probabilities

Fig. 3. Numerical results associated with the graph of Fig. 1a. The abscissa is the parameter ρ defined in (12).

between classes but not the total queue lengths, vary with ρ. In particular, the probability of transitions $=/=$ is maximal when $\rho = \frac{1}{2}$, which may explain why $\rho = \frac{1}{2}$ minimizes the average waiting probability and mean waiting time.

Appendix: Proofs of the Results of Sect. 3

Proof of Corollary 1. Summing (8) over all $\mathcal{A} \in \mathbb{I}$ and rearranging the sum symbols yields

$$\sum_{(i,k) \in \mathcal{I} \times \mathcal{K}} \lambda_i \mu_k \sum_{\substack{\mathcal{A} \in \mathbb{I}: i \in \mathcal{I}(\mathcal{A} \cap \mathcal{K}), \\ k \in \mathcal{K}(\mathcal{A} \cap \mathcal{I})}} \pi(\mathcal{A})$$

$$= \sum_{\substack{(i,k) \in \mathcal{I} \times \mathcal{K}: \\ i \sim k}} \lambda_i \mu_k \sum_{\substack{\mathcal{A} \in \mathbb{I}: \\ i \in \mathcal{A}, k \in \mathcal{A}}} \left(\pi(\mathcal{A}) + \pi(\mathcal{A} \backslash \{i\}) + \pi(\mathcal{A} \backslash \{k\}) + \pi(\mathcal{A} \backslash \{i, k\}) \right). \quad (13)$$

The left-hand side of this equation is the left-hand side of (9). The right-hand side can be rewritten by making changes of variables. For instance, for each $i \in \mathcal{I}$ and $k \in \mathcal{K}$ such that $i \nsim k$, replacing $\mathcal{A}$ with $\mathcal{A} \backslash \{i, k\}$ in the last sum yields

$$\sum_{\mathcal{A} \in \mathbb{I}: i \in \mathcal{A}, k \in \mathcal{A}} \pi(\mathcal{A} \backslash \{i, k\}) = \sum_{\substack{\mathcal{A} \subseteq \mathcal{I} \cup \mathcal{K}: i \notin \mathcal{A}, k \notin \mathcal{A}, \\ \mathcal{A} \cup \{i, k\} \in \mathbb{I}}} \pi(\mathcal{A}) = \sum_{\substack{\mathcal{A} \in \mathbb{I}_0: i \notin \mathcal{A}, k \notin \mathcal{A}, \\ i \notin \mathcal{I}(\mathcal{A} \cap \mathcal{K}), k \notin \mathcal{K}(\mathcal{A} \cap \mathcal{I})}} \pi(\mathcal{A}).$$

The second equality is true only because $i \nsim k$. By applying changes of variables to the other terms, we obtain that the right-hand side of (13) is equal to

$$\sum_{\substack{(i,k) \in \mathcal{I} \times \mathcal{K}: \\ i \nsim k}} \lambda_i \mu_k \left(\sum_{\substack{\mathcal{A} \in \mathbb{I}_0: i \in \mathcal{I}, k \in \mathcal{A}, \\ i \notin \mathcal{I}(\mathcal{A} \cap \mathcal{K}), k \notin \mathcal{K}(\mathcal{A} \cap \mathcal{I})}} \pi(\mathcal{A}) + \sum_{\substack{\mathcal{A} \in \mathbb{I}_0: i \notin \mathcal{I}, k \in \mathcal{A}, \\ i \notin \mathcal{I}(\mathcal{A} \cap \mathcal{K}), k \notin \mathcal{K}(\mathcal{A} \cap \mathcal{I})}} \pi(\mathcal{A}) \right.$$

$$\left. + \sum_{\substack{\mathcal{A} \in \mathbb{I}_0: i \in \mathcal{I}, k \notin \mathcal{A}, \\ i \notin \mathcal{I}(\mathcal{A} \cap \mathcal{K}), k \notin \mathcal{K}(\mathcal{A} \cap \mathcal{I})}} \pi(\mathcal{A}) + \sum_{\substack{\mathcal{A} \in \mathbb{I}_0: i \notin \mathcal{A}, k \notin \mathcal{A}, \\ i \notin \mathcal{I}(\mathcal{A} \cap \mathcal{K}), k \notin \mathcal{K}(\mathcal{A} \cap \mathcal{I})}} \pi(\mathcal{A}) \right)$$

$$= \sum_{(i,k) \in \mathcal{I} \times \mathcal{K}: i \nsim k} \lambda_i \mu_k \sum_{\substack{\mathcal{A} \in \mathbb{I}_0: i \notin \mathcal{I}(\mathcal{A} \cap \mathcal{K}), k \notin \mathcal{K}(\mathcal{A} \cap \mathcal{I})}} \pi(\mathcal{A}).$$

Proof of Proposition 2. Let $i \in \mathcal{I}$. We have $L_i = \sum_{\mathcal{A} \in \mathbb{I}_0} \ell_i(\mathcal{A})$, where

$$\ell_i(\mathcal{A}) = \sum_{(c,d) \in \Pi_{\mathcal{A}}} |c|_i \pi(c, d), \quad \mathcal{A} \in \mathbb{I}_0.$$

Let $\mathcal{A} \in \mathbb{I}$. If $i \notin \mathcal{A}$, we have directly $\ell_i(\mathcal{A}) = 0$ because $|c|_i = 0$ for each $(c, d) \in \Pi_{\mathcal{A}}$. Now assume that $i \in \mathcal{A}$ (so that in particular $\mathcal{A}$ is non-empty). The method is similar to the proof of Proposition 1. First, by applying (1), we have

$$\ell_i(\mathcal{A}) = \sum_{(c,d) \in \Pi_{\mathcal{A}}} |c|_i \frac{\lambda_{c_n}}{\mu(\mathcal{K}(\mathcal{A} \cap \mathcal{I}))} \frac{\mu_{d_n}}{\lambda(\mathcal{I}(\mathcal{A} \cap \mathcal{K}))} \pi((c_1, \ldots, c_{n-1}), (d_1, \ldots, d_{n-1})).$$

Then applying (6) and doing a change of variable yields

$$\mu(\mathcal{K}(\mathcal{A} \cap \mathcal{I})) \lambda(\mathcal{I}(\mathcal{A} \cap \mathcal{K})) \ell_i(\mathcal{A})$$

$$= \lambda_i \sum_{k \in \mathcal{A} \cap \mathcal{K}} \mu_k \left(\sum_{(c,d) \in \Pi_{\mathcal{A}}} (|c|_i + 1) \pi(c, d) + \sum_{(c,d) \in \Pi_{\mathcal{A} \backslash \{i\}}} (0 + 1) \pi(c, d) \right.$$

$$\left. + \sum_{(c,d) \in \Pi_{\mathcal{A} \backslash \{k\}}} (|c|_i + 1) \pi(c, d) + \sum_{(c,d) \in \Pi_{\mathcal{A} \backslash \{i, k\}}} (0 + 1) \pi(c, d) \right),$$

$$+ \sum_{j \in (\mathcal{A} \backslash \{i\}) \cap \mathcal{I}} \sum_{k \in \mathcal{A} \cap \mathcal{K}} \lambda_j \mu_k \left(\sum_{(c,d) \in \Pi_{\mathcal{A}}} |c|_i \pi(c, d) + \sum_{(c,d) \in \Pi_{\mathcal{A} \backslash \{j\}}} |c|_i \pi(c, d) \right.$$

$$\left. + \sum_{(c,d) \in \Pi_{\mathcal{A} \backslash \{k\}}} |c|_i \pi(c, d) + \sum_{(c,d) \in \Pi_{\mathcal{A} \backslash \{j, k\}}} |c|_i \pi(c, d) \right),$$

$$= \lambda_i \sum_{k \in \mathcal{A} \cap \mathcal{K}} \mu_k \left(\ell_i(\mathcal{A}) + \pi(\mathcal{A}) + \pi(\mathcal{A} \backslash \{i\}) + \ell_i(\mathcal{A} \backslash \{k\}) + \pi(\mathcal{A} \backslash \{k\}) + \pi(\mathcal{A} \backslash \{i, k\}) \right)$$

$$+ \sum_{j \in (\mathcal{A} \backslash \{i\}) \cap \mathcal{I}} \sum_{k \in \mathcal{A} \cap \mathcal{K}} \lambda_j \mu_k \left(\ell_i(\mathcal{A}) + \ell_i(\mathcal{A} \backslash \{j\}) + \ell_i(\mathcal{A} \backslash \{k\}) + \ell_i(\mathcal{A} \backslash \{j, k\}) \right).$$

The result follows by rearranging the terms.

Proof of Proposition 3. Equation (11) follows by summing (10) over all $i \in \mathcal{I} \cap \mathcal{A}$ and simplifying the result using (8).

References

1. Adan, I., Busic, A., Mairesse, J., Weiss, G.: Reversibility and further properties of FCFS infinite bipartite matching. Math. Oper. Res. **43**(2), 598–621 (2017)
2. Adan, I., Kleiner, I., Righter, R., Weiss, G.: FCFS parallel service systems and matching models. Perform. Eval. **127–128**, 253–272 (2018)
3. Adan, I., Weiss, G.: Exact FCFS matching rates for two infinite multitype sequences. Oper. Res. **60**(2), 475–489 (2012)
4. Adan, I., Weiss, G.: A skill based parallel service system under FCFS-ALIS - steady state, overloads, and abandonments. Stoch. Syst. **4**(1), 250–299 (2014)
5. Berezner, S., Krzesinski, A.E.: Order independent loss queues. Queueing Syst. **23**(1), 331–335 (1996)
6. Bonald, T., Comte, C., Mathieu, F.: Performance of balanced fairness in resource pools: a recursive approach. Proc. ACM Meas. Anal. Comput. Syst. **1**(2), 41:1–41:25 (2017)
7. Busic, A., Gupta, V., Mairesse, J.: Stability of the bipartite matching model. Adv. Appl. Probab. **45**(2), 351–378 (2013)
8. Caldentey, R., Kaplan, E.H., Weiss, G.: FCFS infinite bipartite matching of servers and customers. Adv. Appl. Probab. **41**(3), 695–730 (2009)
9. Comte, C.: Resource management in computer clusters: algorithm design and performance analysis. Ph.D. thesis. Institut Polytechnique de Paris (2019)
10. Comte, C.: Stochastic non-bipartite matching models and order-independent loss queues. Stochast. Models, 1–36 (2021). https://www.tandfonline.com/doi/full/10.1080/15326349.2021.1962352
11. Droste, M., Kuich, W., Vogler, H. (eds.): Handbook of Weighted Automata. Monographs in Theoretical Computer Science. An EATCS Series, Springer, Heidelberg (2009). https://doi.org/10.1007/978-3-642-01492-5
12. Gardner, K., Righter, R.: Product forms for FCFS queueing models with arbitrary server-job compatibilities: an overview. Queueing Syst. 3–51 (2020). https://doi.org/10.1007/s11134-020-09668-6
13. Gardner, K., Zbarsky, S., Doroudi, S., Harchol-Balter, M., Hyytia, E.: Reducing latency via redundant requests: exact analysis. ACM SIGMETRICS Perform. Eval. Rev. **43**(1), 347–360 (2015)
14. Krzesinski, A.E.: Order independent queues. In: Boucherie, R.J., Van Dijk, N.M. (eds.) Queueing Networks: A Fundamental Approach. International Series in Operations Research and Management Science, vol. 154, pp. 85–120. Springer, Boston (2011). https://doi.org/10.1007/978-1-4419-6472-4_2
15. Mairesse, J., Moyal, P.: Stability of the stochastic matching model. J. Appl. Probab. **53**(4), 1064–1077 (2016)

16. Moyal, P., Busic, A., Mairesse, J.: A product form for the general stochastic matching model. J. Appl. Probab. **58**(2), 449–468 (2021)
17. Tsukiyama, S., Ide, M., Ariyoshi, H., Shirakawa, I.: A new algorithm for generating all the maximal independent sets. SIAM J. Comput. **6**(3), 505–517 (1977)
18. Zhao, Y.: The number of independent sets in a regular graph. Comb. Probab. Comput. **19**(2), 315–320 (2010)

Analysis of Tandem Retrial Queue with Common Orbit and Poisson Arrival Process

Anatoly Nazarov[1], Svetlana Paul[1(✉)], Tuan Phung-Duc[2],
and Mariya Morozova[1]

[1] National Research Tomsk State University, 36 Lenina Avenue,
634050 Tomsk, Russia
[2] University of Tsukuba, 1-1-1 Tennodai, Tsukuba, Ibaraki 305-8573, Japan
`tuan@sk.tsukuba.ac.jp`

Abstract. In this paper, we present a diffusion limit for the time-dependent distribution of the number of customers in the orbit for a tandem queueing system with one orbit, Poisson arrival process of incoming calls and two sequentially connected servers using a characteristic function approach. Under the condition that the mean time of a customer in the orbit tends to infinity, the number of customers in the orbit explodes. Using a proper scaling, we prove that the scaled version of the number of customers in the orbit asymptotically follows a diffusion process. Using the steady-state solution of the diffusion process, we build an approximation for the steady-state distribution of the number of customers in the orbit. We compare this new approximation with the traditional approximation based on the central limit theorem and with simulation. Numerical results show that the new approximation has higher accuracy than that based on the central limit theorem.

Keywords: Tandem retrial queue system · Sequentially connected servers · Asymptotic diffusion analysis

1 Introduction

Retrial phenomena are ubiquitous in service systems. For example, in call centers, customers who cannot immediately connect with the operator may make a phone call later. In some ITC systems, if requests are not processed immediately, some protocol automatically and repeatedly reconnects with the server. The analysis of retrial queues is more challenging in comparison with the counter part models with infinite buffer because the arrival rate of retrial customers depends on the number of retrying customers. As a result, explicit results are obtained in only a few special cases with small number of servers [9]. We refer to Phung-Duc [7] for a review of recent results on the research of retrial queues. In the network context, retrial queueing networks do not possess product form and thus the computation is challenging.

© Springer Nature Switzerland AG 2021
P. Ballarini et al. (Eds.): EPEW 2021/ASMTA 2021, LNCS 13104, pp. 441–456, 2021.
https://doi.org/10.1007/978-3-030-91825-5_27

To the best of our knowledge, only a few works on tandem queues with retrials are available [1,2,4,6]. The papers [1,2] present some exact and approximate analyses for tandem queues with a common orbit of constant retrial rate. In [4], the authors consider a tandem queue without intermediate buffer and thus the blocking phenomenon occurs in the first server, where arriving customers who see the first server either busy or blocked join the orbit and retry to the first server according to a constant retrial rate policy. As for tandem queue with linear retrial rate, Phung-Duc [6] considers a model with two servers in tandem and without an intermediate buffer. Customers who finish service at the first server and sees the second server busy are lost (not join the orbit). Later, Falin [3] studies the same model using an alternative method. For that model, explicit expressions of the stationary queue length distribution are derived.

In this paper, we consider a more complex model than that by Phung-Duc [6] where we assume that customers who finish service in the first server and see the second server busy also joins the same orbit as customers who are blocked at the first server. Although the model can be formulated using a level-dependent quasi-birth-and-death process for which some numerical method is available (see e.g. [8]), an explicit solution for even the stationary distribution cannot be obtained. In this paper, we consider a challenging problem characterizing the time-dependent distribution of the number of customers in the orbit. To this end, we focus to the asymptotic behavior of the distribution of the number of customers in the orbit in a special regime with an extremely small retrial rate (i.e. extremely large mean time in the orbit).

The rest of the paper is organized as follows. In Sect. 2, the description of the model is presented. In Sect. 3, we present a set of Kolmogorov differential equations while Sect. 4 and 5 are devoted to the first and the second order asymptotic analysis of the distribution of the number of customers in the orbit. Section 6 presents the asymptotic diffusion approach where we obtain an approximation for the distribution of the number of customers in the orbit in the stationary regime. Section 7 presents some numerical examples.

2   Mathematical Model

Let us consider a retrial tandem queueing system with Poisson arrival process of incoming calls with rate λ and two sequentially connected servers (see Fig. 1). Upon the arrival of a call, if the first server is free, the call occupies it. The call is served for a random time exponentially distributed with parameter μ_1 and then tries to go to the second server. If the second server is free, the call moves to it for a random time exponentially distributed with parameter μ_2. When a call arrives, if the first server is busy, the call instantly goes to the orbit, stays there during a random time exponentially distributed with parameter σ and then tries to occupy the first server again. After being served at the first server, if the call finds that the second server is busy, it instantly goes to the same orbit, where, after an exponentially distributed delay with parameter σ, tries to move to the first server for service again.

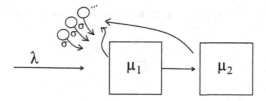

Fig. 1. The model.

We define some notations as follows.

Process $N_1(t)$ - the state of the first server at time t: 0, if the server is free; 1, if the server is busy;
Process $N_2(t)$ - the state of the second server at time t: 0, if the server is free; 1, if the server is busy;
Process $I(t)$ - number of calls in the orbit at the time t.

The goal of the study is two-fold. First, we derive a diffusion limit of the time-dependent distribution of the number of calls in the orbit $I(t)$ and the distribution of the servers' states in our system, under a special regime where the retrial rate is extremely small ($\sigma \to 0$). Second, based on the steady-state behavior of the diffusion solution, we obtain an approximation to the stationary distribution of the number of customers in the orbit.

3   The System of Differential Kolmogorov Equations

We define probabilities

$$P_{n_1 n_2}(i,t) = P\{N_1(t) = n_1, N_2(t) = n_2, I(t) = i\}; n_1 = 0,1; n_2 = 0,1. \quad (1)$$

The three-dimensional process $\{N_1(t), N_2(t), I(t)\}$ is a Markov chain. For probability distribution (1), we can write the system of differential Kolmogorov equations:

$$\frac{\partial P_{00}(i,t)}{\partial t} = -(\lambda + i\sigma)P_{00}(i,t) + \mu_2 P_{01}(i,t),$$

$$\frac{\partial P_{10}(i,t)}{\partial t} = \lambda P_{00}(i,t) + (i+1)\sigma P_{00}(i+1,t) - (\lambda + \mu_1)P_{10}(i,t)$$

$$+ \lambda P_{10}(i-1,t) + \mu_2 P_{11}(i,t),$$

$$\frac{\partial P_{01}(i,t)}{\partial t} = \mu_1 P_{10}(i,t) - (\lambda + i\sigma + \mu_2)P_{01}(i,t) + \mu_1 P_{11}(i-1,t),$$

$$\frac{\partial P_{11}(i,t)}{\partial t} = \lambda P_{01}(i,t) + (i+1)\sigma P_{01}(i+1,t) - (\lambda + \mu_1 + \mu_2)P_{11}(i,t)$$

$$+ \lambda P_{11}(i-1,t). \quad (2)$$

We introduce partial characteristic functions, denoting $j = \sqrt{-1}$

$$H_{n_1 n_2}(u, t) = \sum_{i=0}^{\infty} e^{jui} P_{n_1 n_2}(i, t). \tag{3}$$

So, we have

$$\frac{\partial H_{00}(u, t)}{\partial t} = -\lambda H_{00}(u, t) + \mu_2 H_{01}(u, t) + j\sigma \frac{\partial H_{00}(u, t)}{\partial u},$$

$$\frac{\partial H_{10}(u, t)}{\partial t} = \left(\lambda \left(e^{ju} - 1\right) - \mu_1\right) H_{10}(u, t) + \lambda H_{00} + \mu_2 H_{11}(u, t)$$

$$- j\sigma e^{-ju} \frac{\partial H_{00}(u, t)}{\partial u},$$

$$\frac{\partial H_{01}(u, t)}{\partial t} = \mu_1 H_{10}(u, t) - (\lambda + \mu_2) H_{01}(u, t) + \mu_1 e^{ju} H_{11}(u, t)$$

$$+ j\sigma \frac{\partial H_{01}(u, t)}{\partial u},$$

$$\frac{\partial H_{11}(u, t)}{\partial t} = \left(\lambda \left(e^{ju} - 1\right) - \mu_1 - \mu_2\right) H_{11}(u, t) + \lambda H_{01}(u, t)$$

$$- j\sigma e^{-ju} \frac{\partial H_{01}(u, t)}{\partial u}. \tag{4}$$

Define matrices

$$\mathbf{A} = \begin{bmatrix} -\lambda & \lambda & 0 & 0 \\ 0 & -(\lambda + \mu_1) & \mu_1 & 0 \\ \mu_2 & 0 & -(\lambda + \mu_2) & \lambda \\ 0 & \mu_2 & 0 & -(\lambda + \mu_1 + \mu_2) \end{bmatrix},$$

$$\mathbf{B} = \begin{bmatrix} 0 & 0 & 0 & 0 \\ 0 & \lambda & 0 & 0 \\ 0 & 0 & 0 & 0 \\ 0 & 0 & \mu_2 & \lambda \end{bmatrix}, \mathbf{I}_0 = \begin{bmatrix} 1 & 0 & 0 & 0 \\ 0 & 0 & 0 & 0 \\ 0 & 0 & 1 & 0 \\ 0 & 0 & 0 & 0 \end{bmatrix}, \mathbf{I}_1 = \begin{bmatrix} 0 & 1 & 0 & 0 \\ 0 & 0 & 0 & 0 \\ 0 & 0 & 0 & 1 \\ 0 & 0 & 0 & 0 \end{bmatrix}. \tag{5}$$

Let us write the system (4) in the matrix form

$$\frac{\partial \mathbf{H}(u, t)}{\partial t} = \mathbf{H}(u, t)\{\mathbf{A} + e^{ju}\mathbf{B}\} + j\sigma \frac{\partial \mathbf{H}(u, t)}{\partial u}\{\mathbf{I}_0 - e^{-ju}\mathbf{I}_1\}, \tag{6}$$

where $\mathbf{H}(u, t) = \{H_{00}(u, t), H_{10}(u, t), H_{01}(u, t), H_{11}(u, t)\}$.

Multiplying equations of system (6) by the identity column vector $\mathbf{e}$, we obtain

$$\frac{\partial \mathbf{H}(u, t)}{\partial t} \mathbf{e} = \mathbf{H}(u, t)\{\mathbf{A} + e^{ju}\mathbf{B}\}\mathbf{e} + ju \frac{\partial \mathbf{H}(u, t)}{\partial u}\{\mathbf{I}_0 - e^{-ju}\mathbf{I}_1\}\mathbf{e}. \tag{7}$$

Given $(\mathbf{A} + \mathbf{B})\mathbf{e} = 0$ and $(\mathbf{I}_0 - \mathbf{I}_1)\mathbf{e} = 0$, we get scalar equation

$$\frac{\partial \mathbf{H}(u,t)}{\partial t}\mathbf{e} = (e^{ju} - 1)\left\{\mathbf{H}(u,t)\mathbf{B} + j\sigma e^{-ju}\frac{\partial \mathbf{H}(u,t)}{\partial u}\mathbf{I}_1\right\}\mathbf{e}. \tag{8}$$

From matrix equation (6) and scalar equation (8) we have the form

$$\frac{\partial \mathbf{H}(u,t)}{\partial t} = \mathbf{H}(u,t)\{\mathbf{A} + e^{ju}\mathbf{B}\} + j\sigma\frac{\partial \mathbf{H}(u,t)}{\partial u}\{\mathbf{I}_0 - e^{-ju}\mathbf{I}_1\},$$

$$\frac{\partial \mathbf{H}(u,t)}{\partial t}\mathbf{e} = (e^{ju} - 1)\left\{\mathbf{H}(u,t)\mathbf{B} + j\sigma e^{-ju}\frac{\partial \mathbf{H}(u,t)}{\partial u}\mathbf{I}_1\right\}\mathbf{e}. \tag{9}$$

This system of equations is the basis in further research. We will solve it by a method of asymptotic diffusion analysis under the asymptotic condition $\sigma \to 0$.

4   The First Order Asymptotic: Fluid Limit

In this section, we present the first order asymptotic of the number of customers in the orbit. This type of asymptotic is also called *fluid limit* in the literature [10].

By denoting $\sigma = \varepsilon$ and performing the following substitution in the system (9)

$$\tau = t\varepsilon, \ u = w\varepsilon, \ \mathbf{H}(u,t) = \mathbf{F}(w,\tau,\varepsilon), \tag{10}$$

we rewrite the system (9) as

$$\varepsilon\frac{\partial \mathbf{F}(w,\tau,\varepsilon)}{\partial t} = \mathbf{F}(w,\tau,\varepsilon)(\mathbf{A} + e^{j\varepsilon w}\mathbf{B}) + j\frac{\partial \mathbf{F}(w,\tau,\varepsilon)}{\partial w}(\mathbf{I}_0 - e^{-j\varepsilon w}\mathbf{I}_1),$$

$$\varepsilon\frac{\partial \mathbf{F}(w,\tau,\varepsilon)}{\partial t}\mathbf{e} = (e^{j\varepsilon w} - 1)\left\{\mathbf{F}(w,\tau,\varepsilon)\mathbf{B} + je^{-j\varepsilon w}\frac{\partial \mathbf{F}(w,\tau,\varepsilon)}{\partial w}\mathbf{I}_0\right\}\mathbf{e}. \tag{11}$$

Let us denote

$$\lim_{\varepsilon \to 0} \mathbf{F}(w,\tau,\varepsilon) = \mathbf{F}(w,\tau), \ \mathbf{F}(0,\tau) = \mathbf{r}. \tag{12}$$

The row vector $\mathbf{r}$ defines two-dimensional probability distribution of the states of servers (n_1, n_2). It will be seen below that the row vector $\mathbf{r}$, that satisfies the normalization condition $\mathbf{re} = 1$, is a solution of the matrix equation

$$\mathbf{r}\{\mathbf{A} + \mathbf{B} - x(\mathbf{I}_0 - \mathbf{I}_1)\} = 0. \tag{13}$$

Coefficients of this equation depend on variable x, so, solution $\mathbf{r}$ depends on value of x, therefore denote $\mathbf{r} = \mathbf{r}(x)$. It should be noted that x is a function of τ and $\mathbf{r}$ is a function of x but sometime we omit τ and x for simplicity.

Solving the system (11) under asymptotic condition $\varepsilon \to 0$ $(\sigma \to 0)$, we obtain the following statement.

Theorem 1. *Under the limit condition* $\sigma \to 0$, *the following equality is true*

$$\lim_{\sigma \to 0} \mathbb{E}e^{jw\sigma I\left(\frac{\tau}{\sigma}\right)} = e^{jwx(\tau)}. \tag{14}$$

Here the scalar function $x = x(\tau)$ *is a solution of differential equation*

$$x'(\tau) = \mathbf{r}(x)(\mathbf{B} - x\mathbf{I}_0)\mathbf{e}, \tag{15}$$

where the vector $\mathbf{r}(x)$ *satisfies the normalization condition*

$$\mathbf{r}(x)\mathbf{e} = 1 \tag{16}$$

and is a solution of the matrix equation

$$\mathbf{r}(x)\{\mathbf{A} + \mathbf{B} - x(\mathbf{I}_0 - \mathbf{I}_1)\} = 0. \tag{17}$$

Proof. Let us take the limit $\varepsilon \to 0$ in the system (11). Denoting $\lim_{\varepsilon \to 0} \mathbf{F}(w, \tau, \varepsilon) = \mathbf{F}(w, \tau)$ we obtain

$$\mathbf{F}(w, \tau)(\mathbf{A} + \mathbf{B}) + j\frac{\partial \mathbf{F}(w, \tau)}{\partial w}(\mathbf{I}_0 - \mathbf{I}_1) = 0,$$

$$\frac{\partial \mathbf{F}(w, \tau)}{\partial \tau}\mathbf{e} = jw\left\{\mathbf{F}(w, \tau)\mathbf{B} + j\frac{\partial \mathbf{F}(w, \tau)}{\partial w}\mathbf{I}_0\right\}\mathbf{e}. \tag{18}$$

We find the solution of system (18) in the form

$$\mathbf{F}(w, \tau) = \mathbf{r}(x)e^{jwx(\tau)}, \tag{19}$$

where $x = x(\tau)$ - is a scalar function with argument τ, which has a meaning of asymptotic (while $\varepsilon \to 0$) value of $\sigma I(\tau/\sigma)$, i.e., the number of customers in the orbit normalized by $\varepsilon = \sigma$. Substituting (19) into (18), we obtain

$$\mathbf{r}(x)\{\mathbf{A} + \mathbf{B} - x(\mathbf{I}_0 - \mathbf{I}_1)\} = 0, \tag{20}$$

$$x'(\tau) = \mathbf{r}(x)(\mathbf{B} - x\mathbf{I}_0)\mathbf{e}. \tag{21}$$

Because the scalar function $x(\tau)$ with argument τ is an asymptotic value (while $\varepsilon \to 0$) of the normalized number of calls in the orbit $\sigma i(\tau/\sigma)$, equality (14) is true. So, Theorem 1 is proved.

Let us substitute the solution $\mathbf{r}(x)$ of the system of Eqs. (20) in the scalar equation (21) and we will get

$$a(x) = \mathbf{r}(x)(\mathbf{B} - x\mathbf{I}_0)\mathbf{e}, \tag{22}$$

Function $a(x)$ is very important for the study retrial queuing systems by the method of asymptotic diffusion analysis. Firstly, as we have shown in Theorem 1, $a(x) = x'(\tau)$, therefore, function $a(x)$ characterizes dynamic of the process $x(\tau)$, the limit under $\sigma \to 0$ for the normalized number of calls in the orbit $\sigma(\tau/\sigma)$. Secondly, we will show that function $a(x)$ is the drift coefficient for diffusion process which determines the asymptotic number of customers in the orbit $I(t)$. Using $a(x)$ we will get necessary condition for the existence of steady-state regime in the retrial queuing system under consideration.

5 The Second Order Asymptotic: Diffusion Limit

Substituting the following in the system (9)

$$\mathbf{H}(u,t) = e^{j\frac{u}{\sigma}x(\sigma t)}\mathbf{H}^{(1)}(u,t) \tag{23}$$

and taking into account the Eq. (22), we obtain

$$\frac{\partial \mathbf{H}^{(1)}(u,t)}{\partial t} + jua(x)\mathbf{H}^{(1)}(u,t) = \mathbf{H}^{(1)}(u,t)\left(\mathbf{A} + e^{ju}\mathbf{B} - x\left(\mathbf{I}_0 - e^{-ju}\mathbf{I}_1\right)\right)$$

$$+ j\sigma\frac{\partial \mathbf{H}^{(1)}(u,t)}{\partial u}(\mathbf{I}_0 - e^{-ju}\mathbf{I}_1),$$

$$\frac{\partial \mathbf{H}^{(1)}(u,t)}{\partial t}\mathbf{e} + jua(x)\mathbf{H}^{(1)}(u,t)\mathbf{e} = (e^{ju} - 1)$$

$$\times \left\{\mathbf{H}^{(1)}(u,t)(\mathbf{B} - e^{-ju}x\mathbf{I}_0) + j\sigma e^{-ju}\frac{\partial \mathbf{H}^{(1)}(u,t)}{\partial u}\mathbf{I}_0\right\}\mathbf{e}. \tag{24}$$

We make a substitute (23) with a view to asymptotic centering of random process $I(t)$ because $\mathbf{H}^{(1)}(u,t)$ is a vector characteristic function of a centring random process, where the function $x(\tau)$ was obtained in the first stage of asymptotic analysis.

By denoting $\sigma = \varepsilon^2$ in the system (24) and making substitutions

$$\tau = t\varepsilon^2, \ u = w\varepsilon, \ \mathbf{H}^{(1)}(u,t) = \mathbf{F}^{(1)}(w,\tau,\varepsilon), \tag{25}$$

we can rewrite the system in the following form

$$\varepsilon^2\frac{\partial \mathbf{F}^{(1)}(w,\tau,\varepsilon)}{\partial \tau} + j\varepsilon wa(x)\mathbf{F}^{(1)}(w,\tau,\varepsilon)$$

$$= \mathbf{F}^{(1)}(w,\tau,\varepsilon)\left(\mathbf{A} + e^{j\varepsilon w}\mathbf{B} - x(\mathbf{I}_0 - e^{-j\varepsilon w}\mathbf{I}_1)\right)$$

$$+ j\varepsilon\frac{\partial \mathbf{F}^{(1)}(w,\tau,\varepsilon)}{\partial w}(\mathbf{I}_0 - e^{-j\varepsilon w}\mathbf{I}_1),$$

$$\varepsilon^2\frac{\partial \mathbf{F}^{(1)}(w,\tau,\varepsilon)}{\partial \tau}\mathbf{e} + j\varepsilon wa(x)\mathbf{F}^{(1)}(w,\tau,\varepsilon)\mathbf{e} = (e^{j\varepsilon w} - 1)$$

$$\times \left(\mathbf{F}^{(1)}(w,\tau,\varepsilon)\left(\mathbf{B} - e^{-j\varepsilon w}x\mathbf{I}_0\right) + e^{-j\varepsilon w}j\varepsilon\frac{\partial \mathbf{F}^{(1)}(w,\tau,\varepsilon)}{\partial w}\mathbf{I}_0\right)\mathbf{e}. \tag{26}$$

Denote

$$\lim_{\varepsilon \to 0}\mathbf{F}^{(1)}(w,\tau,\varepsilon) = \mathbf{F}^{(1)}(w,\tau), \lim_{\varepsilon \to 0}\frac{\partial \mathbf{F}^{(1)}(w,\tau,\varepsilon)}{\partial \tau} = \frac{\partial \mathbf{F}^{(1)}(w,\tau)}{\partial \tau} \tag{27}$$

and prove the following statement.

Theorem 2. *Function* $\mathbf{F}^{(1)}(w, \tau)$ *has the following form*

$$\mathbf{F}^{(1)}(w, \tau) = \Phi(w, \tau)\mathbf{r}(x), \tag{28}$$

where the row vector $\mathbf{r}(x)$ *depends on variable* x. *The vector* $\mathbf{r}(x)$ *is determined by Theorem 1, with vector's components* $\mathbf{r}_{n_1 n_2}(x)$, *and the scalar function* $\Phi(w, \tau)$ *is the solution of the partial differential equation:*

$$\frac{\partial \Phi(w, \tau)}{\partial \tau} = a'(x)w\frac{\partial \Phi(w, \tau)}{\partial w} + b(x)\frac{(jw)^2}{2}\Phi(w, \tau). \tag{29}$$

Here the function $a(x)$ *is determined by (15) and the scalar function* $b(x)$ *has the form*

$$b(x) = a(x) + 2\mathbf{g}(x)(\mathbf{B} - x\mathbf{I}_0)\mathbf{e} + 2x\mathbf{r}(x)\mathbf{e}, \tag{30}$$

where vector $\mathbf{g}(x)$ *is determined by system of equations*

$$\mathbf{g}(x)\left(\mathbf{A} + \mathbf{B} + x(\mathbf{I}_1 - \mathbf{I}_0)\right) = a(x)\mathbf{r}(x) + \mathbf{r}(x)(x\mathbf{I}_0 - \mathbf{B}),$$

$$\mathbf{g}(x)\mathbf{e} = 0. \tag{31}$$

Proof. Let us write the first equation of the system (26) up to $O\left(\varepsilon^2\right)$

$$j\varepsilon w a(x)\mathbf{F}^{(1)}(w, \tau, \varepsilon)\left(\mathbf{A} + \mathbf{B} + j\varepsilon w\mathbf{B} - x(\mathbf{I}_0 - \mathbf{I}_1 + j\varepsilon w\mathbf{I}_1)\right)$$

$$+ j\varepsilon\frac{\partial \mathbf{F}^{(1)}(w, \tau, \varepsilon)}{\partial w}(\mathbf{I}_0 - \mathbf{I}_1) = O\left(\varepsilon^2\right). \tag{32}$$

We find the solution of this equation in the following form

$$\mathbf{F}^{(1)}(w, \tau, \varepsilon) = \Phi(w, \tau)\left(\mathbf{r}(x) + j\varepsilon w\mathbf{f}(x)\right) + O\left(\varepsilon^2\right), \tag{33}$$

where $\Phi(w, \tau)$ - is some scalar function whose expression is obtained later. We have

$$j\varepsilon w a(x)\Phi(w, \tau)\mathbf{r}(x) = \Phi(w, \tau)\left\{\mathbf{r}(x)\left(\mathbf{A} + \mathbf{B} - x(\mathbf{I}_0 - \mathbf{I}_1)\right)\right.$$

$$+ j\varepsilon w\left[\mathbf{f}(x)\left(\mathbf{A} + \mathbf{B} - x(\mathbf{I}_0 - \mathbf{I}_1)\right) + \mathbf{r}(x)(\mathbf{B} - x\mathbf{I}_0)\right]\}$$

$$+ j\varepsilon\frac{\partial \Phi(w, \tau)}{\partial w}\mathbf{r}(x)(\mathbf{I}_0 - \mathbf{I}_1) + O\left(\varepsilon^2\right). \tag{34}$$

Taking Eq. (22) into account, dividing Eq. (34) by $j\varepsilon$ and taking the limit $\varepsilon \to 0$, we obtain

$$\mathbf{f}(x)\left(\mathbf{A} + \mathbf{B} - x(\mathbf{I}_0 - \mathbf{I}_1)\right)$$

$$= a(x)\mathbf{r}(x) - \mathbf{r}(x)(\mathbf{B} - x\mathbf{I}_0) + \frac{\partial \Phi(w, \tau)/\partial w}{w\Phi(w, \tau)}\mathbf{r}(x)(\mathbf{I}_0 - \mathbf{I}_1). \tag{35}$$

According to the superposition principle, we can write a solution $\mathbf{f}(x)$ of this equation in the form of sum

$$\mathbf{f}(x) = C\mathbf{r}(x) + \mathbf{g}(x) - \varphi(x)\frac{\partial \Phi(w, \tau)/\partial w}{w\Phi(w, \tau)}. \tag{36}$$

Substituting it into Eq. (35), we obtain equations

$$\boldsymbol{\varphi}(x)\left(\mathbf{A}+\mathbf{B}-x(\mathbf{I}_1-\mathbf{I}_0)\right)=\mathbf{r}(x)(\mathbf{I}_0-\mathbf{I}_1), \tag{37}$$

$$\mathbf{g}(x)\left(\mathbf{A}+\mathbf{B}-x(\mathbf{I}_1-\mathbf{I}_0)\right)=a(x)\mathbf{r}(x)+\mathbf{r}(x)(x\mathbf{I}_0-\mathbf{B}). \tag{38}$$

Notice that Eq. (38) for vector $\mathbf{g}(x)$ coincides with the first expression of Eq. (31), therefore Eq. (31) is true.

Now, consider Eq. (20). Let us differentiate it by x to obtain equation

$$\frac{\partial \mathbf{r}(x)}{\partial x}\left\{\mathbf{A}+\mathbf{B}-x(\mathbf{I}_0-\mathbf{I}_1)\right\}+\mathbf{r}(x)(\mathbf{I}_0-\mathbf{I}_1)=0. \tag{39}$$

Comparing Eq. (39) and Eq. (37) for $\boldsymbol{\varphi}(x)$, we can conclude that:

$$\boldsymbol{\varphi}(x)=\frac{\partial \mathbf{r}(x)}{\partial x}. \tag{40}$$

The additional condition $\boldsymbol{\varphi}(x)\mathbf{e}=0$ is fulfilled because of the normalization condition (16) for vector $\boldsymbol{\varphi}(x)$.

Vector $\mathbf{g}(x)$ is a particular solution of the non-homogeneous system of Eqs. (38), so it should satisfy some additional condition which we choose in the form $\mathbf{g}(x)\mathbf{e}=0$. Then the solution $\mathbf{g}(x)$ of the system (38) is uniquely defined by the system (31).

Now, let us consider the second scalar equation of the system (24). We substitute the expansion (33) in it and rewrite it up to $O(\varepsilon^3)$

$$\varepsilon^2 \frac{\partial \Phi(w,\tau)}{\partial \tau}+j\varepsilon w a(x)\Phi(w,\tau)+(j\varepsilon w)^2 a(x)\Phi(w,\tau)\mathbf{f}(x)\mathbf{e}$$

$$=\Phi(w,\tau)\left\{(j\varepsilon w)^2 \mathbf{f}(x)(\mathbf{B}-x\mathbf{I}_0)+(j\varepsilon w)^2 \mathbf{r}(x)x\mathbf{I}_0\right.$$

$$\left.+\frac{(j\varepsilon w)^2}{2}\mathbf{r}(x)(\mathbf{B}-x\mathbf{I}_0)+(j\varepsilon w)^2 \frac{\partial \Phi(w,\tau)/\partial w}{w}\mathbf{r}(x)\mathbf{I}_0\right\}\mathbf{e}+O(\varepsilon^3). \tag{41}$$

Applying Eq. (22), we obtain

$$\varepsilon^2 \frac{\partial \Phi(w,\tau)}{\partial \tau}+(j\varepsilon w)^2 a(x)\Phi(w,\tau)\mathbf{f}(x)\mathbf{e}$$

$$=\Phi(w,\tau)\left\{(j\varepsilon w)^2 \mathbf{f}(x)(\mathbf{B}-x\mathbf{I}_0)+(j\varepsilon w)^2 \mathbf{r}(x)x\mathbf{I}_0\right.$$

$$\left.+\frac{(j\varepsilon w)^2}{2}a(x)+(j\varepsilon w)^2 \frac{\partial \Phi(w,\tau)/\partial w}{w}\mathbf{r}(x)\mathbf{I}_0\right\}\mathbf{e}+O(\varepsilon^3). \tag{42}$$

Let us divide this equation by ε^2 and take the limit $\varepsilon \to 0$ to obtain

$$\frac{\partial \Phi(w,\tau)/\partial \tau}{\Phi(w,\tau)}=\frac{(jw)^2}{2}\left\{2\mathbf{f}(x)(\mathbf{B}-x\mathbf{I}_0)+2\mathbf{r}(x)x\mathbf{I}_0+a(x)\right.$$

$$\left.-2a(x)\mathbf{f}(x)+\frac{\partial \Phi(w,\tau)/\partial w}{w}\mathbf{r}(x)\mathbf{I}_0\right\}\mathbf{e}. \tag{43}$$

Substituting the Eq. (36) here, we obtain

$$\frac{\partial \Phi(w, \tau)/\partial \tau}{\Phi(w, \tau)} = \frac{(jw)^2}{2} \left\{ 2\mathbf{g}(x)(\mathbf{B} - x\mathbf{I}_0)\mathbf{e} + 2\mathbf{r}(x)x\mathbf{I}_0\mathbf{e} + a(x) \right\}$$

$$- w \frac{\partial \Phi(w, \tau)/\partial w}{\Phi(w, \tau)} \left(\boldsymbol{\varphi}(x)(\mathbf{B} - x\mathbf{I}_0) - \mathbf{r}(x)\mathbf{I}_0 \right) \mathbf{e}. \qquad (44)$$

Denoting

$$b(x) = 2\mathbf{g}(x)(\mathbf{B} - x\mathbf{I}_0)\mathbf{e} + 2\mathbf{r}(x)\mathbf{I}_0 x\mathbf{e} + a(x), \qquad (45)$$

we can rewrite the Eq. (44) in the form

$$\frac{\partial \Phi(w, \tau)}{\partial \tau} = w \frac{\partial \Phi(w, \tau)}{\partial w} \left(\boldsymbol{\varphi}(x)(\mathbf{B} - x\mathbf{I}_0)\mathbf{e} - \mathbf{r}(x)\mathbf{I}_0\mathbf{e} \right) + \frac{(jw)^2}{2} b(x)\Phi(w, \tau). \quad (46)$$

Let us consider the expression individually

$$\boldsymbol{\varphi}(x)(\mathbf{B} - x\mathbf{I}_0)\mathbf{e} - \mathbf{r}(x)\mathbf{I}_0\mathbf{e}. \qquad (47)$$

Using (40), we obtain

$$\frac{\partial \mathbf{r}(x)}{\partial x}(\mathbf{B} - x\mathbf{I}_0)\mathbf{e} - \mathbf{r}(x)\mathbf{I}_0\mathbf{e}. \qquad (48)$$

Let us consider the expression from (22)

$$a(x) = \mathbf{r}(x)(\mathbf{B} - x\mathbf{I}_0)\mathbf{e}. \qquad (49)$$

Differentiating $a(x)$ by x and taking into account that vector $\mathbf{r}(x)$ is a solution of the system (15) and depends on x, we obtain

$$d(x) = a'(x) = \frac{\partial \mathbf{r}(x)}{\partial x}(\mathbf{B} - x\mathbf{I}_0)\mathbf{e} - \mathbf{r}(x)\mathbf{I}_0. \qquad (50)$$

Comparing (50) and (48), we rewrite (46) in the form

$$\frac{\partial \Phi(w, \tau)}{\partial \tau} = a'(x)w \frac{\partial \Phi(w, \tau)}{\partial w} + b(x)\frac{(jw)^2}{2}\Phi(w, \tau), \qquad (51)$$

that coincides with Eq. (29). So, Theorem 2 is proved.

Later we will show that function $b(x)$ is the diffusion coefficient of a diffusion process which has the function $a(x)$ as the coefficient of drift, defined by the Eq. (22).

Thus, we have defined functions $a(x)$ by the Eq. (22) and $b(x)$ by the Eq. (30). These functions are important in the next section where we propose an approximation to the stationary distribution of the number of customers in the orbit.

6 Stationary Distribution of the Diffusion Process and Queue Length Approximation

In this section of the paper, we will consider an implementation of the diffusion limit in Theorem 2 for finding the stationary probability distribution of the number of calls in the orbit $I(t)$ under the asymptotic condition $\sigma \to 0$.

Lemma 1. *Asymptotic stochastic process under the condition* $\sigma \to 0$

$$y(\tau) = \lim_{\sigma \to 0} \sqrt{\sigma} \left\{ I(\frac{\tau}{\sigma}) - \frac{1}{\sigma} x(\tau) \right\}, \tag{52}$$

is a solution of the stochastic differential equation

$$dy(\tau) = a'(x)y d\tau + \sqrt{b(x)} dw(\tau), \tag{53}$$

that depends on continuous parameter x.

Proof. Consider the Eq. (29) from Theorem 2

$$\frac{\partial \Phi(w, \tau)}{\partial \tau} = a'(x) w \frac{\partial \Phi(w, \tau)}{\partial w} + b(x) \frac{(jw)^2}{2} \Phi(w, \tau), \tag{54}$$

with $a(x)$ and $b(x)$ determined by Eqs. (22) and (30).

Solution $\Phi(w, \tau)$ of this equation determines the asymptotic characteristic function for the centered and normalized stochastic process $\sqrt{\sigma} \left\{ I(\frac{\tau}{\sigma}) - \frac{1}{\sigma} x(\tau) \right\}$ of the number $I(t)$ of calls in the orbit under the condition $\sigma \to 0$ and for its probability density distribution.

Let us make an inverse Fourier transform in this equation on argument w. Then for the probability density function $p(y, \tau)$ of process $y(\tau)$, we obtain the equation

$$\frac{\partial p(y, \tau)}{\partial \tau} = -\frac{\partial}{\partial y} \{a'(x) y p(y, \tau)\} + \frac{1}{2} \frac{\partial^2}{\partial y^2} \{b(x) p(y, \tau)\}, \tag{55}$$

which is the Fokker-Planck equation for probability density function $p(y, \tau)$. Hence, the stochastic process $y(\tau)$ is a diffusion process with drift coefficient $d(x)y$ and diffusion coefficient $b(x)$. Therefore, the diffusion process $y(\tau)$ is a solution of the stochastic differential equation (53). So, Lemma 1 is proved.

Let us consider the following stochastic process

$$z(\tau) = x(\tau) + \varepsilon y(\tau),$$

where $\varepsilon = \sqrt{\sigma}$ as before. This process is the sum of the normalized mean and the centered number of calls in the orbit.

Lemma 2. *Stochastic process* $z(\tau)$ *is a solution to stochastic differential equation*

$$dz(\tau) = a(z) d\tau + \sqrt{\sigma b(z)} dw(\tau) \tag{56}$$

with a precision up to an infinitesimal of order ϵ^2.

Proof. Because $x(\tau)$ is a solution of differential equation $dx(\tau) = a(x)d\tau$ and process $y(\tau)$ satisfied Eq. (53), the following equality is true

$$dz(\tau) = d(x(\tau) + \varepsilon y(\tau)) = (a(x) + \varepsilon y a'(x))d\tau + \varepsilon\sqrt{b(x)}dw(\tau). \qquad (57)$$

We can represent its coefficients in the form

$$a(x) + \varepsilon y d(x) = a(x + \varepsilon y) + O(\varepsilon^2) = a(z) + O(\varepsilon^2),$$

$$\varepsilon\sqrt{b(x)} = \varepsilon\sqrt{b(x + \varepsilon y)} + O(\varepsilon) = \varepsilon\left(\sqrt{b(z)} + O(\varepsilon)\right) = \sqrt{\sigma b(z)} + O(\varepsilon^2), \qquad (58)$$

and, so, we can rewrite Eq. (46) as follows:

$$dz(\tau) = a(z)d\tau + \sqrt{\sigma b(z)}dw(\tau) + O(\varepsilon^2). \qquad (59)$$

It coincides with the Eq. (56) with a precision up to infinitesimal $O(\varepsilon^2)$. So, Lemma 2 is proved.

Suppose that the system is in steady-state regime. We consider the stationary probability density function for the process $z(\tau)$

$$s(z, \tau) = s(z) = \frac{\partial P\{z(\tau) < z\}}{\partial z}. \qquad (60)$$

Theorem 3. *Stationary probability density $s(z)$ of the stochastic process $z(\tau)$ has the form*

$$s(z) = \frac{C}{b(z)}\exp\left\{\frac{2}{\sigma}\int_0^z \frac{a(x)}{b(x)}dx\right\}, \qquad (61)$$

where C is a normalizing constant.

Proof. Because $z(\tau)$ is the solution of the stochastic differential equation (56), the process is diffusion with drift coefficient $a(z)$ and diffusion coefficient $b(z)$. Therefore, its probability density function $s(z)$ is the solution of the Fokker-Planck equation

$$-\frac{\partial}{\partial z}\{a(z)s(z)\} + \frac{1}{2}\frac{\partial^2}{\partial z^2}\{\sigma b(x)s(z)\} = 0. \qquad (62)$$

This equation is an ordinary differential equation of the second order

$$(-a(z)s(z))' + \frac{\sigma}{2}(b(z)s(z))'' = 0,$$

$$-a(z)s(z) + \frac{\sigma}{2}(b(z)s(z))' = 0. \qquad (63)$$

Solving it and taking into account the normalization condition and boundary condition $s(\infty) = 0$, we obtain the probability density function $s(z)$ of the normalized number of calls in the orbit in the following form

$$s(z) = \frac{C}{b(z)}\exp\left\{\frac{2}{\sigma}\int_0^z \frac{a(x)}{b(x)}dx\right\}. \qquad (64)$$

So, Theorem 3 is proved.

7   Approximations Accuracy and Their Application Area

On of the goals our paper is to find an approximation of discrete probability distribution $P(i)$ number of calls in the orbit. Using density function $s(z)$ of the stochastic process $z(\tau)$, we construct an approximation for discrete probability distribution $P(i)$. There are different ways to shift from the density function $s(z)$ of continuous stochastic process $z(\tau)$ to discrete distribution $P(i)$ of the discrete stochastic process $i(\tau)$. We will use the following one.

Taking into account Eq. (31), we write a non-negative function $G(i)$ of the discrete argument i in the form

$$G(i) = \frac{C}{b(\sigma i)} \exp \left\{ \frac{2}{\sigma} \int_0^{\sigma i} \frac{a(x)}{b(x)} dx \right\},   \tag{65}$$

Using the normalization condition, we can write

$$P_1(i) = \frac{G(i)}{\sum\limits_{i=0}^{\infty} G(i)}.   \tag{66}$$

This probability distribution $P_1(i)$ we will use as an approximation for the probability distribution $P(i) = P\{I(t) = i\}$ that the number of calls in the orbit. Also, early we had obtained one more approximation using the classical method of asymptotic analysis [5]. Denote it by $P_2(i)$.

Approximations accuracy will be defined and compare by using Kolmogorov range

$$\Delta_{v=1,2} = \max_{k \geq 0} \left| \sum_{i=0}^{k} (P_v(i) - P(i)) \right|,   \tag{67}$$

where $P(i)$ is an empirical probability distribution of the number i of calls in the orbit obtained by the simulation.

The table contains values for this range for various values of σ and ρ (system load) (Table 1):

$$\rho = \frac{\lambda(\mu_1 + \mu_2)}{\mu_1 \mu_2}.   \tag{68}$$

We consider $\mu_1 = 1$ and $\mu_2 = 2$ for all experiments.

Density diagrams of probability distributions are shown in Figs. 2, 3 and 4. The solid line represents the probability distribution of the number i of calls in the orbit obtained by the simulation, the dotted - approximations obtained by method of asymptotic diffusion and analysis (P1) and by the classical method of asymptotic analysis (P2).

Table 1. Kolmogorov range.

σ	ρ = 0.5	ρ = 0.6	ρ = 0.7	ρ = 0.8	ρ = 0.9	
2	0.061	0.041	0.021	0.006	0.009	Δ_1
	0.094	0.158	0.191	0.258	0.363	Δ_2
1.3	0.049	0.025	0.007	0.017	0.018	Δ_1
	0.101	0.134	0.176	0.224	0.305	Δ_2
0.5	0.011	0.027	0.032	0.030	0.021	Δ_1
	0.142	0.125	0.112	0.146	0.198	Δ_2
0.1	0.033	0.023	0.014	0.012	0.008	Δ_1
	0.071	0.049	0.055	0.071	0.097	Δ_2
0.05	0.019	0.012	0.008	0.035	0.003	Δ_1
	0.034	0.039	0.040	0.036	0.074	Δ_2
0.02	0.013	0.009	0.002	0.011	0.004	Δ_1
	0.022	0.024	0.026	0.031	0.049	Δ_2

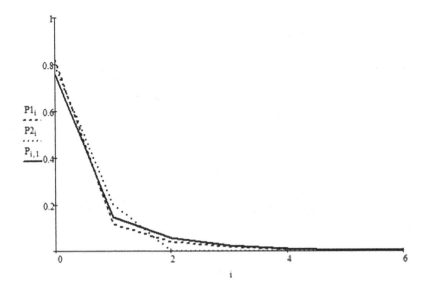

Fig. 2. $\sigma = 2, \rho = 0.5$.

It can be seen, the accuracy of the approximations increases with decreasing parameters ρ and σ. The approximation $P_1(i)$ is applicable for values of $\sigma < 1.3$, where the relative error, in the form of the Kolmogorov distance, does not exceed 0.05. The approximation $P_2(i)$ is applicable for values of $\sigma < 0.02$, that is about 7.5 times less than for approximation obtained through the method of asymptotic diffusion analysis.

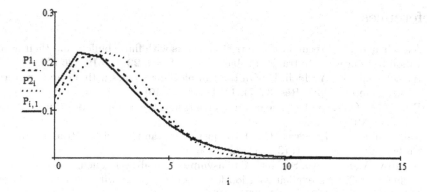

Fig. 3. $\sigma = 0.1, \rho = 0.5$.

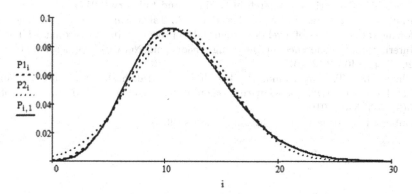

Fig. 4. $\sigma = 0.02, \rho = 0.5$.

8   Conclusion

In this paper, we consider the tandem retrial queueing system with Poisson
arrival process. Using the method of asymptotic diffusion analysis under the
asymptotic condition of the long delay in the orbit, we obtain parameters of the
diffusion process. Probability density distribution of this process has enabled us
to construct an approximation for probability distribution number of calls in the
orbit in the considered RQ-system.

Also we compare the applicability of analytical results obtained by the
method of asymptotic diffusion analysis and by the classical method of asymptotic analysis on the basis of the simulation. It turns out that the method of
asymptotic diffusion analysis is 7.5 times more accurate and can be used for
$\sigma < 1.3$.

References

1. Avrachenkov, K., Yechiali, U.: Retrial networks with finite buffers and their application to internet data traffic. Probab. Eng. Inf. Sci. **22**(4), 519–536 (2008)
2. Avrachenkov, K., Yechiali, U.: On tandem blocking queues with a common retrial queue. Comput. Oper. Res. **37**(7), 1174–1180 (2010)
3. Falin, G.I.: On a tandem queue with retrials and losses. Oper. Res. Int. J. **13**(3), 415–427 (2013)
4. Moutzoukis, E., Langaris, C.: Two queues in tandem with retrial customers. Probab. Eng. Inf. Sci. **15**(3), 311–325 (2001)
5. Nazarov, A., Moiseeva, S.: Method of asymptotic analysis in queuing theory (2006)
6. Phung-Duc, T.: An explicit solution for a tandem queue with retrials and losses. Oper. Res. Int. J. **12**(2), 189–207 (2012)
7. Phung-Duc, T.: Retrial queueing models: a survey on theory and applications. Stochastic Operations Research in Business and Industry (2017)
8. Phung-Duc, T., Masuyama, H., Kasahara, S., Takahashi, Y.: A simple algorithm for the rate matrices of level-dependent QBD processes. In: Proceedings of The 5th International Conference on Queueing Theory and Network Applications (QTNA 2010), pp. 46–52 (2010)
9. Phung-Duc, T., Masuyama, H., Kasahara, S., Takahashi, Y.: State-dependent M/M/c/c+ r retrial queues with Bernoulli abandonment. J. Ind. Manag. Optim. **6**(3), 517–540 (2010)
10. Robert, P.: Stochastic networks and queues (2013)

Queueing Analysis of a Mixed Model of Public and Demand Responsive Transportations

Ayane Nakamura[1], Tuan Phung-Duc[2(✉)], and Hiroyasu Ando[3]

[1] Graduate School of Science and Technology, University of Tsukuba, Tsukuba, Ibaraki 305-8577, Japan
s2020431@s.tsukuba.ac.jp
[2] Faculty of Engineering, Information and Systems, University of Tsukuba, Tsukuba, Ibaraki 305-8577, Japan
tuan@sk.tsukuba.ac.jp
[3] Advanced Institute for Materials Research, Tohoku University, Sendai, Miyagi 980-8577, Japan
hiroyasu.ando.d1@tohoku.ac.jp

Abstract. In this research, we discuss Car/Ride-Share (CRS), which is a novel concept of Demand Responsive Transportation (DRT) aiming at reducing the uneven distribution of cars in traditional carsharing service and the congestion of cars and people. We model a scenario where CRS service is introduced between a spot (e.g., university, company, etc.) and its nearest train station by a bus company using queueing theory. Then, using an approximation model, we derive the probability density function of the required time (the sum of waiting and traveling times) for the customers considering the road congestion. Further, we show some numerical results of the distribution of the required time and discuss the effectiveness of CRS depending on road congestion. We confirm that the excessive introduction of CRS may be ineffective when the road is congested from the perspective of the required time.

Keywords: Transportation systems · Queueing theory · Road congestion

1 Introduction

In this research, we discuss Car/Ride-Share (CRS), which may be a new type of demand responsive transportation (DRT) with good social impacts (see the detailed explanation in Sect. 2). We consider a queueing model of CRS and discuss the effectiveness of CRS in this paper.

CRS is defined as a system where people carry out carsharing (i.e., the car rental for short periods) and ridesharing (e.g., the system where people ride a car together to the destination) simultaneously using private cars [1]. We consider a scenario where a bus company itself introduces CRS between a train station and

© Springer Nature Switzerland AG 2021
P. Ballarini et al. (Eds.): EPEW 2021/ASMTA 2021, LNCS 13104, pp. 457–471, 2021.
https://doi.org/10.1007/978-3-030-91825-5_28

its nearest spot (e.g., university, company) where bus transportation services already exist for both directions and reduces the number of the buses. This CRS system has the following three features [1].

1. Owners of private cars can get financial incentives by sharing their cars.
2. People can carry out carsharing and ridesharing simultaneously, so that the disadvantage of conventional carsharing such as uneven distribution of cars does not occur, i.e., the operator does not have to redistribute the cars.
3. It might be an alternative transportation service with less financial and time burden for existing transportations (bus in this study) in the case of congestion.

About CRS, Ando et al. [1] conducted simulation experiments and showed the decrease of the mean waiting time for customers. Besides, Nakamura et al. [2–4] modeled CRS between a university and a station where a bus transportation already existed using queueing theory and discussed the characteristics of the system. Furthermore, [3,4] considered various scenarios of the price mechanism of CRS, where a third organization or a bus company introduces this service. However, we put several assumptions, for example, Poisson arrival of buses, the state of the station side is the number of the demands of CRS; the buses from the station are always full because of the congestion, and the customers arrive one by one to simplify the model. Besides, we did not incorporate the state of the road between the two points to our model. We did not consider the possibility that the occurrences of CRS cause the road congestion, which is not a good situation for the customers.

Based on the above, we propose analyzing a queueing model of CRS with the following conditions and execute the Monte Carlo simulation aiming at a more realistic discussion in this research.

1. Inter-arrivals of buses are independently and identically distributed (I.I.D) according to Erlang distribution which can be used to approximate the fixed interval (this enables us to discuss the influence of the uncertainty of bus inter-arrivals to the system by adjusting the variance of the distribution).
2. We define the state of the station as the number of people, and do not assume that the buses from the station are always full as in [2–4].
3. The customers at the station side arrive in groups (imaging that people got off the train arrive at all once).

As an extension of [2–4], we incorporated road congestion into our model in [5]. Nevertheless, we only discussed the mean required time (the sum of the waiting time and the traveling time) for customers considering the road congestion, and we did not discuss the distribution of the required time in [5]. In this research, by analyzing a queueing model, we derive the distribution of the required time for the customers in the CRS system using the theory of phase-type distribution. It should be noted that this research considers the distribution of the required time. In contrast, all previous studies [2–5] only considered the mean required time of customers, and thus this paper enables more detailed discussions.

Here, we introduce several related works using some mathematical methods. There are many studies using optimization theory to operate shared transportation properly [6, 7]. However, most of these optimization models assume that the travel demands are known in advance and do not take into account the uncertainty in traffic (e.g., traffic delay due to an accident, the time it takes customers to get on and off the vehicles, fluctuations in traffic demand due to exogenous factors).

As the latest studies using queueing theory, Shuang et al. [8, 9] considered bike-sharing queueing models. However, they assumed that customers arrive one by one according to a Poisson process and also assumed that the travel time of bike-sharing simply follows an exponential distribution, which means the possibility that the occurrence of the bike-sharing induces road congestion (i.e., the travel time depends on the number of the occurrences of the bike-sharing) is not considered.

From a more general perspective on transportation, Daganzo et al. [10] constructed a simple stochastic model of demand-responsive transportation services, including non-shared taxi, dial-a-ride, and ridesharing. They also compared the existing urban transportation modes in scenarios involving different city types and levels of demand. However, the crucial point that must be further studied in [10] is to consider how multiple transportation systems can work together because there is not just one mobility service at a time like the model in [10] in the real world.

Compared to the above previous research, the novelties of our research of CRS are as follows:

I. Proposing a new concept of transportation, in other words, a fusion of multiple shared transportation services (i.e., carsharing and ridesharing).
II. Solving the hassle for the redistribution of cars in carsharing by considering the matching trip demand in the opposite direction, not by a conventional optimization method.
III. Constructing a stochastic model of CRS (note that this approach does not cost money, unlike social experiments), which enables us to consider the uncertainty of people and traffic in this system by conducting numerical experiments in various settings.
IV. Incorporating more realistic and complicated elements into the model compared to the previous work [8–10], considering the batch arrival of customers, the road congestion, and the coexistence of multiple transportation services.

The rest of the paper is structured as follows. In Sect. 2, we state the detailed mechanism of CRS using queueing theory. In Sect. 3, we present a queueing model for CRS. Section 4 shows the derivation of the distribution of the required time for the customers in the CRS system. Section 5 presents some numerical results for the discussion of Sects. 4. Finally, we present concluding remarks in Sect. 6.

2   The CRS System

In this section, we describe the model of CRS.

2.1   Queueing Model

We model a scenario in which CRS is introduced by a bus company between a train station and a spot (e.g., university, company, etc.) where bus transportation has been existing already, and we also assume that there is a parking lot of cars to carry out CRS at the spot side (see Fig. 1).

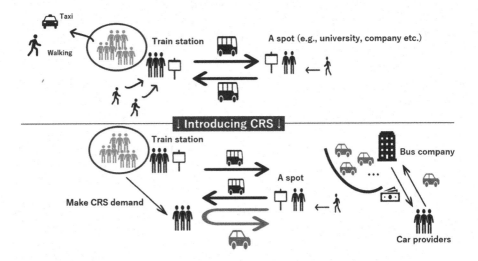

Fig. 1. The schematic illustration of the CRS system.

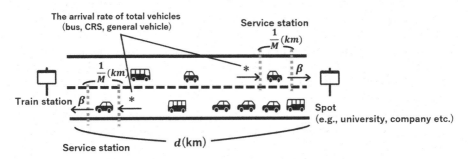

Fig. 2. The modelling of the road.

Table 1. The parameters used in the CRS model.

Parameters	Definitions
λ (people/h)	Arrival rate of customers at the spot
σ (veh/h)	Occurrence rate of CRS
δ (group/h)	Arrival rate of a group of customers at the station
$q_{1(2)}$, $r_{1(2)}$	Shape and Rate parameters of the Erlang distribution for the departure interval of buses from the spot (station)
l (people)	Capacity of buses
m (people)	Minimum number of passengers of cars
n (people)	Maximum number of passengers of cars
K (people)	Maximum number of customers that can exist at the station
M (veh/km)	Maximum traffic density
$\epsilon_{us(su)}$ (veh/h)	Arrival rate of general cars at the service station from the spot to the station (from the station to the spot)
$\alpha_{us(su)}$ (veh/h)	Total arrival rate at the service station on the road from the spot to the station (from the station to the spot). $\rightarrow$ in Sect. 3.2
β (veh/h)	Service rate at the service stations
SN (km/h)	Nominal speed of cars
d (km)	Distance between the spot and the station
s_{us} (km/h)	Effective speed of cars on the road from the spot to the station
s_{su} (km/h)	Effective speed of cars on the road from the station to the spot

As a premise, people who came to the spot by their private cars provide these cars as CRS cars (simply denoted by car) for receiving some financial incentives. Recently, peer-to-peer carsharing [11], where people lend their private cars for some incentives, is becoming popular, and we incorporate this system into our model simply. This paper does not consider the insurance implication and other realistic conditions to simplify the model.

We assume that the minimum and the maximum numbers of passengers for a car are m and n, respectively. Customers can use these cars to move between the spot and the station, but these cars must be returned to the spot.

At the station, groups of customers arrive according to a Poisson process with a rate δ, and the number of people for a group (e.g., people who got off a train arrives at all the once) follows an arbitrary distribution. We define the batch size (i.e., the number of people in a group) X follows an arbitrary distribution, and write as

$$x_k = P(X = k),$$

for $k \in \mathbb{N}$.

To make it easier to analyze, we assume that it is acceptable to have up to K people at the station at the same moment. If there are already K people at the station side, new visitors are blocked and go to the spot by other means of transportation, such as a taxi. We also assume that if the number of free spaces is less than the size of the arriving group at the station, the exceeded number of customers are blocked under the assumption that the probability that each person in the group is blocked is identical. On the spot side, customers arrive according to a Poisson process with rate a λ. Here, we further assume that all the customers have a driving license and do not prefer whether they use the bus or CRS to simplify the discussion.

Buses depart from the spot (the station) to the station (the spot) at intervals following Erlang distribution with rate $r_{1(2)}$ and shape $q_{1(2)}$ (this means the sum of $q_{1(2)}$ exponentially distributed random variables of parameter $r_{1(2)}$). We assume that the capacity of a bus is l, and the spot and the station are the first bus stops, i.e., a bus is empty upon arrival. People lined up at the spot continuously get on a bus on a first-come, first-served basis within the limit of the capacity of the bus.

CRS occurs from the spot at intervals following the exponential distribution with parameter σ when both the spot and the station are crowded above a certain level i.e., when there are more than m people are on the both sides. Here, m is the minimum number of passengers for a car, as mentioned above. In the same way as buses, people get on a car on a first-come, first-served basis within the capacity limit. The interval of the occurrences of CRS follows an exponential distribution because it takes for a group of passengers to reach the point where the car exists in the parking lot and actually starts driving may vary. Furthermore, it takes some matching time for the group of customers in the spot and that in the station. Besides, we assume that the customers who use CRS at the station exit the queue at the moment of the departure of the car (i.e., the car which they will take over) at the spot and that they wait for the arrival of the car at the waiting space for CRS.

A car departing at the spot goes to the station, and after arriving at the station, those in the car get off. Then, people waiting at the station take over the car and drive the CRS to the spot. After the car arrives at the spot, people who got on at the station get off, and the car is returned to its original position. This series of flows, can be considered to meet the demand on the station side under the condition that the cars return to their original point (the spot).

The movement on the road between two points is modeled as shown in Fig. 2 regarding the method of Vandaele et al. [12]. In this method, a part of the road between two points is considered a service station (i.e., a queue with single server), and the service interval follows an exponential distribution of the parameter β. The parameter β of this exponential distribution is derived as β (number of vehicles/unit time) $= SN$ (km/unit time) M (number of vehicles/km). We also assume that other vehicles (defined as general vehicles) except for the buses and CRS cars arrive at the service station according to a Poisson process with parameter $\epsilon_{us(su)}$.

Using the mean sojourn time in the system, the maximum traffic density of the road M (the inverse of M is considered to be the size of the service station), and the distance of the road between the two points d, we can derive the relative speed of the vehicle on the road $s_{us(su)}$ (see [12] for details). Then, we can determine the travel time required for the vehicles on the road using $s_{us(su)}$.

2.2   The Difficulty of this Model

In this model, the travel time of the vehicles on the road is derived by dividing the distance of the road by the relative speed, so it has a constant distribution. As mentioned above, customers always transfer the cars of CRS at the station in this system. Therefore, to understand the exact arrival process of vehicles (bus, car, and general vehicle) at the service station from the station to the spot, it is necessary to keep track the timing when the CRS cars depart from the spot and the travel times required to reach the station side in the model. As a result, this model does not become a simple Markov chain, and it is not easy to analyze the entire system in a simple form. Based on the above, we consider an approximate model that decomposes the whole system into two queueing models, i.e., vehicle model and road model.

3   Approximation Model

This section describes the outline of two queueing models, i.e., vehicle model and road model. We decompose the whole system into these two models, and we use the results of the vehicle model to calculation of the road model approximately.

3.1   Vehicle Model

The vehicle model expresses how buses depart from both points and how cars of CRS depart from the spot. Thereby, this model also expresses the number of people waiting at both points. The vehicle model can be formulated using a GI/M/1-type Markov chain. Let $\mathbb{Z}_+, I, R_1, R_2$, and S denote $\mathbb{Z}_+ = \{0, 1, 2, \dots\}$, $I = \{0, 1, 2, \dots, K\}$, $R_1 = \{0, 1, 2, \dots, r_1 - 1\}$, $R_2 = \{0, 1, 2, \dots, r_2 - 1\}$ and $S = \mathbb{Z}_+ \times I \times R_1 \times R_2$, respectively. Then let $N(t), I(t), R_1(t)$, and $R_2(t)$ express the number of the waiting people at the spot, the number of the waiting people at the station, the progress of the Erlang distribution for the buses from the spot and that for the buses from the station at time t, respectively. It is easy to see that $\{(N(t), I(t), R_1(t), R_2(t)); t \geq 0\}$ forms a Markov chain in the state space S. Our Markov chain is of GI/M/1-type, where $N(t)$ is the level and $\{(I(t), R_1(t), R_2(t))\}$ is the phase.

Assuming that the Markov chain is stable, we define the steady state probabilities as follows:

$$\pi_{(j,\xi,\psi,\omega)} = \lim_{t \to \infty} P(N(t) = j, I(t) = \xi, R_1(t) = \psi, R_2(t) = \omega),$$

where, $j \in \mathbb{Z}_+, \xi \in I, \psi \in R_1, \omega \in R_2$. We can obtain the values of these probabilities and the stability condition using an existing method [13]. Based on the steady state probability, we define some performance measures in Table 1. In Table 1, the throughput is counted by the number of people served per unit time. We omit the detailed derivation of the queueing analysis due to space limitations.

Table 2. Performance measures of the vehicle model.

Parameters	Definitions
$E[W_u]$	Mean waiting time for customers at the spot
$E[W_s]$	Mean waiting time for customers at the station
$E[C]$	Number of occurrences of CRS per unit time
$T_{total(s)}$	Total throughput from the station
$T_{CRS(s)}$	Throughput by CRS from the station

3.2 Road Model

The road model expresses the state of the road, i.e., the road congestion. As described in Sect. 2.2, it is difficult to grasp the exact arrival processes at the service stations. Therefore, we assume that the arrival processes of the service stations follow the Poisson process with rates

$$\alpha_{us} = \frac{q_1}{r_1} + E[C] + \epsilon_{us}, \qquad \alpha_{su} = \frac{q_2}{r_2} + E[C] + \epsilon_{su}.$$

Here, we assume that the arrival rates of the service stations are the sum of the expected values for the numbers of the buses and CRS, and the arrival rate of general vehicles per unit time. In other words, we use the performance measures of the vehicle model to approximate parameters of the road model.

By these approximations, the service stations become M/M/1 queues, and we can derive the mean sojourn times for vehicles easily as follows:

$$E[W_{us}] = \frac{1}{\beta - \alpha_{us}}, \qquad E[W_{su}] = \frac{1}{\beta - \alpha_{su}}.$$

Then, we derive the effective speeds s_{us} and s_{su} (i.e., the mean speeds of cars on the road) as follows [12]:

$$s_{us} = \frac{1}{E[W_{us}] \times M}, \qquad s_{su} = \frac{1}{E[W_{us}] \times M}.$$

From the above results, the mean of the times for a vehicle to travel from the spot to the station and vice versa $E[R_{us}]$ and $E[R_{su}]$ can be derived as

$$E[R_{us}] = \frac{d}{s_{us}}, \qquad E[R_{su}] = \frac{d}{s_{su}}.$$

Besides, we can calculate the mean of the total required time for customers from the spot to the station (i.e., the mean time from when a customer arrives at the spot to when he arrives at the station) and also vice versa $E[A_{us}]$ and $E[A_{su}]$ as follows. Note that we obtain the values of $E[W_u]$ and $E[W_s]$ from the vehicle model (see Table 2).

$$E[A_{us}] = E[W_u] + E[R_{us}], \qquad E[A_{su}] = E[W_s] + E[R_{su}] + P_{CRS}E[R_{us}],$$

where P_{CRS} is the rate of customers from the station to the spot who use CRS as

$$P_{CRS} = \frac{T_{CRS(s)}}{T_{total(s)}}.$$

Here, note that the cars of CRS are located at the spot; as a result, only customers who use CRS from the station have the waiting time for the car from the spot to the station to come before they ride on it.

4 The Required Time Distribution at the Spot

In this section, we present the derivation of the distribution of required time for the customers at the spot using the existing theory of phase-type distribution. Let $\mathbb{Z}_+^*$ denote $\mathbb{Z}_+^* = \{0, 1, 2, \ldots \varphi\}$ and $S^* = \mathbb{Z}_+^* \times \mathbb{Z}_+^* \times I \times R_1 \times R_2$, where

$$\varphi := \inf\{n \in \mathbb{Z}_+ | 1 - \sum_{j=0}^{n} \|\pi_j\| < \epsilon\}, \tag{1}$$

and $\pi_j = (\pi_{(j,0,0,0)}, \pi_{(j,0,0,1)}, \pi_{(j,0,0,2)}, \ldots, \pi_{(j,K,r_1-1,r_2-1)})$, $\|\pi_j\| = \sum_{\xi=0}^{K} \sum_{\psi=0}^{r_1-1} \sum_{\omega=0}^{r_2-1} \pi_{(j,\xi,\psi,\omega)}$ and ϵ is an extremely small value.

We also assume that Q^* is the square matrix truncated Q at level φ. Then, we can compute the steady state probabilities $\pi^*(= (\pi_{(0,0,0,0)}^*, \pi_{(0,0,0,1)}^*, \cdots, \pi_{(\varphi,K,r_1-1,r_2-1)}^*))$ such that $\pi^* Q^* = 0$ with $\pi^* e = 1$. Letting $X(t)$ denote the number of people lined up before a tagged customer (including the tagged customer) at the spot and $N^*(t)$ denote the number of the customers at the spot, respectively, $\{(X(t), N^*(t), I(t), R_1(t), R_2(t)); t \geq 0\}$ forms an absorbed Markov process in the set space S^* with the absorption state as $X(t) = 0$. The infinitesimal generator of this absorbed Markov chain is written as

$$\overline{Q} = \begin{pmatrix} 0 & 0 \\ \widehat{q} & \widehat{Q} \end{pmatrix},$$

where the contents of $\widehat{q}$ and $\widehat{Q}$ are defined in Appendix. Besides, we define the initial distribution a^* of the absorbed Markov chain as

$$a^* = (a_0, a_1, \ldots, a_\varphi) = (a_0, \dot{a}),$$

where $a_0 = 0, \dot{a} = (a_1, \ldots, a_\varphi), a_j = (a_{(j,0,0,0,0)}, a_{(j,0,0,0,1)}, \cdots, a_{(j,\varphi,K,r_1-1,r_2-1)})$.

Then, we can write the distribution function $\widetilde{F}(t)$ and the probability density function $\widetilde{f}(t)$ of the waiting time for a tagged customer as follows:

$$\widetilde{F}(t) = 1 - \dot{a}\exp(\widehat{Q}t)e, \qquad t \geq 0,$$

$$\widetilde{f}(t) = \widetilde{F}'(t) = \dot{a}\exp(\widehat{Q}t)\widehat{q}, \qquad t \geq 0.$$

Letting $\dot{b}$ denote

$$\dot{b}_{(\zeta,\theta,\xi,\psi,\omega)} = (\dot{b}_{(1,0,0,0,0)}, \dot{b}_{(1,0,0,0,0)}, \dot{b}_{(1,0,0,0,0)}, \ldots, \dot{b}_{(\varphi,\varphi,K,r_1-1,r_2-1)}),$$

where, $\dot{b}_{(\dot{\zeta},\zeta,\dot{\xi},\dot{\psi},\dot{\omega})} = 1$ for $\dot{\zeta} \in X \setminus \{0\}, \dot{\xi} \in I, \dot{\psi} \in R_1, \dot{\omega} \in R_2$, 0 for the other elements, the distribution function $\overline{F}(t)$ and the probability density function $\overline{f}(t)$ of the waiting time for any customer are written as follows:

$$\overline{F}(t) = \sum_{\dot{\zeta}=1}^{\varphi} \sum_{\dot{\xi}=0}^{K} \sum_{\dot{\psi}=0}^{r_1-1} \sum_{\dot{\omega}=0}^{r_2-1} \pi^*_{(\dot{\zeta}-1,\dot{\xi},\dot{\psi},\dot{\omega})} \{1 - \dot{b}_{(\dot{\zeta},\zeta,\dot{\xi},\dot{\psi},\dot{\omega})} \exp(\widehat{Q}t)e\}, \qquad t \geq 0,$$

$$\overline{f}(t) = \sum_{\dot{\zeta}=1}^{\varphi} \sum_{\dot{\xi}=0}^{K} \sum_{\dot{\psi}=0}^{r_1-1} \sum_{\dot{\omega}=0}^{r_2-1} \pi^*_{(\dot{\zeta}-1,\dot{\xi},\dot{\psi},\dot{\omega})} \{\dot{b}_{(\dot{\zeta},\zeta,\dot{\xi},\dot{\psi},\dot{\omega})} \exp(\widehat{Q}t)\widehat{q}\}, \qquad t \geq 0.$$

Here, we derive the distribution of the required time (the sum of the waiting time and the travel time) for any customer at the spot. Considering R_{us} is constant for all the vehicles on the road, the distribution function $F(t)$ and the probability density function $f(t)$ of the required time for any customer at the spot are written as

$$F(t) = \overline{F}(t - R_{us}), \qquad t \geq 0,$$
$$f(t) = \overline{f}(t - R_{us}), \qquad t \geq 0.$$

When we conduct numerical experiments, we apply the method of normalization for the matrix $\widehat{Q}$. Letting $\widehat{q}_{ii}$ denote the diagonal elements of $\widehat{Q}$, we can define $\widehat{P}$ as

$$\widehat{P} = I + \frac{1}{\widehat{\nu}}\widehat{Q},$$

where, $\widehat{\nu} = \sup_{0 \leq i \leq \varphi}\{-\widehat{q}_{ii}\}$.

Therefore, we can rewrite $\overline{f}(t)$ as

$$\overline{f}(t) = \sum_{\dot{\zeta}=1}^{\varphi} \sum_{\dot{\xi}=0}^{K} \sum_{\dot{\psi}=0}^{r_1-1} \sum_{\dot{\omega}=0}^{r_2-1} \pi^*_{(\dot{\zeta}-1,\dot{\xi},\dot{\psi},\dot{\omega})} \{\dot{b}_{(\dot{\zeta},\zeta,\dot{\xi},\dot{\psi},\dot{\omega})} e^{-\widehat{\nu}t} \sum_{n=0}^{\upsilon} \frac{(\widehat{\nu}t)^n}{n!} \widehat{P}^n \widehat{q}\}, \qquad t \geq 0,$$

where υ is defined as

$$\upsilon := \inf\{k \in \mathbb{Z}_+ | 1 - \sum_{n=0}^{k} e^{-\widehat{\nu}t} \frac{(\widehat{\nu}t)^n}{n!} < \epsilon\}. \qquad (2)$$

The distribution of the required time for any customer at the station can be derived in the same way. However, we omit it due to space limitations.

5   Numerical Experiments

In this section, we show some numerical results for the probability density function and the distribution function of the waiting time, and the required time of any customer at the spot, which are derived in the previous section. At first, we show the results of the comparison of the numerical experiment results for an approximation model (i.e., the formulae derived in Sect. 4) and the simulation results in Fig. 3 ($\overline{f}(t)$) and Fig. 4 ($f(t)$). Here, we should note that the effective speed of cars at the road s_{us} is determined by reflecting the exact arrival rates of cars at the service station considering the whole system as one queueing network. We set the parameters as $\lambda = 10, \sigma = 30, \delta = 100, m = 2, n = 4, l = 10, K = 10, q_1 = 10, r_1 = 1, q_2 = 10, r_2 = 1, \epsilon_{us} = 10000, SN = 60$, and $M = 200$ for both theoretical analysis and simulation. We also set $\epsilon = 10^{-5}$ in both (1) and (2) for the approximation model. From the results, we can consider that the approximation model can explain a similar tendency. More concretely, we show the 20 times simulation results for the required time distribution function ($F(t)$) in Table 3. The parameter setting is the same as in Fig. 3 and Fig. 4. We can understand that the approximation model can show a similar tendency as the simulation results, although the approximation model evaluates the required time a little longer. That means that the approximation shows safer results for the required time.

Next, we show the numerical results of $f(t)$ in Sect. 3 for two scenarios; the case where the road is not congested (Fig. 5) and the road is congested (Fig. 6). Note that we can change how congested the road is by adjusting the values of ϵ_{us} (the arrival rate of general cars at the service station from the spot to the station) properly. In these graphs, we set $\epsilon_{us} = 10000$ in Fig. 5 and $\epsilon_{us} = 11950$ in Fig. 6. Other parameter settings are the same as in Fig. 3 and Fig. 4.

It turns out that the required time becomes smaller as σ (the occurrence rate of CRS) becomes larger in Fig. 5 (i.e., the introduction of CRS is effective from the perspective of the required time for the customers). On the other hand, interestingly the opposite trend is observed in Fig. 6. That means that the introduction of CRS accelerates the road congestion and makes the required time longer as a result in the case that the road is already congested. To summarize, according to the results in Figs. 5 and 6, our analysis implies that CRS should be actively introduced when the road is not so congested. However, it turned out that the operator has to be careful about introducing CRS when the road is congested.

Table 3. Numerical results of the distribution of the required time of customers $F(t)$.

t		0.30	0.35	0.40	0.45
Approximation		0.5677	0.8273	0.933	0.9744
Simulation	(Mean)	0.6412	0.8607	0.9451	0.9823
	(Standard deviation)	0.1027	0.0306	0.0123	0.0007
	(95% Confidence interval lower limit)	0.5962	0.8472	0.9398	0.9794
	(95% Confidence interval upper limit)	0.6861	0.8741	0.9505	0.0852

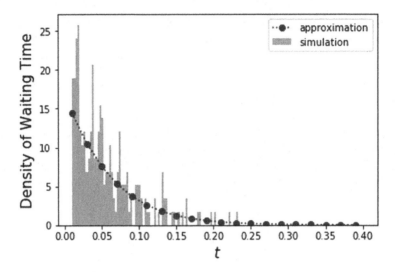

Fig. 3. The density of the waiting time of customers $\overline{f}(t)$ (approximation model and simulation).

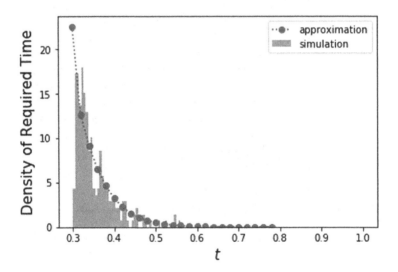

Fig. 4. The density of the required time of customers $f(t)$ (approximation model and simulation).

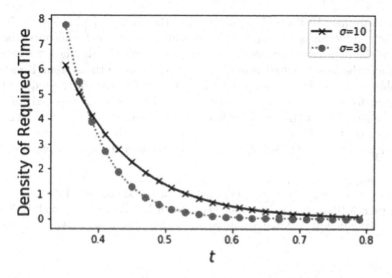

Fig. 5. The density of the required time of customers $f(t)$ (the case where the road is not congested).

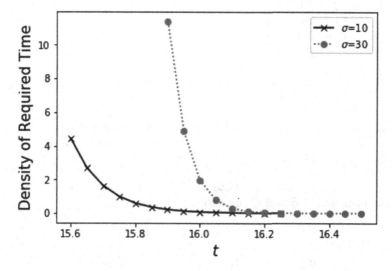

Fig. 6. The density of the required time of customers $f(t)$ (the case where the road is congested).

6 Conclusion

In this paper, we have considered the CRS system considering road congestion using queueing theory. We have analyzed an approximation model which decomposes the whole system into two models; the vehicle model and the road model. Further, we have derived the probability density function of the required time for the customers in the CRS system using the theory of phase-type distribution. By conducting some numerical experiments, we have confirmed that the result of the approximation model generally matches the simulation result that reproduces the actual behavior of the cars on the road in the system. Besides, we have shown the results of the probability density function of the required time for the customers for both cases where the road is not congested and congested and have shown that the introduction of CRS would not be practical when the road is congested. These results have implied that the operator should decide whether to introduce CRS depending on the state of the road congestion.

Acknowledgments. This work is partly supported by JSPS KAKENHI Nos. 19K12198, 18K18006, 21K11765, JST MIRAI No. JPMJMI19B. This study is supported by F-MIRAI: R&D Center for Frontiers of MIRAI in Policy and Technology, the University of Tsukuba and Toyota Motor Corporation collaborative R&D center.

Appendix

We define the elements of $\widehat{q}$ and $\widehat{Q}$ for $\zeta \in X \setminus \{0\}, \theta \in \mathbb{Z}_+^*, \xi \in I, \psi \in R_1, \omega \in R_2$. Here, we write $\widehat{q}$ and $\widehat{Q}$ as

$$
\widehat{q} = \begin{pmatrix} \widehat{q}_{(1,0,0,0,0)} \\ \widehat{q}_{(1,0,0,0,1)} \\ \widehat{q}_{(1,0,0,0,2)} \\ \vdots \\ \widehat{q}_{(\varphi,\varphi,K,r_1-1,r_2-1)} \end{pmatrix},
$$

$$
\widehat{Q} =
\begin{pmatrix}
\widehat{Q}_{(1,0,0,0,0),(1,0,0,0)} & \widehat{Q}_{(1,0,0,0,0),(1,0,0,1)} & \widehat{Q}_{(1,0,0,0,0),(1,0,0,2)} & \cdots & \widehat{Q}_{(1,0,0,0,0),(\varphi,K,r_1-1,r_2-1)} \\
\widehat{Q}_{(1,0,0,0,1),(1,0,0,0)} & \widehat{Q}_{(1,0,0,0,1),(1,0,0,1)} & \widehat{Q}_{(1,0,0,0,1),(1,0,0,2)} & \cdots & \widehat{Q}_{(1,0,0,0,1),(\varphi,K,r_1-1,r_2-1)} \\
\widehat{Q}_{(1,0,0,0,2),(1,0,0,0)} & \widehat{Q}_{(1,0,0,0,2),(1,0,0,1)} & \widehat{Q}_{(1,0,0,0,2),(1,0,0,2)} & \cdots & \widehat{Q}_{(1,0,0,0,2),(\varphi,K,r_1-1,r_2-1)} \\
\vdots & \ddots & \ddots & \ddots & \vdots \\
\widehat{Q}_{(\varphi,\varphi,K,r_1-1,r_2-1),(1,0,0,0)} & \widehat{Q}_{(\varphi,\varphi,K,r_1-1,r_2-1),(1,0,0,1)} & \widehat{Q}_{(\varphi,\varphi,K,r_1-1,r_2-1),(1,0,0,2)} & \cdots & \widehat{Q}_{(\varphi,\varphi,K,r_1-1,r_2-1),(\varphi,K,r_1-1,r_2-1)}
\end{pmatrix}.
$$

Then, the elements of $\widehat{q}$ and $\widehat{Q}$ are defined as follows:

$$
\widehat{q}_{(\zeta,\theta,\xi,r_1-1,\omega)} = q_1, \qquad \zeta \leqq l,
$$

$$
\widehat{q}_{(\zeta,\theta,\xi,\psi,\omega)} = \sigma, \qquad \zeta \leqq n, \theta \geqq m, \xi \geqq m,
$$

$$
\widehat{Q}_{(\zeta,\theta,\xi,\psi,\omega),(\zeta,\theta+1,\xi,\psi,\omega)} = \lambda, \qquad \theta \leqq \varphi - 1,
$$

$$
\widehat{Q}_{(\zeta,\theta,\xi,\psi,\omega),(\zeta,\theta,\xi',\psi,\omega)} = \delta x_{\xi'-\xi}, \qquad \xi + 1 \leqq \xi' \leqq K - 1,
$$

$$
\widehat{Q}_{(\zeta,\theta,\xi,\psi,\omega),(\zeta,\theta,K,\psi,\omega)} = \delta \sum_{k=K-\xi}^{\infty} x_k, \qquad 0 \leqq \xi \leqq K - 1,
$$

$$
\widehat{Q}_{(\zeta,\theta,\xi,\psi,\omega),(\zeta,\theta,\xi,\psi+1,\omega)} = q_1, \qquad 0 \leqq \psi \leqq r_1 - 2,
$$

$$
\widehat{Q}_{(\zeta,\theta,\xi,\psi,\omega),(\zeta,\theta,\xi,\psi,\omega+1)} = q_2, \qquad 0 \leqq \omega \leqq r_2 - 2,
$$

$$\widehat{Q}_{(\zeta,\theta,\xi,r_1-1,\omega),(\zeta-l,\theta-l,\xi,0,\omega)} = q_1, \qquad \zeta \geqq l+1,$$

$$\widehat{Q}_{(\zeta,\theta,\xi,\psi,r_2-1),(\zeta,\theta,\xi-l,\psi,0)} = q_2, \qquad \xi \geqq l+1,$$

$$\widehat{Q}_{(\zeta,\theta,\xi,\psi,r_2-1),(\zeta,\theta,0,\psi,0)} = q_2(1 - \delta_{(\xi,q_2),(0,0)}), \qquad \xi \leqq l,$$

$$\widehat{Q}_{(\zeta,\theta,\xi,\psi,\omega),(\zeta-n,\theta-n,\xi-min(n,\xi),\psi,\omega)} = \sigma, \qquad \zeta \geqq n+1, \theta \geqq m, \xi \geqq m,$$

$$\widehat{Q}_{(\zeta,\theta,\xi,\psi,\omega),(\zeta,\theta,\xi,\psi,\omega)} = -\Big(\sum_{\substack{(\zeta',\theta',\xi',\psi',\omega')\in X\setminus\{0\}\times\mathbb{Z}_+^*\times I\times R_1\times R_2 \\ (\zeta',\theta',\xi',\psi',\omega')\neq(\zeta,\theta,\xi,\psi,\omega)}} \widehat{Q}_{(\zeta,\theta,\xi,\psi,\omega),(\zeta',\theta',\xi',\psi',\omega')}$$

$$+ \sum_{(\theta',\xi',\psi',\omega')\in\mathbb{Z}_+^*\times I\times R_1\times R_2} \widehat{q}_{(\zeta,\theta,\xi,\psi,\omega),(\theta',\xi',\psi',\omega')} \Big),$$

where, $\delta_{(\xi,q_2),(0,0)} = 1$ for $(\xi, q_2) = (0,0)$ and 0 otherwise.

References

1. Ando, H., Takahara, I., Osawa, Y.: Mobility services in university campus. Commun. Oper. Res. Soc. Japan **64**(8), 447–452 (2019). in Japanese
2. Nakamura, A., Phung-Duc, T., Ando, H.: Queueing analysis for a mixed model of carsharing and ridesharing. In: Gribaudo, M., Sopin, E., Kochetkova, I. (eds.) ASMTA 2019. LNCS, vol. 12023, pp. 42–56. Springer, Cham (2020). https://doi.org/10.1007/978-3-030-62885-7_4
3. Nakamura, A., Phung-Duc, T., Ando, H.: Queueing analysis of a car/ride-share system. Ann. Oper. Res. (2021, to appear). https://doi.org/10.1007/s10479-021-04313-8
4. Nakamura, A., Phung-Duc, T., Ando, H.: A stochastic model for car/ride-share service by a bus company. In: Proceedings of The 2020 International Symposium on Nonlinear Theory and Its Applications (NOLTA2020), pp. 312–315 (2020)
5. Nakamura, A., Phung-Duc, T., Ando, H.: A stochastic analysis and price mechanism of car/ride-share system considering road congestion (submitted)
6. Boyaci, B., Zografos, K.G., Geroliminis, N.: An integrated optimization-simulation framework for vehicle and personnel relocations of electric carsharing systems with reservations. Transp. Res. Part B: Methodol. **95**, 214–237 (2017)
7. Agatz, N., Erera, A., Savelsbergh, M., Wang, X.: Optimization for dynamic ridesharing: a review. Eur. J. Oper. Res. **223**(2), 295–303 (2012)
8. Tao, S., Pender, J.: A stochastic analysis of bike sharing systems. Probab. Eng. Inf. Sci. 1–58 (2020)
9. Tao, S., Pender, J.: The impact of smartphone apps on bike sharing systems. SSRN 3582275 (2020)
10. Daganzo, C.F., Ouyang, Y.: A general model of demand-responsive transportation services: from taxi to ride sharing to dial-a-ride. Transp. Res. Part B: Methodol. **126**, 213–224 (2019)
11. Hampshire, R.C., Gaites, C.: Peer-to-peer carsharing: market analysis and potential growth. Transp. Res. Rec. **2217**(1), 119–126 (2011)
12. Vandaele, N., Van Woensel, T., Verbruggen, A.: A queueing based traffic flow model. Transp. Res. Part D: Transp. Environ. **5**(2), 121–135 (2000)
13. Adan, I., Leeuwaarden, V.J., Selen, J.: Analysis of structured Markov processes. arXiv:1709.09060v1 (2017)

An Analytical Framework for Video Quality and Excess Data Distribution in Multiple-Quality Video Under Dynamic Channel Conditions

Mehmet Akif Yazici$^{(\boxtimes)}$ (iD)

Information and Communications Research Group (ICRG), Institute of Informatics,
Istanbul Technical University, Istanbul, Turkey
`yazicima@itu.edu.tr`

Abstract. We present an analytical framework to obtain the distribution of frozen, low quality, and high quality video in a setting with two video quality levels where the channel is dynamic and the data rate the user can achieve varies with time. The presented model, which is based on multi-regime Markov fluid queues, is also capable of producing the distribution of the excess data present in the playout buffer at the end of the video session duration, which will be wasted. The playout control is assumed to be hysteretic, and the effects of the values of thresholds selected for starting playout, switching to low/high quality levels, and pausing/resuming download on the distribution of video quality and excess data is investigated. The presented model can be extended to quality levels more than two.

Keywords: Quality of experience · Markov fluid queue · Video · Wasted video data

1 Introduction

Video content remains one of the significant components of Internet traffic [1]. In the abundance of content, the quality of experience (QoE) a user sees is a decisive factor in the user's behavior in watching the content. Among the objective QoE metrics that can be computed and used as indicators of subjective QoE metrics are video freeze probability, video quality and the share of time a user sees high quality content. In this study, we present an analytical framework for evaluating the distribution of frozen, low quality, and high quality video content where two quality levels are available in a scenario where the data rate the user device achieves is varying with time due to channel conditions. This is a typical wireless (WiFi, cellular, or other) channel scenario, while the method we present is independent of the underlying technology.

This study is supported by İstanbul Technical University Scientific Research Projects Coordination Unit (BAP), grant number MGA-2020-42575.

P. Ballarini et al. (Eds.): EPEW 2021/ASMTA 2021, LNCS 13104, pp. 472–487, 2021.
https://doi.org/10.1007/978-3-030-91825-5_29

A typical playout control mechanism present in most contemporary video providers/players is pausing download when a certain amount of unwatched video is present in the playout buffer. As video, and in general Internet traffic demand is very high and continuously increasing, providers limit the download in order to reduce traffic load on the backbone as well as to reduce energy consumption. To this end, the framework presented in this study also provides the distribution and the mean of the excess (wasted) data present in the playout buffer when the user stops the playout, or seeks to another location within a longer content. This is typical video consumption behavior especially for video-on-demand content. The video session duration considered in this study could be viewed as either the duration of the video, or the duration that the user is willing to watch the content. This may depend on the physical quality of the video as well as the quality of the content with respect to the tastes of the user.

There exists a number of studies in the literature investigating the QoE metrics for video. [18] analyzes the effect of prefetching on the probability of buffer starvation and the distribution of playback intervals. [11] offers an asymptotic solution for a fluid buffer in case of base station caching. [13] proposes an M/D/1 model, whereas [12] employs diffusion approximation to solve a G/G/1/N buffer model. [8] gives a simple two-state Markovian model to estimate the appropriate initial buffering delay to achieve a given buffer depletion probability. In a previous study [19], we had presented a Markov fluid queue-based analytical framework for the computation of freeze probability of video playout in a cellular communication setting, and investigated the effects of initial buffering, initial state, and adaptive playout.

The contribution of this paper can be summarized as follows:

- We present an analytical framework for the distribution of frozen, low quality, and high quality video playout in a dynamical channel. This fills a gap in the literature as we have not encountered any publications on analytical models for video QoE metrics with multiple quality levels.
- The framework presented is also capable of producing the distribution of the excess data, as well as its mean. Existing works in the literature for wasted video data either rely on simulations [7,15] or use measurements [16] and present only average values. Although these are valuable work, they cannot provide the distribution of the excess data in a variety of scenarios.

2 System Model

We consider a system where a user wants to watch video content that has two quality levels: low quality and high quality. The bit rates for low quality and high quality contents are v_L and v_H, respectively. The dynamics of the channel that the user downloads the content is modeled by a continuous-time Markov chain (CTMC) with infinitesimal generator Q_c, with N states. In state i, $1 \leq i \leq N$, the data rate the user gets is $r_i \geq 0$. In this setting, the technology of the communication and the nature of the channel (e.g. wired/wireless) is irrelevant as long as it can be modeled as a CTMC.

474 M. A. Yazici

The switch from low to high and high to low quality video depends on the amount of data in the playout buffer. In order to avoid frequent quality switches, we consider a hysteretic playout control:

(i) Three thresholds are defined: $0 < B_1 < B_2 < B_3$.
(ii) When the request to the video content is made and data starts being downloaded, we require the buffer level to reach B_2 to start the playout.
(iii) Playout starts with low quality when the buffer reaches B_2.
(iv) During playout with low quality, if the buffer gets depleted, video playout freezes. Playout can resume only when the buffer reaches B_2 again.
(v) During playout with low quality, if the buffer reaches B_3, playout switches to high quality video. At the same time, download is paused.
(vi) During playout with high quality, if the buffer drops to B_2, download is resumed and playout continues in high quality.
(vii) During playout with high quality, if the buffer drops to B_1, playout switches to low quality video.

This mechanism is summarized in Fig. 1. Pausing the download when the buffer level reaches a certain level could be either due to a limited buffer, or in order to minimize the excess data downloaded but not played out. When user behavior is considered regarding video content, especially for relatively short on-demand videos, giving up on the video or skipping to a different position is common. Hence, downloading is not continued indefinitely even if the channel conditions are fine. This behavior can be observed in most modern video content providers.

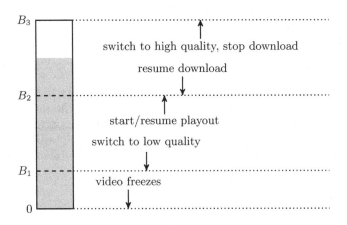

Fig. 1. Summary of the hysteretic playout control.

2.1 Markov Fluid Queues

The playout buffer of user device is modeled as a multi-regime Markov fluid queue. A Markov fluid queue is a two dimensional stochastic process $\{X(t), Z(t)\}$

where $X(t)$ represents the fluid amount in the queue (buffer) and $Z(t)$ represents the background CTMC that determines the net drift into (or out of) the buffer:

$$\frac{d}{dt}X(t) = d(Z(t)), \ t > 0.$$

For ease of notation, we write $d(Z(t)) = d_i$ when $Z(t) = i$.

The steady-state distribution of the fluid level, if it exists, is defined as

$$f_i(x) = \lim_{t \to \infty} \frac{d}{dx} P\{X(t) \le x, \ Z(t) = i\},$$
$$f(x) = [f_1(x), f_2(x), \cdots, f_N(x)].$$

This distribution satisfies the system of differential equations:

$$\frac{d}{dx}f(x)\,R = f(x)\,Q,$$

along with a set of boundary conditions, where Q is the infinitesimal generator of $Z(t)$, and R is the diagonal matrix with the drift rates $d_1, \ldots, d_N$ on its main diagonal [10]. Solution methods for the steady-state distribution are well-known [4].

If the system parameters, namely Q and R matrices, depend on the buffer level $X(t)$ in a piecewise-constant manner, the structure is called a multi-regime Markov fluid queue (MRMFQ) [17]. In a K–regime MRMFQ, there are $K + 1$ regime boundaries: $0 = B_0 < B_1 < \cdots < B_K$. When $B_{i-1} < X(t) < B_i$, the MRMFQ is said to be in regime i, where the state transitions occur according to the infinitesimal generator $Q^{(i)}$, and the drift rates are given in $R^{(i)}$. The behavior of the fluid at the thresholds are also important, and state transitions occur according to $\tilde{Q}^{(i)}$ and net drifts are given in $\tilde{R}^{(i)}$ when $x = B_i$. The steady state probability density of the fluid in this case satisfies

$$\frac{d}{dx}f^{(i)}(x)\,R^{(i)} = f^{(i)}(x)\,Q^{(i)}, \ k \in \{1, 2, \ldots, K\}, \tag{1}$$

along with a set of boundary conditions. The solution for (1) is given in [9]. Probability masses can be observed at the regime thresholds depending on the signs of the drifts on both sides of the threshold. For details, we refer the reader to [9].

2.2 Multi-regime Markov Fluid Queue Model for the Playout Buffer

The MRMFQ model for the playout buffer consists of three regimes, bounded by the thresholds $0 < B_1 < B_2 < B_3$. We will also model the video session duration using a phase-type distribution. This will cover most cases as phase-type distributions are dense in the field of all positive-valued distributions. If the video session duration is deterministic (fixed), Erlangization [14] or ME-fication [3] can be used. Let α represent the initial distribution of the phase-type distribution of the session duration, whereas

$$\begin{bmatrix} T & T^0 \\ \mathbf{0}_{1 \times L} & 0 \end{bmatrix}$$

is the state transition matrix, where T and T^0 blocks are $L \times L$ and $L \times 1$, respectively, and satisfy $T^0 = -T \, \mathbf{1}_{L \times 1}$. In this formulation and the remainder, we denote a matrix of size $m \times n$ of all ones with $\mathbf{1}_{m \times n}$, and a matrix of size $m \times n$ of all zeros with $\mathbf{0}_{m \times n}$.

As the playout is not an indefinite process and expires after some time, the problem inherently becomes a transient one. We employ a modified version of the technique described in [21], which introduces three artificial states that derive the buffer to its initial state after each cycle of the duration. In reference [21], the cycle is assumed to complete when either the duration ends, or the buffer becomes empty, and it is desirable to distinguish between these two situations. Furthermore, the initial buffer content is a non-zero value. In our model, we always start the buffer from 0, and do not stop the cycle when buffer is depleted. Hence, compared to the three artificial states introduced in [21], we use a single "reset" (RST) state. At the instant when the phase-type distribution modeling the session duration expires, the buffer level gives the value of the excess data that will not be played out, i.e. essentially wasted. Then, RST state drives the buffer content to 0 to start a new cycle. In this way, akin to a simulation which is repeated infinitely many times, the steady state distribution of this MRMFQ will give us the distributions at the time when the video duration expires.

Apart from the RST state, each state needs to represent three variables:

(i) the channel state (to determine the data rate),
(ii) the quality level (or frozen video), and
(iii) the phase-type state (for the session duration).

Hence, a state will be represented by a triplet of the form (i, q, s), where $i \in \{1, 2, \ldots, N\}$, $q \in \{F, L, H_p, H_d\}$, and $s \in \{1, 2, \ldots, L\}$. Here, F, L, H_p, and H_d refer to *frozen video*, *low quality video*, *high quality video with playout only*, and *high quality video with download*, respectively.

The following state transitions are observed within each regime:

- (i, q, s) to (j, q, s), $i, j \in \{1, 2, \ldots, N\}$: A change in channel state
- (i, q, s) to $(i, q, s + 1)$, $s \in \{1, 2, \ldots, L - 1\}$: A change in duration state
- (i, q, L) to RST: Duration expires

Enumerating all the states as $\{\text{RST}, (1, F, 1), (2, F, 1), \ldots, (N, F, 1), (1, L, 1), (2, L, 1), \ldots, (N, L, 1), (1, H_p, 1), (2, H_p, 1), \ldots, (N, H_p, 1), (1, H_d, 1), (2, H_d, 1), \ldots, (N, H_d, 1), (1, F, 2), \ldots, (N, H_d, 2), (1, F, 3), \ldots, (N, H_d, L)\}$, the infinitesimal generator of $Z(t)$ in each regime can be written as

$$Q^{(1)} = Q^{(2)} = Q^{(3)} = \begin{bmatrix} 0 & \mathbf{0}_{1 \times 4NL} \\ T^0 \otimes \mathbf{1}_{4N \times 1} & I_L \otimes (I_4 \otimes Q_c) + T \otimes I_{4N} \end{bmatrix}, \quad (2)$$

where I_m is the $m \times m$ identity matrix.

Recalling that the video bit rate is v_L and v_H in low and high quality states, respectively, and download is paused in H_p states, and the RST state takes the buffer to 0, the drift matrices can be written as

$$R^{(1)} = R^{(2)} = R^{(3)} = \begin{bmatrix} -1 & \mathbf{0}_{1 \times 4NL} \\ \mathbf{0}_{4NL \times 1} & I_L \otimes \bar{R} \end{bmatrix}, \tag{3}$$

where

$$R_c = \begin{bmatrix} r_1 & & & \\ & r_2 & & \\ & & \ddots & \\ & & & r_N \end{bmatrix},$$

$$\bar{R} = \begin{bmatrix} R_c & & & \\ & R_c - v_L I_N & & \\ & & -v_H I_N & \\ & & & R_c - v_H I_N \end{bmatrix}.$$

At the regime boundaries, the following state transitions occur:

- At $B_0 = 0$,
  - (i, L, s) to (i, F, s): Low quality video to frozen video
  - (i, F, s) to (j, F, s): A change in channel state
- At B_1,
  - (i, H_d, s) to (i, L, s): High quality video to low quality video
  - (i, F, s) to (j, F, s): A change in channel state
  - (i, L, s) to (j, L, s): A change in channel state
- At B_2,
  - (i, H_p, s) to (i, H_d, s): Download is resumed
  - (i, F, s) to (i, L, s): Playout starts/resumes
  - (i, H_d, s) to (j, H_d, s): A change in channel state
  - (i, L, s) to (j, L, s): A change in channel state
- At B_3,
  - (i, L, s) to (i, H_p, s): Low quality video to high quality video, and download is paused
  - (i, H_d, s) to (i, H_p, s): Download is paused

The transition rate from (i, q, s) to j, q, s is equal to $Q_c(i, j)$ when $i \neq j$. The rates for the other transitions are selected as 1. Note that these transitions in reality are immediate and instantaneous. For instance, when high quality video is playing and the buffer content falls to B_1 from above, it is supposed to immediately switch to low quality video. However, this instantaneous transitions cannot be accommodated in a Markovian model. Hence, we let the buffer spend an exponentially distributed amount of time at level B_1 (during which no other transition is allowed), and then allow the state to change. After the steady state distribution is obtained, the probability masses at the regime boundaries

Table 1. Existence of probability masses on the thresholds and their conditions.

Boundary	RST state	F states	L states	H_p states	H_d states
B_3	–	–	if $d > 0$	–	if $d > 0$
B_2	–	if $d > 0$	–	–	–
B_1	–	–	–	if $d < 0$	if $d < 0$
B_0	Yes	if $d < 0$	if $d < 0$	–	–

due to these spurious periods can be censored out and the distribution can be normalized. Based on these, the infinitesimal generators at the regime boundaries are written as:

$$
\tilde{Q}^{(i)} = \begin{bmatrix} -1 & 1_{1 \times L} \otimes (\alpha \otimes \alpha_c) \\ 0_{4NL \times 1} & I_L \otimes \left(A_1^{(i)} \otimes I_N + A_2^{(i)} \otimes Q_c \right) \end{bmatrix}, \quad i \in \{0, 1, 2, 3\}, \quad (4)
$$

where α_c is the initial distribution of the channel states, and

$$
A_1^{(0)} = \begin{bmatrix} 0,0,0,0 \\ 1,0,0,0 \\ 1,0,0,0 \\ 1,0,0,0 \end{bmatrix}, \quad
A_1^{(1)} = \begin{bmatrix} 0,0,0,0 \\ 0,0,0,0 \\ 0,1,0,0 \\ 0,1,0,0 \end{bmatrix}, \quad
A_1^{(2)} = \begin{bmatrix} 0,1,0,0 \\ 0,0,0,0 \\ 0,0,0,1 \\ 0,0,0,0 \end{bmatrix}, \quad
A_1^{(3)} = \begin{bmatrix} 0,0,1,0 \\ 0,0,1,0 \\ 0,0,0,0 \\ 0,0,1,0 \end{bmatrix},
$$

$$
A_2^{(0)} = \begin{bmatrix} 1,0,0,0 \\ 0,0,0,0 \\ 0,0,0,0 \\ 0,0,0,0 \end{bmatrix}, \quad
A_2^{(1)} = \begin{bmatrix} 1,0,0,0 \\ 0,1,0,0 \\ 0,0,0,0 \\ 0,0,0,0 \end{bmatrix}, \quad
A_2^{(2)} = \begin{bmatrix} 0,0,0,0 \\ 0,1,0,0 \\ 0,0,0,0 \\ 0,0,0,1 \end{bmatrix}, \quad
A_2^{(3)} = \begin{bmatrix} 0,0,0,0 \\ 0,0,0,0 \\ 0,0,1,0 \\ 0,0,0,0 \end{bmatrix}.
$$

The drift matrices at the regime boundaries, $\tilde{R}^{(i)}$, $i \in \{0, 1, 2, 3\}$, are essentially equal to $R^{(i)}$, except for the transitions at the corresponding boundaries. For instance, the drift in state (i, H_d, s) is $r_i - v_H$. However, since there will be a forced transition from (i, H_d, s) to (i, L, s) at B_1, the corresponding rates in the matrix $\tilde{R}^{(1)}$ are changed to 0 to ensure the buffer level stays at B_1 until the state transition. Accordingly, the probability masses indicated in Table 1 will occur in the steady state distribution. Among these, only the probability mass due to F states at 0 are observed in reality. Hence, all masses except for that will be censored out from the steady state distribution.

This completes the definition of the MRMFQ model for the playout buffer. We follow the methodology laid out in [9] to solve this system and leave out the details here.

2.3   The Distribution of Excess Data

The data present in the playout buffer at the end of the session duration will not be played out and hence is wasted. In this study, we quantify the distribution as well as the mean excess data. The end of the session is signified by a transition into the RST state. Hence, the distribution of the buffer occupancy at the beginning of the RST state gives the distribution of the excess data. In a Markov fluid queue, the CDF of the buffer level at the state transition instant into state j is given by [2]

$$
\begin{aligned}
F_b(x, j) &= \lim_{t \to \infty, \Delta t \downarrow 0} P\{X(t) \le x \mid Z(t + \Delta t) = j, \ Z(t) \ne j\} \\
&= \frac{\displaystyle\sum_{i \ne j} F(x, i) \, Q(i, j)}{\displaystyle\sum_{i \ne j} F(\infty, i) \, Q(i, j)},
\end{aligned}
\tag{5}
$$

where $F(x, i) = \lim_{t \to \infty} P\{X(t) \le x, \ Z(t) = i\}$ is the steady state CDF of the buffer level, and Q is the infinitesimal generator. Hence, by setting $j = \mathrm{RST}$, we obtain the CDF of the excess data. To find the mean, one can use the well-known relation between the mean and the CDF of any non-negative random variable:

$$
E[X] = \int_0^\infty (1 - F_X(x)) \, dx.
\tag{6}
$$

3   Numerical Results

We consider a system with the channel state determined by a CTMC whose infinitesimal generator is

$$
Q_c = \begin{bmatrix}
-w & w/3 & w/3 & w/3 \\
0.2 & -1 & 0.4 & 0.4 \\
0.2 & 0.4 & -1 & 0.4 \\
0.2 & 0.4 & 0.4 & -1
\end{bmatrix},
\tag{7}
$$

with the corresponding data rates in Mbps

$$
r_1 = 10^{-3}, \qquad r_2 = 1, \qquad r_3 = 3, \qquad r_4 = 5.
$$

Here, the first state represents the state in which the channel deteriorates and the data rate reduces to a mere 1 Kbps. As seen from (7), the mean holding times for the states are $1\,\mathrm{s}$, except for the state with 1 Kbps data rate, whose mean holding time is parametrized as $1/w$. Furthermore, the transition rates among states 2 to 4 are equal to each other and double the transition rate into state 1. The transition rates from state 1 to the others are selected equal. The initial distribution α_c is selected as the steady state distribution of Q_c, i.e. α_c satisfies $\alpha_c Q_c = \mathbf{0}_{1 \times N}$, $\alpha_c \mathbf{1}_{N \times 1} = 1$.

The video bit rates are selected as $v_L = 2$ and $v_H = 4$ Mbps for the low and high quality content, respectively, in line with YouTube recommendations [5,6]. Therefore, in the first and second states, the channel does not have sufficient data rate for neither quality level; in the third state, the channel does not have sufficient data rate for the high quality content but has sufficient data rate for the low quality content; and the channel can provide the high quality content in the fourth state in this scenario. The duration of the video session is assumed to have an Erlang-2 distribution with a mean of H.

We start by investigating the effect of the size of the playout buffer. In this scenario, B_1 is varied from 1 to 10 Mbps, $B_3 = 2 B_2 = 4 B_1$, and $w = 3$. The distribution of the time the user sees frozen, low quality, and high quality video is given in Fig. 2 for $H = 60$ s, and in Fig. 3 for $H = 300$ s. From these figures, we observe that the value of B_1 has a more profound effect on the probability of frozen video for smaller H, which is due to the fact that the video starts frozen until the buffer content reaches B_2. Moreover, for $H = 60$ s, there seems to be an optimal value of B_1 with respect to the probability of frozen video. On the other hand, the probability of frozen video monotonically decreases with increasing B_1 for $H = 300$ s as expected, since the initial buffering becomes a much smaller part of the playout duration.

When it comes to the probability of high quality video, as the buffer is required to reach B_3 in order to switch to high quality, the probability of high quality video also decreases with increasing B_1. This shows that a joint optimization with respect to the probability of frozen video and the probability of high quality video would be required under this setting.

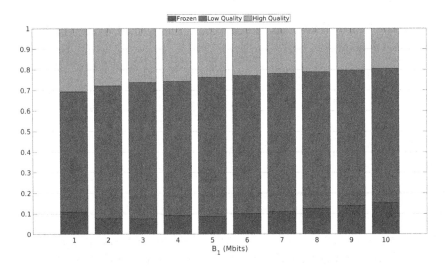

Fig. 2. Video quality distribution for $B_3 = 2 B_2 = 4 B_1$ varied from 4 to 40 Mbits, $H = 60$ s, and $w = 3$

We also look at the amount of data present in the buffer at the end of the session. This is another metric that can be taken into account when designing playout and buffering mechanisms. Under the same scenario with $H = 300$, we plot two of the CDFs for the excess data in Figs. 4 and 5, which are selected as they are representative of the other scenarios. Notice the deflections at B_1

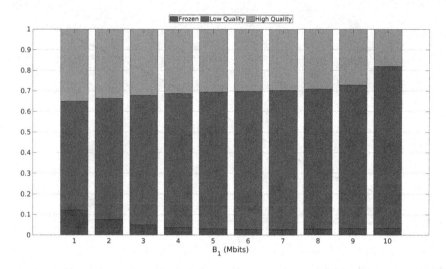

Fig. 3. Video quality distribution for $B_3 = 2\,B_2 = 4\,B_1$ varied from 4 to 40 Mbits, $H = 300\,\text{s}$, and $w = 3$.

and B_2 at each of these plots. Furthermore, we plot the mean of the excess data computed from (6) in Fig. 6, which shows that mean excess data is almost linear in B_1, with very little dependence on session duration.

Next, we look at the effect of the selection of B_2, when B_1 and B_3 are fixed. We select $B_1 = 4$ and $B_3 = 16$ Mbits as this pair of values seem to be reasonable choices based on Figs. 2 and 3. Again, we use $w = 3$. The distribution of the time the user sees frozen, low quality, and high quality video is given in Fig. 7 for $H = 60\,\text{s}$, and in Fig. 8 for $H = 300\,\text{s}$. From these figures, it is evident that increasing B_2 hurts the performance in terms of the probability of frozen video whereas improves the probability of high quality video. The effect of B_2 on the probability of frozen video is mainly due to initial buffering as seen from the comparison of Figs. 7 and 8. Thus, for longer video durations, selecting higher B_2 values could be preferable.

In Fig. 9, we present the mean excess data for again $B_1 = 4$ and $B_3 = 16$ Mbits, B_2 ranging from 5 to 15 Mbits, $H = 300$, and $w = 3$. Although there seems to be an optimum B_2 value for the mean excess data, the absolute change in the amount of the mean excess data is not significant. Our experiments (whose results are omitted here) with other sets of parameters show that the mean excess data has little dependence on B_2 when B_1 and B_3 are fixed.

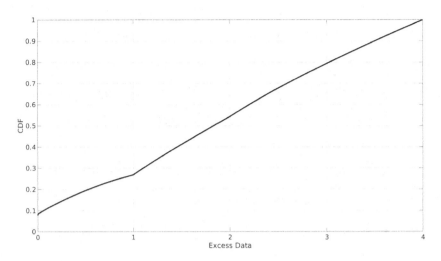

Fig. 4. Excess data CDF for $4B_1 = 2B_2 = B_3 = 4$ Mbits, $H = 300$ s, and $w = 3$.

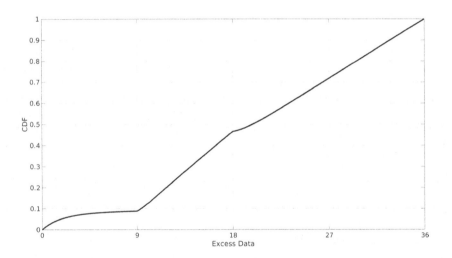

Fig. 5. Excess data CDF for $4B_1 = 2B_2 = B_3 = 36$ Mbits, $H = 300$ s, and $w = 3$.

Lastly, we look at the effect of the parameter w, the transition rate out of state 1. As the data rate in state 1 is very low, we expect improved performance as w increases, which is seen in Figs. 10 and 11. Although not obvious from the figures, the relative improvement in the probability of frozen video (from 0.1115 to 0.0578 for $H = 60$, and from 0.07578 to 0.02158 for $H = 300$) is greater than

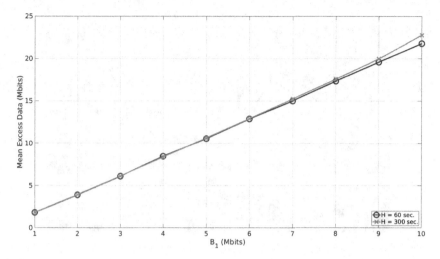

Fig. 6. Mean excess data CDF for $4\,B_1 = 2\,B_2 = B_3$ ranging from 4 to 40 Mbits, and $w = 3$.

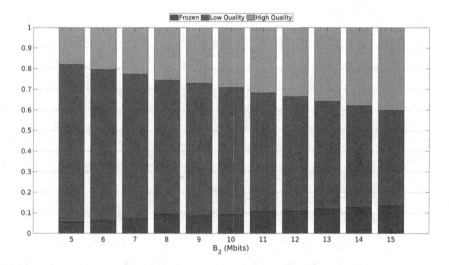

Fig. 7. Video quality distribution for $B_1 = 4$ Mbits, $B_3 = 16$ Mbits, B_2 is varied from 5 to 15 Mbits, $H = 60$ s, and $w = 3$.

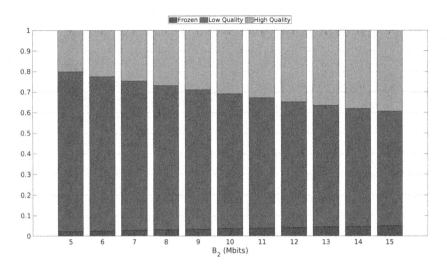

Fig. 8. Video quality distribution for $B_1 = 4$ Mbits, $B_3 = 16$ Mbits, B_2 is varied from 5 to 15 Mbits, $H = 300$ s, and $w = 3$.

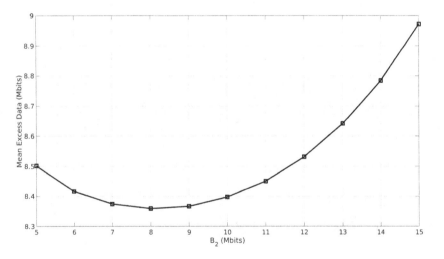

Fig. 9. Mean excess data for $B_1 = 4$ Mbits, $B_3 = 16$ Mbits, B_2 is varied from 5 to 15 Mbits, $H = 300$ s, and $w = 3$.

the relative improvement in the probability of high quality video (from 0.1827 to 0.283 for $H = 60$, and from 0.1946 to 0.3002 for $H = 300$) as w is varied from 1 to 6.

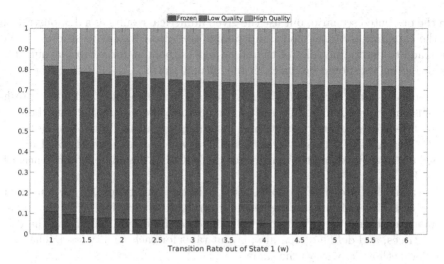

Fig. 10. Video quality distribution for $B_3 = 2\,B_2 = 4\,B_1 = 16$ Mbits, $H = 60$ s, and w is varied.

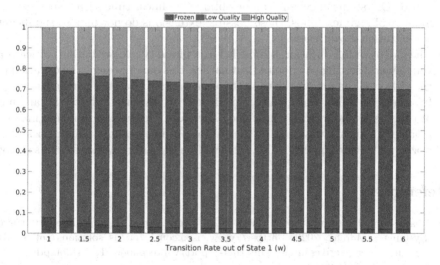

Fig. 11. Video quality distribution for $B_3 = 2\,B_2 = 4\,B_1 = 16$ Mbits, $H = 300$ s, and w is varied.

4   Conclusion

We present an analytical framework based on Markov fluid queues for the distribution of frozen, low quality, and high quality video playout of a video content with two quality levels in a dynamic channel. The framework also produces the distribution of the excess (wasted) data that is downloaded but not played out. We present a number of scenarios based on the presented model and draw conclusions. It should be noted that we do not make any universal inferences based on the presented scenario, but use the given numerical results as a demonstration of the capabilities of the model and sample scenarios. Extensive studies should be made to determine rules of thumb or general policies for video playout. The following points should be made on the current work:

- The roles of the thresholds B_1, B_2, and B_3 can be modified and more thresholds can be accommodated easily. For instance, we use B_2 both as the level required for starting playout, and resuming download. It is possible to use distinct values for these two thresholds, in which case one additional regime should be defined and the matrices of the MRMFQ should be modified accordingly.
- Although we present the model using only two quality levels (low and high) for ease of definition and implementation, more quality levels can be accommodated easily by defining new thresholds as necessary, modifying the drift matrices, and defining separate playout rates for each quality level. The solution methodology stays the same.
- In both of these extensions, the number of regimes increases as new thresholds are defined. However, as pointed out in [20], the numerical solution of MRMFQ systems in general can be obtained in linear time with respect to the number of regimes. Thus, these possible extensions do not impose significant computational overheads.
- The video session duration is modeled as a phase-type distribution, which is a very general setting. As pointed out earlier, this can represent the duration of the video, or the amount of time the user wants to watch a particular video. In a general sense, the quality experienced by the user affects the amount of time he or she watches the video. Therefore, the session duration may be modeled in such a way that the desire to continue watching diminishes with frozen and/or low quality video. This will be the future work after this study.

References

1. Cisco annual internet report (2018–2023) white paper. Technical report, Cisco Systems, March 2020. https://www.cisco.com/c/en/us/solutions/collateral/executive-perspectives/annual-internet-report/white-paper-c11-741490.pdf. Accessed 15 July 2021
2. Akar, N.: Performance analysis of an asynchronous transfer mode multiplexer with Markov modulated inputs. Ph.D. thesis, Bilkent University (1993)
3. Akar, N., Gursoy, O., Horvath, G., Telek, M.: Transient and first passage time distributions of first-and second-order multi-regime Markov fluid queues via ME-fication. Methodol. Comput. Appl. Probab. **23**, 1–27 (2020)
4. Akar, N., Sohraby, K.: Infinite-and finite-buffer Markov fluid queues: a unified analysis. J. Appl. Probab. **41**(2), 557–569 (2004)
5. Google: Live encoder settings, bitrates, and resolutions - YouTube Help. https://support.google.com/youtube/answer/2853702. Accessed 15 July 2021
6. Google: Recommended upload encoding settings - YouTube Help. https://support.google.com/youtube/answer/1722171. Accessed 15 July 2021

7. He, J., Xue, Z., Wu, D., Wu, D.O., Wen, Y.: CBM: online strategies on cost-aware buffer management for mobile video streaming. IEEE Trans. Multimed. **16**(1), 242–252 (2013)
8. Kalman, M., Steinbach, E., Girod, B.: Adaptive media playout for low-delay video streaming over error-prone channels. IEEE Trans. Circuits Syst. Video Technol. **14**(6), 841–851 (2004). https://doi.org/10.1109/TCSVT.2004.828335
9. Kankaya, H.E., Akar, N.: Solving multi-regime feedback fluid queues. Stoch. Models **24**(3), 425–450 (2008)
10. Kulkarni, V.G.: Fluid models for single buffer systems. Front. Queueing: Models Appl. Sci. Eng. **321**, 338 (1997)
11. Liu, A., Lau, V.K.: Exploiting base station caching in MIMO cellular networks: opportunistic cooperation for video streaming. IEEE Trans. Signal Process. **63**(1), 57–69 (2014). https://doi.org/10.1109/TSP.2014.2367473
12. Luan, T.H., Cai, L.X., Shen, X.: Impact of network dynamics on user's video quality: analytical framework and QoS provision. IEEE Trans. Multimed. **12**(1), 64–78 (2010)
13. ParandehGheibi, A., Médard, M., Shakkottai, S., Ozdaglar, A.: Avoiding interruptions-QoE trade-offs in block-coded streaming media applications. In: 2010 IEEE International Symposium on Information Theory, pp. 1778–1782. IEEE (2010)
14. Ramaswami, V., Woolford, D.G., Stanford, D.A.: The Erlangization method for Markovian fluid flows. Ann. Oper. Res. **160**(1), 215–225 (2008)
15. Schwartz, C., Scheib, M., Hoßfeld, T., Tran-Gia, P., Gimenez-Guzman, J.M.: Trade-offs for video-providers in LTE networks: smartphone energy consumption vs wasted traffic. In: 2013 22nd ITC Specialist Seminar on Energy Efficient and Green Networking (SSEEGN), pp. 1–6. IEEE (2013)
16. Sieber, C., Hoßfeld, T., Zinner, T., Tran-Gia, P., Timmerer, C.: Implementation and user-centric comparison of a novel adaptation logic for dash with SVC. In: 2013 IFIP/IEEE International Symposium on Integrated Network Management (IM 2013), pp. 1318–1323. IEEE (2013)
17. da Silva Soares, A., Latouche, G.: Fluid queues with level dependent evolution. Eur. J. Oper. Res. **196**(3), 1041–1048 (2009)
18. Xu, Y., Altman, E., El-Azouzi, R., Elayoubi, S.E., Haddad, M.: QoE analysis of media streaming in wireless data networks. In: Bestak, R., Kencl, L., Li, L.E., Widmer, J., Yin, H. (eds.) NETWORKING 2012. LNCS, vol. 7290, pp. 343–354. Springer, Heidelberg (2012). https://doi.org/10.1007/978-3-642-30054-7_27
19. Yazici, M.A.: Markov fluid queue model for video freeze probability in a random environment. In: 14th International Conference on Queueing Theory and Network Applications (QTNA) (2019)
20. Yazici, M.A., Akar, N.: The workload-dependent MAP/PH/1 queue with infinite/finite workload capacity. Perform. Eval. **70**(12), 1047–1058 (2013)
21. Yazici, M.A., Akar, N.: The finite/infinite horizon ruin problem with multi-threshold premiums: a Markov fluid queue approach. Ann. Oper. Res. **252**(1), 85–99 (2017)

Author Index

Printed in the United States
by Baker & Taylor Publisher Services